Establishing Food Security and Alternatives to International Trade in Emerging Economies

Vasily Erokhin
Harbin Engineering University, China

A volume in the Advances in Public Policy and Administration (APPA) Book Series

IGI Global
DISSEMINATOR OF KNOWLEDGE

Published in the United States of America by
IGI Global
Business Science Reference (an imprint of IGI Global)
701 E. Chocolate Avenue
Hershey PA, USA 17033
Tel: 717-533-8845
Fax: 717-533-8661
E-mail: cust@igi-global.com
Web site: http://www.igi-global.com

Library of Congress Cataloging-in-Publication Data

Names: Erokhin, Vasily, 1980- editor.
Title: Establishing food security and alternatives to international trade in emerging economies / Vasily Erokhin, editor.
Description: Hershey, PA : Business Science Reference, 2017. | Includes bibliographical references and index.
Identifiers: LCCN 2017009417| ISBN 9781522527336 (hardcover) | ISBN 9781522527343 (ebook)
Subjects: LCSH: Food industry and trade--Developing countries. | Food security--Developing countries. | Free trade--Developing countries.
Classification: LCC HD9018.D44 E88 2017 | DDC 338.1/91724--dc23 LC record available at https://lccn.loc.gov/2017009417

This book is published in the IGI Global book series Advances in Public Policy and Administration (APPA) (ISSN: 2475-6644; eISSN: 2475-6652)

British Cataloguing in Publication Data
A Cataloguing in Publication record for this book is available from the British Library.

All work contributed to this book is new, previously-unpublished material. The views expressed in this book are those of the authors, but not necessarily of the publisher.

For electronic access to this publication, please contact: eresources@igi-global.com.

Advances in Public Policy and Administration (APPA) Book Series

ISSN:2475-6644
EISSN:2475-6652

Mission

Proper management of the public sphere is necessary in order to maintain order in modern society. Research developments in the field of public policy and administration can assist in uncovering the latest tools, practices, and methodologies for governing societies around the world.

The **Advances in Public Policy and Administration (APPA) Book Series** aims to publish scholarly publications focused on topics pertaining to the governance of the public domain. APPA's focus on timely topics relating to government, public funding, politics, public safety, policy, and law enforcement is particularly relevant to academicians, government officials, and upper-level students seeking the most up-to-date research in their field.

Coverage

- Government
- Law enforcement
- Political Economy
- Politics
- Public Administration
- Public Funding
- Public Policy
- Resource Allocation
- Urban Planning

IGI Global is currently accepting manuscripts for publication within this series. To submit a proposal for a volume in this series, please contact our Acquisition Editors at Acquisitions@igi-global.com or visit: http://www.igi-global.com/publish/.

The Advances in Public Policy and Administration (APPA) Book Series (ISSN 2475-6644) is published by IGI Global, 701 E. Chocolate Avenue, Hershey, PA 17033-1240, USA, www.igi-global.com. This series is composed of titles available for purchase individually; each title is edited to be contextually exclusive from any other title within the series. For pricing and ordering information please visit http://www.igi-global.com/book-series/advances-public-policy-administration/97862. Postmaster: Send all address changes to above address.

Titles in this Series

For a list of additional titles in this series, please visit: www.igi-global.com/book-series

Impacts of Faith-Based Decision Making on the Individual-Level Legislative Process Emerging Research and Opportunities
Karla Drenner (Kaplan University, USA)
Information Science Reference • copyright 2017 • 186pp • H/C (ISBN: 9781522523888) • US $120.00 (our price)

Handbook of Research on Emerging Business Models and Managerial Strategies in the Nonprofit Sector
Lindy Lou West (Griffith University, Australia) and Andrew Worthington (Griffith University, Australia)
Business Science Reference • copyright 2017 • 509pp • H/C (ISBN: 9781522525370) • US $245.00 (our price)

Ideological Messaging and the Role of Political Literature
Önder Çakırtaş (Bingol University, Turkey)
Information Science Reference • copyright 2017 • 317pp • H/C (ISBN: 9781522523918) • US $175.00 (our price)

Therapeutic Jurisprudence and Overcoming Violence Against Women
Debarati Halder (Centre for Cyber Victim Counselling (CCVC), India & Unitedworld School of Law, India) and K. Jaishankar (Raksha Shakti University, India)
Information Science Reference • copyright 2017 • 344pp • H/C (ISBN: 9781522524724) • US $230.00 (our price)

Public Sector Entrepreneurship and the Integration of Innovative Business Models
Mateusz Lewandowski (Jagiellonian University in Kraków, Poland) and Barbara Kożuch (Jagiellonian University in Kraków, Poland)
Business Science Reference • copyright 2017 • 309pp • H/C (ISBN: 9781522522157) • US $190.00 (our price)

Handbook of Research on Citizen Engagement and Public Participation in the Era of New Media
Marco Adria (University of Alberta, Canada) and Yuping Mao (California State University Long Beach, USA)
Information Science Reference • copyright 2017 • 503pp • H/C (ISBN: 9781522510819) • US $265.00 (our price)

Global Perspectives on Development Administration and Cultural Change
Gbenga Emmanuel Afolayan (Murdoch University, Australia) and Akeem Ayofe Akinwale (University of Lagos, Nigeria)
Information Science Reference • copyright 2017 • 277pp • H/C (ISBN: 9781522506294) • US $170.00 (our price)

Handbook of Research on Managerial Solutions in Non-Profit Organizations
Vojko Potocan (University of Maribor, Slovenia) Mustafa C. Üngan (Sakarya University, Turkey) and Zlatko Nedelko (University of Maribor, Slovenia)
Information Science Reference • copyright 2017 • 609pp • H/C (ISBN: 9781522507314) • US $300.00 (our price)

IGI Global
DISSEMINATOR OF KNOWLEDGE

701 East Chocolate Avenue, Hershey, PA 17033, USA
Tel: 717-533-8845 x100 • Fax: 717-533-8661
E-Mail: cust@igi-global.com • www.igi-global.com

Table of Contents

Detailed Table of Contents

This chapter presents the main issues around globalization and its effect on trade relations of emerging markets. Based on extensive literature review the basic globalization trends with corresponding factors are revealed. The author's proposition states that globalization trends (including political, economic, and social globalization) affect trade relations between and within emerging markets. The analysis covers BRIC and CIVETS countries for the period from 1971 to 2013 (based on KOF Index of Globalization) and from 1960 to 2015 for the exports/imports flows. Several statements and propositions regarding trade relations between and within emerging markets are defined. The proposed implications for future research are focused on: exploring emerging markets as a system which variations can be modeled by stochastic processes, and applying network theory to explore the influences of trade liberalizations and gravitational effects of various regional trade agreements on the economic performance of major emerging markets.

Although free trade is a key factor in promoting economic growth, there are markets, which require specific approaches. The food market is one of them due to its vital importance for securing nutritional requirements of the population. Providing the population with food in sufficient quantity and variety is a challenge, which includes a range of issues of food production and dependence on import of the food market. Food security is increasingly influenced by foreign trade policies implemented by national governments. Many of them are re-examining their strategies for dealing with food security concerns and seeking to protect domestic food producers by restricting presence of foreign competitors on the domestic food market. The chapter addresses recent food security trends in emerging countries, including BRICS, Eurasian Economic Union (EAEU), and Russia, assesses the effects of food trade distortions in order to balance protection and liberalization policies for achieving effective levels of food security.

The chapter examines the practices of state support of agricultural production in various emerging economies in comparison with selected OECD countries. The aim of the research is to discover how agricultural protectionism and support of domestic farmers affect the level of food security on the emerging markets in the conditions of expanding globalization and liberalization of trade in food. The authors focus on the evaluation of the best practices of state support and discovery of opportunities of their utilization on the emerging markets. Content and mechanisms of state regulation are examined based on the data and evaluation methods obtained from the Organization for Economic Cooperation and Development (OECD). The authors find out that the system of state regulation is one of the key determinants of achieving food security in the conditions of turbulent emerging markets. Both the volume and priority directions of state support of agriculture are determined by financial capacities of emerging economies and current goals of their agrarian policies.

"Novel foods" indicates all those foods that do not fall within the traditional western diet. The authors underline that there is an important economic potential linked to this new sector. After having introduced the topic by identifying what is meant by new food, food sustainability, and entrepreneurship in this sector, the analysis focuses on the qualitative aspects of novel foods. The authors emphasize the advantages of these sources, their qualities, and the impact on the environment. In this chapter, the authors demonstrate that novel food sources can be a substitute for other food that require production processes, and thus make a big impact on the environment. After figuring out the advantages of novel food sources, the authors analyze some success stories from the companies that already operate in this area. They also investigate existing business models on the novel food market and the way they operate and create wealth. Summarizing the results of their research, the authors underline the economic potential of the new foods exploitation in the Mediterranean area.

The objective of the chapter is to understand India's role in providing food security through trade with the ASEAN under the India-ASEAN Strategic Partnership. A dynamic panel has been employed to assess and estimate the said objective, given that India and most of the ASEAN countries are considered emerging markets. Econometric investigation points out that while almost more than half the variables in the augmented gravity model hold good, food imports from India, agricultural labour force, common maritime border, distance from India, food price index, food production index, GDP (size) of the importing country, per capita GDP of the importing economy, and trade openness have statistically significant effect on the food trade from India to the ASEAN. Improving maritime infrastructure and agri-logistics, investing in climate change and water management while augmenting the agri-labour productivity are of paramount importance in order to improve food trade between India and the ASEAN.

The costs of the Southeast Asia maritime piracy to the global economy are largely underestimated regarding the issue of the fight against piracy. In particular, the region suffers from severe negative socio-economic impacts due to over-fishing, and the reduction and collapse of the fisheries have led to a widespread loss of income and employment. The attacks on fishers take place regularly in most Southeast Asian countries. Also, in many cases, the pirates are the fishers themselves, and they attack fishers of other nationalities. All those attacks have a serious negative impact on food security in the region by damaging fishing industry and threatening artisanal fishers who risk losing their entire income. Given the fact that large parts of Southeast Asia's fish populations are overfished, fishers have less possibility to avoid areas prone to pirate attacks.

The goal of this chapter is to analyze the socio-economic role of value-added agriculture (VAA) for Central Asian (CA) countries. The agricultural sector of the region provides raw materials for the food, textile, and leather industry. Cotton, wheat, rice, and fruit (fresh and dried) play an important role in the foreign trade of each CA country. These countries have unrealized potential for storing, freezing, processing, and packaging of the wide nomenclature of fruit, food production, and drinks with the organization of their further exports to perspective markets. Adding value to agricultural products lead to increasing the share of finished goods in export, supplying import-substituting products, improving infrastructure in rural areas, providing new jobs, and growing farmers' income. Based on the analysis, the authors recommend using value-added agriculture for the CA countries by attracting domestic and foreign investments to rural areas, establishing tax incentives, and allocating preferential credits for agribusiness.

The impact of economic globalization on sustainable development has been more and more perceptible since the 21st century. Regions of Russian Far East and Northeast China are the important regions affecting the sustainable development of the world because of their rich natural resources, good environment condition, and large potential for economic development. Due to a number of historical and geopolitical reasons, the cooperation between Russia and China in the region of Northeast Asia is progressing, but at a quite slow speed and in an unsustainable manner. The chapter addresses some problems and challenges of sustainable rural development in the Heilongjiang Province, P.R. China and neighbor cross-border regions of Russia (Far East Federal District and Siberian Federal District), investigates similarities and differences of rural way of life in those regions, and discusses measures to promote the cooperation between the two countries and ensure sustainable rural development and food security in the macro-region of Northeast Asia.

The importance of the agricultural sector in the economy of Azerbaijan is high. This sector has always been at the center of economic reforms by the government. This chapter overviews economic and trade policies of the government focused on the development of agricultural production since 1991. Authors carry out analysis of policy changes during separate development stages. The research output presents agricultural policy before the oil boom as mainly devoted to achieving structural transformation from centrally planned economy to the market environment. Within the oil boom period, the government provided substantial fiscal and technical support to the agricultural sector as well as applied tax incentives to farmers but did not pay attention to the transformation from family farming to medium and large-scale production. However, fiscal and macroeconomic challenges of post-oil boom period forced the government to focus on increasing efficiency of the subsidies and incentives and implementation of further agricultural reforms.

The aim of this chapter is to give an overview of the trade agreements existing in Nigeria to find out their contribution to the development of the economy. There are several such agreements, however, the authors investigate how much they really operate and serve the wellbeing of people as well as food security and safety. There have been various initiatives in the countries of Africa to develop the agricultural sector, which has employed millions of people and has had a significant role in keeping the population in the rural areas. However, the authors claim that less progress has been achieved than expected. In this chapter, the authors intend to analyze the situation in Africa, with a special focus on Nigeria, to summarize the achievements, and to list up some recommendations on future measures.

Feeding the quickly growing population in Africa remains a global challenge. As the demand for food increases, climate change, on the other hand, poses more challenges to agricultural productivity, implying that the provision of sufficient quantities and qualities of food is threatened. Traditional leafy vegetables (TLVs) in Africa are resilient to adverse weather conditions and are naturally rich in nutrients including vitamins A & C, iron, protein and other micronutrients. The objective of this chapter is to assess the potential of TLVs improving food security in Africa. TLVs represent a robust local source of food, their consumption is crucial in complementing Africa's diet with micronutrients. TLVs have therefore the potential to play a major role in improving food security and facilitating food sovereignty. Research on seeds, developing seed systems, coupled by preparing and processing methods to TLV products like cakes or flour is needed to increase their consumption, particularly among the young, elite and urban dwellers.

The chapter is aimed at assessing the effect of climate change on crop productivity and prices in Benue State, Nigeria. Time series data on selected output of crops (maize, rice, sorghum, yam, millet, groundnut, beans, and cassava), area planted, price, and climate variables such as rainfall, temperature, and sea level were used. Due to differing periods in data availability, this study used the time period 1995-2009 for analysis, in order to maintain a common period for all the series. First, the trend of productivity, prices, and climate change was analyzed using visual plots and results indicate some level of variability in these

Consumer perception is one of the aspects from which food safety can be considered. This view of food safety, named in the literature as subjective safety, does not necessarily correspond to objective food safety and is also considered as not being easily influenced by education of consumers. Having that in mind, the first goal of this chapter is to examine different approaches to food safety and the results of previous research dealing with consumer perspective of food safety. Besides giving broad literature review of the topic, the goal of this chapter is to implement Consumer confidence in food safety scale in Serbian consumer research. The validity, reliability, and one-dimensionality of this scale, which consists of optimism and pessimism (two dimensions of confidence), are tested and determinants of different rating on it are examined. The research from 2016 includes 300 participants in the main survey. Finally, starting from profiling consumers with different levels of general confidence in food safety, managerial recommendations are provided.

The chapter gives an overview of current issues of achievement of food security through reforming of veterinary service in the light of the social, ecological, and economic development of the society. The authors analyze certain challenges of food security existing in emerging countries, including expansion of market relations within veterinary service. The results of organizational and economic reforms of veterinary service are assessed on the case of Russia in a form of the survey of the heads of regional veterinary centers. Such issues as governmental and public regulations of quarantine operations and the most dangerous animal diseases are also discussed. In order to improve food security, the authors justify the measures for competition and demand stimulation, smoothing consequences of market imperfections, encouraging veterinarians' efforts on reducing the negative impact of livestock breeding on the environment.

Foreword

This book edited by Dr. Erokhin on food security is a must read. It complements the recent volume on this topic by Schmitz and Meyers (*Transition to Agricultural Market Economies*, 2015), and the three volumes by Schmitz, Kennedy, and Schmitz (*Food Security in an Uncertain World: An International Perspective*, 2015; *Food Security in a Food Abundant Word: An Individual Country Perspective*, 2016; and *World Agricultural Resources and Food Security*, 2017). Our research has a heavy emphasis on investing in agricultural research, including the United States, and the effects on poverty and diets of low-income populations. On the other hand, this volume focuses on emerging markets of Asia, Latin America, and Africa. The authors of this volume represent 20 respected institutions, universities, and organizations worldwide.

The objective of the book is the development of conceptual approaches to the establishment of food security and food sovereignty in the growing openness of domestic food markets to the world. The book is structured into 16 chapters that address international trade in agricultural products within and between the various emerging markets of Asia, Latin America, and Africa. It focuses on protection and domestic support of agricultural production in emerging countries. It also deals with the effects of climate change on crop productivity and advanced farming technologies aimed at increasing output; development of value-added chains in the agricultural sector; and the potential of traditional foods, local plants, and novel foods for ensuring food security of vulnerable populations in various parts of the world. Food safety problems specific to emerging countries and threats to traditional ways of life and economic activities in rural areas are also addressed.

Dr. Erokhin notes that "the case of establishing food security in emerging economies is one of the most challenging." Institutional reforms often lag behind global market economic growth and integration. As a result, food shortages become a major problem. Relying on agricultural imports leaves many emerging economies vulnerable to global price fluctuations and unstable export revenue which greatly affects their food security.

He also notes that the "global and national dimensions of food security" are influenced by two opposing economic policies: global liberalization of economic and trade relations versus domestic protection.

Our four volumes on food security mentioned above, as well as this volume, do not detail a very important issue on the link between population dynamics and food security. Is it even possible to "feed the world" without at the same time controlling population growth? The answer to this question is controversial due to the many institutions (social, religious, governmental) that oppose population control. Further work on food security should link the interaction between agricultural productivity growth, food distribution, and population growth.

Andrew Schmitz
University of Florida, USA

Foreword

Providing food security is becoming increasingly challenging in today's fast-changing global economic environment. Degradation of land, unprecedented depletion of ground water and other natural resources, and visible impacts of climate change in shifting weather patterns have made agricultural output and food prices more volatile. Development assistance for agricultural R&D has slowed, leading to a slump in productivity growth. Rapid urbanization has shifted the structure of developing countries from agriculture towards industry and services, converting farm land into urban sprawls. Being far more educated than their parents, today's youth prefer white-collar and even blue-collar jobs rather than work on farms. At the same time, demand for food is growing. Although population growth is expected to be much slower in the coming decades, global food production needs to increase by at least 50% by 2050 to feed the growing population.

Trade plays a vital role in stabilizing food supplies and food prices since abundant foods stocks in some countries coexist with shortages in some others. Unforeseen price surges can push millions of people in developing countries into poverty, aggravating income inequalities and threatening social cohesion. Malnutrition can result in maternal mortality and stunted growth of children, reducing their learning ability and lowering their productivity in adulthood.

Emerging economies are pursuing various policy options for ensuring food security: expanding investment in agriculture; encouraging climate-friendly technology; restoring degraded farmland; improving post-harvest storage and supply chains; and even indulging in the promotion of niche products. They face special challenges as their growing middle class shifts from traditional staples to more nutritious products such as meat, fish and dairy whose higher resource intensity requires expansion of domestic agricultural capacity or greater reliance on imports. But importing countries also worry about unreliability of world markets in times of need and promote self-sufficiency in food along with trade policy restrictions.

This volume rightly develops the theme of food security with a critical focus on the importance of balancing trade liberalization vis-à-vis protectionism. It probes into many of the choices that link national, regional and global policies extensively with the provision of food security for all. I congratulate the contributors for this timely compilation which should be of great interest to policymakers, academics and students.

Shikha Jha
Asian Development Bank, Philippines

Preface

Food security concerns are growing as the scale of hunger in the world expands. In the era of globalization and booming international trade, food security is a subject of international regulation. However, being a combination of physical availability of food on the particular domestic market and economic access to adequate supply by all people of a country, food security is increasingly influenced by foreign trade policies implemented by national governments. In the conditions of globalizing market, many of them seek to protect domestic food producers by restricting the presence of foreign competitors on the domestic food market, i.e. struggle for making the country self-sufficient in food. In some of the cases, such restrictions lead to improving self-sufficiency, but in most of the cases, they substantially decrease food security in the internationally-recognized sense of this term.

Sufficient nutrition is the major human need and a guarantee of a right to life enshrined in the legislation of most of the countries. In a context of globalization, food problem is a focus of much attention of each particular country seeking to secure its self-sufficiency and ensure the well-being of its people. Food security is one of the dimensions of the provision of economic sovereignty of a state, a major component of demographic policy, a necessary condition of implementation of the strategic national priority, which is the improvement of the quality of life of citizens by means of guaranteeing them high standards of living and nutrition.

The scale of hunger in the world is expanding. In 2005, there were 850 million starving people worldwide, whereas in 2015 – over one billion. Some countries, primarily leading food producers introduce export bans in an effort to decrease domestic food prices and create food reserves in the country. In such a way, food becomes a tool of economic, trade and even political pressure in international relations.

In the past, the main reasons for food shortages used to be droughts and other natural disasters. Contemporary problem is that food became so expensive that many people can not afford to buy it. World prices for food and agricultural commodities have become very volatile because of objective natural circumstances, but also due to the speculative trading and artificial scarcity of available food.

Price instability is detrimental to poor countries, where deteriorating living conditions of people may raise food conflicts. Even now, over two billion poorest people in the world spend up to 70% of their disposable income on food. Demand for food is especially increasing in developing countries. China, India, Brazil, and other emerging markets, which aggregated population is now over 40% of world's total population have succeeded to increase living standards of their people. People have started to consume more; demand outreached supply; the situation requested an increase of domestic production or imports of food.

However, the case of establishing food security in emerging economies is one of the most challenging. In such countries, institutional reforms often lay behind the paces of economic growth and integration to the global market. Providing the population with food in sufficient quantity and variety becomes a problem, which includes a range of issues of food production, import dependence and export orientation of the food market, solvency and dietary patterns of the population. Many emerging economies have to rely on agricultural imports, leaving them vulnerable to global price fluctuations and affecting their export revenues, which tremendously threat food independence and food security of those nations.

Both global and national dimensions of food security of particular emerging country are developed under the influence of two major types of economic policy, which are often countered with one another. On the one side, there is a global trend towards liberalization of economic and trade relations, which is facilitated in the framework of international organizations (for instance, the World Trade Organization) and various alliances of countries. On the other side, emerging countries tend to protect domestic producers, support purchasing power of consumers, especially those with low income, and promote domestic products abroad.

Although free trade is a key factor in promoting economic growth through international competition and efficient allocation of resources, the food market is very specific. Developed countries implement a wide range of tools that affect the competitiveness of domestic farmers and character of international trade in food directly and indirectly. Such policies support the effective elimination of price disparity and growth of farmers' incomes. The off-loading of agricultural surpluses of developed countries on the world market brings down prices and creates disincentives for local producers in many emerging countries where agriculture is the main source of livelihood for a major part of the population. Trade integration and liberalization froze the support and trade policies and left emerging economies with very few policy instruments to protect themselves from food imports and to subsidize their agriculture. This book seeks to explain influences of trade liberalizations, gravitational effects of various regional trade blocks, sustainable development of agricultural production and rural territories, food security issues in the emerging part of the world. The objective of the book is the development of conceptual approaches to the establishment of food security and food sovereignty in the conditions of growing openness of domestic food markets to the world.

Since the research is carried out on the cases of emerging economies of Asia, Latin America, and Africa, its outcomes may be shared with a wide international network of stakeholders, including research institutions, universities, and individual researchers. The target audience of this book is composed of professionals and researchers working in the spheres of food security, international trade, rural development, agricultural production, and marketing. The book is appropriate for government officials, policy makers, and businesses of most of the emerging countries. Adaptation of research outcomes to the conditions of various regions (Southeast, Central, and Northeast Asia, Central and Eastern Europe, Latin America, and Africa) increases the omnitude of the publication and elaborates new practices and solution in the sphere of achieving food independence and food security, applicable in the rapidly changing international environment.

The book is structured in 16 main chapters, which address international trade in agricultural products and food within and between various emerging markets of Asia, Latin America, and Africa; protection and domestic support of agricultural production in emerging countries; effects of climate change on crop productivity and advanced farming technologies aimed at increasing of output; development of value-added chains in agricultural sector; potentials of traditional foods, local plants, and novel foods for ensuring food security of vulnerable populations in various parts of the worlds; food safety problems specific for emerging countries; threats to traditional way of life and economic activities in rural areas.

The authors of these chapters include scholars from 16 countries of Europe, Asia, Latin America, and Africa representing 20 respected research institutions, universities, and organizations. This variety of talent, expertise, and experience and the wide scope of topics make this book a unique collection of information and analysis, and a valuable resource for anyone studying emerging economies and specifics of establishing food security and alternatives to international trade on such markets.

Chapter 1 titled "Trade Relations Within and Between the Various Emerging Markets of Asia, Latin America, and Africa: How Globalization Affects Emerging Markets" is written by Bistra Vassileva. It presents the main issues around globalization and its effect on trade relations of emerging markets. In this chapter, based on extensive literature review the basic globalization trends with corresponding factors are revealed. The author's proposition states that globalization trends (including political, economic, and social globalization) affect trade relations between and within emerging markets. The analysis covers BRIC and CIVETS countries for the period from 1971 to 2013 (based on KOF Index of Globalization) and from 1960 to 2015 for the exports/imports flows. Several statements and propositions regarding trade relations between and within emerging markets are defined. The proposed implications for future research are focused on: exploring emerging markets as a system which variations can be modeled by stochastic processes, and applying network theory to explore the influences of trade liberalizations and gravitational effects of various regional trade agreements on the economic performance of major emerging markets.

Although globalization and free trade are the key factors in promoting economic growth, there are markets, which require specific approaches. The food market is one of them due to its vital importance for securing nutritional requirements of the population. Providing the population with food in sufficient quantity and variety is a challenge, which includes a range of issues of food production and dependence on import of the food market. Food security is increasingly influenced by foreign trade policies implemented by national governments. Many of them are re-examining their strategies for dealing with food security concerns and seeking to protect domestic food producers by restricting the presence of foreign competitors on the domestic food market. In the Chapter 2 titled "Trade in Agricultural Products and Food Security Concerns on Emerging Markets: How to Balance Protection and Liberalization" Vasilii Erokhin addresses recent food security trends in emerging countries, including BRICS, Eurasian Economic Union (EAEU), and Russia, and assesses the effects of food trade distortions in order to balance protection and liberalization policies for achieving effective levels of food security.

Titled "State Support of Agricultural Production in Emerging Countries as a Tool to Ensure Food Security" Chapter 3, written by Marina Lescheva, Anna Ivolga, and Oleksandr Labenko, examines the practices of state support of agricultural production in various emerging economies in comparison with selected OECD countries. The aim of their research is to discover how agricultural protectionism and support of domestic farmers affect the level of food security on the emerging markets in the conditions of expanding globalization and liberalization of trade in food. The authors focus on the evaluation of the best practices of state support and discovery of opportunities of their utilization on the emerging markets. The authors found out that the system of state regulation is one of the key determinants of achieving food security in the conditions of turbulent emerging markets, however, both the volume and priority directions of state support of agriculture are determined by financial capacities of emerging economies and current goals of their agrarian policies.

Chapter 4 by Vittorio D'Aleo, Francesco D'Aleo, and Roberta Bonanno titled "New Food Industries Toward a New Level of Sustainable Supply: Success Stories, Business Models, and Strategies" discusses the novel foods which are all those foods that do not fall within the traditional western diet. The authors underline that there is an important economic potential linked to this new sector. After having introduced the topic by identifying what is meant by new food, food sustainability, and entrepreneurship in this sector, they focus on the qualitative aspects of novel foods. The authors emphasize the advantages of these sources, their qualities, and the impact on the environment. In this chapter, the authors demonstrate that novel food sources can be a substitute for other food that requires production processes and thus make a big impact on the environment. After figuring out the advantages of novel food sources, the authors analyze some success stories from the companies that already operate in this area. They also investigate existing business models on the novel food market and the way they operate and create wealth. Summarizing the results of their research, the authors underline the economic potential of the new foods exploitation in the Mediterranean area.

Several chapters of the book address emerging economies of Asia. Ishita Ghosh and Ishita Ghoshal study food security issues in India in the light of the trade relations with the Association of Southeast Asian Nations (ASEAN). In Chapter 5 titled "Implications of Trade Liberalization for Food Security Under the ASEAN-India Strategic Partnership: Gravity Model Approach" the authors focus on understanding India's role in providing food security through trade with the ASEAN under the India-ASEAN Strategic Partnership. The authors employ a dynamic panel in order to assess and estimate the said objective, given that India and most of the ASEAN countries are considered emerging markets. Econometric investigation points out that while almost more than half the variables in the augmented gravity model hold good, food imports from India, agricultural labour force, common maritime border, distance from India, food price index, food production index, GDP (size) of the importing country, per capita GDP of the importing economy, and trade openness have statistically significant effect on the food trade from India to the ASEAN. The authors conclude that improving maritime infrastructure and agri-logistics, investing in climate change and water management and, and augmenting the agri-labour productivity are of paramount importance in order to improve food trade between India and the ASEAN.

In Chapter 6 titled "Maritime Piracy in Southeast Asia: Consequences on Economic and Political Integration" Giacomo Morabito, Vittorio D'Aleo, Davide Di Fatta, Roberto Musotto, and Walter Vesperi assess the effects of the maritime piracy for the food security of emerging countries of Southeast Asia. According to the authors, the region suffers from severe negative socio-economic impacts due to over-fishing, and the reduction and collapse of the fisheries have led to a widespread loss of income and employment. The problem is that the attacks on fishers take place regularly in most Southeast Asian countries and, in many cases, the pirates are the fishers themselves, and they attack fishers of other nationalities. All those attacks have a serious negative impact on food security in the region by damaging fishing industry and threatening artisanal fishers who risk losing their entire income.

Khabibullo Pirmatov, Jana Galova, and Elena Horska study food security issues in the emerging economies of Central Asia from a perspective of value-added agriculture. In their chapter titled "Value-Added Agriculture for Central Asian Countries" the authors analyze the socio-economic role of value-added agriculture for the countries of Central Asia. The agricultural sector of the region provides raw materials for the food, textile, and leather industry. Cotton, wheat, rice, and fruit play an important role in the foreign trade of each Central Asian country. These countries have unrealized potential for storing, freezing, processing, and packaging of the wide nomenclature of fruit, food production, and drinks with the organization of their further exports to the perspective markets. Adding value to agricultural products lead to increasing the share of finished goods in export, supplying import-substituting products, improving infrastructure in rural areas, providing new jobs, and growing farmers' income. Based on the analysis, the authors recommend using value-added agriculture for the Central Asian countries by attracting domestic and foreign investments to rural areas, establishing tax incentives, and allocating preferential credits for agribusiness.

The theme of establishing food security in Asian countries is continued by Gao Tianming. In his chapter titled "Food Security and Rural Development on Emerging Markets of Northeast Asia: Cases of Chinese North and Russian Far East" he analyses the impact of economic globalization on food security and sustainable rural development in the regions of Russian Far East and Northeast China which are the important regions affecting the sustainable development of the world because of their rich natural resources, good environment condition, and large potential for economic development. The author complains that due to a number of historical and geopolitical reasons, the cooperation between Russia and China in the region of Northeast Asia is progressing, but at a quite slow speed and in an unsustainable manner. The chapter addresses some problems and challenges of sustainable rural development in the Heilongjiang Province, P.R.China and neighbor cross-border regions of Russia (Far East Federal District and Siberian Federal District), investigates similarities and differences of rural way of life in those regions, and discusses measures to promote the cooperation between the two countries and ensure sustainable rural development and food security in the macro region of Northeast Asia.

In emerging countries, the agricultural sector has always been at the center of reforms. Khatai Aliyev and Ilkin Gasimov in their chapter titled "Retrospective of Economic and Trade Policies Focused on Agricultural Development: Case of Azerbaijan" overview economic and trade policies focused on the development of agricultural production since 1991. Authors carry out analysis of policy changes during separate development stages. The research output presents agricultural policy before the oil boom as mainly devoted to achieving structural transformation from centrally planned economy to the market environment. Within the oil boom period, the government provided substantial fiscal and technical support to the agricultural sector as well as applied tax incentives to farmers but did not pay attention to the transformation from family farming to medium and large-scale production. The authors conclude that new macroeconomic challenges experienced by emerging economies force the governments focusing on increasing efficiency of the subsidies and incentives and implementation of further agricultural reforms.

Three chapters of the book address food security issues in Africa. In Chapter 10 titled "Effects of International Trade Agreements on the Economy and Society of Africa: Special Focus on Nigeria" Henrietta Nagy, József Káposzta, Gyorgy Ivan Neszmelyi, and Omokheka Gregory Obozuwa give an overview of the trade agreements existing in Nigeria to find out their contribution to the development of the economy. There are several such agreements, however, the authors try to see how much they really operate and serve the wellbeing of people as well as food security and safety. There have been various initiatives in the countries of Africa to develop the agricultural sector, which has employed millions of people and has had a significant role in keeping the population in the rural areas. However, the authors claim that less progress has been achieved than expected. In this chapter, the authors intend to analyze the situation in Africa, with a special focus on Nigeria, to summarize the achievements, and to list up some recommendations on future measures.

Feeding the quickly growing population in Africa remains a global challenge. As the demand for food increases, climate change, on the other hand, poses more challenges to agricultural productivity, implying that the provision of sufficient quantities and qualities of food is threatened. Praxedis Dube, Wim Heijman, Rico Ihle, and Justus Ochieng in their Chapter 11 titled "The Potential of Traditional Leafy Vegetables for Improving Food Security in Africa" assess the potential of traditional leafy vegetables in improving food security in Africa. Traditional leafy vegetables in Africa are resilient to adverse weather conditions and are naturally rich in nutrients including vitamins A and C, iron, protein and other micronutrients. Traditional leafy vegetables represent a robust local source of food, their consumption is crucial in complementing Africa's diet with micronutrients. Traditional leafy vegetables have therefore the potential to play a major role in improving food security and facilitating food sovereignty.

Effect of climate change on crop productivity and food prices are assessed by Goodness Aye and Ruth Harunah in the chapter "Effects of Climate Change on Crop Productivity and Prices in Benue State, Nigeria: Implications for Food Security". On the case of Benue State, Nigeria, the authors use the time series data on selected output of crops (maize, rice, sorghum, yam, millet, groundnut, beans, and cassava), the area planted, price, and climate variables such as rainfall, temperature, and sea level. Due to differing periods of data availability, their study used the time period 1995-2009 for analysis, in order to maintain a common period for all the series. First, the trend of productivity, prices, and climate change was analyzed using visual plots and results indicate some level of variability in these series over time. Second, a three-stage least square regression was used to simultaneously analyze the effect of climate change on productivity and prices. The authors show that climate change had a significant impact on the productivity of millet, sorghum, cassava, and groundnut while it had a significant impact only on the price of maize.

These findings have important implications for food security not only in Nigeria but also in other countries of Africa and other emerging economies worldwide. As long as agriculture is among the sectors strongly affected by climate changes and droughts, which simultaneously influence food security through the volume of production and safety of products, the book pays much attention to such issues as farming technologies and innovations. Marko Jeločnik and Jovan Zubović in their chapter titled "Irrigation and Food Security: Case of Soybean Production in Serbia" investigate the ways of increasing soybean output by means of irrigation. According to the authors, in the climate conditions of the Republic of Serbia soybean is predominantly produced in the system of dry land farming. The authors analyze the economic importance of soybean production for the Republic of Serbia (with a special focus on sustainability of national food security). One of their goals is the assessment of cumulative losses incurred in the soybean production in the climatic conditions of the Republic of Serbia (dominantly caused by drought) in the last several years. They also estimate the possible positive impact of irrigation on the current state of food security on the macro level, as well as the higher profitability of soybean production on the micro level.

Jesus Gonzalez-Feliu, Laura Palacios-Arguello, Carlos Osorio-Ramírez, and Carlos Talamantes in their chapter titled "Local Production-Based Dietary Supplement Distribution in Emerging Countries: Bienestarina Distribution in Colombia" address the production and distribution of Bienestarina to the vulnerable population of Colombia as one of the strategies of the Colombian Institute of Familiar Wellness to fight malnutrition, especially among children. This case is a good example of establishing food security and social improvement logistics that merits particular attention. The chapter presents an analysis of the Bienestarina supply chain based on the four elements: steering, organization, development, and financial issues. First, an overview of social improvement logistics and the Bienestarina context is provided. Second, theoretical frameworks related to the case are presented. Third, the case is described on the basis of the proposed analysis framework. Finally, generalization issues and conclusions allow the authors proposing the first characterization of social improvement logistics.

The book is finalized with two chapters discussing food security in emerging economies from a point of view of food safety. In such a light, food safety can be considered in various dimensions, one of which is consumer perception. This view of food safety, named in the literature as subjective safety, does not necessarily correspond to objective food safety and is also considered as not being easily influenced by education of consumers. Having that in mind, Ivana Domazet and Nenad Djokic in Chapter 15 titled

"Food Safety from Consumer Perspective: Consumer Confidence in Food Safety" examine different approaches to food safety and the results of previous research dealing with consumer perspective of food safety. Besides giving broad literature review of the topic, the goal of this chapter is to implement Consumer confidence in food safety scale in Serbian consumer research. The validity, reliability, and one-dimensionality of this scale, which consists of optimism and pessimism (two dimensions of confidence), are tested and determinants of different rating on it are examined. The research from 2016 includes 300 participants in the main survey. Finally, starting from profiling consumers with different levels of general confidence in food safety, the authors provide managerial recommendations.

Another specific sphere of food safety is the quality of animal products and animal health. The book is concluded with Chapter 16, titled "Improvement of Food Security Through Reforming of Domestic Veterinary Service: Case of Russia", written by Anna Ivolga, Vladimir Trukhachev, Natalia Bannikova, and Anzhelika Baicherova. Using the case of Russia, the authors give an overview of current issues of achievement of food security through reforming of veterinary service in the light of the social, ecological, and economic development of the society. The authors analyze certain challenges of food security existing in emerging countries, including expansion of market relations within veterinary service. The results of organizational and economic reforms of veterinary service are assessed on the case of Russia. Such issues as governmental and public regulations of quarantine operations and the most dangerous animal diseases are also discussed. In order to improve food security, the authors justify the measures for competition and demand stimulation, smoothing consequences of market imperfections, encouraging efforts on reducing the negative impact of livestock breeding on the environment.

This book attempts to develop a long-term outlook of those challenges the emerging economies face on their way to the establishment of food security. There are various factors influencing food security on the often-turbulent emerging markets. Direct-acting ones are those related to administration of availability of food on the domestic market, including regulation of domestic food production through support and imports through various forms of protectionism, including food import bans. Indirect-acting factors are those, which determine consumption behavior on the domestic market, i.e. dietary habits, the level of life, inflation, the purchasing power of population, etc. From the above reasoning, assessing food security should be undertaken with consideration of both administrative regulations and economic patterns of domestic food market of any emerging country.

Vasilii Erokhin
Harbin Engineering University, China

Acknowledgment

This book is a product of research and support of many individuals to whom I would like to express my gratitude.

I thank all the chapter contributors representing 16 countries and 20 respected universities, research institutes, and organizations. My acknowledgment is made to the members of the Editorial Advisory Board and international team of reviewers for sharing their precious time, professional experience, valuable advices, and insights. I give my special thanks to Prof. Andrew Schmitz from the University of Florida (USA) and Dr. Shikha Jha from the Asian Development Bank (Philippines) who both kindly prepared their outstanding forewords.

I wish to express my sincere thanks to the School of Economics and Management, Harbin Engineering University, China, for providing me with all the necessary facilities. I would like to personally thank Dr. Gao Tianming, Director of the Polar Development and Northeast Asian Economic Research Center, China, for his full support and encouragement throughout my work.

I sincerely thank the entire IGI Global team for assistance and support throughout the book development process.

I am truly indebted to my wife, Anna, who shared her continuing support, both morally and physically, throughout the work on this book.

Chapter 1
Trade Relations Within and Between the Various Emerging Markets of Asia, Latin America, and Africa:
How Globalization Affects Emerging Markets

Bistra Vassileva
University of Economics – Varna, Bulgaria

ABSTRACT

This chapter presents the main issues around globalization and its effect on trade relations of emerging markets. Based on extensive literature review the basic globalization trends with corresponding factors are revealed. The author's proposition states that globalization trends (including political, economic, and social globalization) affect trade relations between and within emerging markets. The analysis covers BRIC and CIVETS countries for the period from 1971 to 2013 (based on KOF Index of Globalization) and from 1960 to 2015 for the exports/imports flows. Several statements and propositions regarding trade relations between and within emerging markets are defined. The proposed implications for future research are focused on: exploring emerging markets as a system which variations can be modeled by stochastic processes, and applying network theory to explore the influences of trade liberalizations and gravitational effects of various regional trade agreements on the economic performance of major emerging markets.

INTRODUCTION

Globalization is signified as a quintessence of the last decade of the 20th century. Its effect on international trade is outstanding since the globalization contributes a great deal to the international markets interconnectedness. The shifting balance of power on the global markets has been indicated as a transition from Globalization 2.0 (Western-dominated) to Globalization 3.0 (China-dominated) (Walker, 2007).

DOI: 10.4018/978-1-5225-2733-6.ch001

However, the optimistic forecasts for continuing high growth of emerging markets (especially for the abbreviated blocks of countries, such as BRICS (Brazil, Russia, India, China, and South Africa), MIST (Mexico, Indonesia, South Korea, and Turkey), and CIVETS (Colombia, Indonesia, Vietnam, Egypt, Turkey, and South Africa)) started to fade away during the last few years. The average growth rate in the emerging world fell back to 4% in 2013 (Sharma, 2014). Some of the reasons include the following: too much focus on one factor – demography, linear extrapolation of the trends, political myopia, and ignorance of global factors.

The business environment nowadays is characterized by increasing complexity, uncertainty, and discontinuity. Changing market conditions, intensified global competition, and increasingly shorter product life cycles mean that countries and companies have to re-examine the traditional methods and strategies for doing business in international markets (Bartlett & Ghoshal, 1987; Ohmae, 1989). Global competition is growing and will continue to increase. According to the competitiveness roadmap 2007-2050 (Garelli, 2007)[1] among the main issues with a high impact on world competitiveness environment during the next three decades are the following ones:

- Protectionism is on the rise;
- Service and integration are becoming key competitiveness factors;
- Labor cost differences shrink;
- Technological divide disappears;
- China, India, and Russia emerge as technological powers;
- Climate change affects economic resources.

It is argued that protectionist reprisal (including environmental protection, corporate governance, social protection, or intellectual property) will increasingly confront acquisitions pursued by some of emerging economies due to perceptions of a potential loss of economic power and national image. Emerging economies, as an opposite reaction, will significantly increase pressure to gain access to decision-making in international institutions by emphasizing their predominant economic weight and their financial capabilities to fund such institutions (Garelli, 2007). Analysis of the situation from these two contrasting points of view provides a background for researchers (Le Pere, 2005, p. 36) to conclude that there are two opposite dimensions of globalization. The first dimension is characterized with bringing the economies and societies closer together because of the flourishing interrelations of trade, investment, technology, cross-border production systems, and information and communication flows. The second dimension corresponds to the new transnational dynamics which marginalize and exclude a great proportion of the world's countries and people.

BACKGROUND

The world is changing at a speed which has never been seen before. Years of research done by the McKinsey Global Institute (MGI) and McKinsey's Strategy Practice (Dobbs, Ramaswamy, Stephenson, & Viguerie, 2014) reveal three major economic forces the global economy has never previously experienced: the collision of technological disruption, rapid emerging-markets growth, and widespread aging. Much bigger shifts in each of these areas are expected which will tremendously affect the economy, social life, and personal behavior worldwide.

First, technology and connectivity have disrupted industries and transformed the lives of billions of people in their different roles as workers, consumers, and citizens. The KPMG report on complexity (KMPG, 2011) shows that technology is changing business models, improving processes, and opening new markets, but also creating volumes of new data that must be managed, supported, and secured. More transactions are taking place across more borders. The changing global regulatory environment is forcing businesses to react to ensure compliance while at the same time managing new risks. We are witnessing an extraordinary growth in computing capacity, power, and speed of penetration of information and communication technologies (ICT)[2]. This acceleration in the scope, scale, and economic impact of technology will be supplemented by a new age of artificial intelligence, consumer products and services, instant communication, and unlimited information which in turn will distress business in unthinkable ways. With instant information and communication, virtually everything is available to anyone, anywhere. Markets are now global and many corporations are often richer and more powerful than many countries.

Second, the world's economic center of gravity has continued shifting from the West to the East, with China being at the center of the trend. This shifting locus of economic activity and dynamism to emerging markets and to cities within those markets, will give rise to a new class of global competitors, both companies and brands. The global urban population is growing by 65 million a year, and nearly half of global GDP growth between 2010 and 2025 will come from 440 cities in emerging markets, 95% of them being almost unknown small and medium-sized cities in emerging markets. This shifting balance of power has been indicated as a transition from Globalization 2.0 (Western-dominated) to Globalization 3.0 (China-dominated) (Walker, 2007)[3]. Globalization 3.0 is characterized by the fact that the West no longer dominates the world's savings, and as a result, no longer dominates global investment and finance. The erosion of Western power is accompanied by the erosion of the authority of the grand institutions of Globalization 2.0, which sustained power by enforcing the implicit rules of Western economic orthodoxy. This situation is confirmed by the 2016 FDI Confidence Index ranking and scores[4]. The first and second places are occupied by the United States and China with maintained ranking from 2015 (respectively, 2.02/2.10 for the United States and 1.82/2.00 for China out of maximum 3.0)[5].

Third, rapid aging of the world's population will create a massive set of economic pressures. The baby boomers have begun retiring. Aging has been evident in developed economies for a few years, with Japan and Russia seeing their populations decline and the trend is spreading slowly to China. It is expected that during the next few years it will "reach" Latin America.

It is suggested that during the collision of these three forces, the resulting change will be so significant that much of the management expertise, know-how, and intuition that has survived in the past will become irrelevant. Both countries and companies will be faced with more discontinuity and volatility, with long-term charts no longer looking like smooth upward curves, with outdated long-held assumptions, and useless formerly powerful business models (Dobbs et al., 2014). The aggregated effect of their activities will affect in turn trade relations on a global scale, including the agricultural sector. There is no doubt that globalization is a fact, and that it will continue to affect companies and their trade activities providing global market opportunities and/or raising global competitive threats. Combined with global market uncertainty, globalization can be considered to be a global business driver (Table 1).

Several case studies (Hartzenberg, 2011; Hopewell, 2014; Gavin & Sindzingre, 2009) illustrate the effects of the abovementioned globalization shifts on regional development with a focus on the agricultural sector. The implications of the theory of dependent development are demonstrated by Hopewell (2014, p. 292) based on the situation in Brazil where the emergence of a highly competitive export-oriented agribusiness sector has prompted the expansion and internationalization of domestic capital. The rise of

Table 1. Global business drivers

Driver	Description
Globalization	■ Realignment of global supply chain. ■ Economic shift from global North to South. ■ Changing global governance. ■ Elevated financial volatility and risk. ■ More inclusive globalization.
Consumer behaviour	■ New models of consumer engagement. ■ The insatiable consumer. ■ New era of squeezed profitability.
Demographics	■ War for talent. ■ Growing infrastructure needs.
Regulation and activism	■ Changing role of government. ■ Rising geopolitical instability.
Environment and natural resources	■ Depleting natural resources. ■ Responding to sustainability challenge.
Technology	■ New paradigms in product design and manufacturing. ■ New and innovative R&D models.

Source: Adapted by A.T.Kearney (n.d.)

Brazilian agribusiness triggered the state and capital to allied together to aggressively pursue the expansion of markets for Brazilian exports, even to set negotiations and dispute settlement with World Trade Organization (WTO). As a result, Brazil continues to be a major exporter of agricultural products among top six exporters (third place after the EU-28 and the USA according to the World Trade Organization (2014a, p. 54). These top six places are equally distributed among developed countries (the EU-28, the United States, and Canada) and emerging markets (Brazil, China, and India). The top six exporters of agricultural products saw growth in 2013 (except the EU-28 and the United States), with India showing the greatest increase (11%).

Based on her comparative research on regional integration in Africa, Hartzenberg (2011, p. 10) concluded that regional integration could be a rational response to the difficulties faced by Africa, a continent with many small national markets and landlocked countries. According to the results of her research, on average, only about 10% to 12% of African trade takes place amongst African nations. Africa's exports are still targeted mainly (above 80%) to international markets, with the EU and the United States accounting for more than 50% of total Africa's export. China is also an important export market for African countries. The emergence of China on global trade scene provoked the attention of researchers (Blázquez-Lidoy, Rodríguez, & Santiso, 2006) on Chinese impact on global export flows which is considered to be both a threat and opportunity.

International trade relations are affected by many and various actors on the global scene which pursue different goals using numerous means to achieve them. These actors are located at different levels of the pyramid of international trade relations, including the following:

1. International organizations and institutions (the United Nations (UN), the World Bank, the World Trade Organization (WTO), the International Monetary Fund (IMF), the Food and Agriculture Organization of the United Nations (FAO), etc.);

2. Regional and sub-regional blocks (Mercosur, the European Union (EU), the Eurasian Economic Union (EAEU), the Caribbean Single Market and Economy (CSME), three regional integration arrangements (the Economic Community of West African States (ECOWAS), the Preferential Trade Area (PTA) – the precursor of the Common Market for Eastern and Southern Africa (COMESA); and the Economic Community of Central African States (ECCAS) supported by the Economic Commission for Africa (ECA)), etc.);
3. Customs and monetary unions (the Economic and Monetary Community of Central Africa (CEMAC), the Andean Community (CAN), the East African Community (EAC), etc.);
4. Multilateral free trade area (the Central European Free Trade Agreement (CEFTA), the Common Market for Eastern and Southern Africa (COMESA), the European Free Trade Association (EFTA), the Greater Arab Free Trade Area (GAFTA), the North America Free Trade Agreement (NAFTA), etc.);
5. Governments (states) and state-owned agencies;
6. Multinational corporations (mncs) and transnational corporations (tncs);
7. Companies;
8. Farmers;
9. Households and individuals.

The homogenization of policies and institutions is considered as a side effect of globalization (Le Pere, 2005) which interconnects the abovementioned two opposite dimensions of globalization. It is characterized by trade and capital market liberalization, the establishment of international standards for labor and the environment, implementation of regulatory agreements on international trade, and other standardization activities. Policy liberalization is studied by Sally (2011, p. 5) as one of the main engines of modern globalization, technology being the second one. These policies and measures toward homogenization generate the framework of international rules, norms, and practices to regulate international trade relations. The topic about the effects of harmonization and liberalization is also controversial, especially regarding the politics. According to Le Pere (2005, p. 37), globalization of the world economy and related markets creates new instabilities, insecurities, and inequalities. Heribert (2004, p. 40) pointed out that markets cannot carry the task of "shaping the globalization since it is a task for a political sphere". Based on the abovementioned research results and conclusions, it could be proposed that the overall resulting effect (positive or negative) of Globalization on emerging markets and their trade relations depends on the capability of political forces to overcome the negative consequences of international volatility of trade, capital flows, and production by reasonable political action to offset the risk and uncertainty. Otherwise, the global insecurity will be given a rise (Kaplinsky, 2001).

GLOBALIZATION AND TRADE RELATIONS OF BRIC AND CIVETS MARKETS

Issues, Controversies, Problems

Trade interdependence between continents, regions, and countries has rapidly increased in recent years. Several geo-economic shifts affect trade interdependence between continents and between countries, namely:

1. Globalization vs. Deglobalization;
2. North vs. South;
3. Protectionism vs. Liberalism;
4. Divergence vs. convergence between the West and emerging markets.

As it was presented above, these shifts create controversies which are further amplified by the increasing tension between political, economic, and social globalization. The analysis of selected emerging markets below takes into account these three components of the globalization which are measured by the KOF Index of Globalization.

Trade relations depend substantially on the interactions between states and international markets which in the scope of the four abovementioned geo-economic shifts reflect on the increasingly interdependent relationship between the roles of the state and the market. The role of the state is utilized through politics by the means of currency regimes, capital movements, standards, and the financial foundations of systemic stability. Because of the strong interrelations between politics and economy, economic globalization becomes very much power-centered. It is debatable if global trade governance declines due to the proliferation of the free trade agreements (FTAs). According to the World Trade Organization (2014b), there are over 200 FTAs in function today. Three-quarters of world trade will soon be governed by free trade agreements that are negotiated outside of the WTO.

Trade interdependence between continents is especially evident for the Europe–Asia trade flows which now constitute a "third link" in the global economy (Gavin & Sindzingre, 2009). South-South trade flows have also increased substantially. The share of export as a percentage of GDP is used to present the role of emerging markets in trade relations on the global market.

Comparative Analysis of Globalization Trends in Emerging Markets

Over the years, researchers and practitioners have debated country classification issues (Nielsen, 2011, p. 7). Several classification systems have been developed by international organizations and have been widely recognized. The country classification system by the United National Development Programme is built around the Human Development Index (HDI) launched together with the Human Development Report (HDR) in 1990. The classification systems in the World Bank are developed both for operational and analytical purposes. The operational classification satisfies the needs of the World Bank's International Bank for Reconstruction and Development (IBRD) which has a statutory obligation to lend only to creditworthy member countries. That is why objective criteria are needed to assess the countries which need a credit. Under this system, countries that borrow from the IBRD and exceed a certain income threshold engage in a process that moves the country to non-borrowing status. The analytical country classification was constructed by the World Bank in 1978. It has been modified in 1989 when the countries were divided into categories based on the income level (Table 2).

Similar to the World Bank, classification systems in the International Monetary Fund (IMF) are used for both operational and analytical purposes. Several analytical classifications have been developed during the years starting from 1948. According to the IMF (Abiad, Bluedorn, Guajardo, & Topalova, 2012), emerging market economies (EMs) and low-income countries form a single category which is named Emerging Market and Developing Economies (EMDEs). According to the criteria established by the IMF, an emerging market is defined by a GDP-per-capita ratio that ranges between \$2,000 and \$12,000.

Table 2. The World Bank's classification of countries

Type of countries	Number	GDP, $ mln	Population, million people	GDP per capita, $
Low income countries	59	1,165,000	3,094	380
Lower middle income countries	69	1,635,000	1,099	1,490
Upper middle income countries	42	2,135,000	498	4,320
High-income countries	39	19,304,000	834	23,150

Source: World Bank, 2000

The comparative analysis focuses on BRIC[6] and CIVETS[7] countries as the most popular emerging markets. For the purposes of analysis, several secondary sources of information are used (Vassileva & Nikolov, 2016, p. 300). Countries are comparatively analyzed based on the composite elements of the KOF Index of Globalization, namely economic globalization, actual flows restrictions, social globalization, personal contact, information flows, cultural proximity, political globalization, and the overall globalization index itself.

According to the overall globalization index, there is a strong positive tendency toward globalization for BRIC and CIVETS countries (Figure 1) which trend is quite visible after the 1990s.

As the author presents in Figure 2 through Figure 8, there are obvious differences in the structure of the globalization process between the analyzed countries, especially for the elements "personal contact" and "cultural proximity".

Two "jumps" in cultural proximity (1989 and 1997) are evident in China. Personal contact remains at a low level with values quite similar to those of India (Figure 3).

India presents an appealing case with a level of political globalization which differs substantially from the other elements of the KOF composite index. Russian Federation is an interesting and challenging market because of its instability. As it shown in Figure 2, there is a serious "leap" in cultural proximity in 1998. During the last two years, few negative trends are observed. There is a slight decline in information flows, actual flows, and restrictions. As it was mentioned in the introductory part of the chapter, protectionism is on the rise nowadays which causes problems, especially for investment strategies.

There is a group of scholars and researchers (Ryder, 2005; Laudicina, 2012) who suggest that the locus of global economic, political and demographic power has been shifting with growing intensity from the Global North (developed countries) to the Global South (developing countries) as well. Both trends affect global supply chains decisions, especially their sourcing locations as well as expansion strategies of the companies. Brazil could serve as a good example (Figure 2) with few fluctuations during the years. India (Figure 3) differs a lot compared to the other BRIC countries. Political globalization shows a permanent trend to increase while the growth rate for the rest of the factors is almost flat. There is a substantial rise in cultural proximity in 1995 but the personal contact did not change during the analyzed period.

Detailed data about the analyzed markets in regard to the components and indicators comprising KOF Index of Globalization are provided in Table 3.

Comparing BRIC country profiles for 2013, it is evident that all countries are at the same (almost maximum) level for political globalization (Figure 4). Regarding cultural proximity and personal contact, the following subdivision can be done: China and Russia share the same level (around 80) of cultural proximity while the value of this index component for India and Brazil is approximately 40. As for the personal contact, Russia differs substantially from the rest (Brazil, India, and China).

Figure 1. Comparison of BRIC and CIVETS by the Overall Globalization Index (KOF Index of Globalization) in 1971–2013
Note: Data on Russia are not available from 1971 to 1989
Source: Swiss Federal Institute of Technology Zurich, 2016

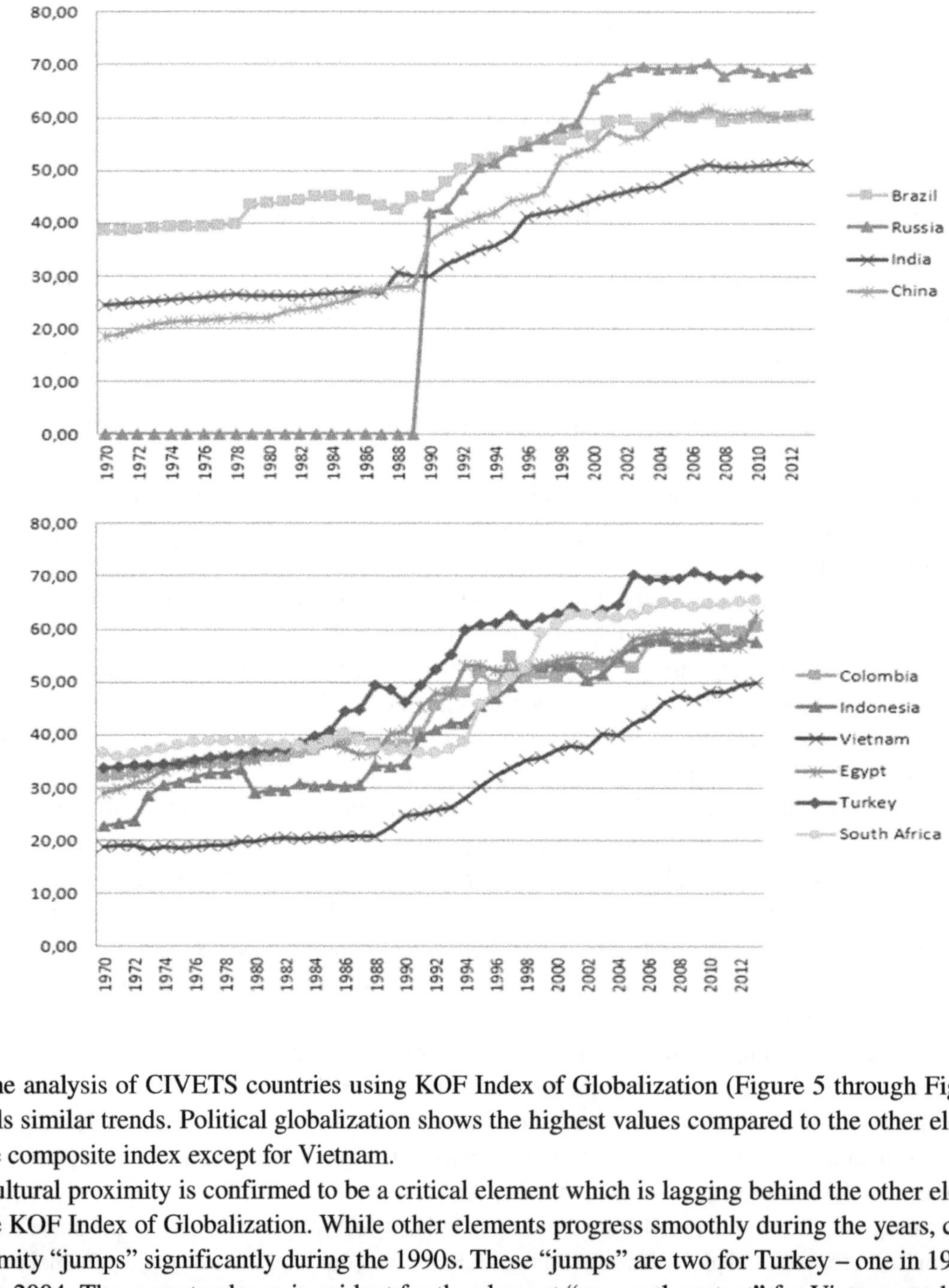

The analysis of CIVETS countries using KOF Index of Globalization (Figure 5 through Figure 8) reveals similar trends. Political globalization shows the highest values compared to the other elements of the composite index except for Vietnam.

Cultural proximity is confirmed to be a critical element which is lagging behind the other elements of the KOF Index of Globalization. While other elements progress smoothly during the years, cultural proximity "jumps" significantly during the 1990s. These "jumps" are two for Turkey – one in 1986 and one in 2004. The same tendency is evident for the element "personal contact" for Vietnam and Egypt. Cultural proximity, personal contact, and information flows draw together the element "social globalization".

Figure 2. KOF Index of Globalization for Brazil and Russia in 1971–2013
Source: Swiss Federal Institute of Technology Zurich, 2016

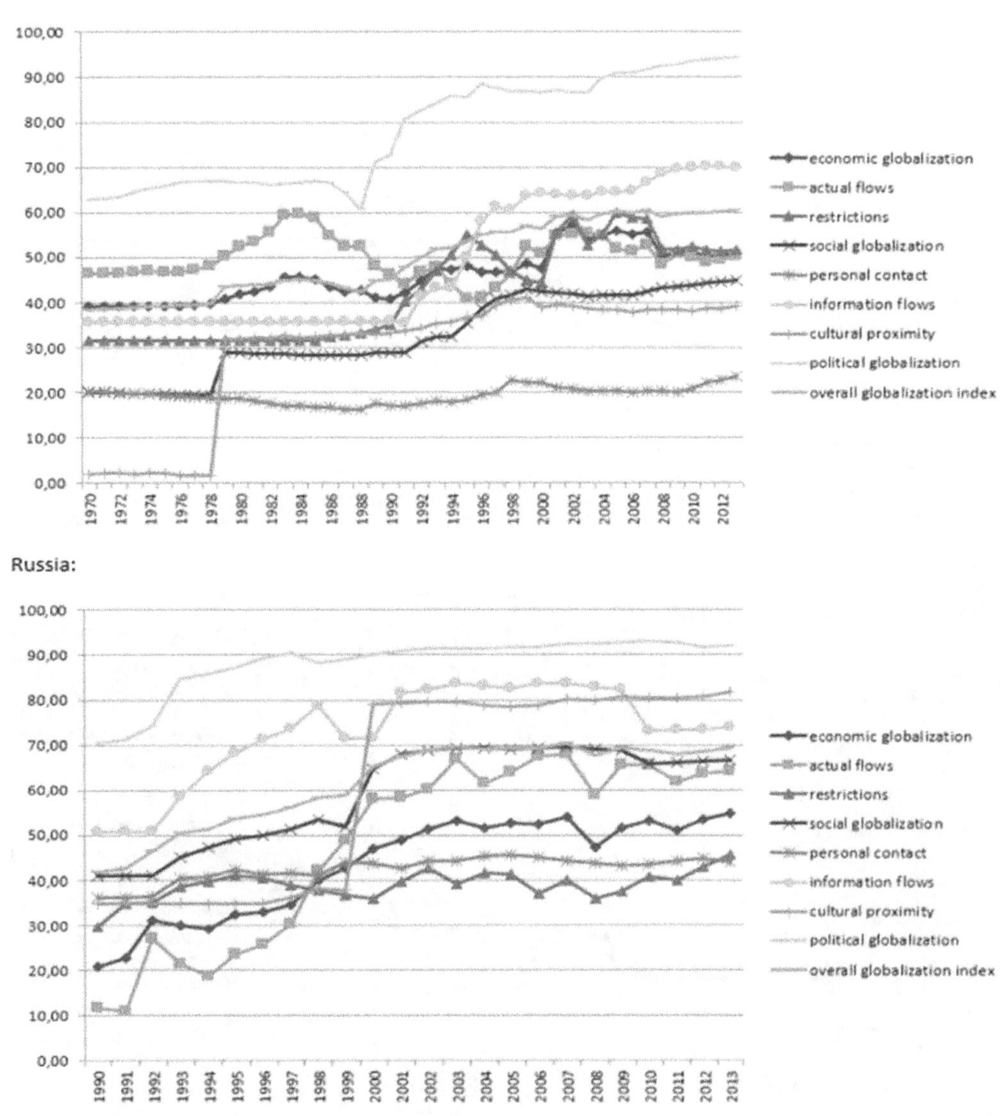

The level of social globalization is quite below the economic and political globalization for the analyzed countries. Vietnam presents an extreme example of the discrepancy between the main components of the KOF Index of Globalization. Quite similar is the situation with Egypt (Figure 6).

Regarding South Africa (Figure 7), a sharp improvement in cultural proximity in 1994 is observed along with a steady growth of political globalization after 1994. Information flows and social globalization are raising as well.

Comparing CIVETS country profiles for 2013 (Figure 8), the same conclusion regarding political globalization as for BRIC countries can be made. Only Vietnam differs with a value of 55.27. The average value of this element of the KOF Index of Globalization for Vietnam is quite low (37.94) especially

Figure 3. KOF Index of Globalization for India and China in 1971–2013
Source: Swiss Federal Institute of Technology Zurich, 2016

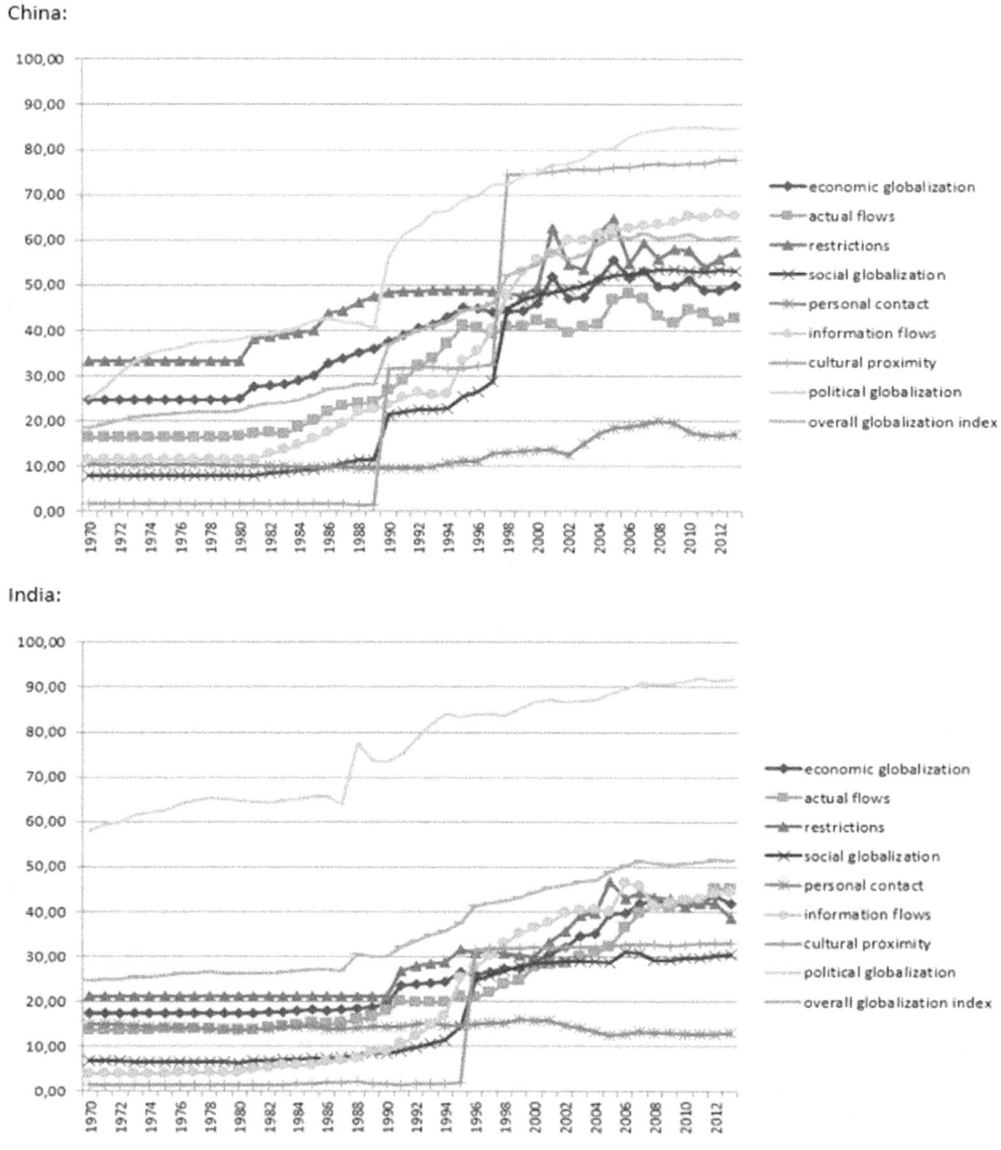

with a comparison to the other countries under analysis. Vietnam profile contrasts considerably from the globalization pattern of the rest of the countries. Cultural proximity can be used as a benchmark to segment the emerging country markets. Indonesia, Colombia, and South Africa share the same level (around 40) of cultural proximity while the value of this index component for Turkey and Egypt is approximately 80.

Based on the results of the above analysis highlighting the importance of social globalization, especially cultural proximity and personal contact, a comparative analysis of BRIC and CIVETS countries using data on Hofstede cultural dimensions is performed (Figure 9).

Table 3. Comparative analysis of BRIC and CIVETS countries, KOF Index of Globalization, 1970-2013, average values

	Brazil	Russia	India	China	Colombia	Indonesia	Vietnam	Egypt	Turkey	South Africa
Economic globalization	46.05	43.29	25.83	38.11	39.27	43.85	36.30	37.49	45.42	59.62
Actual flows	49.88	47.69	22.81	30.15	36.88	48.91	42.19	45.01	33.09	58.30
Restrictions	42.22	38.89	28.85	46.07	41.66	38.79	30.27	29.96	57.74	60.94
Social Globalization	32.83	59.25	16.39	26.83	36.96	19.65	17.37	31.12	40.57	34.18
Personal contact	19.38	42.48	14.13	12.46	25.24	14.71	13.80	30.45	42.01	45.43
Information flows	48.59	72.89	20.43	33.64	62.69	25.86	33.99	42.85	47.32	36.47
Cultural proximity	29.16	61.45	14.20	34.18	20.29	17.83	1.44	18.64	31.47	19.86
Political Globalization	77.91	88.18	76.28	58.05	62.25	68.25	37.94	75.87	75.75	49.87
Overall Globalization index	49.82	61.50	36.09	39.35	44.69	41.50	29.70	45.60	51.89	47.48

Note: Average values for Russia are calculated for the period 1990-2013
Source: Swiss Federal Institute of Technology Zurich, 2016

Figure 4. BRIC comparative country profiles by KOF Index variables in 2013
Source: Swiss Federal Institute of Technology Zurich, 2016

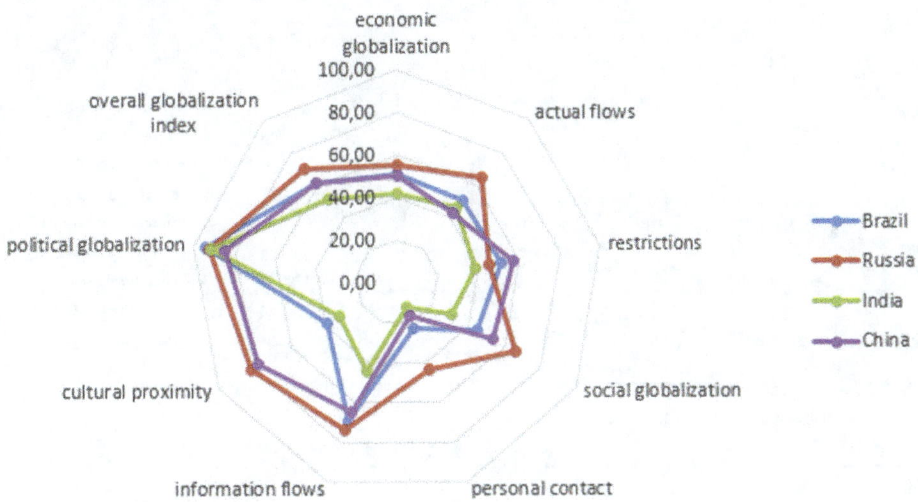

Russia, Indonesia, India, and China are quite similar on Power Distance (PDI) dimension (93–77), followed by Vietnam, Egypt, Brazil, and Turkey (70–66). South Africa (49) differs substantially from the rest. Power Distance deals with the fact that the individuals in societies are not equal. Such countries are quite centralized which increases macro risk levels. Macro risk levels extensively affect trade relations between countries. The score on Indulgence (IND) and Long Term Orientation (LTO) dimensions fluctuate considerably among analyzed countries. The following three segments are defined based on these two dimensions:

Figure 5. KOF Index of Globalization for Colombia and Indonesia in 1971–2013
Source: Swiss Federal Institute of Technology Zurich, 2016

Colombia:

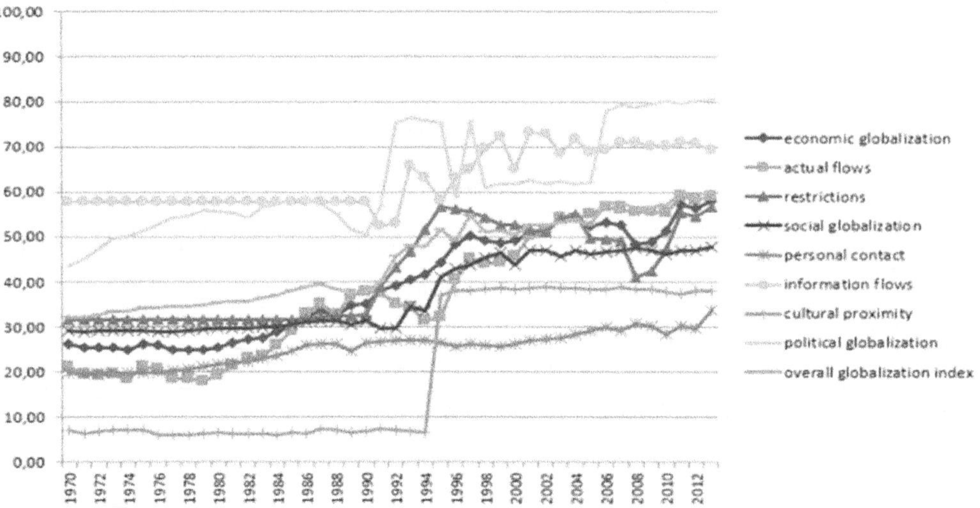

Indonesia:

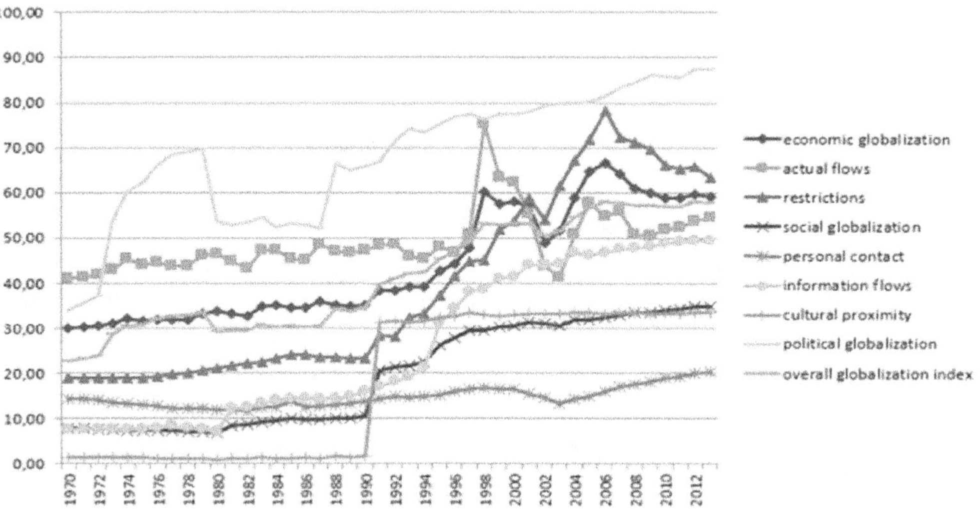

1. Russia and China (LTO ≈ 80 / IND ≈ 20);
2. Indonesia and Vietnam (LTO ≈ 60 / IND ≈ 35);
3. Brazil and Turkey (LTO ≈ 45 / IND ≈ 50).

The high score on Long Term Orientation suggests that Russia and China are definitely the countries with a pragmatic mindset, i.e. people in these countries show an ability to adapt traditions to changed conditions easily. This might have a positive effect on trade negotiations. At the same time, a low score on IND suggests that these societies suppress gratification of needs and apply regulation by using strict social norms. Russia and China can be defined as "restrained" cultures according to the low score on

Figure 6. KOF Index of Globalization for Vietnam and Egypt in 1971–2013
Source: Swiss Federal Institute of Technology Zurich, 2016

Vietnam:

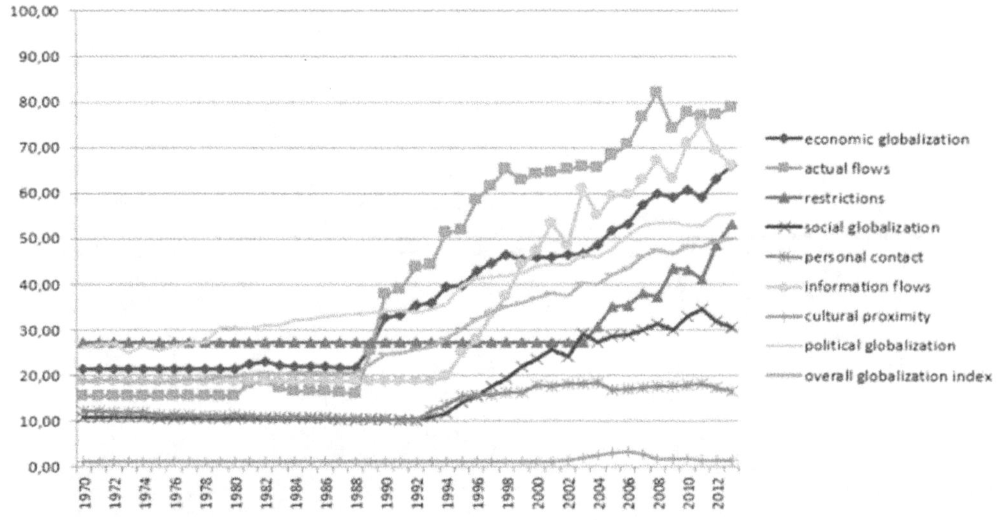

Egypt:

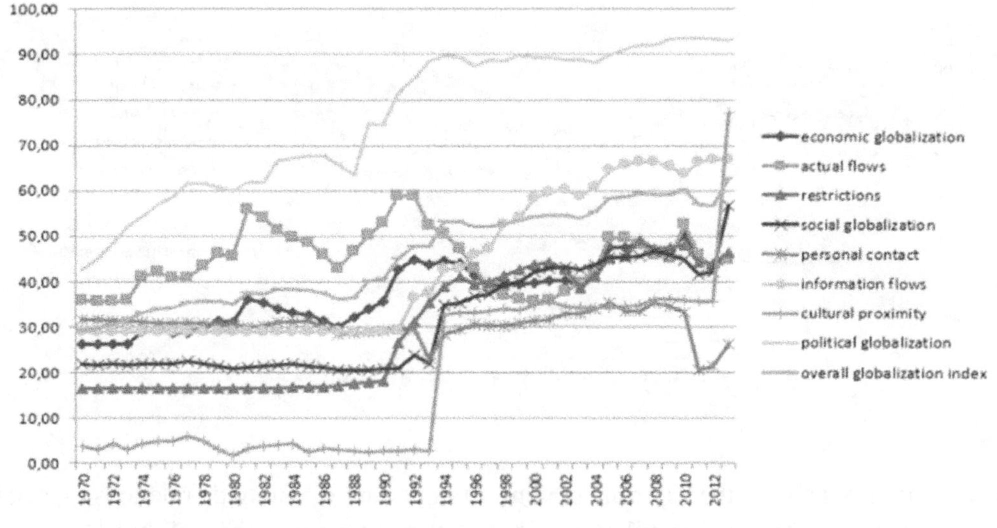

Indulgence dimension. Societies with a low score in this dimension have a tendency to cynicism and pessimism because of the perception that human actions are restrained by social norms.

Export and Import Flows of Emerging Markets

According to the statistics of the United Nations (Nicita, 2015), world trade in goods has increased intensely over the last decade, rising from less than $8.0 trillion in 2003 to more than $18.5 trillion in 2013. Trade in services has also greatly increased between 2003 and 2013 (from about $2.0 trillion to about $4.7 trillion). Trade flows are disproportionately distributed among the regions in the world.

Figure 7. KOF Index of Globalization for Turkey and South Africa in 1971–2013
Source: Swiss Federal Institute of Technology Zurich, 2016

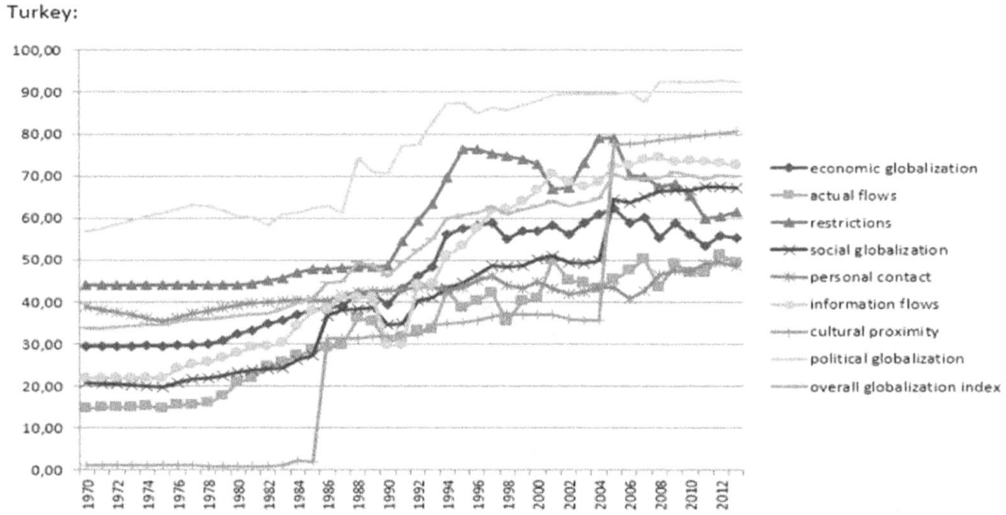

As it was discussed above, the FTAs are playing an important role in trade relations especially with and between emerging markets. The milestones in the process are presented in Table 4.

According to the Organization for Economic Cooperation and Development [OECD] (2015), exports of agricultural commodities are projected to become concentrated in fewer countries, while imports become more dispersed over a large number of countries. The importance of relatively few countries in supplying global markets for some key commodities increases market risks, including those associated with natural disasters or the adoption of disruptive trade measures. Overall, trade is expected to increase more slowly than in the previous decade but maintaining a stable share relative to global production and consumption.

Trade flows cannot be analyzed independently because they affect economic situation in emerging markets. With regard to the impact of globalization on income inequality. Stiglitz (2003) suggests an increase in income inequality both within and across emerging markets (Table 5).

Figure 8. CIVETS comparative country profiles by KOF Index variables in 2013
Source: Swiss Federal Institute of Technology Zurich, 2016

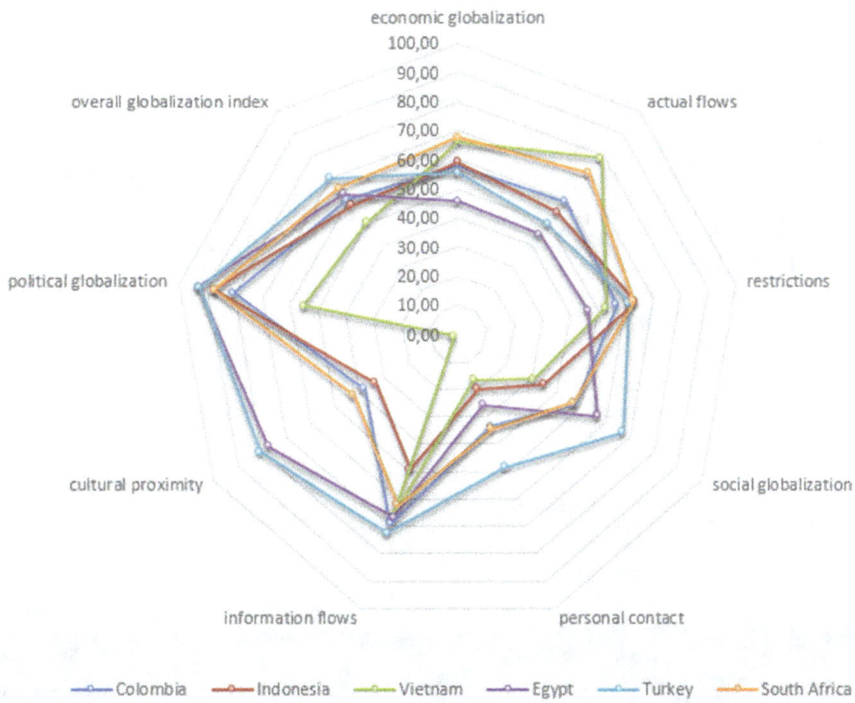

Table 4. The milestones of trade negotiation process with emerging markets

Year	Trade negotiation	Focus
1994	Uruguay Round	Unfair rules of international trade; developing-country membership of the WTO
2001	Doha development round	A code that emphasized the centrality of developing country concerns
2003	Cancun	The advent of an emerging market developing country coalition in the form of the G-20+
2004	Meeting of the WTO's General Council	The restoration of the integrity of the Doha consensus
2005	6th WTO ministerial meeting in Hong Kong	Improving market access in key areas such as agriculture, textiles, and cotton; reform of developed countries' protectionist domestic support and export subsidy regimes

Source: Adapted by Le Pere, 2005; Sally, 2001

Europe–Asia bilateral trade flows have increased by more than 50% in recent years. Prior to that, both regions directed their largest trade flows to North America. As it was pointed out above, there has been a significant shift away from the previous pattern of US-centric trade interdependence with flows redirection Europe–Asia (the "third link") and South–South. The European Union (EU) and East Asia[8] have become the two largest regional trade blocks in the global economy and what drives their trade is intensive intraregional trade. Shifts in trade flow geographically and shifts in the core of the trade flow – from manufacturing (with a focus on tariffs) to services and investments.

Figure 9. Comparative profiles of BRIC and CIVETS countries by Hofstede cultural dimensions
Source: Data were extracted from Hofstede, n.d.

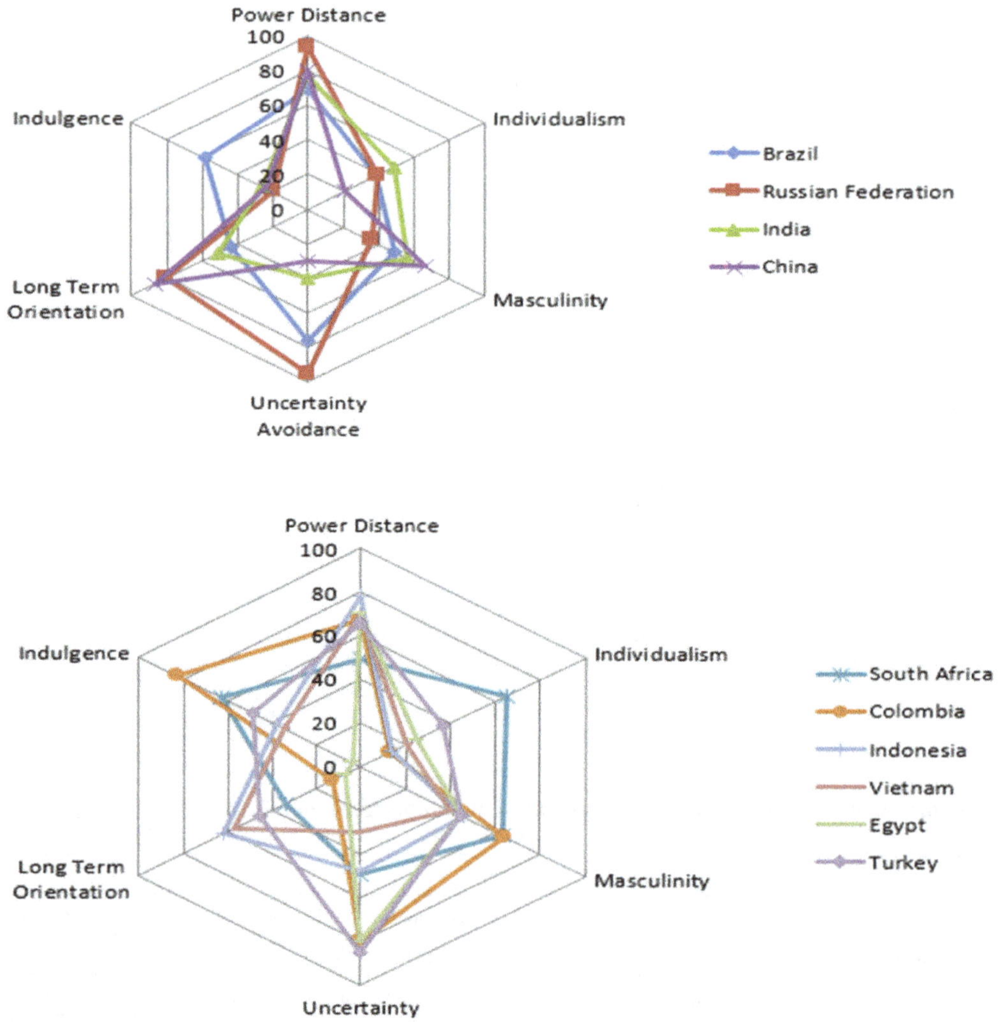

As it is shown in Figure 10 and Table 6, during the last decade analyzed countries follow a similar pattern of their export with a similar rate except for Vietnam which export and import levels increased tremendously.

Despite that, the trend of the export of goods and services as a percentage of GDP shows a comparable pattern during the last two decades among analyzed countries (Figure 10), it is quite dissimilar when the period is longer (1960–2015) as shown in Figure 11 to Figure 13.

Relying on a single indicator (export as a percentage of GDP) could be misleading especially for forecasting purposes since the inherently cyclical nature of both political and economic development will be ignored. Sharma (2014) explains the shortcomings of single-factor theories with the nowadays situation with short economic cycles, typically running just three to five years from the peak to the trough. The competitive landscape can shift completely in that time, whether through technological innovation

Table 5. Comparative macroeconomic profiles of BRIC and CIVETS countries in 2000 and 2014, % of GDP

	Household final consumption expenditure		General government final consumption expenditure		Gross capital formation		Exports of goods and services		Imports of goods and services		Gross savings	
	2000	2014	2000	2014	2000	2014	2000	2014	2000	2014	2000	2014
Brazil	65	62	19	19	19	21	10	11	12	14	13	16
China	47	37	16	14	35	46	21	24	19	21	36	49
Colombia	69	61	17	18	15	26	16	16	17	21	n.a.	19
Egypt, Arab Republic	76	83	11	12	20	14	16	14	23	23	18	12
India	64	58	13	11	24	34	13	23	14	26	25	33
Indonesia	61	57	7	9	22	35	41	24	30	24	25	31
Russian Federation	46	54	15	19	19	21	44	28	24	21	36	24
South Africa	62	61	18	20	16	20	27	31	24	33	16	15
Turkey	71	69	12	15	21	20	20	28	23	32	18	15
Vietnam	70	64	6	6	27	27	50	86	53	83	28	30
World	59	58	16	17	24	24	26	30	25	30	25	24
East Asia & Pacific	53	50	15	16	29	32	27	32	25	31	31	35
Europe & Central Asia	57	57	18	20	23	20	36	42	34	39	24	22
Latin America & Caribbean	66	64	15	16	20	22	19	20	20	22	17	18
Middle East & North Africa	48	46	18	17	23	27	39	46	28	38	29	n.a.
North America	65	67	15	15	23	20	14	15	17	18	21	18
South Asia	66	62	12	11	24	32	14	22	15	26	25	32
Sub-Saharan Africa	61	67	16	15	15	21	38	28	31	31	20	16
Low income	80	76	12	14	16	27	18	22	29	40	n.a.	*17*
Lower middle income	66	65	11	11	21	28	27	26	26	30	25	29
Upper middle income	55	51	15	15	26	32	27	27	23	25	28	32
High income	59	60	17	18	24	21	26	32	26	31	23	21

Note: Value in italic presents the Most Recent Value if data for the specified year or full period are not available
Source: World Bank, 2016

or political transformation. Quite often political cycles are neglected even though they are as important to a nation's prospects as economic ones.

China and India follow more or less a similar pattern of export flows with a stable growth (Figure 12) while export levels of Colombia and Egypt fluctuate tremendously over the analyzed period (Figure 13). This could be explained with the "hidden" political cycles.

China is already a much more open economy than most of emerging markets. Vietnam goes along the same way. Because of the stable positive trade flows trends of China and India (see Figure 12) the OECD Development Centre launched an initiative to study their impact ("Asian Drivers" is used as a term for China and India) on other developing countries in Africa, Asia, and Latin America. The emergence of

Table 6. Exports of goods and services, % of GDP, BRIC, CIVETS, Mexico, and Pakistan 2010-2015

	2010	2011	2012	2013	2014	2015
India	22.0	24.5	24.5	25.3	22.9	n.a.
Brazil	10.7	11.5	11.7	11.7	11.2	13.0
Colombia	15.9	18.7	18.3	17.6	16.0	14.7
China	26.5	26.8	25.7	24.8	23.9	22.4
Egypt, Arab Republic	21.3	20.6	16.6	17.2	14.4	13.2
Indonesia	24.3	26.3	24.6	23.9	23.6	21.1
Mexico	29.9	31.3	32.6	31.8	32.3	35.3
Pakistan	13.5	14.0	12.4	13.3	12.3	10.9
Russian Federation	29.2	28.3	27.4	26.6	27.5	29.5
South Africa	28.6	30.4	29.7	31.0	31.3	30.9
Turkey	21.2	24.0	26.3	25.6	27.9	28.0
Vietnam	72.0	79.4	80.0	83.6	86.4	89.8

Source: World Bank, 2016

Figure 10. Export of goods and services, BRIC and CIVETS markets, 1990-2015, % of GDP
Source: World Bank, 2016; OECD, 2016

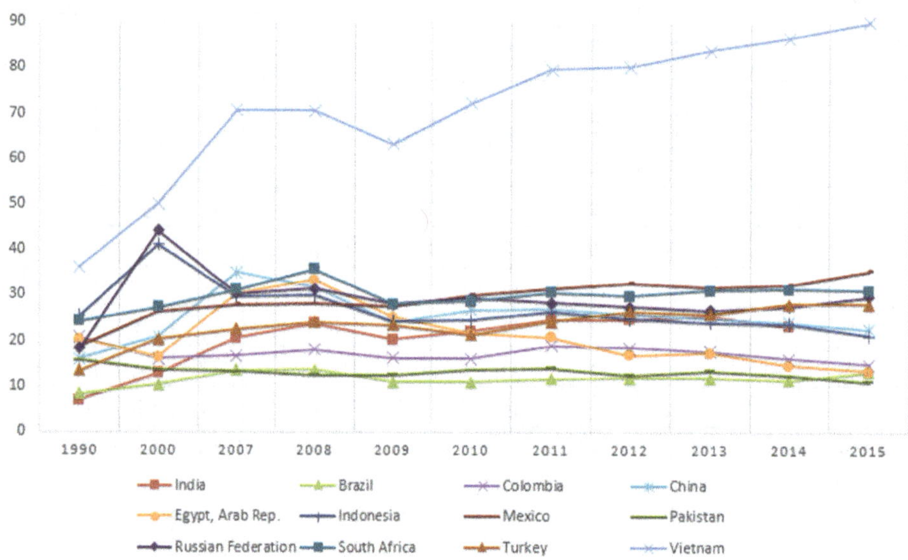

China is viewed as a challenge, especially in terms of trade relations, for Latin American countries. The growing impact of China on Latin America has raised the interest of major institutions involved in Latin America (Blázquez-Lidoy et al. 2006, p. 11). It reinforces the urgent case for more reforms, in particular in the area of infrastructure, to maintain the continent's comparative advantage.

The export positions and trends of BRIC (Table 7) and CIVETS (Table 8) countries on the global market for total cereals are further explored.

Figure 11. Exports of goods and services as a percentage of GDP, 1960-2015, World and Brazil
Source: Data were extracted from World Bank, 2016

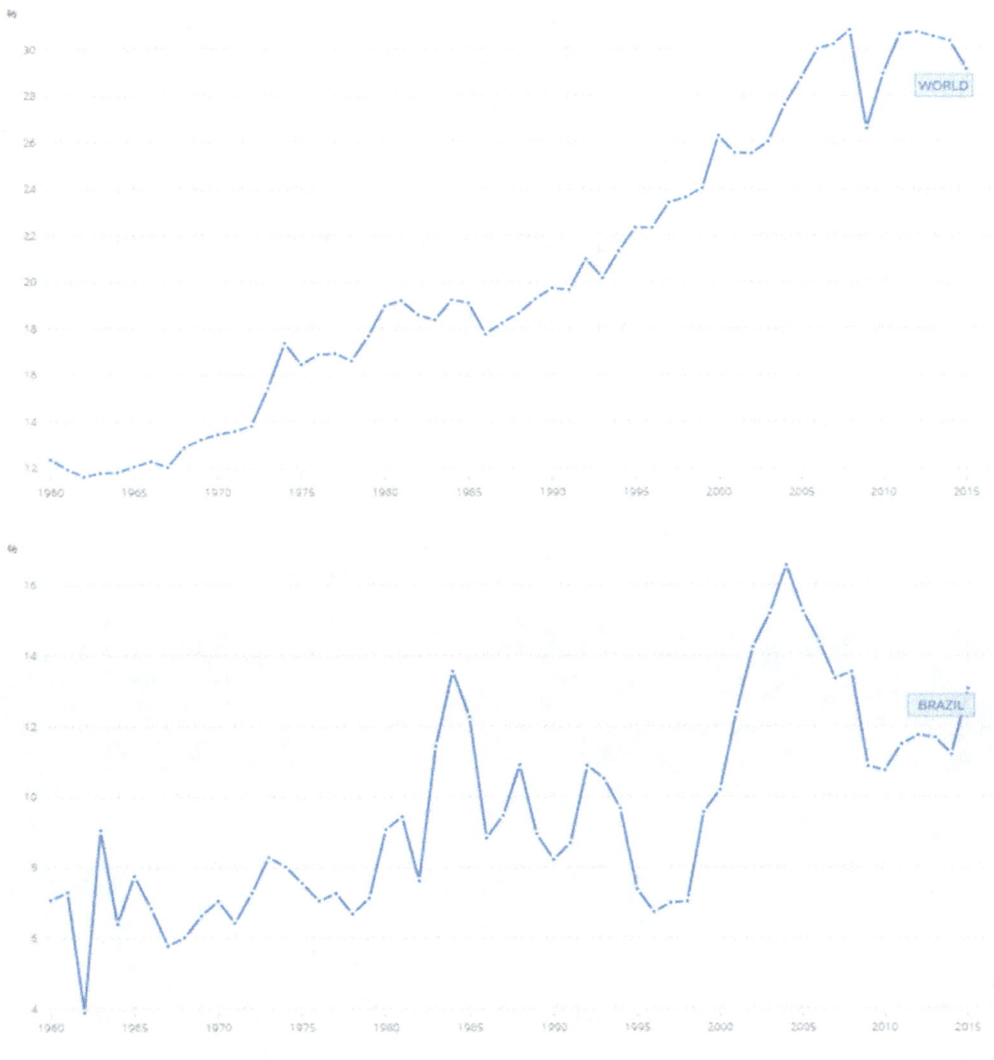

Table 7. Exports (ITY) of total cereals, million tons, BRIC, 2000/2001 –2016/2017

Countries	2000 / 2001	2001/ 2002	2002 / 2003	2003 / 2004	2004 / 2005	2005 / 2006	2006 / 2007	2007 / 2008	2008 / 2009	2009 / 2010	2010 / 2011	2011 / 2012	2012 / 2013	2013 / 2014	2014 / 2015	2015 / 2016	2016 / 2017
Brazil	2.32	5.04	2.44	7.62	2.97	2.28	6.15	12.17	7.82	8.07	15.37	11.45	29.00	24.44	23.22	37.60	18.61
China	12.41	9.82	19.28	15.21	7.42	8.39	8.55	5.38	1.42	1.46	1.28	1.05	1.13	0.96	0.70	0.77	0.75
India	4.08	9.36	8.14	8.16	7.98	5.66	6.89	6.89	6.00	5.15	8.01	16.44	24.72	21.83	14.93	12.09	12.45
Russia	1.26	7.02	16.24	6.26	9.15	12.08	12.28	13.30	23.56	20.96	4.53	27.63	15.79	25.52	30.80	34.69	37.79

Source: Agricultural Market Information System [AMIS], n.d.
Note: ITY – International Trade Year

Figure 12. Exports of goods and services as a percentage of GDP, 1960-2015, China and India
Source: Data were extracted from World Bank, 2016

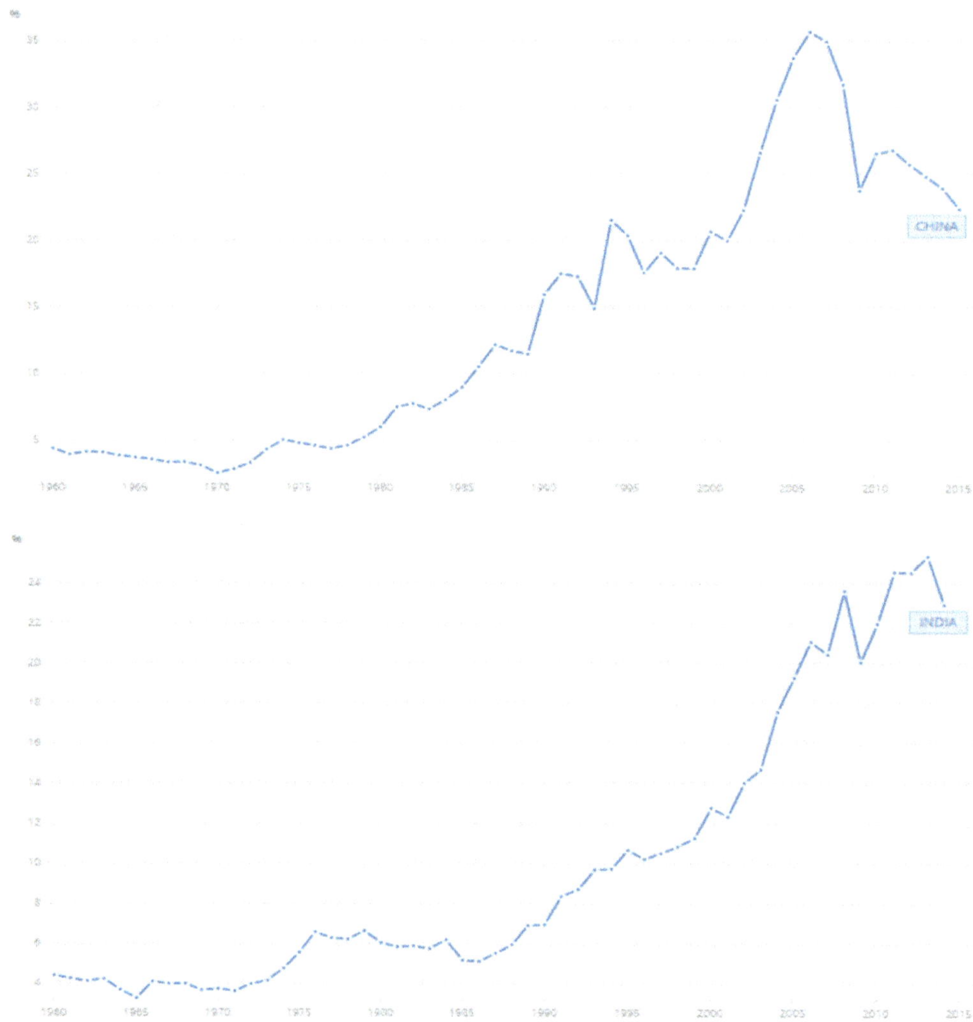

Russia dominates the export market for cereals. Since 2012/2013 nowadays, the trend is positive and constantly increasing. The situation around Brazil and India is promising to be interesting during the next few years (Figure 14).

CIVETS countries cannot be considered as competitors on the global market of cereals (Table 8).

Until 2008/2009, the import of cereals in BRICS countries is comparatively similar (Figure 15).

Chinese import climbed sharply in 2010/2011 and kept its rise until 2015/2016.

Presented analysis of global positions of BRIC and CIVETS markets and their export-import flows, including agricultural products raises a question about the future sustainability of global agribusiness. In the light of the five goals of the Common Agricultural Policy (CAP) assigned by the Treaty of Rome, on the one hand, and the complex and sometimes-volatile environment, from the other hand, countries, companies, farmers, and customers are confronted by many challenges. These challenges are summarized by Boulanger & Messerin (2009) as long term – connected with climate change, water and energy, and

Table 8. Exports (ITY) of total cereals, million tons, CIVETS selected countries, 2000/2001 – 2016/2017

Countries	2000 / 2001	2001/ 2002	2002 / 2003	2003 / 2004	2004 / 2005	2005 / 2006	2006/ 2007	2007/ 2008	2008 / 2009	2009/ 2010	2010 / 2011	2011 / 2012	2012/ 2013	2013 / 2014	2014 / 2015	2015 / 2016	2016/ 2017
Egypt	0.72	0.49	0.59	0.84	1.11	0.98	1.22	0.36	0.64	0.54	0.10	0.37	0.40	0.42	0.40	0.40	0.40
Indonesia	0.11	0.11	0.03	0.05	0.10	0.08	0.03	0.18	0.14	0.08	0.08	0.08	0.08	0.11	0.25	0.23	0.20
South Africa	2.04	1.54	1.20	1.25	1.23	2.02	0.68	0.83	2.63	1.72	2.63	2.51	2.40	2.30	2.27	1.04	0.91
Turkey	1.72	1.03	1.15	1.14	2.46	3.00	2.10	1.56	2.33	4.57	3.20	3.29	2.79	3.59	3.34	4.77	4.71
Vietnam	3.47	3.25	3.82	4.07	5.26	4.65	4.59	4.75	5.96	7.09	7.37	8.64	8.32	8.17	8.61	3.61	7.44

Source: AMIS, n.d.

Note: ITY – International Trade Year

Figure 13. Exports of goods and services as a percentage of GDP, 1960-2015, Colombia and Egypt
Source: Data were extracted from World Bank, 2016

Figure 14. Exports of total cereals, BRIC, million tons, NMY, 2006/2007 – 2016/2017
Source: Data were extracted from AMIS, n.d.
Note: NMY – National Marketing Year

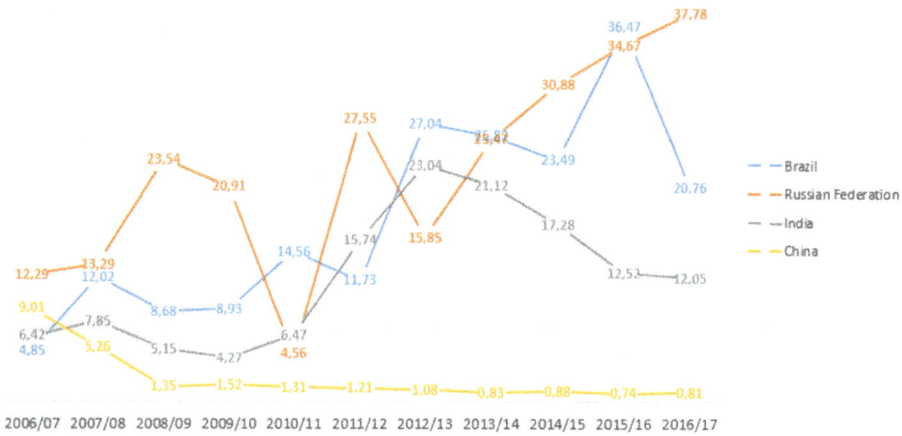

Figure 15. Imports of total cereals, BRICS, million tons, NMY, 2000/2001 – 2016/2017
Source: Data were extracted from AMIS, n.d.
Note: NMY – National Marketing Year

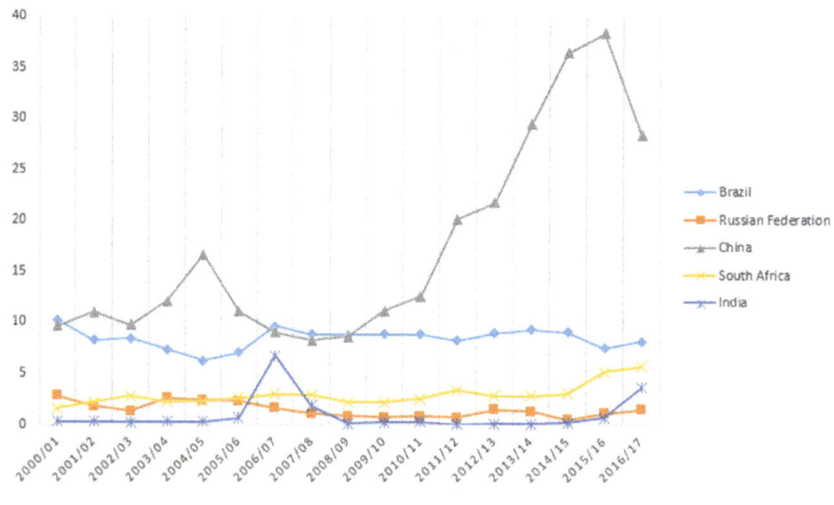

policy-based challenges. Long-term challenges require proactive international trade processes to increase the global resilience of agriculture. Policy-based challenges demand targeted and complex interventions in research and development of agriculture on the global and domestic scale.

FUTURE RESEARCH DIRECTIONS

Emerging markets, often classified as middle powers, have become critical players on the global scene which started a process of power diffusion in the international system. In order to look at the trends which are presented in this chapter, emerging markets should be explored as a system of markets. The main assumption should be that there are variations in the factors which affect the distribution of the trade volumes (import and export flows) as well as that these variations can be modeled by stochastic processes. The selection of a proper stochastic process instead of the model equation of factors affecting the economic performance of emerging markets (in terms of international trade flows) should serve as a starting point of future research methodology. The main research goals should be the following:

- If the state of equilibrium of the ranked emerging markets fits the power law;
- If the deviations are characterized by the index of sheppard.

Additionally, the influences of trade liberalizations and gravitational effects of various regional trade agreements on the economic performance of major emerging markets should be explored by the means of network theory.

CONCLUSION

Globalization is signified as a quintessence of the last decade of the 20th century. Its effect on international trade is outstanding since the globalization contributes a great deal to the international markets interconnectedness. The shifting balance of power on the global markets has been indicated as a transition from Globalization 2.0 (Western-dominated) to Globalization 3.0 (China-dominated). However, optimistic forecasts for continuing the high growth of emerging markets (especially for the abbreviated blocks of countries, such as BRICS, MIST, and CIVETS) started to fade away during the last few years. The gap between the West and the emerging markets has narrowed but there is a notion that the West suffered from the financial crisis while emerging markets suffered from a trade or deGlobalization crisis.

Global markets nowadays are characterized by multilateral trade system, raising regionalism in the global economy, trade liberalization, and controversial geo-economic shifts. Emerging markets as "middle power" are building new trade models through promising and structural relationship between the continents Asia–Latin America and Asia–Africa which destroy the West-dominant trade model. "Asian Drivers" are a major change in the scale of these relations. For Europe and the United States, this should serve as a message of alert.

REFERENCES

Abiad, A., Bluedorn, J., Guajardo, J., & Topalova, P. (2012). *The Rising Resilience of Emerging Market and Developing Economies.* IMF Working Paper, WP/12/300. Washington, DC: International Monetary Fund.

Agricultural Market Information System. (n.d.). *AMIS Market Database.* Retrieved February 8, 2017, from http://statistics.amis-outlook.org/data/index.html#COMPARE

Bartlett, C., & Ghoshal, S. (1987). Managing Across Borders: New Strategic Requirements. *Sloan Management Review*, *29*, 7–17.

Blázquez-Lidoy, J., Rodríguez, J., & Santiso, J. (2006). *Angel or Devil? China's Trade Impact on Latin American Emerging Markets*. Working paper No. 252. Paris: OECD Development Centre.

Boulanger, P., & Messerin, P. (2009). *The 2020 European Agriculture: Long Term Challenges, New Public and Private Policies*. Paper presented at the conference "The 2020 European Agriculture: Long Term Challenges, New Public and Private Policies", Paris, France.

Dobbs, R., Ramaswamy, S., Stephenson, E., & Viguerie, P. (2014). *Management Intuition for the Next 50 Years*. Retrieved November 24, 2016, from http://www.mckinsey.com/business-functions/strategy-and-corporate-finance/our-insights/management-intuition-for-the-next-50-years

Dreher, A. (2006). Does Globalization Affect Growth? Evidence from a New Index of Globalization. *Applied Economics*, *38*(10), 1091–1110. doi:10.1080/00036840500392078

Garelli, P. S. (2007). *The Competitiveness Roadmap: 2007–2050*. Lausanne: IMD World Competitiveness Center.

Gavin, B., & Sindzingre, A. (2009). EU Trade Relations with Emerging Asia: Identifying the Issues. *Asia Europe Journal*, *7*(1), 9–22. doi:10.1007/s10308-008-0220-0

Hartzenberg, T. (2011). *Regional Integration in Africa*. Retrieved January 26, 2017, from https://www.wto.org/english/res_e/reser_e/ersd201114_e.pdf

Heribert, D. (2004). The Stability of International Financial Markets: A Global Public Good? In S. Schirm (Ed.), *New Rules for Global Markets: Public and Private Governance in the World Economy* (pp. 23–43). London: Palgrave Macmillan.

Hofstede, G. (n.d.). *Geert Hofstede*. Retrieved November 24, 2016, from https://geert-hofstede.com/

Hopewell, K. (2014). The Transformation of State-Business Relations in an Emerging Economy. The Case of Brazilian Agribusiness. *Critical Perspectives on International Business*, *10*(4), 291–309. doi:10.1108/cpoib-03-2014-0019

Kaplinsky, R. (2001). Globalization and Economic Insecurity. Institute of Development Studies Bulletin (Sussex), 32(2), 13–24.

Kearney, A. T. (n.d.). *Global Business Policy Council*. Retrieved November 23, 2016, from https://www.atkearney.com/gbpc

KPMG. (2011). *Confronting Complexity*. Amsterdam: KPMG International Cooperative.

Laudicina, P. (2012). *Beating the Global Odds: Successful Decision-making in a Confused and Troubled World*. Hoboken, NJ: John Wiley & Sons, Inc.

Le Pere, G. (2005). Emerging Markets – Emerging Powers: Changing Parameters for Global Economic Governance. *IPG*, *2*, 36–51.

Nicita, A. (2015). *Key Statistics and Trends in International Trade 2014*. Geneva: *United Nations Conference on Trade and Development*.

Nielsen, L. (2011). *Classifications of Countries Based on Their Level of Development: How It Is Done and How It Could Be Done*. IMF Working Paper, WP/11/31. Washington, DC: International Monetary Fund.

Ohmae, K. (1989). The Global Logic of Strategic Alliances. *Harvard Business Review*, (March-April), 143–154.

Organization for Economic Cooperation and Development. (2015). *OECD-FAO Agricultural Outlook 2015*. Paris: OECD Publishing.

Organization for Economic Cooperation and Development. (2016). *National Accounts*. Retrieved November 24, 2016, from http://www.oecd.org/std/na/

Ryder, I. (2005). Issues and Patterns in Global Branding. In C. Distler & B. Nivolett (Eds.), *Securing the Business Benefits of Globalization: A European Perspective* (pp. 205–226). Blue Bell, PA: Unisys.

Sally, R. (2001). *International Trade and Emerging Markets since the Crisis*. Johannesburg: South African Institute of International Affairs.

Sharma, R. (2014). The Ever-Emerging Markets. Why Economic Forecasts Fail. *Foreign Affairs*. Retrieved November 23, 2016, from https://www.foreignaffairs.com/articles/2013-12-06/ever-emerging-markets

Stiglitz, J. (2003). *The Roaring Nineties: A New History of the World's Most Prosperous Decade*. New York, NY: W.W. Norton & Co.

Swiss Federal Institute of Technology Zurich. (2016). *KOF Index of Globalization*. Retrieved November 24, 2016, from http://globalization.kof.ethz.ch/

Vassileva, B., & Nikolov, M. (2016). Market Entry Strategies to Emerging Markets: A Conceptual Model of Turnkey Project Development. *Serbian Journal of Management*, *11*(2), 291–310. doi:10.5937/sjm11-10177

Walker, M. (2007). Globalization 3.0. Prospects for a New World Order. *The Wilson Quarterly*, (Autumn): 16–24.

World Bank. (2000). *Economic Review, 11*. Washington, DC: World Bank.

World Bank. (2016). *World Bank Open Data*. Retrieved November 24, 2016, from http://data.worldbank.org/

World Trade Organization. (2014a). *International Trade Statistics 2014*. Retrieved January 10, 2017, from https://www.wto.org/english/res_e/statis_e/its2014_e/its2014_e.pdf

World Trade Organization. (2014b). *Trade and Development: Recent Trends and the Role of the WTO*. Retrieved December 15, 2016, from https://www.wto.org/english/res_e/booksp_e/world_trade_report14_e.pdf

ADDITIONAL READING

Aswathappa, K. (2008). *International Business* (4th ed.). Columbus, OH: McGraw-Hill Education.

Dieter, H. (2003). Reshaping Globalization: A New Order for International Financial Markets. *IGD Occasional Paper, 38*, 40.

Mathews, J. (1997). Power Shift. *Foreign Affairs, 76*(1), 50–66. doi:10.2307/20047909

Morgan Stanley Capital International. (n.d.). *MSCI Market Classification Framework.* Retrieved November 24, 2016, from https://www.msci.com/market-classification

Nualart, D. (n.d.). *Stochastic Processes.* Retrieved November 24, 2016, from https://www.math.ku.edu/~nualart/StochasticCalculus.pdf

KEY TERMS AND DEFINITIONS

Common Agricultural Policy (CAP): In 1952 the Treaty of Rome creates the European Economic Community (a precursor of today's EU), between six western European countries. CAP is foreseen as a common policy, with the objectives to provide affordable food for EU citizens and a fair standard of living for farmers.

Emerging Market: An emerging market is a country that has some characteristics of a developed market, but does not meet standards to be a developed market.

Exports of Goods and Services as a Percentage of GDP: Exports of goods and services represent the value of all goods and other market services provided to the rest of the world. They include the value of merchandise, freight, insurance, transport, travel, royalties, license fees, and other services, such as communication, construction, financial, information, business, personal, and government services. They exclude compensation of employees and investment income (formerly called factor services) and transfer payments.

FDI Confidence Index: The Foreign Direct Investment Confidence Index®, established in 1998 by A.T.Kearney, ranks countries based on how changes in their political, economic, and regulatory systems are likely to affect foreign direct investment inflows in the coming years.

Globalization 2.0: The driving force in Globalization 2.0 is the globalizing companies. Globalization 1.0 and Globalization 2.0 were primarily driven by European and American people and businesses.

Globalization 3.0: The world entered into the new era of Globalization 3.0 around the year 2000. The driving force in Globalization 3.0 is the newfound power of the individuals to collaborate and to compete globally provided by the software and the Internet. Globalization 3.0 is going to be powered by a non-western and non-white group of individuals.

KOF Index of Globalization: The KOF Index of Globalization measures the three main dimensions of globalization: economic, social and political. The overall index of globalization and sub-indices are calculated referring to actual economic flows, economic restrictions, data on information flows, data on personal contact, and data on cultural proximity. Data are available on a yearly basis for 207 countries over the period 1970–2012.

Stochastic Process: A stochastic process with state space *S* is a collection of random variables $\{X_t, t \in T\}$ $\left\{X_t, t \in T\right\}$. dined on the same probability space (*Ω, F, P*). The set *T* is called its parameter set. If $T = N = \{0, 1, 2, \ldots\}$ $T = N = \left\{0,1,2,\ldots\right\}$. ,he process is said to be a discrete parameter process. If *T* is not countable, the process is said to have a continuous parameter.

ENDNOTES

1 The "Competitiveness Roadmap" is an attempt to describe and assess the main issues that will affect the world competitiveness landscape. Issues are shown along two axes, degree of impact and time-scale. It is a subjective assessment which aims to bring some coherence to the multitude of issues that are said to be having an impact on the competitiveness landscape. The issues which are described are 45 with different levels of impact.

2 According to Moore's Law, the overall processing power of computers doubles every two years.

3 The exact moment of the shift is considered to be the accession to WTO membership of China on December 11, 2001.

4 The Foreign Direct Investment Confidence Index®, established in 1998 by A.T.Kearney, ranks countries based on how changes in their political, economic, and regulatory systems are likely to affect foreign direct investment inflows in the coming years.

5 There are scholars (Dreher, 2006) who question the statistical significance of FDI Confidence Index because of the following reasons: it covers only 67 countries, there is no clear explanation about the weights and cultural factors are excluded. They propose KOF Index of Globalization which measures the three main dimensions of Globalization: economic, social and political, and includes sub-indices referring to: actual economic flows, economic restrictions, data on information flows, data on personal contact, and data on cultural proximity. An alternative perspective to measuring Globalization from the perspective of nation-states can be found in UNCTAD's "Transnationality Index" (TNI). Although ostensibly a measure of how internationalised MNCs are, the TNI can also be construed as reflecting organisational responses to Globalization.

6 Brazil, Russia, India, and China as the first original grouping as BRIC (BRICs) before the induction of South Africa on April 11, 2010. South Africa is included in CIVETS analysis.

7 Colombia, Indonesia, Vietnam, Egypt, Turkey, and South Africa.

8 East Asia is usually defined as ASEAN + 3 comprising North and South East Asian countries, but India is increasingly linked to this grouping and is a member of the East Asian Summit.

Chapter 2
Trade in Agricultural Products and Food Security Concerns on Emerging Markets:
How to Balance Protection and Liberalization

Vasilii Erokhin
Harbin Engineering University, China

ABSTRACT

Although free trade is a key factor in promoting economic growth, there are markets, which require specific approaches. The food market is one of them due to its vital importance for securing nutritional requirements of the population. Providing the population with food in sufficient quantity and variety is a challenge, which includes a range of issues of food production and dependence on import of the food market. Food security is increasingly influenced by foreign trade policies implemented by national governments. Many of them are re-examining their strategies for dealing with food security concerns and seeking to protect domestic food producers by restricting presence of foreign competitors on the domestic food market. The chapter addresses recent food security trends in emerging countries, including BRICS, Eurasian Economic Union (EAEU), and Russia, assesses the effects of food trade distortions in order to balance protection and liberalization policies for achieving effective levels of food security.

INTRODUCTION

In the early XX century, developed countries came to an understanding of the necessity of foreign trade liberalization for effective development of international trade and easier access to their commodities to foreign markets. Contrariwise, developed countries when practicing efficient protectionist policy were reluctant to abandon completely the protectionist practices and kept securing their domestic markets from foreign competitors (Erokhin, Ivolga, & Heijman, 2014).

DOI: 10.4018/978-1-5225-2733-6.ch002

Principles of trade liberalism are widely accepted by the World Trade Organization (WTO) and various alliances of countries as the state of things most advantageous for everyone, however, it is rather challenging to find the proper balance between protectionism and liberalization especially between developed countries and emerging economies. Too much liberalization could lead to less protection for domestic producers against international corporations, while too much protectionism could create an artificial market regime and stifle competition, the very driver of effectiveness and productivity.

Countries depending on their economic power and position on the global market apply varying policies in terms of the relationship between protectionism and liberalization in the sphere of international trade. Developed countries are within their full economic and political power to implement a wide range of tools to protect their national producers and promote their influence over the "weaker" nations on the global market. Developing economies are still in some or other way in a dependent position on the global market in comparison with developed countries. They supply raw materials, mineral resources, and agricultural commodities and consume products of higher conversion. Evidently, being dependent, developing economies are to a far greater degree concerned that the major part of benefits of trade liberalization goes to the developed states. Shah (2010) claims that often those nations that promote free trade for all, want protectionism for themselves. The major concern shared also by Smith (2005) and Akyüz (2005) is that developed nations are pressing developing nations to abandon any protectionism before their economies are ready to enter the global markets.

That is why, many developing economies continue to protect and regulate various sectors of the domestic market in order to ensure their ability to provide standards for the full range of population, especially those related to agricultural production and food consumption. As of Shah (2010), developing economies consider their food security as a crucial parameter of independence, and even if other countries may be more efficient producers of some agricultural products or food items, a country may decide to grow some of their own and support their farmers.

Those developing countries which have embarked on economic development and reform programs, and have begun to open up their markets and emerge onto the global scene are referred to as emerging economies. Most of the emerging economies are characterized as transitional, meaning they are in the process of moving from a closed economy to an open market economy, such as Russia and other countries of the former Soviet Union. Another criterion for a country to be characterized as emerging is its rate of economic growth. Emerging economies are considered to be fast-growing economies, such as China, India, and other countries of Asia and Latin America.

Emerging countries have played a significant role in increasing global food availability in the past decade (Henneberry & Diaz, 2014). They have increased their agricultural production in order to meet their growing demand, but still, one in eight people in emerging markets suffer from hunger and malnutrition. As of The Swiss Re Group (2013), out of the 850 million people suffering from hunger worldwide, 98% are located in emerging markets. The Asia-Pacific region has the greatest number (528 million), followed by sub-Saharan Africa (237 million).

Food security is a holistic concept that addresses a wide range of dimensions, including availability, access, utilization, and stability of food (Food and Agriculture Organization of the United Nations [FAO], 2006). In terms of availability, in order for a country to achieve food security, the total food supply needs to equal its actual food demand, not only in quantitative but also in a qualitative way (variety, nutrient content, and safety). Food supply involves elements such as production, imports, exports, and also changes in national stocks. Similarly, demand involves the amount of food needed to feed, seed, manufacture, and the amount wasted to post-harvest loss (Henneberry & Diaz, 2014). However, steady population growth, insufficient farm investment, socio-economic instability, and adverse weather events remain key challenges to the world's food supply, often barring people's access to food. Moreover, dietary preferences and nutritional demand are changing and are also putting pressure on food production and availability. Recently, rising food prices have become a major concern due to a combination of complex factors (The Swiss Re Group, 2013).

For the emerging economies, where the economic reforms are still in the progress, market institutions are very fragile, and many people suffer from malnutrition, food security issues are vitally important. For such countries, ensuring food security has become one of the 21st century's greatest challenges. On the one hand, governments try to increase the competitiveness of the national economy by integrating it into the global market, on the other hand, they tend to secure food sovereignty of their countries by implementing a wide range of tools, including hidden measures of protectionism in the sphere of agricultural production and trade in food. In this chapter, the author will investigate the tendencies in trade in agricultural products and food in the countries, which comprise the two alliances of emerging economies, i.e. BRICS (Brazil, Russia, India, China, and South Africa) and EAEU (Eurasian Economic Union, including former countries of the Soviet Union, namely Armenia, Belarus, Kazakhstan, Kyrgyzstan, and Russia). The author will also analyze various policies those countries implement for the purpose of achieving food security levels, with a special focus on extremities – the ban on import of certain categories of agricultural products and food imposed by Russia against some developed countries.

BACKGROUND

The concept of food self-sufficiency means the extent to which a country can satisfy its food needs from its own domestic production (FAO, 1999). The practical application of this concept is that food self-sufficiency means that a country produces a proportion of its own food needs that approaches or exceeds 100% of its food consumption (Clapp, 2015). Ritson (1980) explored the relationship between food security and food self-sufficiency in the case of the UK. He understood food self-sufficiency as a proportion of imported food consumption, while in terms of security, he suggested that an efficient system for the international trading of food was a vital component of world food security and that absolute national self-sufficiency was not commensurate with this. Luan, Cui, and Ferrat (2013) define self-sufficiency as the ability of a region to sustain its own requirement for food. Therefore, self-sufficiency depends on two factors: production and consumption.

While food self-sufficiency is mainly focused on the provision of food security worldwide through the import of food products, their self-sufficiency, or optimum balance between domestic food production and import (Meskhia, 2016), food security is a wider concept. Food security is usually categorized as a non-trade concern within trade policy as it incorporates factors other than those directly relevant to the operation of an international market system for the production and exchange of agricultural goods. FAO (1992) defines food security as physical, social and economic access by all people at all times to sufficient, safe and nutritious food which meets their dietary needs and food preferences for an active and healthy life.

Food security exists when people have access to sufficient amounts of safe and nutritious food and are therefore consuming the food required for normal growth and development, and for an active and healthy life. Food insecurity happens due to the unavailability of food or insufficient purchasing power (FAO, 1992). Thus, FAO's approach to food security is based on two pillars which are the physical availability of food and agricultural products on the domestic market (domestic production plus import) and economic access to adequate supply by all people (purchasing power, food inflation, distribution, etc.).

Physical food availability is the level of saturation of the domestic market with both domestically produced and imported foodstuffs when the population in every locality of the country can purchase them in quantities and assortments no less than the established food consumption rates. Such consumption rates are various foodstuffs in quantities and combinations that meet the modern scientific principles of the optimum nutrition taking into consideration the existing nutritional structure and traditions of the country.

The second pillar, economic access to adequate supply, is a possibility to purchase foodstuffs at prevailing prices in quantities and assortments no less than the established consumption rates supported by the appropriate income level.

According to this two-pillar line of argumentation, food self-sufficiency is just a component part of food security, specifically, physical availability, since it is directly related to a balance between domestic production and imports, not to economic access to supply.

Anderson (1990), who defines food security as an access by all members of a household at all times to enough food for an active, healthy life, has the alike understanding of food security pillars. In his opinion, they are the availability of nutritionally adequate and safe foods and assured the ability to acquire acceptable foods in socially acceptable ways. The implication of such approach is that food insecurity means uncertain (market environment) or limited (administrative regulations) availability of food on the domestic market, or ability to acquire acceptable foods by means of domestic production or import.

The liberalization of international trade has become a significant source of tension in contemporary agricultural change with the incorporation of agriculture into the world trading system (Lee, 2007). In the conditions of globalization, where liberalization of food trade and the reduction of administrative protection of food producers are mandated by the rules of the WTO, many countries have lost a part of their sovereignty over food policies (Lawrence & McMichael, 2012). Some of them (primarily, developing ones) have become food dependent, others managed to benefit from easier access to foreign markets and unified framework of global trade in food. In general, globalization has refocused attention from trade-based food self-sufficiency to availability-based food security.

WTO officials argue that in order to achieve food security nations should encourage international trade in food (Lee, 2007). Trade liberalization can affect the availability of certain foods by removal of barriers to imports, but also to foreign investments in the development of domestic production. Thow and Hawkes (2009) found that availability of processed food has risen in developing countries after foreign direct investment by multinational food companies. Thus, changes in trade policies have facilitated the rising availability and consumption of foods. Dorosh (2004) found the positive contribution of trade liberalization to food security and stabilization of supply and food market in the case of Bangladesh. He concluded that food self-sufficiency was not a prerequisite for food security in the country.

Protection policies, including administrative restrictions of food imports with a view to increasing food self-sufficiency of the country, on the contrary, may cause unnecessary social costs and place food self-sufficiency into conflict with the goals of food security and poverty reduction. Warr (2011) studied the effects of an import ban on rice introduced in Indonesia in an effort to increase food self-sufficiency of the country on this staple food. He concluded that such a policy let to achieve the required level of food self-sufficiency, but only at the expense of reducing food security of the most vulnerable people – the poorest net consumers.

Food security can be severely decreased not by administrative regulations only, but by high food costs, poor availability of healthy food, and low purchasing power of population (Socha, Zahaf, Chambers,

Abraham, & Fiddler, 2012). Samaratunga (2011) studied food security issues in the case of Sri Lanka and concluded that the problem of food security in developing countries rested on many parameters apart from domestic production and import, such as growth and distribution of income, population growth, and food consumption patterns. Power (2005), Sonnino, Faus, and Maggio (2014), Kalugina (2014), Erokhin, Ivolga, and Heijman (2014) support that food insecurity is directly related to low income, not to dependence on imports of foodstuffs.

MAIN FOCUS OF THE CHAPTER

Food Security on Emerging Markets: Major Trends

Indeed, the level of economic development is an essential ingredient of food security. In developed countries, where peoples' incomes are high, food security is also sufficient. In those countries, which grow faster than their neighbors do, economic development raises the incomes of the poor and improves their ability to gain access to food.

From 2005 to 2015, the share of emerging markets in the World GDP grew from 9% to 25%. China itself tripled its own contribution from 5% in 2005 to around 15% in 2015. Economic growth in these emerging countries, in general, has been strongly related to agricultural development (Henneberry & Diaz, 2014). However, the level of economic development and speed of economic growth do not mean food security is achieved automatically. Hunger in the world still exists, especially in those countries which tend to demonstrate higher rates of economic growth than the developed nations, i.e. developing economies. On the fast-growing emerging markets, food insecurity remains rooted in disparities in income, access to food, its availability to all citizens of a country, and its safety and quality.

Henneberry and Diaz (2014) state that food security may be determined by the ability of each country to include the poorest population in the nation's development. This can be achieved by creating the proper infrastructure and providing the poor with new or better income opportunities. Resting on such broad categories as "proper infrastructure" and "better income opportunities", it is rather challenging to measure the progress of a single country in achieving food security. Throughout the years, there have been several approaches to measuring food security within countries with different levels of economic development. One of the most widely accepted tools is the Global Food Security Index (GFSI), which examines food security comprehensively across the three internationally established dimensions of affordability, availability, and quality. GFSI is a scoring model, constructed from 28 indicators (Table 1), that measures the drivers of food security across countries irrespective of the level of their economic development, i.e. developed, developing, emerging, economies in transition, etc.

The overall goal of the study is to assess which countries are most and least vulnerable to food insecurity (The Economist Intelligence Unit, 2016). The overall score for the GFSI is calculated from a simple weighted average of the evaluated indicators and scaled from 0-100, where 100 equals the most favorable (Henneberry & Diaz, 2014). Developed countries of North America and Europe are on the top, led by the United States while developing and emerging economies have lower rankings (Table 2).

Not everyone agrees entirely on which countries are emerging markets. The International Monetary Fund (IMF), Morgan Stanley Capital International (MSCI), Standard and Poor's (S&P), Russell, and Dow Jones each have their lists of emerging markets. There are 16 countries that each organization classifies as emerging markets as of 2016: Brazil, Chile, China, Colombia, Hungary, Indonesia, India, Malaysia,

Table 1. Indicators within the GFSI

Category	Indicator	Content
Affordability	Food consumption as a share of household expenditure	Measures the average percentage of household expenditure that is spent on food
	Proportion of population under global poverty line	Measures the percentage of the population living under $3.10/day in purchasing power parity
	Gross domestic product per capita	Measures individual income and, hence, the ability to afford food in $US purchasing power parity terms
	Agricultural import tariffs	Measures the average applied most favored nation (MFN) tariffs on agricultural imports
	Access to financing for farmers	A qualitative indicator that measures the availability of financing to farmers from the government, multilateral, and private sectors
	Presence of food safety net programs	A qualitative indicator that measures public initiatives to protect the poor from food-related shocks
Availability	Sufficiency of supply	A composite indicator that measures the food availability through the supply of kcal/capita/day and levels of food aid
	Public expenditure on agricultural R&D	Measured as a percentage of agricultural GDP. It is a proxy for agricultural innovation and technology that increases market efficiency and access
	Agricultural infrastructure	Composite indicator that measures the ability to store and transport crops to market
	Volatility of agricultural production	Measured as the standard deviation of the annual growth in agricultural production
	Political instability	Qualitative indicator that measures the presence of general political instability
	Corruption	A measure of the pervasiveness of corruption in a country, by measuring the risk of corruption
	Urban absorption capacity	A measure of a country's capacity to absorb the stresses placed by urbanization and still ensure food security
	Food loss	A measure of post-harvest and pre-consumer food loss as a ratio of the total domestic supply of crops, livestock and fish commodities
Quality and safety	Diet diversification	Measures the share of non-starchy foods in total dietary energy consumption
	Nutritional standards	A composite indicator that measures nutrition governance
	Micronutrient availability	A composite indicator that measures the availability of iron and vitamin A in the food supply
	Protein quality	This indicator measures the grams of quality protein consumed using the methodology of the Protein Digestibility Corrected Amino Acid Score
	Food safety	A composite indicator that measures the enabling environment for food safety

Source: The Economist Intelligence Unit, 2016

Mexico, Peru, Philippines, Poland, Russia, South Africa, Thailand, and Turkey. GFSI rankings for those countries along with the leading developed economies are presented in Table 2.

Over the past five years, the GFSI has shown improvements in food security. Overall global economic growth has led to improvements in the structural areas that are essential to improving people's access to a wide range of affordable, nutritious foods, This is particularly evident in emerging-market countries, which have reached the economic and development threshold necessary to enable them to focus on improving government programmes to enhance food security (The Economist Intelligence Unit, 2016).

Most of the emerging economies succeeded to increase their food security level during 2012-2016, with three countries of Latin America (Chile, Colombia, and Peru) and China gaining the biggest increase.

Table 2. GFSI rankings of top ten developed countries and selected emerging economies in 2012-2016

Ranking	Country	2012	2013	2014	2015	2016	Average in 2012-2016	5-year change
1	United States	86.1	85.6	86.2	85.8	86.6	86.1	+0.5
2	Ireland	82.1	81.4	81.8	82.3	84.3	82.4	+2.2
3	Singapore	82.5	80.8	83.4	83.0	83.9	82.7	+1.4
4	Australia	81.7	81.0	80.5	81.8	82.6	81.5	+0.9
5	Netherlands	82.9	82.4	82.8	82.2	82.6	82.6	-0.3
6	France	83.5	83.0	81.5	81.6	82.5	82.4	-1.0
7	Germany	81.2	80.9	81.5	81.6	82.5	81.5	+1.3
8	Canada	82.3	81.0	81.4	81.5	81.9	81.6	-0.4
9	United Kingdom	78.8	79.0	79.4	79.3	81.9	79.7	+3.1
10	Sweden	80.7	80.1	80.3	80.0	81.3	80.5	+0.6
...								
24	Chile	70.6	70.8	72.4	72.9	74.4	72.2	+3.8
...								
29	Poland	72.6	71.6	71.7	72.1	72.4	72.1	-0.2
...								
34	Hungary	71.9	71.6	69.6	68.5	69.3	70.2	-2.6
35	Malaysia	66.8	66.6	68.6	68.9	69.0	68.0	+2.2
...								
39	Mexico	66.8	65.9	65.7	67.2	68.1	66.7	+1.3
...								
41	Brazil	67.8	67.7	67.6	66.8	67.6	67.5	-0.2
42	China	62.3	62.3	63.5	64.3	65.5	63.6	+3.2
...								
45	Turkey	62.3	63.0	62.5	64.1	63.6	63.1	+1.3
...								
47	South Africa	60.8	61.0	62.3	62.3	62.9	61.9	+2.1
48	Russia	66.7	65.6	62.8	61.5	62.3	63.8	-4.4
49	Colombia	57.7	57.6	57.8	58.8	61.0	58.6	+3.3
...								
51	Thailand	58.7	59.4	59.5	59.0	59.5	59.2	+0.8
...								
55	Peru	54.7	55.0	55.6	56.9	57.7	56.0	+3.0
...								
71	Indonesia	47.7	48.2	48.0	47.9	50.6	48.5	+2.9
...								
74	Philippines	49.8	48.6	48.8	49.1	49.5	49.2	-0.3
75	India	48.7	48.2	47.9	48.9	49.4	48.6	+0.7

Source: The Economist Intelligence Unit, 2016

These positive developments have largely been driven by rising incomes in most emerging countries and general improvements in the global economy. Falling food prices have also positively impacted food security (The Economist Intelligence Unit, 2016). Russia was the worst performer with a decline of 4.4 points, and the reasons of that are discussed below.

Production and Trade in Agricultural Products in BRICS

The alliance of five major emerging national economies, Brazil, Russia, India, China, and South Africa, is BRICS. The BRICS members are countries, distinguished by their large, fast-growing economies and growing influence on regional and global affairs. As of 2015, five BRICS countries represent almost 3 billion people, which are 40% of the world population, with a combined nominal GDP of $17 trillion, which is 20% of the world GDP.

Three of these countries, except Brazil, India, and South Africa, entered the WTO in 1995 when the organization was established. Two decades of liberalization of trade in agricultural products and food influenced agricultural production increased imports of food and benefited food security in those countries. Particularly, Brazil has increased its agricultural production in almost three times, while export increased in 5.3 times, and import increased by 53% (Table 3).

India also has been able to ensure a positive trade balance in food and agricultural products during 1995-2015. The share of import in GDP is growing but still remains very low (Table 4).

South Africa is the smallest economy among the five members of BRICS and the most dependent on import of food. Import trade quota is 0.80 in 2014, which means that the country imports almost as much as it produces domestically. However, South Africa keeps its trade balance positive (Table 5).

China entered the WTO in 2002. By 2015, the country has become the leading producer and consumer of food and agricultural products among BRICS countries, with almost $1 trillion worth of agricultural production, and $102 billion of agricultural import. Trade balance in agricultural products has been remaining negative since 2008 (Table 6).

Table 3. Agricultural production and foreign trade in agricultural products in Brazil in 1995-2015, $ million

Indicators	1995*	2000	2005	2010	2011	2012	2013	2014	2015
GDP by kind of economic activity: agriculture, hunting, forestry, and fishing	37,791	30,519	41,435	91,132	113,921	107,736	114,542	111,497	n.a.
Export of agricultural products	13,258	12,808	30,432	60,778	77,389	77,235	82,108	78,845	70,981
Import of agricultural products	5,756	3,665	3,210	8,373	10,244	10,626	11,713	11,224	8,817
Trade balance	7,502	9,143	27,222	52,405	67,145	66,609	70,395	67,621	62,164
Import Trade Quota (Import/GDP)	0.15	0.12	0.08	0.09	0.09	0.10	0.10	0.10	n.a.

Source: Author's development based on the United Nations Conference on Trade and Development [UNCTAD] (2016)
Note: * - the year of accession to the WTO

Table 4. Agricultural production and foreign trade in agricultural products in India in 1995-2015, $ million

Indicators	1995*	2000	2005	2010	2011	2012	2013	2014	2015
GDP by kind of economic activity: agriculture, hunting, forestry, and fishing	91,022	100,517	145,375	289,880	322,598	312,268	321,028	321,895	n.a.
Export of agricultural products	5,912	5,418	9,001	18,199	27,009	30,534	37,429	36,003	29,954
Import of agricultural products	1,675	3,216	6,332	13,612	16,556	18,399	17,418	19,980	20,777
Trade balance	4,237	2,202	2,669	4,587	10,453	12,135	20,011	16,023	9,177
Import Trade Quota (Import/GDP)	0.02	0.03	0.04	0.05	0.05	0.06	0.05	0.06	n.a.

Source: Author's development based on the UNCTAD (2016)
Note: * - the year of accession to the WTO

Table 5. Agricultural production and foreign trade in agricultural products in South Africa in 1995-2015, $ million

Indicators	1995*	2000	2005	2010	2011	2012	2013	2014	2015
GDP by kind of economic activity: agriculture, hunting, forestry, and fishing	5,499	4,084	6,160	8,961	9,446	8,556	7,608	7,801	n.a.
Export of agricultural products	2,779	2,232	3,990	8,230	9,012	8,742	9,294	9,455	7,977
Import of agricultural products	1,409	1,254	2,451	5,483	7,009	7,448	6,777	6,272	4,896
Trade balance	1,370	978	1,539	2,747	2,003	1,294	2,517	3,183	3,081
Import Trade Quota (Import/GDP)	0.26	0.31	0.40	0.61	0.74	0.87	0.89	0.80	n.a.

Source: Author's development based on the UNCTAD (2016)
Note: * - the year of accession to the WTO

Table 6. Agricultural production and foreign trade in agricultural products in China in 1995-2015, $ million

Indicators	1995	2000	2002*	2005	2010	2011	2012	2013	2014	2015
GDP by kind of economic activity: agriculture, hunting, forestry, and fishing	145,301	180,496	199,759	273,518	598,526	734,708	829,468	919,435	979,224	n.a.
Export of agricultural products	12,299	13,559	16,164	24,635	44,152	54,168	56,318	59,983	63,495	63,236
Import of agricultural products	9,236	9,043	9,891	21,541	59,556	75,455	90,650	98,646	105,263	102,787
Trade balance	3,063	6,273	6,273	3,094	-15,404	-21,287	-34,332	-38,663	-41,768	-39,551
Import Trade Quota (Import/GDP)	0.06	0.05	0.05	0.08	0.10	0.10	0.11	0.11	0.11	n.a.

Source: Author's development based on the UNCTAD (2016)
Note: * - the year of accession to the WTO

China is an ultimate leader among BRICS countries in agricultural production. The country has been increasing the volume of domestic agricultural production starting from 2003, right after its accession to the WTO (Figure 1).

Due to a huge domestic market, China is not able to export much of its agricultural products. Brazil is the leader in the volume of agricultural export. India is also increasing its export (Figure 2).

China is the most distinct importer among the four countries, with over $100 billion of import in 2015. Other countries are really low, with only India reaching a level of $20 billion (Figure 3).

South Africa, being the smallest economy and food producer among the BRICS countries, is very dependent on import of food, and the dependency is rising. China, India, and Brazil are quite stable in terms of Import/GDP ratio (Figure 4).

The fifth member of BRICS, Russia, is a special case in terms of its recent trade policy and influences on food security. According to the GFSI, Russia was the only country among emerging economies which lost several ranking points in terms of all three dimensions of food security, i.e. affordability, availability, and quality and safety, during 2012-2016. At the same time, Russia is the only emerging country which abandoned trade liberalization pathway and banned imports of a variety of food products from the developed countries, including the EU, the USA, Canada, Australia, and a number of countries. In an effort to make up losses in terms of its influence on global affairs, Russia stresses on regional integration in the form of the emerging alliance of countries which used to be the republics of the Soviet Union and gained independence in 1991. The alliance is the Eurasian Economic Union which originated from the Commonwealth of Independent States in 2000 and was transformed into the EAEU in 2014. Russia is a founding member and a major driving force of this alliance established for the effective promotion of free movement of goods, services, capital, and labor between participating countries (Armenia, Belarus, Kazakhstan, Kyrgyzstan, and Russia). That is why, before the investigation of the Russia's food security case, contemporary trends in trade in food in the macro-region of Eurasia has to be addressed.

Figure 1. Agricultural production of BRICS countries in 1995-2014, $ million
Source: Author's development based on the UNCTAD (2016)

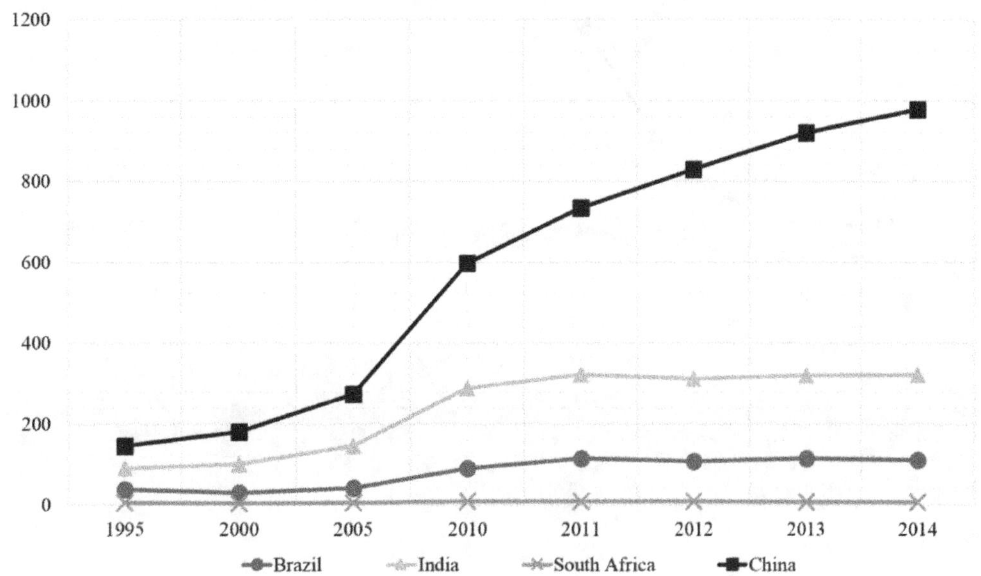

Figure 2. Export of agricultural products and food in BRICS countries in 1995-2015, $ million
Source: Author's development based on the UNCTAD (2016)

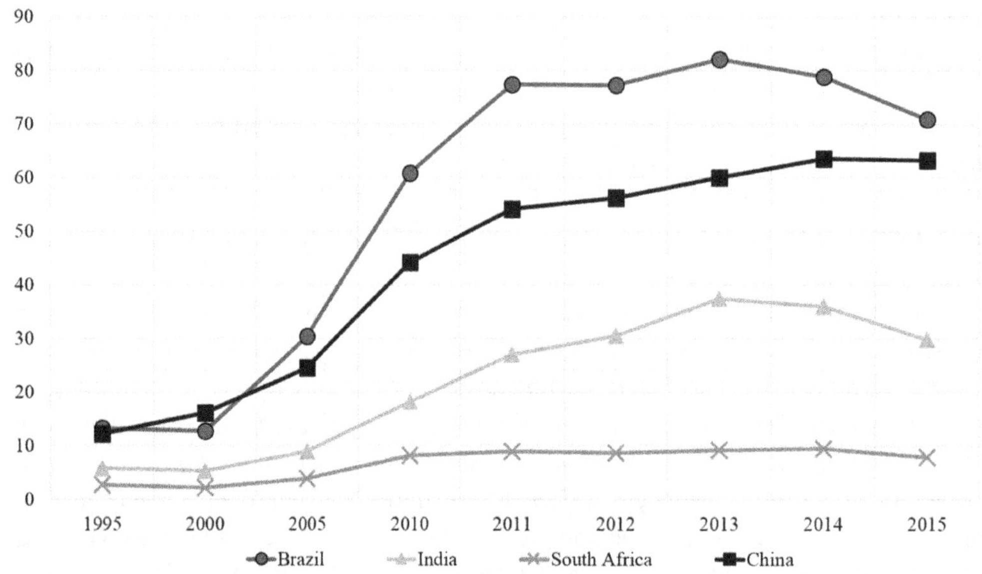

Figure 3. Import of agricultural products and food in BRICS countries in 1995-2015, $ million
Source: Author's development based on the UNCTAD (2016)

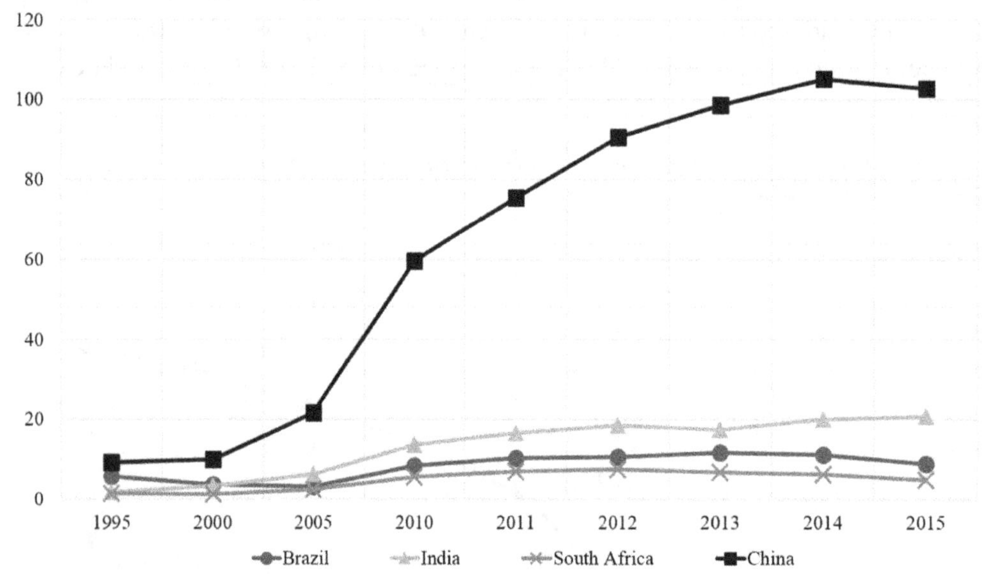

Figure 4. Import Quota of BRICS countries in 1995-2014
Source: Author's development based on the UNCTAD (2016)

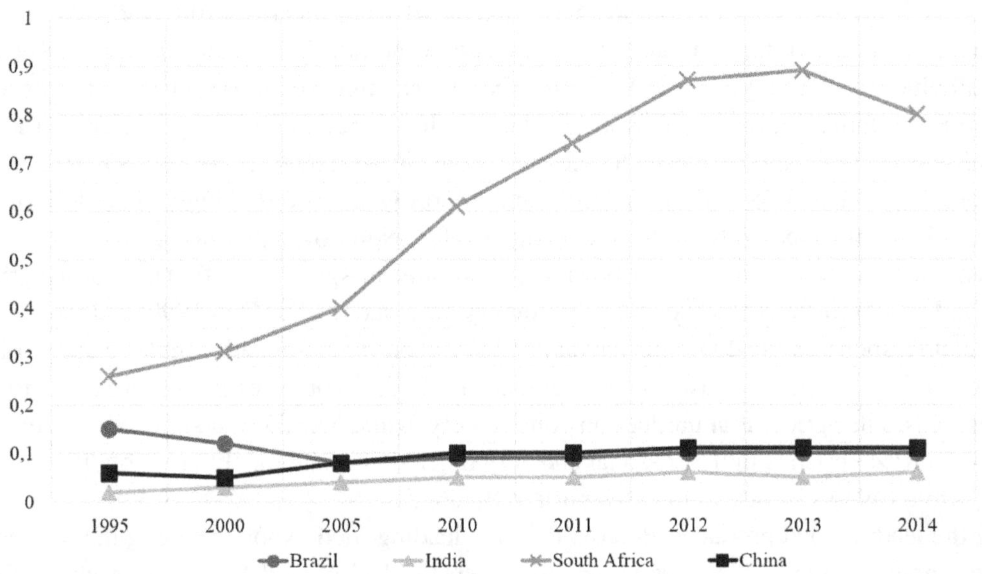

Production and Trade in Agricultural Products in the EAEU

While the EAEU may appear small in terms of its members, it still represents about 183 million people spread amongst the vast territory from the Baltic Sea in the West to the Pacific Ocean in the East. Participating countries are Armenia, Belarus, Kazakhstan, Kyrgyzstan, and Russia. Agricultural production in Armenia is growing, but the country is not self-sufficient. Trade balance in food and agricultural products is negative. Import Quota is 0.37 in 2014 which means that import comprises 37% of the total agricultural production of the country (Table 7).

Table 7. Agricultural production and foreign trade in agricultural commodities in Armenia in 1995-2015, $ million

Indicators	1995	2000	2003*	2005	2010	2011	2012	2013	2014	2015
GDP by kind of economic activity: agriculture, hunting, forestry, and fishing	524	444	605	936	1,574	2,061	1,901	2,050	2,149	n.a.
Export of agricultural products	34	32	84	111	164	235	346	404	408	383
Import of agricultural products	210	210	213	299	662	769	798	830	789	644
Trade balance	-176	-178	-129	-188	-498	-534	-452	-426	-381	-261
Import Trade Quota (Import/ GDP)	0.40	0.47	0.35	0.32	0.42	0.37	0.42	0.40	0.37	n.a.

Source: Author's development based on the UNCTAD (2016)
Note: * - the year of accession to the WTO

Belarus is the country which focuses on agricultural production the most among other countries of the EAEU. The volume of agricultural production has been growing rapidly starting from 1995, and especially during the last 2-3 years, and reached almost $6 billion in 2014. Starting from 2009, the Belarussian trade balance in food and agricultural products became positive and remains like that until now. The country substantially increased its export in 2013-2014, when Russia imposed a ban on import of agricultural products from certain countries of the world, and deliveries started to go through Belarus (Table 8).

Kazakhstan, on the contrary, turned to become a consumer-oriented country – import deliveries of food are growing, the trade balance is negative, and Import Quota is rather high – 0.43 in 2014, which means that 43% of the food on the domestic market is of foreign origin (Table 9).

Kyrgyzstan is more or less in the same position as Kazakhstan – growing deficit in trade in agricultural products and food, growing import quota. The country accessed the WTO in 1999, and the author sees the immediate response of the domestic market in 2000 and onwards – negative trade balance (Table 10).

Belarus and Kazakhstan are two leaders among the EAEU countries on agricultural production. Belarus increases its agricultural production quite rapidly, while Kazakhstan slows down. Agricultural production in Kazakhstan is rather seasonal and very much dependent on climatic conditions and yield of wheat, while Belarus is rather more diversified (Figure 5).

Being the leading food producer, Belarus is also a leading food exporter among the four countries under consideration. Its export is growing steadily, except the decline in 2015, while Kazakhstan's export is fluctuating in parallel with the production of wheat (Figure 6).

Table 8. Agricultural production and foreign trade in agricultural products in Belarus in 1995-2015, $ million

Indicators	1995	2000	2005	2010	2011	2012	2013	2014	2015
GDP by kind of economic activity: agriculture, hunting, forestry, and fishing	2,185	1,277	2,573	5,096	4,950	5,446	5,188	5,935	n.a.
Export of agricultural products	89	495	1,319	3,190	3,792	4,519	5,349	5,333	4,169
Import of agricultural products	665	1,027	1,562	2,748	3,099	3,509	3,972	4,656	4,278
Trade balance	-576	-532	-243	442	693	1,010	1,377	677	-109
Import Trade Quota (Import/GDP)	0.30	0.80	0.61	0.54	0.63	0.64	0.77	0.78	n.a.

Source: Author's development based on the UNCTAD (2016)

Table 9. Agricultural production and foreign trade in agricultural products in Kazakhstan in 1995-2015, $ million

Indicators	1995	2000	2005	2010	2011	2012	2013	2014	2015*
GDP by kind of economic activity: agriculture, hunting, forestry, and fishing	2,535	1,484	3,639	6,678	9,610	8,920	10,657	9,143	n.a.
Export of agricultural products	518	600	669	1,934	1,826	3,087	2,675	2,605	1,346
Import of agricultural products	374	431	1,135	2,203	3,172	3,823	4,125	3,961	2,709
Trade balance	144	169	-466	-269	-1,346	-736	-1,450	-1,356	-1,363
Import Trade Quota (Import/GDP)	0.15	0.29	0.31	0.33	0.33	0.43	0.39	0.43	n.a.

Source: Author's development based on the UNCTAD (2016)
Note: * - the year of accession to the WTO

Table 10. Agricultural production and foreign trade in agricultural products in Kyrgyzstan in 1995-2015, $ million

Indicators	1995	1999*	2000	2005	2010	2011	2012	2013	2014	2015
GDP by kind of economic activity: agriculture, hunting, forestry, and fishing	607	436	468	701	837	1,207	1,100	1,074	1,098	n.a.
Export of agricultural products	89	92	71	84	340	306	239	213	239	156
Import of agricultural products	96	82	81	166	544	708	780	842	869	567
Trade balance	-7	10	-10	-82	-204	-402	-541	-629	-630	-411
Import Trade Quota (Import/GDP)	0.16	0.19	0.17	0.24	0.65	0.59	0.71	0.78	0.79	n.a.

Source: Author's development based on the UNCTAD (2016)
Note: * - the year of accession to the WTO

Figure 5. Agricultural production in the EAEU in 1995-2014, $ million
Source: Author's development based on the UNCTAD (2016)

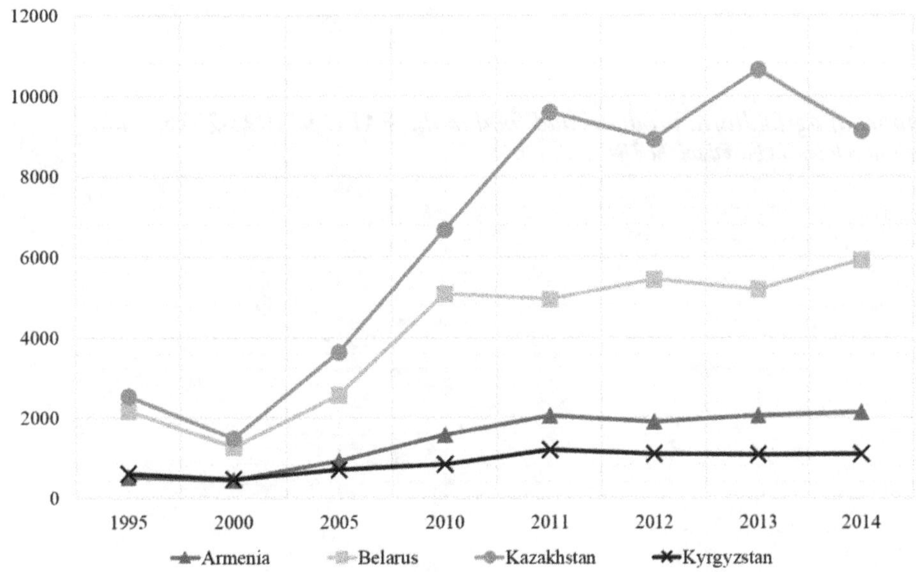

The two countries are also leaders in import. Belarus increases its import from the EU, process raw agricultural commodities, or just repacks food products, labels them and then export them to Russia – this is the way how food products of the EU origin come to Russia in circumvention of the food ban, being formally produced in Belarus (Figure 7).

Belarus is the best performer among other EAEU countries in relation to food independence. Being heavily dependent on import deliveries of food products and agricultural commodities in the mid-1990s, the country was able to decrease its dependency (Figure 8).

Figure 6. Export of agricultural products and food in the EAEU in 1995-2015, $ million
Source: Author's development based on the UNCTAD (2016)

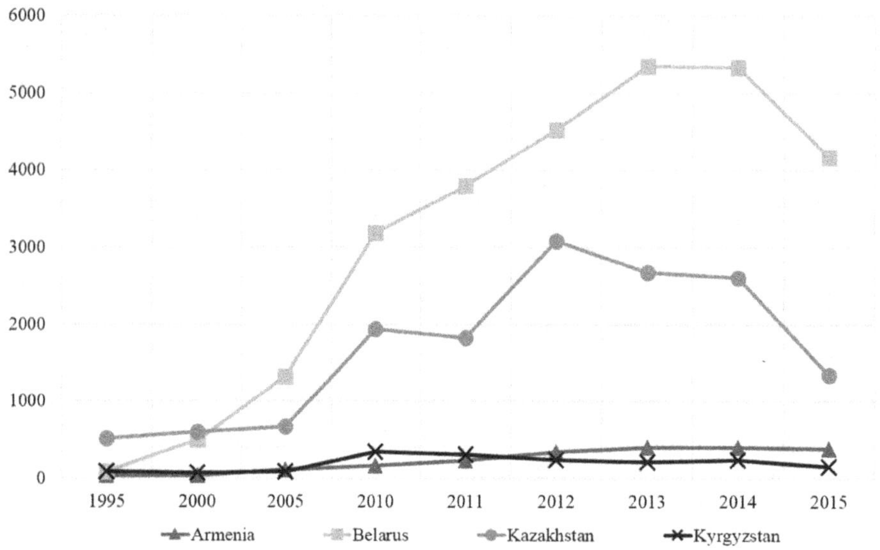

Figure 7. Import of agricultural products and food in the EAEU in 1995-2015, $ million
Source: Author's development based on the UNCTAD (2016)

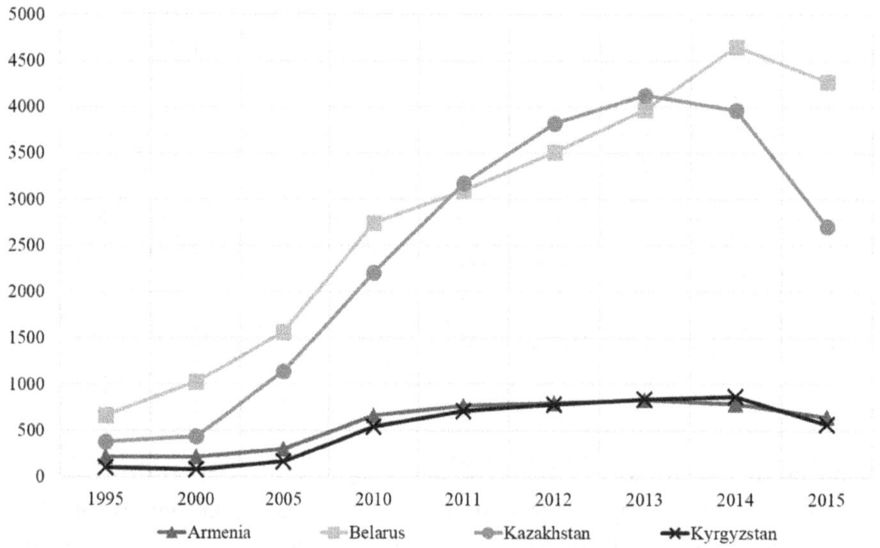

Protectionism and Food Security Issues: Case of Russia

As we have seen on the example of some EAEU and BRICS countries, domestic production of staple agricultural products and food in many countries, including such big agricultural producers as Russia, fails to meet demand. Such countries have to rely on agricultural imports, leaving them vulnerable to

Figure 8. Import Quota of the EAEU countries in 1995-2014
Source: Author's development based on the UNCTAD (2016)

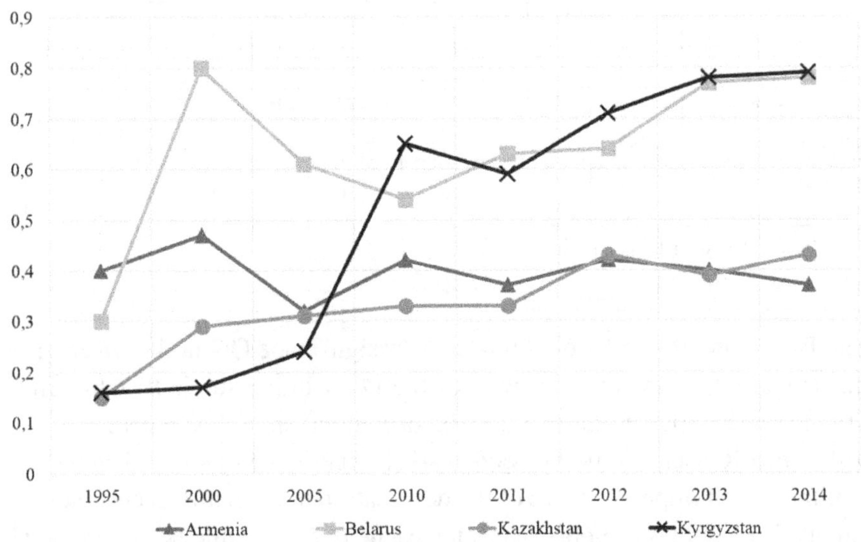

global price fluctuations and affecting their export revenues, which tremendously threat food security of those nations.

Regardless of the world's biggest territory and high volumes of domestic agricultural production, Russia remains a net importer of agricultural products. Currently, over 41% of Russia's agricultural GDP is produced by imports. Russia received up to 55% of its agricultural imports from the countries it has so far sanctioned, including the EU.

Until recently, Russia was not among the WTO members. Russia's accession to this global trading system in 2012 and membership of its major EAEU and BRICS trade partners in the WTO are considered as a precondition of essential structural changes in interregional and even international trade.

However, trade sanctions imposed on both sides between Russia and the Western countries, demonstrated, that trade liberalization in general and the WTO, in particular, had not much to do with those changes. The real threats for Russian agricultural market and trade are unilateral actions, which completely distort trade patterns and destroy trade links between countries and regions. Trade restrictions threat sustainable development not only because of their direct effects (market volatilities, destruction of trade linkages, etc.) but largely because of the hidden ones (distortion of natural economic order, the establishment of artificial market environment and overprotection).

Considering the retrospectives of the Russia's food market, the author may emphasize the rapid decrease of agricultural production (and its share in the GDP) in the beginning of 1990s, after the fall of the Soviet Union and during the early period of liberal economic reforms (Table 11). That sort of the liberalization effect is common to most of the economies in transition, when the opening of the domestic market causes an immediate increase of import deliveries, while local producers do not have facilities to respond swiftly to such a change of external competitive environment.

Russia currently provides 1.17% of the world food export and 1.77% of the world food import (UNCTAD, 2016). Over the recent decades, Russia has been a net importer of agricultural commodities and food. The share of the country in the total world food import has been growing since the very fall of

Table 11. Major indicators of Russia's agriculture in 1990-2015

Indicators	1990	1995	2005	2010	2011	2012	2013	2014	2015
Share of agricultural production in GDP, %	17.0	6.3	4.5	3.3	3.4	3.4	3.4	3.3	3.5
Agricultural land, million hectares	213.8	209.6	193.0	191.9	191.7	191.9	191.8	191.8	191.8
Rural population, million people	39.5	39.4	38.8	37.5	37.3	37.2	37.1	37.0	38.02
Share of agriculture in employment of population, %	13.2	15.7	10.2	7.9	7.9	7.3	7.0	7.0	7.0

Source: Author's development based on Trading Economics (2016)

the Soviet Union. Food Foreign-Trade-to-GDP ratio (Foreign Trade Quota) has been growing over two decades onwards (Figure 9). The outrun growth of the Food Import-to-GDP ratio (Import Quota) in comparison with the Food Export-to-GDP ratio (Export Quota) should be noted.

Even in 2014, when Russia's domestic food market experienced two tendencies, unfavorable for import (depreciation of food import because of the devaluation of the national currency and import ban, introduced in August 2014), Import Quota spiked and reached 47.1%, while export predictably decreased, as Russian producers reoriented on the domestic market.

The negative trade balance has been the permanent trend of Russia's trade in agricultural products over the past two decades (Figure 10). Falling import and introduced import ban can hardly break the tendency. Simple linear trend model shows the expected $25 billion of negative trade balance by 2020.

Figure 9. Food export, import, and foreign trade ratios to Russia's GDP (agriculture) in 1995-2014 and forecast for the period until 2020, %
Source: Author's development based on UNCTAD (2016)

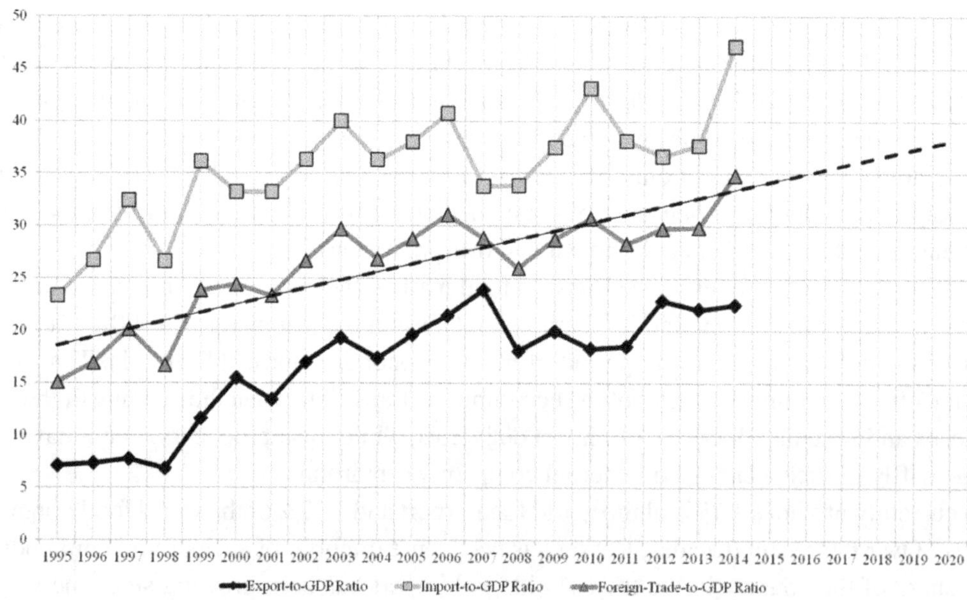

Figure 10. Food export, import, and trade balance of Russia in 1995-2015 and forecast for the period until 2020, %
Source: Author's development based on the UNCTAD (2016)

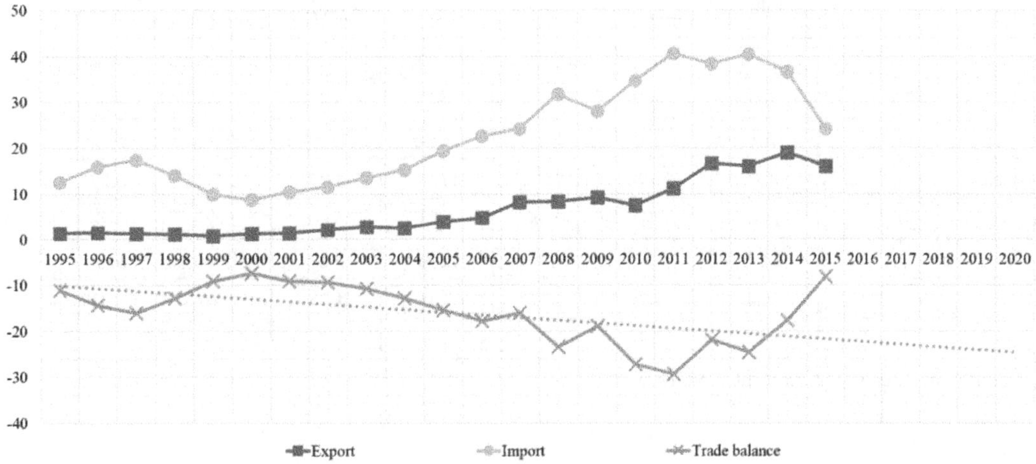

Russia's major food import commodities are fruit, meat and meat products, vegetables, and cheese. There is a decrease of import deliveries in 2015 in comparison with 2013 and 2014 as a result of both introduced import ban and depreciation of import because of the devaluation of the Russian ruble (Table 12).

Analyzing the structure and volumes of food import to Russia, it is worthwhile to note the high share of the EU – about one-fifth of total import. However, if in 2010 the EU supplied almost 30% of Russia's food import, in 2015 its share decreased down to 21.7% (Table 13). Banned foodstuffs from Europe have been replaced on the Russian food market by deliveries from Asia, Latin America, Africa, and some of the CIS countries, particularly, Belarus and Kazakhstan.

Top ten leading food suppliers of Russia are Belarus, Brazil, Germany and other countries of the EU, South America, and Asia. Geographic structure of Russia's food import is rather more diversified in comparison with export since no country has any serious advantage in its market share.

Table 12. Import of selected agricultural and food products to Russia in 1990-2015, $ billion

Product groups	1990	1995	2000	2005	2010	2011	2012	2013	2014	2015
Meat and meat products	0.51	0.62	0.48	1.44	2.82	3.16	3.23	2.94	2.70	1.54
Cheese	0.12	0.21	0.20	0.91	1.93	2.07	2.00	2.33	1.65	0.77
Milk and dairy products	0.19	0.24	0.20	0.40	1.06	0.85	1.00	1.51	1.30	0.81
Fish (fresh, chilled and frozen)	0.09	0.17	0.18	0.98	1.75	2.02	1.98	2.33	1.83	0.92
Vegetables	0.30	0.34	0.28	0.75	2.20	2.88	2.23	2.59	2.60	1.72
Fruit	0.57	0.93	0.72	2.20	5.06	5.71	5.48	5.71	4.71	3.58
Eggs	0.08	0.13	0.41	0.58	0.14	0.18	0.20	0.26	0.33	0.28
Sugar and honey	1.04	1.23	0.96	1.23	1.98	2.44	0.96	0.91	0.84	0.61

Source: Author's development based on the UNCTAD (2016)

Table 13. Russia's food import by regions in 1995-2015

Regions	1995		2000		2010		2015	
	$ billion	**share in total, %**	**$ billion**	**share in total, %**	**$ billion**	**share in total, %**	**$ billion**	**share in total, %**
World, total	14.90	100.00	6.98	100.00	33.62	100.00	24.10	100.00
CIS	3.71	24.89	1.67	23.93	3.41	10.14	4.61	19.13
EU	6.76	45.37	1.84	26.36	10.04	29.86	5.23	21.70
North America	1.37	9.19	0.77	11.03	1.77	5.26	0.57	2.37
Latin America	1.00	6.71	0.68	9.74	6.80	20.23	5.27	21.87
Asia	1.67	11.21	0.86	12.32	5.87	17.46	6.65	27.59
Africa	0.22	1.48	0.21	3.01	1.38	4.10	1.38	5.73
Other	0.17	1.15	0.96	13.61	4.35	12.94	0.39	1.62

Source: Author's development based on the UNCTAD (2016)

Today with escalating political conflict between Russia and Ukraine, economic sanctions imposed by the EU and the USA, and Russia's reorientation on Asia and Latin America, it is not clear how further progress can be achieved between Russia and its trade partners.

Russia banned beef, pork, poultry, fish, fruit, vegetables, cheese, milk and other dairy products from the USA, Canada, the EU, Norway, and Australia. Russia imported about €33 billion worth of food products in 2014, of which the banned goods accounted for nearly €7 billion, or 21.2%. Top ten suppliers to Russia include the countries of Latin America, Europe, and Asia. Five out of the top ten are banned, which are the Netherlands, Poland, Norway, France, and the USA (Table 14). The ban hit about 7% of the EU agricultural exports to Russia. For certain EU countries, i.e. Poland, France, and Norway, losses were essential. Russia used to be a large market for Polish fruits, French cheese, and Norwegian fish.

The desire for the ban was that, by banning those products that Russia had become dependent on, the country would start to become a stronger domestic producer of agricultural goods, thus diminishing the need for heavy reliance on imports (Girard, 2015). However, the degree of dependence is higher than official statistics reports.

Wegren (2013) argues that since trade integration is to lead to more food imports, it is unlikely that Russia's targets for domestic food production can or will be met completely. However, when food security is primarily influenced by such direct-acting factor like food import ban, and imports plunge, how does the domestic Russian market become saturated? As a matter of fact, it does not, since consumption levels of staple foods in Russia are below the WHO recommendations. Per capita consumption of milk and dairy products decreased by 4 kg in Russia in 2015 compared to 2014. The same is true for the Russia's market of meat and meat products: despite some increase in production and decrease in imports, the domestic market is not saturated, and consumption of meat is below the WHO recommendation of healthy nutrition.

Due to the decreased purchasing power, nutritional habits of Russian consumers changed from high-fat animal products to starches (Lunze, Yurasova, Idrisov, Gnatienko, & Migliorini, 2015). Russian citizens are being forced to switch to cheaper products and spend more on food. According to the Analytical Center at the Government of the Russian Federation (2016), already from 2014, when the food import ban was launched, the share of food products in the consumer basket in Russia resumed

Table 14. Import of selected food and agricultural products to Russia from selected countries in 2013 and 2015, $ million

Commodity group	EU		USA		Norway		Canada		Australia	
	2013	2015	2013	2015	2013	2015	2013	2015	2013	2015
Meat of bovine animals, fresh, chilled or frozen	139.91	0.11	0.87	0.00	0.01	0.01	0.58	0.00	178.04	0.00
Other meat and edible meat offal	1,582.72	10.35	349.48	0.60	0.02	0.02	258.82	0.04	59.70	9.98
Meat, edible meat offal, salted, dried; flours, meals	20.95	1.06	0.01	0.00	0.01	0.00	0.00	0.00	0.00	0.00
Meat, edible meat offal, prepared, preserved	174.73	28.17	4.30	0.00	0.01	0.00	0.04	0.00	0.06	0.00
Milk, cream and milk products (excluding butter, cheese)	303.42	12.46	4.45	0.81	0.54	0.17	0.00	0.00	2.30	0.00
Butter and other fats and oils derived from milk	183.08	0.05	0.02	0.00	0.00	0.00	0.00	0.00	67.14	0.00
Cheese and curd	1,247.67	20.59	0.06	0.03	4.42	0.00	0.00	0.00	10.71	0.25
Fish, fresh (live or dead), chilled or frozen	158.41	11.30	75.39	0.00	1,140.51	7.70	34.25	0.00	0.67	0.00
Fish, dried, salted or in brine; smoked fish	1.68	0.01	0.56	0.00	0.17	0.03	0.01	0.00	0.00	0.00
Fish, aqua. invertebrates, prepared, preserved	65.66	23.31	1.98	0.02	0.25	0.07	0.52	0.12	0.00	0.00
Vegetables	1,008.52	80.54	7.40	1.12	0.09	0.00	2.85	0.11	0.11	0.02
Vegetables, roots, tubers, prepared, preserved	358.18	250.13	3.30	7.87	0.00	0.00	0.01	0.01	0.18	0.05
Fruits and nuts (excluding oil nuts), fresh or dried	1,528.03	88.09	235.31	3.20	0.02	0.00	0.11	0.00	9.82	0.00
Fruit, preserved, and fruit preparations (no juice)	147.40	81.30	10.29	14.97	0.00	0.00	2.67	0.02	0.83	0.01

Source: Author's development based on the UNCTAD (2016)

growth and reached 38%. This is much higher than the levels typical for developed countries (10-20%) and even some emerging countries (in Brazil, food costs account for 17.8%). Morozova, Glushakova, and Fadeikina (2016) discovered that Russian people were to redistribute family means in favor of food, reducing the consumption level at the same time. The import ban drastically decreased the available variety of foods, leaving shoppers with fewer options at the higher prices. Food prices rose, in some cases increasing more than by 70% (Girard, 2015). Krivonos (2015) also confirms that import restrictions naturally contributed to overall reduced food imports, but also to higher consumer prices and a lower level of consumption in Russia. Due to the recent food ban, Russia is suffering from a critical lack of diversity in its food selection (Girard, 2015). Along with the decrease in consumption, there is a trend that counterfeit products, such as fake cheese, are appearing on the market.

In fact, liberalization of trade in the form of the Russia's accession to the WTO in 2012 did not cause any essential increase in imports of agricultural products and food. The situation is rather unnatural since the removal of barriers is likely to bring more imports to the country (see the examples of BRICS and EAEU countries). In Russia, quite the opposite, food imports went down during 2014-2015 as a result of a series of macroeconomic events overlapping any effects of trade liberalization: reciprocal sanctions between Russia and Western countries, direct food import ban, devaluation of the ruble, food inflation, and erosion of purchasing power. In such conditions, there are no any preconditions for market capacity growth – people do not increase consumption of domestically produced food staples, as Russian government expected when launching the import ban, but decrease it due to shrinking financial capacity.

Therefore, the country is still heavy dependent on imports. Russia is unlikely to achieve domestic self-sufficiency in food, namely meat, milk, and high-value crops in the near future (Petrick, 2015). The ban makes things worse rather than contributes to achieving of food security of the country in terms of availability, affordability, and even quality and safety of food. Russia must look to abandon distorting restrictions and follow the market way of regulation of foreign trade activities if it wants its population to be adequately fed.

SOLUTIONS AND RECOMMENDATIONS

In emerging countries, food security is influenced by two groups of factors. Direct-acting ones are those related to administration of availability of food on the domestic market, including regulation of domestic food production through support and imports through various forms of protectionism, including food import bans. Liberalization of trade and growing openness of domestic market to international competition are preconditions to ensurance of food security. Lower trade barriers reduce domestic food prices, increase the purchasing power of consumers and afford them a greater variety of food products. Domestic markets reach saturation point through openness as domestic agricultural production grows and the remaining deficit is covered by imports. Freer trade can thus help enhance the availability and affordability of food and contribute to a better-balanced diet. Indirect-acting factors are those, which determine consumption behavior on the domestic market, i.e. dietary habits, the level of life, inflation, the purchasing power of population, etc.

From the above reasoning, assessing food security should be undertaken with consideration of both administrative regulations and economic patterns of domestic food market of any country.

FUTURE RESEARCH DIRECTIONS

Despite the food production growth in many emerging countries, their dependence on international trade is strengthening. This is because more open trade and WTO-supported reduction in trade barriers will encourage specialization and exports according to the comparative advantage of each country (The Economist Intelligence Unit, 2016). Emerging countries often lack basic infrastructure, and smaller incomes inhibit access to and affordability of nutritious food. Such countries also have less power to protect themselves and their domestic food markets against competition from the developed countries. Greater progress requires that country-based food-security policies shift from addressing the entire population to targeting the most vulnerable groups. One of the most powerful tools in terms of improving food security is to contain food price inflation. In such cases, as Russia's food import embargo coupled with economic decline, it is important to understand any changes that may have occurred in the food consumption pattern and to observe any nutritional implications of these changes, with a particular focus on the supply of minimum amounts of staple foods required for healthy human life. It is important to determine the actual level of security of domestic food market with an account of the internationally recognized reference nutrient intakes. There should be monitoring of the level of security of certain market over the certain period of time, which allows comparing it with both the official statistics reports and levels of food self-sufficiency and assessing influences of various foreign trade policies on food security.

CONCLUSION

The growth in international agricultural trade has led to increasing interest in the notion of food security (Lee, 2007). As the author's research shows, trade has an important role to play in increasing food self-sufficiency, which is a proportion of imported food in domestic consumption, but also in improving food security. Transforming paradigm of international economic relations causes rising influence of integration and trade liberalization on domestic food markets of developing and emerging countries. Experiences of some of the emerging economies constituting BRICS and the EAEU show direct links between trade integration (accession to the WTO) and growth of dependency of domestic food markets on import. The absence of such effect in the case of Russia is the result of "overlapping" macroeconomic issues, not related to the WTO: sanctions, currency devaluation, decreasing consumer purchasing power, etc. In the mid-term and long-term perspective effects of trade and integration show up. Internal bottlenecks may become accelerators of food security and economic sovereignty problems when trade liberalization goes on.

What has to be done in such a situation? Adaptation of domestic food markets of emerging countries in the conditions of trade integration and liberalization should have two major directions: promotion of external opportunities and overcoming of internal bottlenecks. Most of the emerging economies are dependent on import deliveries of agricultural products and food (food dependency), in addition to the existing (and even growing, in case of Russia) nonconformance of food consumption volumes to international standards (food insecurity), that is why the attention has to paid to the quality of domestic consumption, not only its quantity. Promotion of external opportunities should include the development of food export potential in the conditions of trade integration and liberalization and should be founded on the discovery of comparative advantages and their promotion on the global and domestic food markets.

REFERENCES

Akyüz, Y. (2005). *Trade, Growth and Industrialization: Issues, Experiences and Policy Challenges*. Penang, Malaysia: Third World Network.

Analytical Center at the Government of the Russian Federation. (2016). *The Food Embargo: The Results of 2015*. Moscow: Government of the Russian Federation.

Anderson, S. A. (1990). Core Indicators of Nutritional State for Difficult-to-Sample Populations. *The Journal of Nutrition*, *120*, 1557–1600. PMID:2243305

Clapp, J. (2015). *Food Self-Sufficiency and International Trade: A False Dichotomy?* Rome: Food and Agriculture Organization of the United Nations.

Dorosh, P. A. (2004). Trade, Food Aid and Food Security: Evolving Rice and Wheat Markets. *Economic and Political Weekly*, *36*(39), 4033–4042.

Erokhin, V., Ivolga, A., & Heijman, W. (2014). Trade Liberalization and State Support of Agriculture: Effects for Developing Countries. *Agricultural Economics*, *60*(11), 524–537.

Food and Agriculture Organization of the United Nations. (1992). *Food, Nutrition, and Agriculture*. Rome: Food and Agriculture Organization of the United Nations.

Food and Agriculture Organization of the United Nations. (1999). *Implications of Economic Policy for Food Security: A Training Manual*. Retrieved February 3, 2017, from http://www. fao.org/docrep/004/x3936e/x3936e03.htm

Food and Agriculture Organization of the United Nations. (2006). *Food Security*. Policy Brief. Retrieved February 1, 2017, from ftp://ftp.fao.org/es/ESA/policybriefs/pb_02.pdf

Girard, M. (2015). *Trade and Transition: Russia's Agricultural Pitfalls and Prospects for the 21st Century*. Orono, ME: University of Maine.

Henneberry, S. R., & Diaz, C. C. (2014). Global Food Security: Emerging Economies and Diverging Food Markets. *Journal of Food Distribution Research*, *45*(1), 29–34.

Kalugina, Z. (2014). Agricultural Policy in Russia: Global Challenges and the Viability of Rural Communities. *International Journal of Sociology of Agriculture and Food*, *1*(21), 115–131.

Krivonos, E. (2015). *Russian Restrictions on Agri-food Imports: Effects on Domestic and World Markets*. Paper presented at the International Conference of Agricultural Economists, Milan.

Lawrence, G., & McMichael, P. (2012). The Question of Food Security. *International Journal of Sociology of Agriculture and Food*, *2*(19), 135–142.

Lee, R. (2007). *Food Security and Food Sovereignty*. Newcastle upon Tyne, UK: Centre for Rural Economy, University of Newcastle upon Tyne.

Luan, Y., Cui, X., & Ferrat, M. (2013). Historical Trends of Food Self-Sufficiency in Africa. *Food Security*, *3*(5), 393–405. doi:10.1007/s12571-013-0260-1

Lunze, K., Yurasova, E., Idrisov, B., Gnatienko, N., & Migliorini, L. (2015). Food Security and Nutrition in the Russian Federation – a Health Policy Analysis. *Global Health Action*, *8*(1), 10. doi:10.3402/gha.v8.27537

Meskhia, I. E. (2016). Food Security Problems in Post Soviet Georgia. *Annals of Agrarian Science*, *2*(14), 46–51. doi:10.1016/j.aasci.2016.03.004

Morozova, E., Glushakova, O., & Fadeikina, N. (2016). Food Consumption as an Indicator of the Quality of Life of the Population in Regions. *Foods and Raw Materials*, *4*(1), 171–180. doi:10.21179/2308-4057-2016-1-171-180

Petrick, M. (2015). *The Prospects for Russian Food Self-Sufficiency*. Paper presented at the International Conference of Agricultural Economists, Milan, Italy.

Power, E. M. (2005). Determinants of Healthy Eating Among Low-Income Canadians. *Canadian Journal of Public Health*, *96*, 537–542. PMID:16042163

Ritson, C. (1980). *Self-sufficiency and Food Security*. Reading, UK: Centre for Agricultural Strategy, University of Reading.

Samaratunga, P. A. (2011). Multiple Facets of Food (In) Security in Sri Lanka: An Input to Food Policy. In S. Mittal & D. Sethi (Eds.), *Policy Options to Achieve Food Security in South Asia* (pp. 51–76). New Delhi: Foundation Books.

Shah, A. (2010). *Deregulation or Protectionism?* Retrieved June 21, 2016, from http://www.globalissues.org/article/43/deregulation-or-protectionism

Smith, J. W. (2005). *Economic Democracy: The Political Struggle for the 21st Century*. Sun City, AZ: Institute for Economic Democracy.

Socha, T., Zahaf, M., Chambers, L., Abraham, R., & Fiddler, T. (2012). Food Security in a Northern First Nations Community: An Exploratory Study on Food Availability and Accessibility. *Journal of Aboriginal Health*, *2*(8), 5–14.

Sonnino, R., Faus, A. M., & Maggio, A. (2014). Sustainable Food Security: An Emerging Research and Policy Agenda. *International Journal of Sociology of Agriculture and Food*, *1*(21), 173–188.

The Economist Intelligence Unit. (2016). *Global Food Security Index 2016. An Annual Measure of the State of Global Food Security*. London: The Economist Intelligence Unit Limited.

The Swiss Re Group. (2013). *New Swiss Re Sigma Study Puts the Spotlight on the Role that Insurance Can Play in Improving Food Security for over 850 Million People Globally*. Retrieved February 1, 2017, from http://www.swissre.com/media/news_releases/nr_20130116_improving_food_security.html

Thow, A. M., & Hawkes, C. (2009). The Implications of Trade Liberalization for Diet and Health: A Case Study from Central America. *Globalization and Health*, *28*(1), 5. doi:10.1186/1744-8603-5-5 PMID:19638196

Trading Economics. (2016). Statistics Database [Data file]. Retrieved from http://ru.tradingeconomics.com/countries

United Nations Conference on Trade and Development. (2016). Statistics Database [Data file]. Retrieved from http://unctad.org/en/Pages/Statistics.aspx

Warr, P. (2011). *Food Security vs. Food Self-Sufficiency: The Indonesian Case*. Canberra: The Australian National University.

Wegren, S. K. (2013). *Return to Putin's Russia. Past Imperfect, Future Uncertain*. Plymouth: Rowman & Littlefield Publishers.

ADDITIONAL READING

Anderson, K., Jha, S., Nelgen, S., & Strutt, A. (2013). Re-examining Policies for Food Security in Asia. *Food Security*, *5*(2), 195–215. doi:10.1007/s12571-012-0237-5

Bairoch, P., & Kozul-Wright, R. (1996). *Globalization Myths: Some Historical Reflections on Integration, Industrialization and Growth in the World Economy*. Geneva: UNCTAD.

Boehringer, C., & Rutherford, T. (1999). *Decomposing General Equilibrium Effects of Policy Intervention in Multi-Regional Trade Models: Method and Sample Application.* Retrieved April 21, 2014, from http://EconPapers.repec.org/ RePEc:zbw:zewdip:5249.

Bokusheva, R., Hockmann, H., & Kumbhakar, S. (2012). Dynamics of Productivity and Technical Efficiency in Russian Agriculture. *European Review of Agriculture Economics*, *39*(4), 611–637. doi:10.1093/erae/jbr059

Bremmer, I. (2015). The New World of Business. *The Fortune*. Retrieved July 23, 2015, from http://fortune.com/2015/01/22/the-new-world-of-business/

Bussière, M., Delle Chiaie, S., & Peltonen, T. A. (2014). Exchange Rate Pass-Through in the Global Economy: The Role of Emerging Market Economies. *IMF Economic Review*, *62*(1), 146–178. doi:10.1057/imfer.2014.5

Calderón, C., & Kubota, M. (2013). Sudden Stops: Are Global and Local Investors Alike? *Journal of International Economics*, *89*(1), 122–142. doi:10.1016/j.jinteco.2012.05.010

Capello, R., & Perucca, G. (2013). Globalization and Growth Patterns in Eastern European Regions: From the Transition Period to the Economic Crisis. Milan: Politecnico di Milano; GRINCOH.

Chang, H.-J., & Nolan, P. (1995). Europe versus Asia: Contrasting Paths to the Reform of Centrally Planned Systems of Political Economy. In H.-J. Chang & P. Nolan (Eds.), *The Transformation of the Communist Economies* (pp. 3–45). London: Macmillan. doi:10.1007/978-1-349-23916-0_1

Dawson, T. C. (2003). *Globalization, the Transition Economies, and the IMF*. Vienna: Joint Vienna Institute.

Devereux, M. (1999). Growth and the Dynamics of Trade Liberalization. *Journal of Economic Dynamics & Control*, *23*(5-6), 773–795. doi:10.1016/S0165-1889(98)00043-8

Erokhin, V., Heijman, W., & Ivolga, A. (2014). Trade Tensions between EU and Russia: Possible Effects on Trade in Agricultural Commodities for Visegrad Countries. *Visegrad Journal on Bioeconomy and Sustainable Development*, *3*(2), 52–57. doi:10.2478/vjbsd-2014-0010

Erokhin, V., & Ivolga, A. (2012). How to Ensure Sustainable Development of Agribusiness in the Conditions of Trade Integration: Russian Approach. *International Journal of Sustainable Economies Management*, *2*(1), 12–23. doi:10.4018/ijsem.2012040102

Estevadeordal, A., Freund, C., & Ornelas, E. (2008). Does Regionalism Affect Trade Liberalization Toward Nonmembers? *The Quarterly Journal of Economics*, *124*(4), 1531–1575. doi:10.1162/qjec.2008.123.4.1531

Gill, I. S., Izvorski, I., Van Eeghen, W., & De Rosa, D. (2014). *Diversified Development: Making the Most of Natural Resources in Eurasia*. Washington, DC: World Bank. doi:10.1596/978-1-4648-0119-8

Henneberry, S. R., & Diaz, C. C. (2015). Food Security Issues: Concepts and the Role of Emerging Markets. In A. Schmitz, P. L. Kennedy, & T. G. Schmitz (Eds.), *Food Security in an Uncertain World* (pp. 63–79). Bingley: Emerald Group Publishing Limited. doi:10.1108/S1574-871520150000015005

Ivolga, A., & Erokhin, V. (2013). Current State and Post-Crisis Developments of EU-CIS Trade: Perspective Tools to Ensure Sustainability. *Visegrad Journal on Bioeconomy and Sustainable Development, 1*, 38–42.

Ivolga, A., Labenko, O., & Ivolga, I. (2013). Vector of the EU and Russia Integration in the Sustainable Development. *Visegrad Journal on Bioeconomy and Sustainable Development, 2*(2), 53–56. doi:10.2478/vjbsd-2013-0010

Khanna, T., & Palepu, K. G. (2010). *Winning in Emerging Markets: A Road Map for Strategy and Execution*. Boston, MA: Harvard Business Press.

Kierzkowski, H. (2000). Challenges of Globalization: The Foreign Trade Restructuring of Transition Economies. *Russian & East European Finance and Trade, 36*(2), 8–41.

Kolapo, I. A., Olufemi, A. J., & Olusola, A. (2012). Globalization and the Transitional Economy: Impacts and Effects on Africa Regional Growth. *European Journal of Globalization and Development Research, 2*(1), 117–126.

Kose, M. A., & Prasad, E. (2010). *Emerging Markets: Resilience and Growth amid Global Turmoil*. Washington, DC: Brookings Institution Press.

McMillan, J., & Naughton, B. (1992). How to Reform a Planned Economy: Lessons from China. *Oxford Review of Economic Policy, 8*(1), 130–143. doi:10.1093/oxrep/8.1.130

Raynolds, L., Murray, D., & Wilkinson, J. (2007). *Fair Trade: The Challenges of Transforming Globalization*. London: Routledge.

Roman-Alcala, A. (2013). *From Food Security to Food Sovereignty?* Retrieved June 22, 2016, from http://civileats.com/2013/05/29/from-food-security-to-food-sovereignty/

Santacreu, A. M. (2015). *The Economic Fundamentals of Emerging Market Volatility*. St. Louis, MO: Federal Reserve Bank of St. Louis.

Schmitz, A., & Meyers, W. H. (Eds.). (2015). *Transition to Agricultural Market Economies. The Future of Kazakhstan, Russia and Ukraine*. Boston, Oxfordshire: CABI. doi:10.1079/9781780645353.0000

Vos, R., & Koparanova, M. (Eds.). (2011). *Globalization and Economic Diversification: Policy Challenges for Economies in Transition*. London: Bloomsbury Academic. doi:10.5040/9781849665742

Windfuhr, M., & Jonsen, J. (2005). *Food Sovereignty: Towards Democracy in Localized Food Systems*. Heidelberg: FIAN ITDG Publishing. doi:10.3362/9781780441160

Wittman, H., Desmarais, A., & Wiebe, N. (2010). *Food Sovereignty: Reconnecting Food, Nature and Community*. Oakland, CA: Food First Books.

Zelenko, O. (2012). *Trade Policy Options of CIS Countries: Twenty Years After*. Rotterdam: Erasmus University.

KEY TERMS AND DEFINITIONS

BRICS: The association of five major emerging countries (Brazil, Russia, India, China, and South Africa), distinguished by their large, fast-growing economies and growing influence on regional and global affairs.

Emerging Economies: The countries, progressing toward becoming advanced, but still lacking market efficiency and sustainable market mechanisms, having greater domestic and external risks due to political instability, infrastructure problems, and currency volatility.

Eurasian Economic Union: The union of Armenia, Belarus, Kazakhstan, Kyrgyzstan, and Russia established in 2014 for effective promotion of free movement of goods, services, capital, and labor between participating countries.

Food Security: A combination of physical food availability on the domestic market by means of domestic production and import, and economic access to sufficient amounts of foodstuffs in accordance with the international standards of adequate food consumption and healthy life.

Food Self-Sufficiency: An extent to which a country can satisfy its food needs from its own domestic production.

Foreign Trade: The system of international commodity-money relations composed of foreign trade activities of all countries worldwide.

Trade Liberalization: A process of reduction or elimination of constraints in the sphere of international trade, including reduction or removal of customs tariffs, import quotas, abolishment of multiple exchange rates, and simplification of administrative requirements to import and export operations.

Trade Protectionism: A set of measures to limit imports or promote exports by putting up barriers to trade that countries use to limit unfair competition from foreign industries.

Chapter 3
State Support of Agricultural Production in Emerging Countries as a Tool to Ensure Food Security

Marina Lescheva
Stavropol State Agrarian University, Russia

Anna Ivolga
Stavropol State Agrarian University, Russia

Oleksandr Labenko
National University of Life and Environmental Sciences of Ukraine, Ukraine

ABSTRACT

The chapter examines the practices of state support of agricultural production in various emerging economies in comparison with selected OECD countries. The aim of the research is to discover how agricultural protectionism and support of domestic farmers affect the level of food security on the emerging markets in the conditions of expanding globalization and liberalization of trade in food. The authors focus on the evaluation of the best practices of state support and discovery of opportunities of their utilization on the emerging markets. Content and mechanisms of state regulation are examined based on the data and evaluation methods obtained from the Organization for Economic Cooperation and Development (OECD). The authors find out that the system of state regulation is one of the key determinants of achieving food security in the conditions of turbulent emerging markets. Both the volume and priority directions of state support of agriculture are determined by financial capacities of emerging economies and current goals of their agrarian policies.

DOI: 10.4018/978-1-5225-2733-6.ch003

INTRODUCTION

International experience of development of agriculture confirms that performance is only possible based on the effective state support. Bearing in mind special role of agricultural production in ensurance of food security and sustainable social and economic development, governments of the leading countries of the world forward substantial resources on the support of domestic agricultural producers, food market regulation, rural social programs, and environment protection. Such measures undertaken as much as possible by most every country affect domestic food markets, agricultural production, and food security.

While international trade in recent decades has been gradually liberalized, trade in agricultural products and food has remained among the ones most influenced by international regulations and national policies (Božić, Bogdanov, & Ševarlić, 2011). As of Markovic and Markovic (2014), agricultural protectionism is a part of the agricultural policy of almost every country. It is focused on the selection of measures of foreign trade and economic policies to achieve the protection of agricultural sector and domestic food market from foreign competition. In a broader dimension, protectionist policies in the form of state support of agricultural production are aimed at the development of sustainable food production, ensurance of food security of a country, and sustainable management of natural resources.

As of Erokhin, Ivolga, and Heijman (2014), developed countries implement a wide range of tools that affect the competitiveness of domestic farmers and food security both directly and indirectly. Such policies support the effective elimination of price disparity and growth of farmers' incomes. The offloading of agricultural surpluses of developed countries on the world market brings down prices and creates disincentives for local producers in many developing countries where agriculture is the main source of livelihood for a major part of the population.

The case of emerging economies is different. In such countries, institutional reforms often lay behind the paces of economic growth and integration to the global market (Erokhin, 2015). Sustainability of domestic food market requires the development of institutions, but such institutional development can be extremely costly on emerging markets. In the early 1990s, due to the difficult economic situation, many emerging countries decreased the level of state support of agriculture and undertaken reductions in the protection of their domestic food markets, at considerable pain and effort, largely with a view to enhancing the supply of food products for their populations. Trade liberalization was successfully implemented in countries where this process was sustained for a long period. In the short run, trade liberalization as a way to increase food security is often painful and may be even damaging for emerging economies, which cannot compete against free trade. Liberalization froze state support and left emerging economies with very few policy instruments to protect themselves from food imports and to subsidize their agriculture.

Despite the certain progress in economic growth during the 2000-2010s, most of the emerging economies still fail to support domestic farmers on a level comparable with the developed states. In many cases, volumes of domestic support gained by farmers in emerging countries are tenfold lower than those in the developed states (Erokhin et al., 2014). Domestic production of basic agricultural products and food in many emerging countries, including such big agricultural producers as Russia, China, Brazil, India, and Argentina fails to meet demand. Providing the population with food in sufficient quantity and variety is a challenge, which includes a range of issues of food production, import dependence and export orientation of the food market, solvency and dietary patterns of the population. Many emerging economies have to rely on agricultural imports, leaving them vulnerable to global price fluctuations and affecting their export revenues, which tremendously threat food security of those nations.

BACKGROUND

State support of agriculture and its influence on food security have been investigated by many authors. In relation to the state support policies implemented in the developed countries, Josling, Anderson, Schmitz, and Tangerman (2010) focused on linkages between state support, international trade in agricultural products, and food security. Schmitz, Moss, Schmitz, Furtan, and Schmitz (2010) investigated current agricultural policies in the USA and other developed countries and elaborated prognosis of agricultural policies for the next decades. Issues of state support of agriculture in the light of liberalization of international trade in food were researched by Devereux (1999), Boehringer and Rutherford (1999), and Estevadeordal, Freund, and Ornelas (2008).

Experiences of emerging countries in the sphere of state support of agriculture have been studied by Petrikov (2012, 2016), Filippov (2014), Malozemov (2014), Visser, Mamonova, Spoor, and Nikulin (2015), and others. Anderson, Jha, and Nelgen (2013) impacted into research of political issues of agricultural protectionism and disarrays on international food market in relation to developing countries and emerging economies of Asia. Wittman, Desmarais, and Wiebe (2010) studied the specifics of food sovereignty in emerging countries in the conditions of growing degree of trade liberalization. As regards to the influences of trade liberalization on food security in emerging countries, Olson (2003) investigated the alternatives to the global trade order and claimed that an expansion of the Agreement on Agriculture of the World Trade Organization (WTO) would make all countries' food security increasingly uncertain and dependent on volatile international market prices and far-flung distribution chain. Liefert and Swinnen (2002) investigated changes in agricultural markets in transition economies, and later Liefert (2004) studied food security issues in Russia in the conditions of economic growth during the early 2000s. In this study, he concluded that the main food security problem was inadequate access to food by certain socioeconomic groups, which lacked sufficient purchasing power to afford a minimally healthy diet (Liefert, 2004). It is true not only for Russia but for many other emerging economies.

The specifics of the emerging markets is that they are very turbulent. There is still no clear understanding of how state support influences food security on the emerging markets in terms of four pillars of food security, i.e. availability, access, stability, and utilization. How limited tools of state support and growing openness to import of food in the conditions of trade liberalization influences domestic agricultural production, farmers and rural dwellers in the emerging countries? In all variety of conducted research, what is needed now is a rethink of international experience. The goal of the given research is an aggregation of international practices and investigation of opportunities for development of state regulation of agriculture in emerging countries with the aim to ensure food security in the conditions of import substitution and an increase of domestic agricultural production.

In this chapter, content and mechanisms of state regulation of agriculture are investigated based on the data and evaluation parameters obtained from the Organization for Economic Cooperation and Development (OECD). According to the OECD's approach, state support of agriculture includes the following elements:

- **Producer Support Estimate (PSE):** Annual monetary value of gross transfers from consumers and taxpayers to support agricultural producers, arising from policy measures, regardless of their nature, objectives or impacts on farm production or income.

- **General Services Support Estimate (GSSE):** Annual monetary value of gross transfers to services provided collectively to agriculture and arising from policy measures which support agriculture.
- **Consumer Support Estimate (CSE):** Includes explicit and implicit transfers associated with compensation to consumers of high market prices for agricultural commodities.
- Aggregated support of agriculture includes the overall amount of transfers received by agricultural producers at the expense of taxpayers and consumers (support of market price); gross transfers from taxpayers to consumers of agricultural products, and transfers at the expense of taxpayers to services provided collectively to agriculture.

OECD unites 34 countries which aggregated share in the global GDP is about 80%. Institutional and structural shifts in the evolution of basic approaches and content of state support of agriculture in those countries reflect specific features of state regulation of agriculture in the modern era of globalization in the conditions of aggravation of global challenges. They have to be considered when correcting the measures of state support of agriculture in emerging countries in order to ensure food security. In the OECD countries, almost one-third of farmers' income on average is not actually earned in agricultural markets, but rather comes from a range of government subsidies and other support measures that restrict agricultural trade and distort markets (Secretariat of the Convention on Biological Diversity, 2005). Effects of such distortions, both positive and negative ones, have to be carefully assessed in order to balance state support, protection, and liberalization policies. The research objective is to find out the key regularities of state support of agriculture in emerging countries and to discover the opportunities of using of accumulated experience for the development of a national mechanism of state regulation of agriculture in the conditions of import substitution policy and an increase of domestic agricultural production.

For the purposes of the present research, the authors used the approach of the United Nations Department of Economic and Social Affairs (DESA) and the United Nations Conference on Trade and Development (UNCTAD), i.e. considered 17 economies (Albania, Armenia, Azerbaijan, Belarus, Bosnia and Herzegovina, Georgia, Kazakhstan, Kyrgyzstan, Macedonia, Moldova, Montenegro, Russia, Serbia, Tajikistan, Turkmenistan, Ukraine, and Uzbekistan) as emerging ones. Additionally, the authors considered seven countries (China, Brazil, India, Vietnam, Laos, Cambodia, and South Africa) recognized as emerging ones by the World Bank and the International Monetary Fund. The research is based on the system methodology. It combines various methods of scientific knowledge, such as monographic, economic and statistical analysis, grouping, trend and comparative analysis.

MAIN FOCUS OF THE CHAPTER

Trade in Food in the Emerging Economies

Despite the resource-oriented model of export, most of the emerging economies are engaged in international trade in other commodities, apart from oil, including food and agricultural products. According to Liefert and Swinnen (2002), during the 1970-1980s, most of the agricultural exports went to the Soviet Union (exports of meat by Hungary, Romania, Ukraine and Kazakhstan; grain by Hungary, Ukraine, and Kazakhstan; sugar by Ukraine; and cotton by Uzbekistan and Turkmenistan). During 1995-2015,

Figure 1. Developed and developing countries and emerging economies by foreign trade turnover in food and agricultural products in 1995-2015, $ billion
Source: Authors' development based on the United Nations Conference on Trade and Development [UNCTAD] (2016)

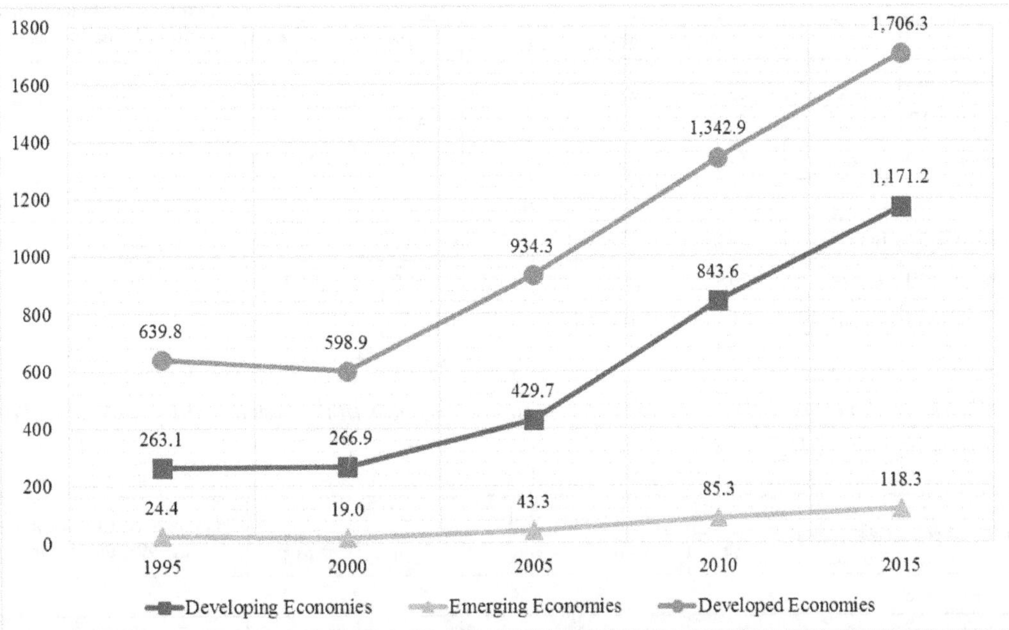

foreign trade turnover of the emerging economies continued to grow steadily and increased fivefold ($118.3 billion in 2015 in comparison with only $24.4 billion in 1995) (Figure 1).

As of Liefert and Swinnen (2002), the pre-reform trade in agriculture for the emerging economies was not market driven but rather was an integral part of countries' economic planning, i.e. resulted from politics, not economics. When most of the emerging economies liberalized trade in the early 1990s, they experienced a collapse in the export of their agricultural products to the world and dramatic rise of import flows to their domestic food markets. Export/import ratio in foreign trade turnover (FTT) in food and agricultural products in 1995 was 25/75, while developed countries had equal proportions of export and import flows in their foreign trade (Table 1). During two decades, emerging economies succeeded in balancing their foreign trade turnover to 44/56 in 2015, but import still prevails.

The share of emerging economies in world trade in food and agricultural products was 3.9% in 2015. It increased by 1.3 percentage points during two decades of transition (Table 2). Share in world import was higher than the one in world export (3.5% and 4.4%, consequently).

Emerging economies are oriented on the world market rather than on the internal markets. In 2015, over 68% of the total trade of the emerging economies allotted to trade with developed and developing countries, and only 31.7% – to intra-group trade. Import of food and agricultural products is even more unbalanced – over 79% of deliveries come from the developed and developing economies, not from the emerging markets (Table 3).

Such a high dependency on import deliveries from developed countries along with the drop in domestic food production during the 1990-2010s have raised concerns about food security and food sovereignty in certain emerging economies in transition, especially Russia. The level of food security and changes during transition have differed importantly between the countries. However, domestic food markets in

Table 1. Proportion of export and import flows in total trade in food and agricultural products among developed countries and emerging economies in 1995-2015, %

Group of countries	year					Change in 2015 in comparison with 1995, percentage points
	1995	2000	2005	2010	2015	
Developed economies						
share of export in FTT in food	48.5	47.3	47.2	49.6	50.6	+2.1
share of import in FTT in food	51.5	52.7	52.8	50.4	49.4	-2.1
Emerging economies						
share of export in FTT in food	25.0	26.8	31.2	34.3	44.1	+19.1
share of import in FTT in food	75.0	73.2	68.8	65.7	55.9	-19.1

Source: Authors' development based on the UNCTAD (2016)

Table 2. Shares of developed countries and emerging economies in the world trade in food and agricultural products in 1995-2015, %

Group of countries	year					Change in 2015 in comparison with 1995, percentage points
	1995	2000	2005	2010	2015	
Developed economies						
share in world trade	69.0	67.7	66.4	59.1	57.0	-8.0
share in world export	67.6	66.2	64.3	58.9	57.5	-10.1
share in world import	70.3	69.0	68.3	45.4	56.5	-13.8
Emerging economies						
share in world trade	2.6	2.1	3.1	3.8	3.9	+1.3
share in world export	1.3	1.2	2.0	2.6	3.5	+2.2
share in world import	3.9	3.0	4.1	3.8	4.4	+0.5

Source: Authors' development based on the UNCTAD (2016)

most of the emerging economies are to a greater extent influenced by internal factors (low competitiveness of food producers, their financial instability, outdated facilities, infrastructure, lowering effective demand, etc.), than external ones. Despite certain progress of reforms, volumes of GDP and foreign trade of the emerging economies are well below than those of the developed countries, while the share of foreign trade turnover in food and agricultural products in GDP is the highest among the groups of countries under consideration (4.5% for emerging economies in comparison with 4.0% for developing and 3.7% for developed countries) (Figure 2).

Consequently, emerging economies have to concentrate on solving internal problems and increasing efficiency of domestic agricultural production. The goal is to create a profitable, market-driven agricultural economy with productivity levels and supporting infrastructure that allows it to compete effectively on the world market. In order to increase their competitiveness on the global market most of the emerging economies implement various tools of state support of agriculture.

Table 3. Proportion of intra-group and extra-group trade in food and agricultural products among developed countries and emerging economies in 1995-2015, %

Group of countries	year					Change in 2015 in comparison with 1995, percentage points
	1995	2000	2005	2010	2015	
Developed economies						
Export: intra-group trade	73.9	76.3	79.1	73.2	70.6	-3.3
Export: extra-group trade	26.1	23.7	20.9	26.8	29.4	+3.3
Import: intra-group trade	69.2	69.4	70.7	69.6	69.4	+0.2
Import: extra-group trade	30.8	30.6	29.3	30.4	30.6	-0.2
Total: intra-group trade	71.5	72.7	74.6	71.3	70.0	-1.5
Total: extra-group trade	28.5	27.3	25.4	28.7	30.0	+1.5
Emerging economies						
Export: intra-group trade	60.7	58.8	53.3	46.4	36.2	-24.5
Export: extra-group trade	39.3	41.2	46.7	53.6	63.8	+24.5
Import: intra-group trade	23.5	26.6	25.5	25.5	28.1	+4.6
Import: extra-group trade	76.5	73.4	74.5	74.5	79.1	-4.6
Total: intra-group trade	32.8	35.3	34.4	32.7	31.7	-1.1
Total: extra-group trade	67.2	64.7	65.6	67.3	68.3	+1.1

Source: Authors' development based on the UNCTAD (2016)

Figure 2. Share of foreign trade turnover in food and agricultural products in GDP in developed, developing and emerging economies in 1995-2015, %
Source: Authors' development based on the UNCTAD (2016)

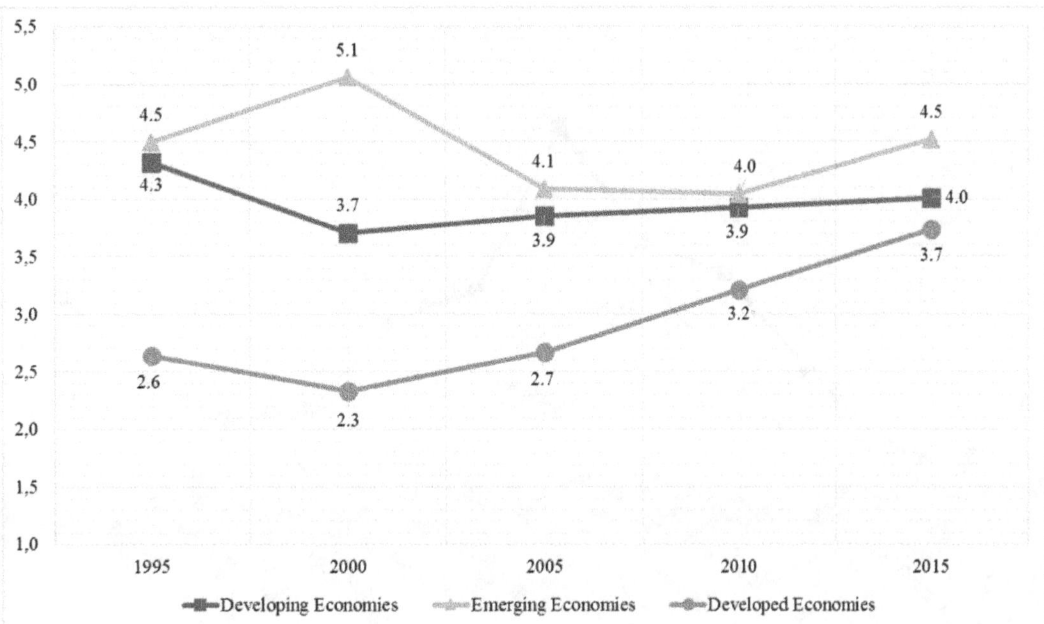

State Support of Agriculture and Effects for Food Security

Conducted analysis reveals that a system of state regulation is a key determinant of the development of agriculture. The strategic importance of support is proved by a complete synchronization of its amount with dynamics of gross agricultural production (GAP) and aggregated support of agriculture (ASA) (Figure 3).

The increase of state support provides outrunning growth of gross agricultural production. When state support decreases and the index is below 1, gross agricultural production index changes correspondingly. It can be observed on the examples of Japan and Switzerland, where a decrease in state support (I = 0.7) was accompanied by a decrease in gross production (I = 0.8).

Analysis of state support in relation to the GDP allowed the authors to discover the tendency of its contraction in the majority of the countries over the previous 20 years (Figure 4).

Thus, in the EU, share of state support of agriculture in the GDP decreased from 1.5% down to 0.8%; in the USA, from 0.6% down to 0.5%; in Canada, from 0.8% down to 0.4%; in Russia, from 2.6% down to 0.7%. Exceptions to this tendency are China and Indonesia, where the relative level of support increased from 1.4% up to 3.2% and from 0.8% up to 3.6%, respectively. In Brazil, in the mid-1990s, part of incomes of agricultural producers was withdrawn in the favor of the state, but currently, agriculture benefits from the support amounted to 0.4% of the GDP. In general, among OECD countries, state support of agriculture decreased twofold (from 1.5% of the GDP in 1995 down to less than 0.8% in 2015). The most critical decrease happened in those countries, where the level of support was higher, i.e. Turkey, South Korea, Iceland, and Switzerland. However, it remains high, above 1% of the GDP.

Figure 3. GAP and ASA growth indices in selected countries (1995-1997 to 2012-2015)
Source: Authors' development based on the UNCTAD (2016)

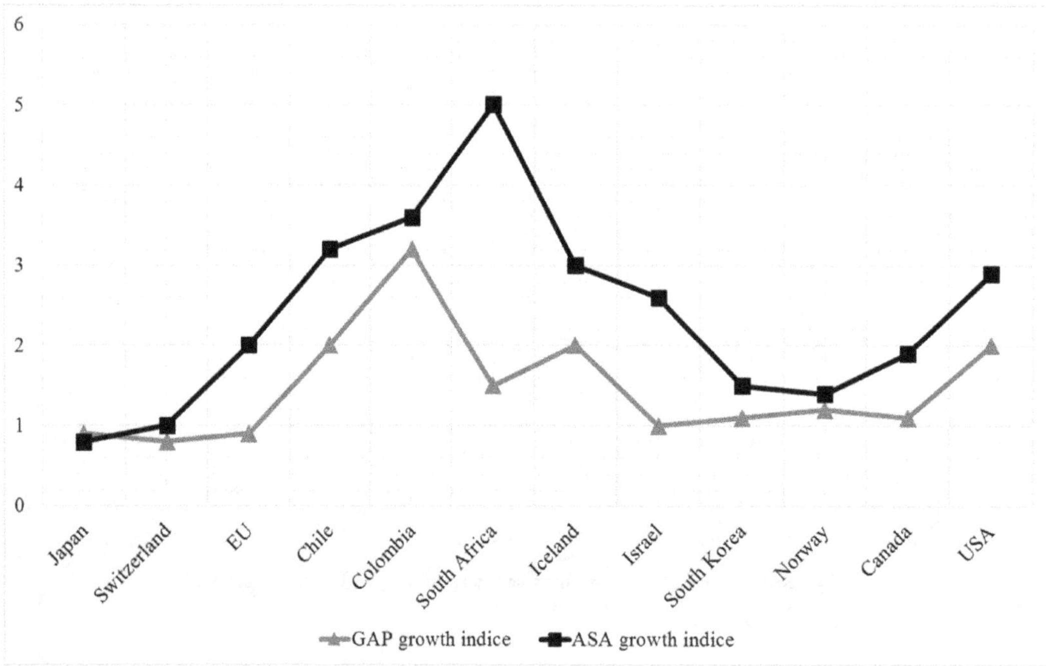

Figure 4. State support of agriculture in selected countries in 1995 and 2015, percentage of GDP
Source: Authors' development based on the UNCTAD (2016)

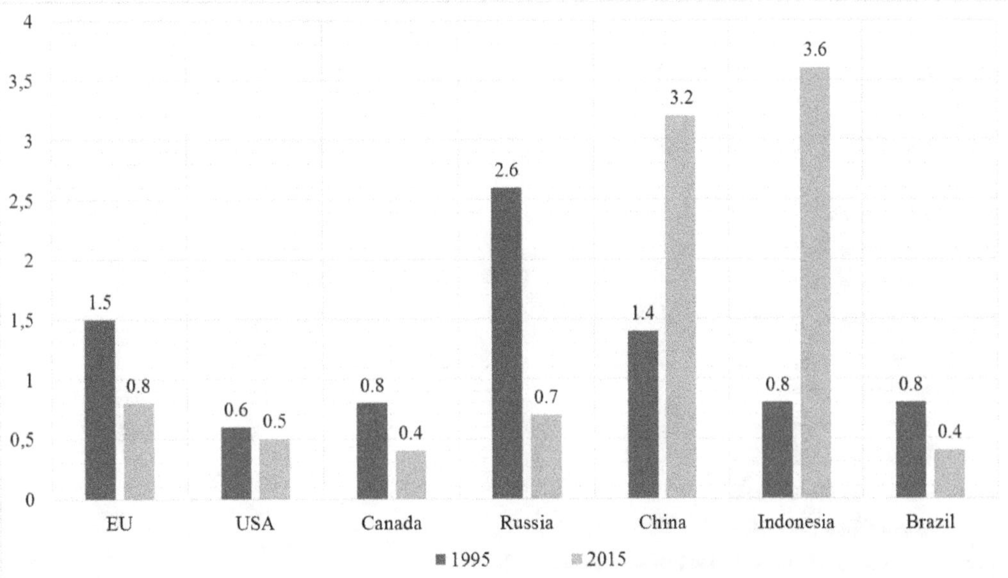

The decrease of the relative level of support does not mean the agriculture now receives less. Support in absolute terms is growing. Thus, during 1995-2015, aggregated amount of state support of agriculture among the OECD countries increased by $23 billion (7.2%), including in the USA, by 195%; in Canada and Turkey, by 170%; in Norway, by 140%. Even more significant growth of the absolute amount of support occurred in the emerging economies, which are not the OECD members. In Brazil, withdrawal of incomes of agricultural producers in the amount of $4.2 billion in 1995 changed to state support in the amount of $9.4 billion in 2015. In China, the amount of support during the same period increased from only $13 billion up to $287 billion.

The variance of the importance of state support for farmers can be estimated by means of matching its volume with the gross product in agriculture (Figure 5). The analysis reveals that percentage ratio between the volume of state support and the gross product in agriculture in the OECD countries decreases, but fluctuates over a wide range, from negative values in Ukraine to almost 100% in Switzerland and Norway. Revealed differences between the countries are determined by the availability of natural resources, the role of agricultural sector in the national economy, the level of development, and goals of agrarian policy.

State regulation of agriculture is a complicated mechanism, which includes the tools of pressure on the structure of agricultural production, agricultural market, social sphere of rural territories, inter-industry, and inter-farm relations. In relation to this, the structure of state support and discovery of trends are the subjects of much attention. The structure of aggregate support of agriculture in the selected OECD countries and emerging economies in accordance with the methodical approach described above is presented in Table 4.

In 1995, support of producers dominated in the aggregated support of agriculture. Among the OECD countries, its relative share was 78%, including in the EU – 87%, in Canada and Japan – 75%, in the USA – over 50%. As for the emerging economies, this type of support also played a critical role. Over

Figure 5. Aggregated support of agriculture in selected countries in 1995 and 2015, percentage of the gross product in agriculture
Source: Authors' development based on the UNCTAD (2016)

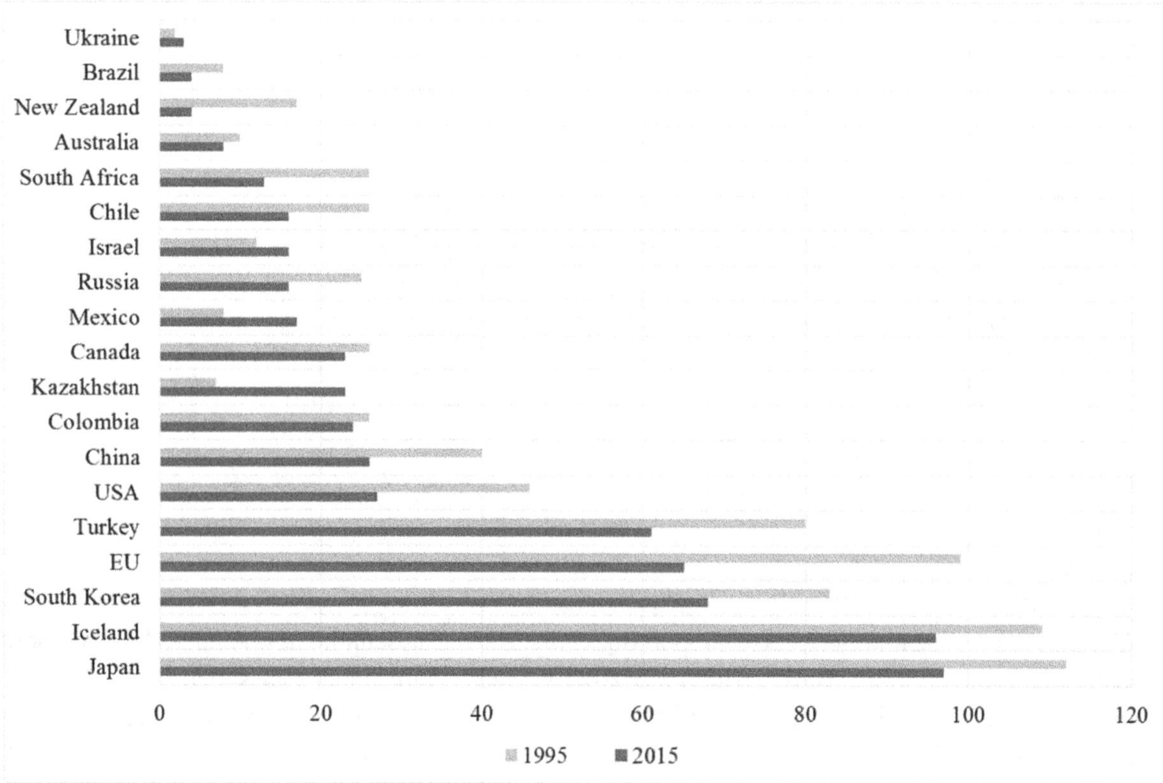

the past 20 years, the importance of the support of producers essentially decreased. Being the members of the WTO, OECD countries are obliged to limit support of producers, which distorts market patterns. That is why they have to redirect budget transfers to support of services or support of consumers. Such a tactic gives them an opportunity to save or even increase the level of support without breaking the WTO rules. Some countries almost redirected state support on the development of infrastructure, research, and development, education in the sphere of agriculture, marketing, promotion of products, and other services. Thus, in 2015, in Australia the share of such expenses in the aggregated support of agriculture is over 50%, in Chile – 48%, in New Zealand – 73%, in Canada – 28%, in Kazakhstan – 23%, and in Israel – 20%.

It is evident that usage of such capacities of support of services not restricted by the WTO is relevant for emerging countries since many of them lose a substantial portion of actual agricultural production because of poor road networks, lack of storage and other infrastructure facilities, which is resulted in the cost of production.

As for the support of consumers, this tool is implemented massively in the USA only, where the agrifood policy is focused on the expansion of domestic and external markets of agricultural raw materials and food. Over a half of the agrarian budget of the USA is spent on food aid to low-income sections of the population in purpose to promote domestic demand. The aid is implemented in the framework of the programs of distribution of food coupons (Supplemental Nutrition Assistance Program – SNAP),

Table 4. Structure of aggregate state support of agriculture in the selected countries in 1995 and 2015, %

Countries	Support of producers		Support of services		Support of consumers	
	1995	2015	1995	2015	1995	2015
Australia	77	50	23	50	0	0
Brazil	176	83	-75	17	0	0
Canada	75	72	25	28	0	0
Chile	83	52	17	48	0	0
China	47	89	50	11	3	0
Columbia	91	85	9	15	0	0
EU	87	85	9	14	4	1
Iceland	87	94	9	5	4	1
Indonesia	73	89	26	6	1	5
Israel	87	80	13	20	0	0
Japan	75	85	25	15	0	0
Kazakhstan	95	76	5	23	0	0
Mexico	64	82	14	11	22	7
New Zealand	31	27	69	73	0	0
Norway	92	93	5	5	3	2
Russia	79	80	21	16	0	4
South Africa	65	57	35	43	0	0
South Korea	86	87	13	13	1	0
Switzerland	82	89	6	11	12	0
Turkey	72	85	28	15	0	0
Ukraine	142	157	-42	-57	0	0
USA	53	38	9	9	38	53

Source: Authors' development based on the UNCTAD (2016)

free lunches in schools (The National School Lunch Program – NSLP), food for women, infants, and children (The Special Supplemental Nutrition Program for Women, Infants, and Children – WIC). The state budget also finances public awareness campaign on nutrition and research in the sphere of healthy nutrition. Each dollar of food aid generates two dollars in economic activity. Each billion of dollars in the program allows creating or supporting 18,000 of workplaces, including 3,000 workplaces in agriculture.

This experience is worthwhile implementing for the development of organizational and economic mechanisms of state regulation of agriculture in emerging countries, since the tools currently implemented in the majority of them are reciprocal and directly or indirectly focused on the increase of output, i.e. on supply, without any regulations of demand. However, as the authors referred to Liefert (2004) in the beginning of the chapter, the main food security problem for emerging countries is not the low volume of output but inadequate access to food by certain socioeconomic groups. The result is that the increase of output in agriculture is not fulfilled even at the extensive support of agriculture. One of the reasons is sensibility to domestic demand. Low effective demand does not drive production increase.

Identifying the overall tendency of decrease of support in relation to the gross agricultural production (in average, in the countries under consideration it decreased from 21% in 1995 down to 17% in 2015), it is necessary to bear in mind that such average indicators submerge differences existing between the OECD countries and emerging economies. While in the former ones the level of protectionism decreased gradually, the latter ones moved up from taxation of agriculture (it still exists in Ukraine) to ensurance of an essential level of support, which just started to catch up with the OECD countries in 2014-2015.

Change in the overall level of support of producers is followed by a change of priorities in the implementation of direct and indirect subsidies. In the late 1990s, support for domestic prices was the major measure of indirect support of agricultural producers in the OECD countries. Its share was over 45% of the overall support in the EU, the USA, Canada, Israel, Turkey, and Australia, 35% in Russia, and over 90% in Japan, Korea, Kazakhstan, and South Africa.

Support of market prices ensures higher incomes for farmers because of keeping high prices for food and making consumers support farmers by paying these high prices. In the case of agricultural surpluses because of high prices, it is possible to promote export using export subsidies, which are paid from the budget, not at the consumers' expense.

The degree of support of farmers by means of such tool is characterized by the Producer Nominal Protection Coefficient (NPC), calculated as a ratio between the average price received by producers on the domestic market and the average world market price. It shows that domestic prices for agricultural products in the USA, Canada, Mexico, and Israel are close to international ones, and did not change essentially during 1995-2015 (Figure 6).

The coefficient decreased substantially in the developed countries, where it was traditionally high. Nevertheless, domestic prices in the EU, Norway, Japan, Turkey, and South Korea surpassed international price in a range from 20% to 200% in 2015. In the emerging economies, on the contrary, the coefficient rises, which is reflected in surpassing of domestic prices over the international price in a range from 2% to 20%

Figure 6. Producer Nominal Protection Coefficient in selected countries in 1995 and 2015
Source: Authors' development based on the UNCTAD (2016)

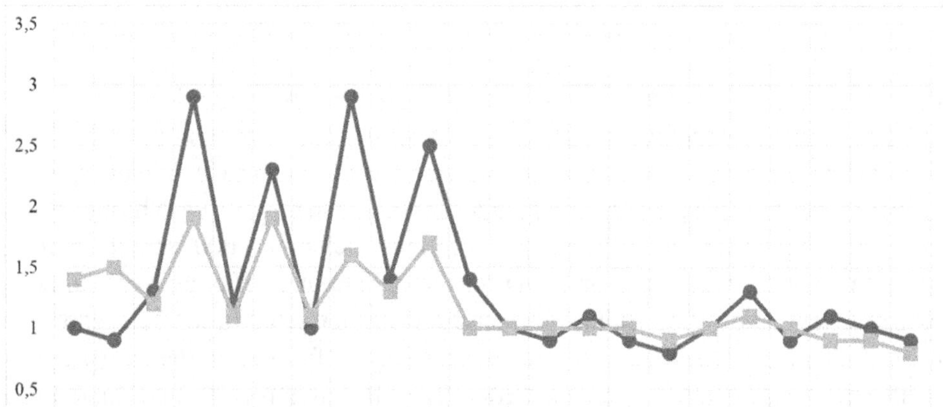

Those emerging countries with financial capacities replace indirect measures of support with direct ones, which do not affect consumer prices. They are implemented by the means of direct state equalization fees; compensation of disaster damages; compensation of redeployment damage (compensation of land diversion, forced cattle slaughtering, etc.); subsidies per unit area or livestock population; payments in the form of financing of procurements (for example, subsidies for purchasing of fertilizers, agricultural chemicals, and fodder); financing of target programs. Across the OECD, direct payments to farmers increased from 19% up to 37% during 1995-2015. Payments per hectare of crop area and livestock unit reached almost a half of state support of agriculture in the USA and 2/3 of the respected payments in the EU. They are increasingly used in order to stimulate implementation of certain technologies of production, including the ones with environmental restrictions. The traditional measure of support is compensation of initial costs. Its relative share in the structure of support of producers across the OECD increased from 9.5% up to 12.0%, including in the EU from 7.0% up to 14.0%.

Agricultural insurance is also one of the major tools of support of farmers' income. International experience shows that subsidizing of insurance premium and reinsurance of risks are the common and effective ways of state support of economic stability in agriculture.

In general, the share of measures, which distort trade patterns the most, decreased by 20 percentage points in the OECD countries during 1995-2015, while in the USA and the EU – more than twofold.

Along with the support of operating activities of farmers, more and more resources of agrarian budgets of the OECD countries are spent on the realization of long-term goals, such as environmental sustainability, development of infrastructure, and implementation of innovations (Table 5). Expenses on infrastructure are essentially financed from the state budgets in Turkey, Japan, Brazil, Chile, and South Korea. Kazakhstan and Canada spend the biggest amount of support on inspection and control services. Improvement of agricultural education, development and implementation into the agricultural production of innovations are priority directions for Australia, Norway, Israel, Switzerland, the EU, and the USA. In those countries, expenses for research and education comprise from 30% up to 64% of total resources spent on support services.

Facilitating organization, financing and implementation of research and development for agriculture at the expense of state budgets, governments of the most of the countries encourage innovation activities in the private sector using direct and indirect financial incentives, providing information on the results of state-funded research, developing state-private partnerships, and organizing work on information and extension support of farmers.

Over the recent years, development of information and extension services is facilitated in a way of decentralization of state services and the introduction of private entities and intermediaries in this field. Innovation brokers appeared in the EU and other countries. They formulate needs of farmers in the sphere of research, help them to get an access to technologies, or develop links in value chains. Some of them are financed from the state funds and governed by a state, through regional organizations. There are pure private systems (for example, in the Netherlands and New Zealand), where farmers pay for the services and select service providers on a commercial basis. There are mixed systems, where services provided by the state entities and private consultants are paid by the farmers either in part or in whole. Finally, there are systems managed by the farmers' organizations (for example, in France and Finland) and financed by the government, farmers' organizations, and private farmers (Kadomtseva, 2010).

Table 5. Structure of state support of services in agriculture in selected countries in 1995 and 2015, %

Countries	Research and education		Inspection and control services		Infrastructure		Other	
	1995	2015	1995	2015	1995	2015	1995	2015
Australia	75	64	5	9	14	25	5	1
Brazil	23	15	4	7	58	51	15	27
Canada	41	39	22	40	12	10	24	11
Chile	28	24	1	20	64	51	7	5
China	8	29	5	6	23	37	64	28
Columbia	25	23	3	7	71	69	0	1
EU	31	37	3	6	24	30	43	27
Iceland	35	12	10	41	20	2	35	45
Indonesia	22	11	5	3	73	72	0	5
Israel	40	44	14	16	3	30	43	10
Japan	5	14	0	1	91	82	4	3
Kazakhstan	0	12	89	62	11	22	0	4
Korea	11	25	2	6	74	49	12	20
Mexico	54	48	6	11	17	38	24	3
New Zealand	65	49	24	32	11	20	14	0
Norway	56	54	18	24	11	16	14	5
Russia	12	42	8	26	15	13	64	18
South Africa	85	41	7	16	8	36	0	7
Switzerland	28	47	3	16	14	36	56	7
Turkey	1	1	3	2	75	82	21	15
Ukraine	25	45	8	30	63	11	4	14
USA	35	29	13	16	1	26	51	30

Source: Authors' development based on the UNCTAD (2016)

SOLUTIONS AND RECOMMENDATIONS

Differences in structural content of state support of agriculture are developed under the influence of historical, political, natural, and economic conditions of various countries. They are determined by their financial capabilities and current tasks of agrarian policies, some of which are: fulfillment of economic potential of agriculture as a sector, which ensures economic growth, and employment of population; increase of environmental sustainability of production in the conditions of natural scarcity; ensurance of food security of a country; expansion of export opportunities. Nevertheless, the analysis let the authors revealing some common tendencies in a change of role and directions of state support of agriculture in the emerging countries in the light of ensurance of food security, in particular:

- While the share of support of agriculture in the GDP decreases, its absolute value increases.

- Growth rates of state support of agriculture are synchronized with the growth rates of the gross product in agriculture. During critical periods of market slump increase of direct support of agriculture helps to stabilize production and domestic food market in order to ensure food security.
- Under an influence of globalization, the structural content of state support changes: tied price support is replaced by the untied support of producers' income.
- Importance of support of services grows, primarily of innovation activities and infrastructure.

Emerging countries have limited capabilities to provide the level support of their domestic farmers which would ensure achievement of target parameters of food security. Involvement into the international trade forces emerging countries opening their domestic food markets for imports. Effective protection of domestic farmers in emerging countries is impeded by the low import tariffs, which facilitate an easier market access for foreign agricultural products and food, lead to a reduction of domestic production, and in such a way threaten food security. Acting in the framework of the common directions of development of state support of agriculture, emerging countries should provide constant diversified aid to the direct producers of agricultural products. Measures of such support are increasingly frequently emerging into the tools of "precise pointing". They are aimed at smoothing negative consequences of unfavorable market environment because of objective and subjective circumstances. Account of those tendencies in the development of tools for state support in the emerging countries will let make the support more focused and effective and will help to adopt the measures of regulation of agriculture to the goals of ensurance of food security.

FUTURE RESEARCH DIRECTIONS

The vital issue for emerging countries in terms of achieving food security is how to secure the sustainable development of national agriculture and agribusiness in the conditions of a growing market openness and liberalization of international trade in food, taking into consideration the incomparably lower financial capabilities. Future research in this field has to be focused on exploring the ways in which emerging countries would be able to ensure the sustainable development of agricultural production and trade. The effects of the following measures on food security have to be investigated: state support of import substitution agricultural production; provision of environmental safety of domestic food and agricultural commodities; agricultural and food export increase once the domestic is saturated; logistical costs saving; optimization of all factors that affect competitiveness of domestic agricultural and food commodities in compliance with rational geographic distribution and specialization of agricultural production.

CONCLUSION

Many economies considered as emerging ones originated from the common socialist-type model of the economy (post-Soviet countries, Eastern European states, some of the Asian countries), where the dominant state and collective ownership of the means of production minimized the role of natural competitive factors (Kolodko, 1998). Foreign trade was controlled by the state, which resulted in various distortions, low level of diversification, specialization of export destinations and import channels due to political preferences. During early stages of transition, most of the economies of that type experienced a decline in output and export, significant increase in inflation, and decrease in living standards of the domestic population. Economic and trade liberalization rapidly embedded emerging economies into the international market, however, being considerably non-competitive, they had to specialize on low value-added exports and high value-added technological imports. For at least two decades after the beginning of market reforms in the 1980-1990s, most of the economies currently recognized as emerging ones have not been able to break the existing trade patterns.

Things are changing. Unimaginable growth of China, certain economic successes of Russia and Kazakhstan, the rapid increase in agricultural production and exports in Brazil and Argentina, and improvements in the region of the Eastern Europe demonstrated the potentials of the emerging economies on the global market. By the 2010s, particular countries and their strengthening alliances had been rather effective in increasing their shares in the global agricultural output and trade in agricultural products and food. New economic powers have arisen: China, Russia, Brazil, and other Eastern European and South-East Asian nations. The countries accelerated integration processes between themselves in order to ensure their food security, aggregate existing competitive advantages, and mitigate market risks. However, despite the evident achievements, there are still old problems and new challenges in terms of achieving food security. Unlike the past two-three decades of transition, coming years will manifest new factors: increasing competition between emerging countries and their alliances for better conditions of production and distribution of agricultural products and food; intensifying role of trade regulations (unification in the WTO framework and protectionism of particular domestic markets) and state support in achieving certain levels of food security; growing tensions between major global powers as to shape a new multi-polar world order.

The long-term outlook for agricultural development, trade in food and agricultural products, and food security in the emerging economies will depend on effective structural reforms, diversification and technical modernization of domestic production and exports, utilization of existing comparative advantages, and benefiting from effective state support of agriculture. As Von Braun, Serova, Seeth, and Melyukhina (1996) stated in relation to Russia, much would depend on how well incentives to increase agricultural production provided by a state are transmitted to the agriculture and food-processing sector and on the opening up of international trade opportunities.

REFERENCES

Anderson, K., Jha, S., Nelgen, S., & Strutt, A. (2013). Re-examining Policies for Food Security in Asia. *Food Security*, *5*(2), 195–215. doi:10.1007/s12571-012-0237-5

Boehringer, C., & Rutherford, T. (1999). *Decomposing General Equilibrium Effects of Policy Intervention in Multi-Regional Trade Models: Method and Sample Application*. Mannheim: Center for European Economic Research.

Božić, D., Bogdanov, N., & Ševarlić, M. (2011). *Ekonomika Poljoprivrede*. Belgrade: University of Belgrade.

Devereux, M. B. (1999). Growth and the Dynamics of Trade Liberalization. *Journal of Economic Dynamics & Control*, *23*(5-6), 773–795. doi:10.1016/S0165-1889(98)00043-8

Erokhin, V. (2015). Structural Changes in International Trade in Food: Competitive Growth Models for Economies in Transition. In *Proceedings of the 3rd International Conference "Economic Scientific Research – Theoretical, Empirical and Practical Approaches"*. Bucharest: Academia Romana.

Erokhin, V., Ivolga, A., & Heijman, W. (2014). Trade Liberalization and State Support of Agriculture: Effects for Developing Countries. *Agricultural Economics*, *60*(11), 524–537.

Estevadeordal, A., Freund, C., & Ornelas, E. (2008). Does Regionalism Affect Trade Liberalization Toward Nonmembers? *The Quarterly Journal of Economics*, *123*(4), 1531–1575. doi:10.1162/qjec.2008.123.4.1531

Filippov, R.V. (2014). International Experience of Subsidizing of Agriculture as Basis of Food Security of Russia. *Naukovedenie, 20*(1).

Josling, T., Anderson, K., Schmitz, A., & Tangerman, S. (2010). Understanding International Trade in Agricultural Products: One Hundred Years of Contributions by Agricultural Economists. *American Journal of Agricultural Economics*, *92*(2), 424–446. doi:10.1093/ajae/aaq011

Kolodko, G. W. (1998). *Equity Issues in Policymaking in Transition Economies*. Washington, DC: International Monetary Fund.

Liefert, W. (2004). *Food Security in Russia: Economic Growth and Rising Incomes are Reducing Insecurity*. Washington, DC: Economic Research Service, USDA.

Liefert, W., & Swinnen, J. (2002). *Changes in Agricultural Markets in Transition Economies*. Washington, DC: Economic Research Service, USDA.

Malozemov, S. I. (2014). Experience of State Support of Agriculture in Foreign Countries. *Science and Modernity*, *32*(1), 136–141.

Markovic, I., & Markovic, M. (2014). Agricultural Protectionism of the European Union in the Conditions of International Trade Liberalization. *Economics of Agriculture*, *61*(2), 423–440.

Olson, D. R. (2003). *Towards Food Sovereignty: Constructing an Alternative to the World Trade Organization's Agreement on Agriculture*. Minneapolis, MN: Institute for Agriculture and Trade Policy.

Petrikov, A. V. (2012). It Is Necessary to Increase Adaptation of Russian Agrarian Sector to WTO Conditions. *Economy of Agricultural and Processing Enterprises*, *6*, 6–8.

Petrikov, A. V. (2016). Major Directions of Implementation of Modern Agrifood and Rural Policy. *International Agricultural Journal*, *1*, 3–9.

Schmitz, A., Moss, C., Schmitz, T., Furtan, W., & Schmitz, H. (2010). *Agricultural Policy, Agribusiness, and Rent-Seeking Behaviour*. Toronto: University of Toronto Press.

Secretariat of the Convention on Biological Diversity. (2005). *The Impact of Trade Liberalization on Agricultural Biological Diversity, Domestic Support Measures and their Effects on Agricultural Biological Diversity*. Montreal: SCBD.

United Nations Conference on Trade and Development. (2016). Statistics Database [Data file]. Retrieved from http://unctad.org/en/Pages/Statistics.aspx

Visser, O., Mamonova, N., Spoor, M., & Nikulin, A. (2015). Quiet Food Sovereignty as Food Sovereignty without a Movement? Insights from Post-socialist Russia. *Globalizations*, *12*(4), 1–16. doi:10.1080/14747731.2015.1005968

Von Braun, J., Serova, E., Seeth, H., & Melyukhina, O. (1996). *Russia's Food Economy in Transition: What Do Reforms Mean for the Long-term Outlook?* Washington, DC: International Food Policy Research Institute.

Wittman, H., Desmarais, A., & Wiebe, N. (2010). *Food Sovereignty: Reconnecting Food, Nature and Community*. Halifax, Canada: Fernwood Publishing and Food First Books.

ADDITIONAL READING

Anderson, K., Dimaran, B., Francois, J., Hertel, T., Hoekman, B., & Will, M. (2001). The Cost of Rich (and Poor) Country Protection to Developing Countries. *Journal of African Economies*, *10*(3), 227–257. doi:10.1093/jae/10.3.227

Ayres, J. M., & Bosia, M. (2011). Beyond Global Summitry: Food Sovereignty as Localized Resistance to Globalization. *Globalizations*, *8*(1), 47–63. doi:10.1080/14747731.2011.544203

Boyer, J. (2010). Food Security, Food Sovereignty, and Local Challenges for Transnational Agrarian Movements: The Honduras Case. *The Journal of Peasant Studies*, *37*(2), 319–351. doi:10.1080/03066151003594997

Claeys, P. (2012). The Creation of New Rights by the Food Sovereignty Movement: The Challenge of Institutionalizing Subversion. *Sociology*, *46*(5), 844–860. doi:10.1177/0038038512451534

Dollar, D. (2001). Globalization, Inequality, and Poverty since 1980. Washington, DC: Development Research Group, the World Bank.

Dornbusch, R. (1992). The Case for Trade Liberalization in Developing Countries. *The Journal of Economic Perspectives*, *6*(1), 69–85. doi:10.1257/jep.6.1.69

Edelman, M. (2014). Food Sovereignty: Forgotten Genealogies and Future Regulatory Challenges. *The Journal of Peasant Studies*, *41*(6), 959–978. doi:10.1080/03066150.2013.876998

Hospes, O. (2009). Food Sovereignty: The Debate, the Deadlock, and a Suggested Detour. *Agriculture and Human Values*, *31*(1). doi:10.1007/s10460-013-9449-3

Khanna, T., & Palepu, K. G. (2010). *Winning in Emerging Markets: A Road Map for Strategy and Execution*. Boston, MA: Harvard Business Press.

Mamonova, N. (2017). *Rethinking Rural Politics in Postsocialist Settings. Rural Communities, Land Grabbing and Agrarian Change in Russia and Ukraine*. Enschede: Ipskamp Drukkers.

Patel, R. (2009). Food Sovereignty. *The Journal of Peasant Studies*, *36*(3), 663–706. doi:10.1080/03066150903143079

Spanu, V. (2003). *Liberalization of the International Trade and Economic Growth: Implications for both Developed and Developing Countries*. Cambridge, MA: Harvard University.

Ushachev, I. (2012). Measures to Secure Competitiveness of Russia's Agricultural Production in the Conditions of its Accession to WTO. *Economics of Agricultural and Processing Enterprises, 6*, 1–5.

Wehrheim, P., & Wobst, P. (2005). The Economic Role of Russias Subsistence Agriculture in the Transition Process. *Agricultural Economics, 33*(1), 91–105. doi:10.1111/j.1574-0862.2005.00136.x

KEY TERMS AND DEFINITIONS

Agriculture: A pool of establishments engaged in growing crops, raising animals, and harvesting fish and other animals.

Developed Country: A country which has a highly developed economy in terms of bigger gross domestic product, gross national product, and per capita income, and advanced technological infrastructure relative to other countries.

Developing Country: A country which has a less developed economy in terms of smaller gross domestic product, gross national product, and per capita income relative to other countries.

Emerging Economy: An economy that is not as advanced as developed countries but progressing toward becoming advanced.

Food Market: A medium that allows buyers and sellers of agricultural raw materials, agricultural products, and food to interact in order to facilitate an exchange.

Food Security: An availability and adequate access at all times to sufficient, safe, nutritious food to maintain a healthy and active life.

State Support: A set of protective measures provided by a state to a certain industry or sector.

Trade Liberalization: A removal or reduction of restrictions or barriers on the free exchange of goods between nations.

Chapter 4
New Food Industries Toward a New Level of Sustainable Supply:
Success Stories, Business Models, and Strategies

Vittorio D'Aleo
University of Messina, Italy

Francesco D'Aleo
University of Messina, Italy

Roberta Bonanno
University of Messina, Italy

ABSTRACT

"Novel foods" indicates all those foods that do not fall within the traditional western diet. The authors underline that there is an important economic potential linked to this new sector. After having introduced the topic by identifying what is meant by new food, food sustainability, and entrepreneurship in this sector, the analysis focuses on the qualitative aspects of novel foods. The authors emphasize the advantages of these sources, their qualities, and the impact on the environment. In this chapter, the authors demonstrate that novel food sources can be a substitute for other food that require production processes, and thus make a big impact on the environment. After figuring out the advantages of novel food sources, the authors analyze some success stories from the companies that already operate in this area. They also investigate existing business models on the novel food market and the way they operate and create wealth. Summarizing the results of their research, the authors underline the economic potential of the new foods exploitation in the Mediterranean area.

DOI: 10.4018/978-1-5225-2733-6.ch004

INTRODUCTION

Feeding is needful but the relationship between people and food is different from country to country, depending on the culture of the population and evolving changes in the different historical periods. The term "new foods" or "novel foods" concerns the use of non-conventional Western and Mediterranean sources of food (Burlingame, Charrondiere, & Mouille, 2009), which are already used since ancient times in other countries of Southeast-Asia area, for example, India, China, and Thailand. It is statistically estimated that in 2050 the world population could grow to about 9 billion people (Brookes, 2007), consequently, the food request could rise enormously. Therefore, there is a need to spread information about the existence of the alternative food sources, which will enrich those which are currently in use. In particular, the economic experts think (World Economic Forum and Food and Agriculture Organization of the United Nations (FAO)) that the food request could increase so high, that a new farmland larger than the territory of Brazil would be needed. The paucity of food resources could be a big problem for whole humanity in the imminent future, and novel foods could provide a possible solution. An example: the oceans cover about the 70% of the Earth even if actually they supply only a smaller quantity of food in comparison to the land surface. Therefore, from Southeast Asian Sea we could import Jellyfishes as a new potential future worldwide food. In fact, actually, the economists estimated (FAO and World Economic Forum) that the potential of this food is enormous. Some countries could import insects or breed them like a new source of food proteins, while African countries could use algae like a supplement of proteins. Spirulina microalgae is an example of food used like a nutrient supplement against the food shortages in many African countries. Hopefully, new farms concentrated in South America and other countries like China and Thailand could be established in other countries as soon as possible

BACKGROUND

Food Market Trends

Before evaluating the aspects of the new supply in detail, it is important to underline the food market trends to better understand the perspective of the global market. Therefore, the authors emphasize the following trends:

- Healthier food. More responsibility regards to animal welfare and land sustainability and more attention to the environment (environmentally friendly).
- Consumer needs access to detailed and clear information.
- Necessity to adapt the alimentation to the lifestyle of everyone.
- More and more consumers would like to feel connected to nature using some natural products.
- Homemade products named "0-kilometer foods" which are provided from their homeland.
- New consumption experience and opportunity to test new flavors like artisanal treats.
- Increase in home-cooking and home entertaining.

World economic traders could try to approach to these new trends and to expand existing market across new consumption occasions. New start-up companies related to "novel foods" should adapt their products to new market requirements.

Novel Foods Market

According to the ISPRA (Brookes, 2007), the global market for novel foods is actually about €35 billion (sales volume), most of which derived from food ingredients with the new primary molecular structure or genetically modified. The European Union (EU) represents about 40% of global novel foods market, an equivalent of €14 billion in sales. Enterprises related to the novel foods sector employ about 90,000 people worldwide. Research and development (R&D) are the most important costs, and novel foods enterprises usually spend an average of 3-5% of their turnover on R&D (range between 2% and 8%). Even if an enterprise wants to increase its profit it has to wait for a longer period of time (4-10 years) compared to the companies in traditional foods sector (1-3 years), because novel foods business is related to innovations. Life expectancy is better for new products (5-15 years) than those which are on the market already (1-3 years) (Hepburn et al., 2008).

European Regulation

The European Parliament has ratified that novel foods might be marketed in all countries of the EU. Receiving an authorization from the individual states will not take long, but the European Food Safety Authority will have to authorize the definitive marketing. What exactly are novel foods? Larvae, algae, scorpions, and spiders are considered novel. These products are already marketed in the Netherlands, Denmark, Belgium, and Great Britain. Within this group of foods, insects and algae are not the only potential resources of food. There are other supplies as well that could be used in purpose to eradicate malnutrition in the world. The FAO researchers all agreed that this could be a new possibility. To support this, people have to remember that there are many hundreds of tons of jellyfishes in the Mediterranean Sea that represents an important source of water and proteins. Actually, Professor Stefano Piraino of the University of Salento (Italy) coordinates the European project named "Med-Jellyrisk", which scope is exploitation of jellyfishes as a source of food. In Europe, jellyfishes and some other products were not considered safe food before 1997 (European Parliament, 1997) and therefore were not subject to any analysis by the European Community which usually advised by the EFSA before to be approved for marketing and consumption. Novel foods and all related products consumed inside the EU are governed by the Regulation 258/97 approved in 1997 (European Parliament, 1997).

This Regulation is a safety tool allowing the entry of novel food only after detailed controls and inspections and the subsequent final authorization in order to drive away unsafe products and foodstuffs.

According to the Regulation 258/97, a product is recognized as novel when it relates to one of the four categories:

- Foods and food ingredients with a new primary molecular structure or genetically engineered.
- Foods and food ingredients which are constituted or isolated from microorganisms (bacteria, fungi, or algae).
- Foods and food ingredients which are composed of plants, animals, or parts of them.
- Products and food ingredients which could perform a change in the nutritional value of the current diet and produced with an innovative or not currently used manufacturing process.

Procedures for requesting the authorizations and for marketing of novel foods within the EU as well as in certain member states are rather complex (Hermann, 2009). Companies willing to deal with novel foods have to submit an application to the related body of the member state, which scientifically reviews it, completes an initial assessment report, and then sends it to the EU commission. The commission members draw up another report in which provide assessments and objections. Applying companies must satisfy all requests of the commission before obtaining authorization for their new products. The final act is the decision for authorization (or denial) published in the Community Official Gazette. Actually, the procedure does not distinguish among innovative genuine products and those used outside the EU which could be imported under different names from the countries of their origin. Whenever a company advances a product which has already been authorized, it is also required to make another submission by the commission of a member state. However, in some cases, the procedure is simplified because of the acknowledgment of "substantial equivalence" of the product compared to the one already authorized. The same procedure may also be followed to certify an equivalence on the market of novel food compared to traditional food.

European Start-Ups' Market

Actually, 28 markets altogether represent the European start-ups' market (D'Aleo & Morabito, 2015). Europe offers funding for entrepreneurs, supports them, and grants network opportunities, but the clutches between 28 submarkets of the EU member states cause a loss of homogeneity unlike other markets (the United States are the most obvious example) (Darkow, Foerster, & von der Gracht, 2015). However, the market of the EU is more complex and diversified compared to the USA, because every member state has its recipes and characteristic variations, which could result in a loss of a market. Even if Europe is in the fifth place among the top 10 most innovative countries worldwide, multiple data indicate that the conditions of European economic could not be ideal for entrepreneurs and for fast-growing companies. This appears due to the fragmentation of the European market, denied access to the capital sources, and a supportive shortage of the initiatives. Furthermore, the absence of the European law on data protection complicates the expansion and displacement of companies that would like to enter into commercial negotiations with European regions. There are European countries wherein venture capital and bank loans are easily accessible: Ireland, Sweden, Luxembourg, the Netherlands, Germany, France, Belgium, Austria, the United Kingdom, Slovak Republic, and Bulgaria. Meanwhile, there are other European countries wherein the capital is more difficult to obtain: Portugal, Romania, Poland, Denmark, Lithuania, Italy, Greece, Slovenia, Spain, Croatia, and Hungary. During 2003-2013, Europe produced 30 "unicorns" (slang term which refers to the companies that have been evaluated billion dollars), compared with 39 in the USA. There are some sectors in Europe that encourage the development of new business; examples are music, financial technologies, and games. The most recent successes have been reported recently: Tradeshift, Soundcloud, Supercell. Start-ups operating in the food sector represent only a small percentage, but the EXPO could symbolize a new stimulus.

Special Focus: Italian Environment for Start-Ups

In Italy, there is a special department for start-up companies that identifies new innovative start-ups. In March 2015, the segment of "innovative start-ups" (Law decree 179/2012) (Government of Italy, 2012) counted about 3,711 companies with an increase of 530 units compared to 2014. Innovative start-ups represent 0.25% of the total number of companies in Italy. The total value of these new companies is estimated to be €192 million. However, when the authors analysed that segment of business, they have observed that most of the innovative start-ups provide business services or consultancy (73%), while others operate in the spheres of industrial manufacturing or new products creation (18.2%), and only a few start-ups work in distribution of new products (Table 1).

New start-ups are dominantly managed by men, while women head only 477 companies, which is equivalent to 12.9% of the total number of start-ups. Furthermore, young people are the founders of only 879 new companies (23.7% of the total). There are only 88 start-ups founded by foreigners in Italy (2.4% of the total). Start-ups are mainly located in the north of Italy. Lombardia is the Italian region that holds the most of the new companies (808, or 21.8% of the total) followed by Emilia Romagna with 451 new companies (12.2% of the national total). Lazio is the most important region in Central Italy with 347 new companies (9.4% of the total). Veneto and Piemonte have 274 and 266 new companies, equal to 7.4% and 7.2%, respectively. Italian regions with the smallest number of new companies are Basilicata with 20 start-ups, Molise with 16, and Valle d'Aosta with 10. Among the southern regions, Campania is the first with 214 new companies (5.8%) followed by Sicily and Puglia each of them hosts 156 new companies (4.20%). Calabria has only 91 new companies or 2.5% of the total number of a start-up in Italy (Table 2).

Big cities are more attractive for new companies. Milan is the main city in Italy chosen to start a new company. 533 companies are now located in Milan, a which is an equivalent to 14.4% of the total number of companies in the entire region. Meantime, Roma, Torino, Bologna, and Naples are also important in terms of presence of start-ups.

New companies have four partner founders and employ 2.50 people in average; return on investment (ROI) and return on equity (ROE) indicators of profit for innovative start-ups record negative values.

Typically, the evolution of start-ups in food sector occurs in two phases:

- Innovations of "incremental", an occurrence that introduces a new tool or channel for the marketplace.
- Creation of unusual service which is substantially discontinuous with the recent past.

Table 1. Breakdown of start-ups by sectors in Italy

Sector	Number of companies
Consulting and business services	2,708
New products production	677
Sells and commerce	153
Building	42
Agricultural and connected activity	13
Tourism	9
Logistics	10
Insurance and finance	6
Other sectors	93
Total	3,711

Source: Camere di Commercio d'Italia, 2015

Table 2. Geographic dispersion of start-ups in Italy

Region	Number of start-ups
Lombardia	808
Emilia Romagna	451
Lazio	347
Veneto	274
Piemonte	266
Toscana	238
Campania	214
Sicilia	156
Puglia	155
Marche	152
Trentino Alto Adige	128
Friuli Venezia Giulia	110
Sardegna	105
Calabria	91
Abbruzzo	70
Liguria	57
Umbria	43
Basilicata	20
Molise	16
Valle D'Aosta	10

Source: Camere di Commercio d'Italia, 2015

When a company starts to change existing product, it tries to create new products and services based on new unpublished experience.

In order to extend the marketplace, new companies should interconnect the real and the virtual market and produce new devices or apply innovations to traditional tools.

Nowadays, the market of "new products" is a subject to an increase of availability of new foods or "niche foods". Therefore, new companies could provide a wide variety of new products and incorporate them into the traditional distribution channel.

MAIN FOCUS OF THE CHAPTER

Insects

There are about sixty thousands vertebrate species on the Earth, and even if the humans (Homo sapiens) apparently dominate the world in term of their impact on the environment and use of resources, insects are the true rulers of the world with about one million of species. Biologically, the great diversity of insects is due to their co-evolution with plants and environment. The insects class is further divided into thirty orders such as Coleoptera, Diptera, or Hemiptera, so at last, it is estimated that about 75% of all animals that live on the Earth are insects. These organisms belong to a bigger systematic group; they are the most different species of animals living on the Earth, except open oceans, and occur inside all ecological niches. As pollinators, they play an important role in the reproduction of plants, biodegradation of organics substrates, waste, and humans' food. Various species of insects serve as carriers for pollens and concur to reproduction system of plants. Ingram, Nabhan, and Buchmann (1996) identified one hundred thousand pollinator species, and almost all of these (98%) are insects. Over 90% of 250.000 species of flowering plants depend on pollinators. An important role is also played by insects in biodegradation of waste (Simon, 2008). Beetles (4,000 species play a significant role in decomposition of dung), flies, ants, and termites break down organic matter until it is fit to be consumed by fungi and bacteria (Van Huis, 2003). Honey and silk are the most commonly known insect products. Insects provide to humans a wide variety of valuable products, for example, about 1.2 million tons of commercial honey is made by bees and 90.000 tons of silk is produced by silkworms every year (Van Huis, 2003; Yong-Woo, 1999). Insects are important for the production of colour red carmine and other colorants used to colour foods, textiles, and pharmaceuticals and for medical applications including maggot therapy (Sherman & Pechter, 1988; Wollina, Karte, Herold, & Looks, 2000), where they are used to treat infected wounds and burns (Van Huis, 2003). However, some of the insects are pests which inflict damage to humans, farm animals, and crops. Some insects pose a serious threat to humans' health, for example, mosquito, a vector of the malaria parasite, kills one million people every year. In agriculture, insects inflict severe damages and cause failure of one-fifth of the world's total crop production every year (Shelton, Zhao, & Roush, 2002; World Health Organization, 2014). The introduction of alien pests into new habitats due to the global exchange causes another severe damage (Snowden, 2010). Bringing a pest to a new geographical area, in most cases, may lead to critical complications (Snowden, 2010). Very particular example is Red palm weevil (Rhynchophorus ferrugineus Olivier, 1790). In Europe and the USA, this pest destroys all ornamental plants belonging to Phoenix species' palms, and it seems to be also involved in the destruction of other palms species. However, in some countries, people consume larvae of these species, and Red palm weevil is the most popular edible beetle in the tropics. Insects belong to Arthropoda's phylum along with crabs; lobsters and shrimps are common foods worldwide. In 2013, Jongema published a list (updated in June 2015) of edible insects of the world, where he presented about 2,000 of species of edible insects that are known to be consumed from the native populations in many countries (Jongema, 2015). In 2013, Van Huis et al. (2013) in FAO/INFOOD conference in Rome published a database of the biological food and included insects. This database is a collection of analytical data on biodiversity food from the published and unpublished sources. It reports 471 entries on about 5,700 values. Even if this database represents the largest compilation of data for this animal group, it seems to be incomplete so providing a definitive idea on the number of edible insect species seems to be difficult. The largest number of edible insects seems to come from North America, where Ramos-Elorduy, Moreno, and Camacho (2009), reported 549 species of edible insects from Mexico (Vantomme, 2015). The true problem about edible insects is that some species cannot be identified especially at the larval stage. Among the thirty orders of insects, only eighteen are considered edible, the most of these belonging to Coleoptera (31%) order followed by Lepidoptera (18%) and Hymenoptera (15%) (Vantomme, 2015).

Entomophagy, the word that indicates consumption of insects as food, is common in the world, mostly in Africa and Asia (Katayama et al., 2008; Aiking, 2011). It is also accepted in North America, but not well accepted among the Western European populations. In 2050, the global population is estimated to be about 9 billion people, and populations from developed countries will have to adapt to alternative sources of animal proteins because traditional breeding will become unsustainable (Aiking & De Boer, 2004; Nowak, Persijn, Rittenschober, & Charrondiere, 2014). In 2013, Caparros Megido et al. (2013) published the first study and results on a possible acceptance of entomophagy among Belgian people. The results were that even if 46.6% of respondents had a negative attitude to eating insects, about 77.7% of them were willing to try. Surprisingly, respondents in the age group from 19 to 45 years after trying some insects declared their readiness to cook them at home. After Caparros Megido et al. (2013) published the paper, a new brand smart-farm for insects breeding (Insect Europe) was established in the Netherlands. The company breeds, sells, and promotes edible insects, mainly crickets and grasshoppers, as human and animal food. In France, three people opened a shop (Europe-entomophagy) selling edible bugs. In January 2014, Jarrod Goldin with his brothers launched a new start-up company (Next Millennium Farms) located in Ontario. He was inspired by a landmark United Nation report, which recommended insects for human consumption. Next Millennium Farms (now Entomo Farms) began in 2014 with 5,000 square feet for mealworms and finished with $65,000 in sales. Insects is a future source of animal proteins. Establishing of new brand and start-ups for breeding and selling insects in Europe is the first step to increasing the number of potential consumers (Rumpold & Schlüter, 2013; De-Magistris, Pascucci, & Mitsopoulos, 2015).

Some business success stories for insect breeding are provided below.

1. Insect Europe BV. The company breeds and elaborates edible insects like crickets, mealworms, and grasshoppers for the food industry. Insect Europe BV undertook the edible insects business some years ago. Nowadays, products of the company are distributed and marketed worldwide. The target of Insect Europe BV is the output of "premium" insects for the food industries, with a focus on high quality and safety standards. Insect Europe BV could get the leadership among the edible insect's trade and could get the top of edible insects culture. Insect Europe BV created two different brands to sell its products: "Bugs for Pets" and "Delibugs". The two brands have two different markets: "Bugs for Pets" is for animal consumption (pet animals); "Delibugs" are products for human consumption. "Delibugs" spread and sell products derived by worms, crickets, grasshoppers, ants, scorpions, and butterflies.
2. Entomo Farms. The company elaborates, harvests, and distributes insects and their derivatives of the highest quality and provide alternatives to farmers producing healthier and certifiable organic fish and chicken. Entomo Farms has been the first business that used the "Entomophagy" term. They follow a way for the dissemination of insect farming methods and marketing of insect protein. Entomo Farms aims to be the world leader in providing protein alternatives in the form of the finest organic and excellent insect proteins with great tasting products and recipes.
3. Entonote was is founded in April 2014 by Giulia Maffei (biologist) and Giulia Tacchini (freelance) with a precise target to offer services and introduce entomophagy in Italy. This is the first experience in Italy. Even if it is not a start-up, it represents the only entity that actually organizes events and provides consultancy in the sphere of entomophagy.

Jellyfishes

Jellyfishes are aquatic, pluricellular organisms (salt and sweet waters) that belong to Cnidaria or Ctenophora phylum. They appeared in the oceans about six hundred million years ago (before the appearance of dinosaurs) and still populate the oceans today (Uye, 2008). Scientists classified about two hundred species of jellyfishes (distributed in 19 families and 50 genera) (Hsieh, Leong, & Rudloe, 2001). The edible jellyfishes are consumed in many countries of Southeast Asia but mostly in China, where domestic demand is increasing. China is the first country to process jellyfish for human consumption. Chinese people have been catching jellyfishes for 1,700 years, and consider them as a delicacy. Jellyfishes are fished in different methods (Iversen & Hale, 1992). Despite the fact that jellyfishes are consumed since ancient time, only recently this business has become commercially active. After catching, jellyfishes are sent to the local farm and then exported as soon as possible. Kitamura and Omori (2010) quote a short list of the six major edible jellyfishes harvested in Southeast Asia. They provide an identification key for a correct taxonomic identification. Fresh jellyfish readily spoil at ambient temperature. Therefore, processing of jellyfish is carried out preferably within a few hours of capture while the animals are still alive. The body of jellyfish consists of a hemispherical transparent umbrella. The mouth is on the under surface of the umbrella and is protected by fused oral arms, commonly known as legs. The umbrella and oral arms of jellyfish are separated immediately after catching. Jellyfishes are cleaned with seawater, scraped to remove mucus membranes and gonadal material. Both umbrella and oral arms are used in processing. Traditional methods of processing involve a step-wise reduction of the water content using salt and alum. A salt mix containing about 10% alum is used for initial salting of jellyfish using about 1 kg salt-alum mix for 8–10 kg of jellyfish. Salted jellyfishes are then left in the brine for 3–4 days, followed by several transfers to another container salted with a fresh mix containing a smaller amount of alum. Salted jellyfish can then be heaped and left dry on a draining rack at room temperature for 2 days, and the heap is turned upside down several times during that period to allow excess water to drain out through compression by its own weight. The entire process requires 20–40 days to produce a salted final product with 60–70% moisture and 16–25% salt (Subasinghe, 1992). The processed jellyfish has a yield of about 7–10% of the raw weight depending on the species and processing formula. Preservation of jellyfish in a mixture of salt and alum is necessary to obtain products of desirable structure and texture. Alum reduces pH, acts as a disinfectant and hardening agent, giving and maintaining a firm texture by precipitating protein. Salt aids in reducing water content and keeping the product microbial stable. Salt or alum used singularly in the processing of jellyfish does not produce a product of satisfactory properties. Extensive liquidation of tissue occurs in the absence of salt, while disagreeable odors develop in the absence of alum. Lowered pH greatly reduces the possibility of microbial growth and enhances the shelf life of the product. Colour of freshly processed jellyfish is creamy white but gradually turns yellowish as the product ages. In the Asian jellyfish market, a whiter colored product has higher retail value. Yellow, but not brown color, is acceptable. The longer the sample remains around, the darker it becomes. If it remains too long, it turns brown and the product is unacceptable. The salted jellyfish has a stable shelf life up to one year at room temperature. The shelf life can be increased to more than two years if the product is kept cool. Chinese wedding or formal banquet is rarely completed without a jellyfish salad. The Chinese have various methods for preparing jellyfish, cooked or uncooked. Fresh unprocessed jellyfishes are rich in minerals such as Na, Ca, K, and Mg, but processed product is depleted of salts after desalting in fresh water. Lipid content of the whole jellyfish ranges from 0.0046% to 0.2% on a wet weight basis (Subasinghe, 1992). Chinese products contain an average of 5.5 and 6.8 g of protein per 100 g of salted product, respectively (Morikawa, 1984).

Consumption in Italy: unofficial sources suggest that Asian restaurants located in Italy serve local jellyfish. Animals, caught by small boats, are treated with a substance similar to flour and carried to Chinese restaurants. Jellyfishes are also dried and exported.

Market characteristics: analysis of the industries (few) allowed detecting the strengths and weaknesses of the most important international companies. AS.MP Jellyfish Processing & Trading Com is a company for processing and trading salted jellyfish. The company is on the market since 1970, however, it sells its products in local markets in the Philippines. Actually, the company tries to sell its products on the international market with over 20 tons of jellyfishes. There is no another company of such caliber existing in Europe.

Algae

Algae were the first "plants-like" to appear on the Earth. Billions of years ago, they transformed the atmosphere allowing to other forms of life to evolve. The term "algae" is an imprecise word. In biology, with the term "algae" one can define a heterogeneous group of generally simple unicellular or multicellular eukaryotic or prokaryotic organisms, some smaller than a cell, others bigger than a football field. Algae are complex organisms fixing carbon dioxide and releasing oxygen. They are also among the biggest producers of organic substance. Even if all algae contain chlorophylls, pigments distinctly color them. The variety of their coloration is so wide that pigmentation may play an important role in algae classification (Buschmann, Troell, & Kautsky, 2001). Algae belong to one of three groups of organisms that biologists divided in red-algae, brown-algae, and green-algae (Kawashima, 1993; Lüning, Yarish, & Kirkman, 1990). The estimated total value of cultivated algae was \$7.8 billion in 2008, which was equivalent to approximately 15.8 million tons of fresh weight, 99.6% of which is dominated by seaweeds. The total of farmed algae in 2010 was estimated at 19 billion tons with a value of \$5.7 billion, in comparison to wild harvested algae, which was equivalent to approximately 800 thousand tons (Neori et al., 2004). Red algae are very delicate organisms and often appear graceful, with a branched structure and with a color range from pink to red made by a phycoerythrin pigment that hides chlorophylls. Even if some organisms inside this group are microscopic, many of them grow up to several centimeters. Some of them grow as thin filaments or leaflike structures. Red Algae includes several edible organisms as Porphyra (known as Nori) genera. There are approximately two thousand and five hundred species of red algae. Brown Algae, approximately about 1,800 species, are the smallest group of algae. This group includes macroscopic members only, with the one big seaweeds known (Kelp's greenwood). Their color ranges from brown to yellow because of the unique pigment, fucoxanthin. There are different brown algae, but only a few are consumed by humans, the main is Saccharina japonica known as Kombo. Green Algae is the most known kind of algae in the human diet. They grow as filaments or branching fronds in the oceans and seas worldwide. The most commonly eaten seaweeds in this algae group are marine lettuce (Ulva lactuca) and Enteromorpha intestinalis (Aonori). Most of the green algae dwell in freshwater, but the most consumed ones come from the sea. There are approximately ten thousand species of green algae, five thousand of which of marine origin (Lüning et al., 1990).

Cyanobacteria are sometimes called blue-green algae, but they are prokaryotic organisms and are not true algae. However, Arthrospira platensis, a cyanobacterium, is the main microalgae provider in terms of commercial production. Its cultivation is particularly attractive for several reasons. Humans eat some varieties of algae for ages. Edible algae, that the ancients used as food for centuries raise from coastal regions or internal lakes. Only a small part of these species of freshwater macro and micro-algae are used for diet because many of them are poisonous. Seaweeds are traditionally consumed as food in North and Southeast Asia. In the USA, some companies began cultivating seaweeds inside tanks for human consumption, and their markets grew so much that they exported these products to Asian countries.

However, in the last few years, there was a small spread of seaweed and freshwater microalgae into the European and occidental cuisine, where the interest for tropical and non-conventional food is now growing.

Red Algae for Food Industry

Among the Red Algae group, there are some species interesting for economic development. Chondrus crispus is a common organism which can be found on rock surfaces. Red seaweed was found on both sides of the Atlantic Ocean, along the Canadian line and from Iceland to Spain and Portugal. Chondrus crispus was also known as Irish Moss is a red macro-alga with a dark red-brown color, but bleaching may occur. Irish Moss extract has been used in Ireland since the beginning of the 19th century as a folk remedy for some illnesses. The name "carrageen" seems to be introduced around 1829; it probably came from Carrigan Head in Co. The use of Irish Moss or Carrageen Moss spread from Ireland, and its use was exported to the USA by emigrates. Small amounts are still collected in Ireland for cooking and as a health drink. Historically, carrageenan was extracted from wild seaweeds Chondrus crispus. However, supply from wild seaweed was insufficient. This has prompted a cultivation of carrageenan-containing seaweeds in tropical oceans. In 2008, about thirty countries launched carrageenan biomass production. Nowadays, according to FAO statistics, red seaweed production is 21% of the total production of all cultivated seaweeds, an equivalent of 9 million tons. The main red seaweed species under cultivation are Kappaphycus and Eucheuma (primarily raw materials for carrageenan), Gracilaria (used to extract agar), and Porphyra, or Nori (mainly for direct human consumption) (Buschmann, Correa, Westermeier, del Carmen Hernandez-Gonzalez, & Norambuena, 2001).

Agar and carrageenan are called hydrocolloids and are used as food additives and in medical or research laboratory. In the past decade, demand for hydrocolloids has grown with an increased consumption of processed food. Nevertheless, there was an insufficient supply of wild seaweed. For this reason, there was a motivated experimentation of cultivating carrageenan seaweeds in tropical waters. This resulted in a rapid expansion of Kappaphycus and Eucheuma cultivation from 944.000 wet tons in 2000 (48% of total red seaweed cultivation) to 5.6 million wet tons in 2010 (63%). Among various carrageenan-containing seaweeds, only Eucheuma has been mainly cultivated. Carrageenan has been used as a natural food additive for over 600 years, and today it is recognized as a harmless food additive without nutritional value. A vast amount of carrageenans is applied for dairy and water products. Nowadays, it is used for stabilization, thickening, and gelation of processed foods. It was used by the food industry in the USA and Europe and for new food products. It was used worldwide to enhance production of milk, toothpaste, flans, puddings, custards, jellies, confections, ice creams, chocolate, and nutritional beverage mixes. Carrageenan is also an approved food additive in many countries for humans and animals. With the development of the agar industry, interest for Gracilaria increased further (Armisen, 1995). The annual yield of agar is about 7,000 tons in the world, the most part constituted by Gracilaria species as a raw material. In many species, the content of agar is about 20-30%, or even more, and the quality is good. Experiments on artificial cultivation have been carried out in many countries (Kapraun, 1999). China, India, and Taiwan, the foremost countries in Gracilaria culture. It is carried out in China since 1984 (Guiry & Blunden, 1991).

Artificial cultivation of Gracilaria is carried out by two methods according to its biological characteristics. The first method is used for the species that can be propagated by the vegetative form and can be harvested to time during the whole growing period (examples are Gracilaria parvaspora and Gracilaria blodgettii). The second method is used for the species that may propagate through spores only (examples are Gracilaria asiatica and Gracilaria tenuistipitata). Agar was the first phycocolloid discovered and prepared as a purified extract. It has an ability to form gels upon cooling of a hot solution to 30–40°C and to melt to sols upon heating to 90–95°C. Agar has a wide variety of uses, other than microbiological use. It is engaged in the food industry, dairy industries, meat packing, pharmaceuticals, and in biomedicine. Agar is used in food industry predominantly for its stabilizing, thickening, and gelling characteristics (Armisen, 1995; Dawes, 1995; Friedlander & Levy, 1995).

Nori or Porphyra spp. is a red algae that grows as a very thin, flat, reddish blade. It is found in the most temperate zones worldwide, where it is consumed by the local populations. It is among the most nutritious seaweeds, with a protein content of about 50% (with the large amounts of three amino acids: alanine, glutamic acid, and glycine) and low content of sugars (below 1%). In addition, vitamin content is very high. During processing, most salt is washed away, so the sodium content is low. The manufactured products are Nori sheets each of about 20×20 cm size and 3.5–4.0 g weight. In ancient times, Porphyra could be collected by hand from natural sources, but now the most are derived from artificial cultivation. Good quality Nori is in demand in the Republic of Korea, where production methods differ between the northern and southern parts of the country. In the Republic of Korea, annual production volume is 270,000 wet tons (Armisen, 1995; Oliveira, Alveal, & Anderson, 2000). There are no data for North Korea. In the Republic of Korea, Nori lightly fried with a little oil is consumed as a popular snack with beer. The culture of Porphyra in non-Asian countries, in particular in the western and eastern coasts of the USA, used to be successful but became unviable commercially when residents of the shore areas objected to the presence of seaweed farms and access to sufficient space to expand the pilot farm was refused. On the east coast of Maine, cultivation of Porphyra had many problems with indigenous species that slowed the progress. Therefore, in the USA the biggest company in the sphere of food biotechnology was reorganized and started developing other businesses. Cultivation of Porphyra (Nori) in Japan yields about 500,000 wet ton per year, or $1.5 billion. Dried nori is oversupplied in Japan, and producers and dealers encourage nori consumption outside national borders, primarily in the USA. Production and markets in China are expanding, although the quality of the product is not always as good as that coming from the Republic of Korea and Japan. In China, Nori industry produces 210,000 wet tons per year (Oliveira et al., 2000). In many countries, processing of wet Porphyra into dried sheets of Nori are highly mechanized, rather like an adaptation of the paper-making process. Wet Porphyra is rinsed, chopped, and stirred in a paste. It is then poured onto a cloth and run through a dryer. Adjusting of the temperature carefully controls the rate of drying. The sheets are removed from the clothes and packed in about ten pieces to be distributed. This product is called "hoshi-nori" which distinguishes it from "yaki-nori", which is toasted and distributed in sealed packages. Nori is used as a luxury food mainly. In Japan, it is consumed with rice balls and sushi, a typical Japanese consisting food, or cut into small pieces and eaten with noodles. It can be incorporated into soy sauce to make a luxury sauce. In China, it is mostly used in soups and for seasoning fried foods.

Brown Algae for Food Industry

There are about 1,800 species of brown algae and their brown color results from the dominance of the xanthophyll pigment fucoxanthin, which masks other pigments, like Chlorophylls. There are no known unicellular organisms inside this group of algae. However, the biggest algae are known, for example, some members of the Macrocystis genus, may grow up to 70 meters and form prominent submerged kelp forests. Brown algae are adapted to a wide variety of marine ecological niches including rocky marine pools and intertidal zones. Only four species are restricted to dwell in freshwater. Among the Brown Algae group, there are some species perspective for economic development. Brown Giant Kelp algae Macrocystis pyrifera, Ascophyllum nodosum, and Laminaria hyperborean are harvested on the coasts of Atlantic coast, and they are used for alginate extraction (Lüning et al., 1990). Alginates are used mainly for commercial use as toothpaste, soap, and other applications. Its primary function is as a binder, stabilizer, and emulsifier agent. Saccharina japonica (Laminaria) is produced in Japan, China, and Southeast Asia for food and alginate production. "Kombu" is the Japanese name for the dried seaweed that is derived from a mixture of diverse Laminaria/Saccharina species. "Haidai" is the Chinese name for Saccharina japonica, the most cultivated species. There are three techniques for cultivation of kelp algae: vertical, horizontal, and mix method. These are similar in each of the countries that produce kelp products.

Vertical culture method is a technique which assumes exploitation of water space. It is a simple, easily managed procedure. Among the kelp line, there is a space that allows the light pass through them, thus stimulating sporophyte growth. However, this method has a drawback, indeed, the light decreases with depth, and the plants at the lower level may not get enough light to meet their requirements for good growth.

Horizontal culture method can be used in lower sea areas. Horizontally positioned culture line give greater and more distributed light to all of the attached plants compared to the vertical method. The disadvantage of this model is interwoven of algae line. This usually happens during storms (Oliveira et al., 2000).

Mixed culture method combines vertical and horizontal systems. Harvesting is an intensive procedure and depends on the seawater temperature. Kelp has to be harvested only when the temperature reaches 21°C. Naturally growing plants are biennial and are ready for harvesting after 20 months. Harvesting is made by manual operations: unfastening the algae line from the floating barges, loading the kelp into boats, and transporting them to the seashore for sun-drying or inside farm to process them. There are two main processing methods: immediate sun-drying or salting. Saccharina japonica species contain about 10% proteins and 2% percent fat, as well as an amount of minerals and vitamins. The cost of seedlings is currently about $0.50-0.60 per seedling-curtain (each with about 40,000–50,000 seedlings). No data on total production costs are available. According to Oliveira et al. (2000), global production of Saccharina japonica was about 5.9 million tons in 2013. Despite Saccharina japonica naturally grows and cultivated in the Republic of Korea, the demand in the country is lower because Koreans prefer consuming of Undaria pinnatifida. This species inhabits the intertidal zone to a depth of about 20 meters. It is introduced in Australia, by not consumed there. Undaria is highly invasive species; it grows rapidly and has the potential threat to native algal species. Harvesting is about 800,000 wet tons of three different species, and about 50% of this is for wakame. Produced from different brown seaweed, Undaria pinnatifida, it is grown in a similar fashion to Saccharina in China where this species was not as popular as Undaria (Yamanaka & Akiyama, 1993). Japan has developed different hybrids with higher growth and nutritional characteristics.

Green Algae for Food Industry

Green algae are the biggest group of algae and they may be unicellular (one cell), multicellular (many cells), and colonial (existing like an aggregation of cells). Most green algae are aquatic and are found commonly in freshwater and marine habitats, even if some are terrestrial. Some species live inside other organisms, like a symbiotic cell, an example is lichens. There are about 4,500 species, including about

2,500 Chlorophyceae (mostly freshwater organisms) and 250 marine Ulvophyceae, or seaweeds. Three important macroscopic edible Green Algae genera are marketed: Monostroma spp., Enteromorpha spp., and Ulva spp. First two genera are now cultivated in Japan and other countries of Southeast Asia, where they are consumed as Aonori (dried seaweed). They are cultivated in calm waters like bays and estuaries, but they may also be grown in deeper waters using floating lines (Monostroma spp.), like some species of Red Algae. They contain about 20% proteins and little fat and have a high content of vitamins and minerals (Blunden, 1991).

It is collected from the wild because it grows naturally, usually attached to rocks, but it may be cultivated using different methods. Ulva spp. has an important role for many coastal communities, and are currently used for numerous purposes. Various coastal communities utilized this resource since ancient time. Ulva is consumed worldwide mainly in soups and salads. Ulva spp. has higher protein content (proportion varies up to 40%) and lower lipid content than other two marketed Green macroscopic algae. Cultivation of Ulva spp. on a commercial scale has a low economic value, due to the high natural abundance (Guiry & Blunden, 1991). However, some cultivation method is selected. Open water cultivation is the oldest, but this genera may be cultivated with a suspended technique which is attained through the attachment of the seaweed to ropes, lines, floating rafts, and nets, or enclosed in cages. Semi-suspended cultivation, basket culture, raft culture, and long line cultivation are other techniques of cultivation. Among microscopic Green Algae, Chlorella genus belonging to the phylum Chlorophyta is the most important for food supply (Iversen & Hale, 1992). Cells of this microscopic species are spherical, subspherical or ellipsoid, single or forming colonies. This genus of freshwater green algae is composed of organisms which are the incredibly sustainable food source offering many important nutrients coupled with the high and complete value of proteins. Chlorella is another alga with high protein content (60%); it provides large quantities of folate, vitamin B-12, iron, and other minerals. However, the cell wall of Chlorella is rigid and indigestible. Therefore, harvested chlorella cells need to be processed in order to break the cell wall before using it as a dietary supplement for its nutrient content.

Green-Blue Algae (Cyanophyta)

Spirulina originated about 3.5 billion years ago. It is a primitive organism which established an ability to use carbon dioxide dissolved in seawater as a nutrient source for its reproduction. Spirulina is a Cyanophyta, or blue-green algae and is not an algae but bacterium. It is found in soil, marshes, freshwater, brackish water, and thermal springs. Alkaline, high temperatures (Spirulina shows an optimum growth between 35°C and 39°C), saline water (>30 g/l) with high pH values (11.0) and high level of solar radiation allowing it to grow rapidly, reaching a maximum density. Spirulina's main photosynthetic pigment is phycocyanin, but it also contains chlorophyll and carotenoids. Characteristic of the genus is a helical shape of the filaments (or trichomes). Spirulina naturally grew in the lakes of Texcoco, Chad, and Niger and under normal water conditions. This organism is one of many algal species, but it thrives in the alkaline lakes where it is difficult or impossible for other microorganisms to survive. Most commercial production systems are based on shallow raceways (open or closed), in which Spirulina cultures are mixed by a paddle wheel. However, there are still some examples of Spirulina which are harvested commercially from natural sites and local populations. Aztecs living in the Valley of Mexico collected Spirulina from a lake and made a like-paste "techuitlatl". Since 1967, Sosa-Texcoco Ltd has been harvesting Spirulina maxima from an area of Texcoco lake. The first pilot plant that produced 150 tons of dry spirulina biomass per year started production in 1973. Sosa-Texcoco Ltd has invested about $5 million in the development of production since 1977. In 1995, Sosa-Texcoco ceased production of Spirulina. Four volcanic lakes with natural Spirulina blooms in Myanmar have been reported to be used as a production site. Production began in 1988, and since 1999 increased to 100 tons/year. About 70% is harvested from boats on the surface of the lake, and about 30% is grown in outdoor ponds alongside the lake. Habib, Parvin, Huntington, and Hasan (2008) suggested that the mass culture of algae (labeled as one of the "future foods") would help to overcome global protein shortages. However, two of the main barriers to the small-scale culture of Spirulina are the costs and availability of inorganic nutrients. Gami, Naik, and Patel (2011) investigated the use of a medium formulated for mass production of Spirulina that is different from the laboratory medium. This newly formulated medium contains single super phosphate (1.25 g/litre), sodium nitrate (2.50 g/litre), sodium chloride (0.50 g/litre), magnesium sulfate (0.15 g/litre), calcium chloride (0.04 g/litre), and sodium bicarbonate (8 g/litre). There are some alternatives for Spirulina culturing, especially from waste effluents, which may be available in rural locations. Mass cultivation of Spirulina is usually carried out in shallow raceway ponds, equipped with paddle wheels to mix the culture. Two types of open raceway ponds are typically used: the first is lined by concrete and is, therefore, expensive, the second is a shallow earthen tunnel lined with polyvinyl-chloride (PVC) or some other durable plastic material. The first type even being expensive for manufacturers could bring bacterial or other contaminations. The surface of commercial raceways varies from 0.1 to 0.8 hectares, and culture depth is usually maintained at 10–25 cm (D'Aleo & D'Aleo, 2015a, 2015b). The paddle wheel, large or small with a speed from 10 rpm is the most common stirring device. If the paddle wheel is not sufficient for shifting the culture or the culture is over-growing, self-shading governs the light availability to the single cell of the culture. Thus a certain part of the culture always fails to receive enough light to fulfill photosynthesis needs, the light will be limited. Torzillo, Carlozzi, Pushparaj, Montaini, and Materassi (1993) showed the Spirulina growing in polyethylene tubes in Italy. Spirulina is produced in at least 22 countries: Benin, Brazil, Burkina Faso, Chad, Chile, China, Costa Rica, Côte d'Ivoire, Cuba, Ecuador, France, India, Madagascar, Mexico, Myanmar, Peru, Israel, Spain, Thailand, Togo, the USA, and Vietnam. According to Van Huis et al. (2013), total industrial production of Spirulina in China was 19.1 million tons in 2003 and 41.60 million tons in 2004 ($7.6 million and $16.6 million, respectively). China started industrial production of Spirulina in 1990, and now the country has over 200 factories. Other countries with significant Spirulina production facilities the USA and France. The USA have the largest intensive farms in the world, mainly located in Hawaii and California. Earthrise has a plot of about 50 hectares located in the Sonoran Desert that produced about 450 tons in 2012. It uses water from the mineral-rich Colorado river. Cyanotech located in Hawaii produces 350 tons of Spirulina annually and use a combination of fresh water and deep ocean water (Guiry & Blunden, 1991; Iversen & Hale, 1992).

Some business success stories for Spirulina production are provided below:

1. Micro-Life. In Italy, Spirulina is produced by Micro-Life farm only which sells different algae products and has approximately €1 million in revenues. Micro-Life's products do not contain any kind of animal ingredients and hence are 100% vegan. Micro-Life is the first Italian company that may produce organic Spirulina. In 2014, it obtained organic certification. High revenues are derived from cosmetics made with Spirulina.

2. There are two new start-ups in southern Italy: Spirulina-Microfarm in Calabria and Apulia Mundi in Puglia. Spirulina Microfarm is specialized in consulting and in the creation of innovative and sustainable systems for Spirulina production. The company is equipped with a laboratory producing small amounts of high-quality product.

Business Model for Novel Food Start-Ups

The aim is to identify the general business model for the entire novel foods, even if the data are scarce. The companies which operate in this field should own some common characteristics. Most of these companies are founded by researchers, biologists, and life cycle experts. It is not possible to imagine a new start-up that operates without a scientist (Aceleanu, Serban, & Burghelea, 2015). These companies, at least initially, have a close relationship with the university or research center. There are no suppliers of such novel food as jellyfishes in Europe, hence there are foreign companies that supply the market (Alvarez, Mackalski, Loeb, & Mazzanti, 2015). However, there are some European suppliers for algae. These new start-ups (Belz & Binder, 2015) are able to connect profits and environmental resources to preserve the nature and contain some of the most important food industry trends: healthier foods, environmentally friendly, natural products, new consumption experience and the opportunity. Initially, the only diffusion channels were the Internet (the potential market is infinite) or small shops dedicated to the sustainable foods. The companies that operate in the algae production may sell their products in pharmacies or health food stores which are very broad distribution channels (D'Aleo, 2016). The Mediterranean products are considered high-quality commodity (Frewer et al., 2001). The researches are important sources for these companies; they provide counsels, trainings, and information. Usually, these companies offer packages, meetings, or guides for all consumers who would like to become small producers. The typical consumer of these products is a man/woman aged between 20 and 40 years with higher education and high disposable income (Hallström, Carlsson-Kanyama, & Börjesson, 2015). These consumers know that novel foods contain higher amounts of nutritional compounds like vitamins, antioxidants, and polyunsaturated fatty acids and smaller amounts of unwanted nutritional compounds such as heavy metals, mycotoxins, and pesticide residues. Therefore, novel foods are better for health and environment than conventional ones in terms of the essential elements and active environmental protection (Frewer, Risvik, & Schifferstein, 2001). Initial costs to start a new company in this area are usually high. Such high costs are because of required permissions (such costs present in all food sectors), high skilled human resources, and machinery. These high initial costs allow getting in active about after 3/4 years or collapsing about at the first months of the second year of activity.

We are sure that in common with all food startups in this sector there are the essential and fundamental elements to give more chance to a business project:

1. Solve a logistical limit.
2. Involve more suppliers.
3. Offer a flexible service to customers.
4. Give guarantees on quality.
5. Propose different methods of taste and try.
6. Reach agreements with local entities.
7. Reduce purchasing choice stress.

8. Draw aggregation tools Win/Win.
9. Strong marketing strategy to create awareness.

SOLUTIONS AND RECOMMENDATIONS

It is proved, therefore, that food tastes evolve from genetic factors such as the preference for sweet or salty and the refusal to bitter or sour, and it is also known that there is an innate and automatic regulation of appetite. The neuroendocrine mechanism behind these individual variations expressed through the inter-relationship between hormones and neurotransmitters so totally unique and genetically predetermined for each of us. However, from birth (in truth already in utero), genetics in its psychophysical manifestations (such as body size, perceptions, behavior) is directed and shaped by life experiences. Experiences will determine whether caloric intake gradually matches the energy expenditure. It is essential, in order to perceive the new foods as "normal", since childhood, even in nations with strong culinary traditions and food, that these new foods are gradually introduced into the daily diets. Therefore, the families must be the first promoters of these new sources, along with schools with proper food education. Certainly, it is appropriate that the state invests in specific information campaigns so that the adult population, skeptical about these new foods, can enjoy all the necessary information. The potential economic impact of these new sources must be conveyed by the state that must invest in measures able to stimulate investment in this sector (Marsden & Smith, 2005). Large industries that operate in the food sector have no interest in this type of novel food that may create distortions in supply chain and demand side. Thus, only small startups supported by government subsidies and an adequate information campaign that starts on the level of families and school altogether represent the driving force for this still embryonic sector.

FUTURE RESEARCH DIRECTIONS

In order to establish a complete vision of the issue, it is necessary to understand the nature of evolution in food requirements for target populations. Indeed, the researchers must be able to tell if it is better to focus on the marketing and commercial strategies, to specific segments of the population and consumers (Kim & Mauborgne, 2015), who may be more sensitive to the trial, or if a general marketing strategy is most appropriate. Researchers should focus future studies to understand how to make these new foods more acceptable. In fact, the way of presenting, serving, preparing, and packaging of novel food become crucial variables.

CONCLUSION

Novel foods could be subjects to boycott by new consumers who could get wrong information from mass media or Internet. These new products are priced higher in comparison to the conventional ones. They possess greater organic value but because consumers get little information through the conventional channels, novel foods have a small market. Professionals and experts in this field promote the reputation of novel foods. Therefore, this knowledge is deeply niched and restricted to regular consumers of novel food only. Food market in Europe and Mediterranean countries is saturated by traditional food products; it is difficult to enter this mature market with novel foods. Therefore, legislation in the sphere of promotion of novel foods is needed. It should help developing logistic and distribution, establishing of sector agreements and ensuring growth in the novel food sphere along with adequate quantitative controls and awareness of potential consumers. Furthermore, related national institutions should provide special funds for start-ups which would like to enter to the novel foods market, a marketplace of high economic potential.

REFERENCES

Aceleanu, M. I., Serban, A. C., & Burghelea, C. (2015). Greening the Youth Employment – A Chance for Sustainable Development. *Sustainability*, *7*(3), 2623–2643. doi:10.3390/su7032623

Aiking, H. (2011). Future Protein Supply. *Trends in Food Science & Technology*, *22*(2), 112–120. doi:10.1016/j.tifs.2010.04.005

Aiking, H., & De Boer, J. (2004). Food Sustainability: Diverging Interpretations. *British Food Journal*, *106*(5), 359–365. doi:10.1108/00070700410531589

Alvarez, J.B., Mackalski, R., Loeb, A., & Mazzanti, L. (2015). Leading Sustainable Change: An Organizational Perspective. *Strategic Management Journal*, *34*(13), 1209–1231.

Armisen, R. (1995). Worldwide Use and Importance of Gracilaria. *Journal of Applied Phycology*, *7*(3), 231–243. doi:10.1007/BF00003998

Belz, F. M., & Binder, J. K. (2015). Sustainable Entrepreneurship: A Convergent Process Model. *Business Strategy and the Environment*. Retrieved November 26, 2016, from http://onlinelibrary.wiley.com/doi/10.1002/bse.1887/abstract

Blunden, G. (1991). Agricultural Uses of Seaweeds and Seaweed Extracts. In M. D. Guiry & G. Blunden (Eds.), *Seaweed Resources in Europe: Uses and Potential* (pp. 65–81). Chichester, UK: John Wiley & Sons Ltd.

Brookes, G. (2007). *Economic Impact Assessment of the Way in which the EU Novel Foods Regulatory Approval Procedures Affect the EU Food Sector*. Retrieved November 26, 2016, from http://www.pgeconomics.co.uk/archive.php

Burlingame, B., Charrondiere, R., & Mouille, B. (2009). Food Composition is Fundamental to the Cross-Cutting Initiative on Biodiversity for Food and Nutrition. *Journal of Food Composition and Analysis*, *22*(5), 361–365. doi:10.1016/j.jfca.2009.05.003

Buschmann, A. H., Correa, J. A., Westermeier, R., del Carmen Hernandez-Gonzalez, M., & Norambuena, R. (2001). Red Algal Farming in Chile: A Review. *Aquaculture (Amsterdam, Netherlands)*, *194*(3), 203–220. doi:10.1016/S0044-8486(00)00518-4

Buschmann, A. H., Troell, M., & Kautsky, N. (2001). Integrated Algal Farming: A Review. *Cahiers de Biologie Marine*, *42*(1–2), 83–90.

Caparros Megido, R., Sablon, L., Geuens, M., Brostaux, Y., Alabi, T., Blecker, C., & Francis, F. et al. (2013). Edible Insects Acceptance by Belgian Consumers: Promising Attitude for Entomophagy Development. *Journal of Sensory Studies*, *29*(1), 14–20. doi:10.1111/joss.12077

D'Aleo, F., & D'Aleo, V. (2015a). *Preliminary Study in Open Raceway for Production of Spirulina Biomass in Southern Italy. First Experience in Calabria.* Poster presented at the 110° Congresso della Società Botanica Italiana, Pavia.

D'Aleo, F., & D'Aleo, V. (2015b). *Start-up Company for Production of Spirulina Biomass in Southern Italy. First Experience in Calabria.* Poster presented at the 43° Congresso Società Italiana di Microbiologia, Napoli.

D'Aleo, V., & Morabito, G. (2015). Verso il prototipo di start-up mediterranea. Confronto fra modelli di politiche pubbliche, incubatori e acceleratori. *Proceedings of the BIC-Italia Net Congress & EBN AGM*. Milan: BIC; EBN AGM.

DAleo, V. (2016). Supply Chain Management: Overview, Competition and Competences, How to Exploit the Hidden Capabilities. *International Journal of Management and Network Economics*, *3*(4), 336–346. doi:10.1504/IJMNE.2016.079878

Darkow, I. L., Foerster, B., & von der Gracht, H. A. (2015). Sustainability in Food Service Supply Chains: Future Expectations from European Industry Experts Toward the Environmental Perspective. *Supply Chain Management: An International Journal*, *20*(2), 163–178. doi:10.1108/SCM-03-2014-0087

Dawes, C. P. (1995). Suspended Cultivation of Gracilaria in the Sea. *Journal of Applied Phycology*, *7*(3), 303–313. doi:10.1007/BF00004004

De-Magistris, T., Pascucci, S., & Mitsopoulos, D. (2015). Paying to See a Bug on my Food: How Regulations and Information Can Hamper Radical Innovations in the European Union. *British Food Journal*, *117*(6), 1777–1792. doi:10.1108/BFJ-06-2014-0222

di Commercio d'Italia, C. (2015). *Infocamere.* Retrieved December 12, 2016, from http://www.camcom.gov.it/

European Parliament. (1997). *Regulation No 258/97 on Novel Foods and Novel Food Ingredients*. Retrieved December 12, 2016, from http://eur-lex.europa.eu/legal-content/EN/TXT/PDF/?uri=CELEX:31997R0258&from=EN

Frewer, L. J., Bergmann, K., Brennan, M., Lion, R., Meertens, R., Rowe, G., & Vereijken, C. et al. (2011). Consumer Response to Novel Agri-Food Technologies: Implications for Predicting Consumer Acceptance of Emerging Food Technologies. *Trends in Food Science & Technology*, *22*(8), 442–456. doi:10.1016/j.tifs.2011.05.005

Frewer, L. J., Risvik, E., & Schifferstein, H. (2001). Food, People and Society: A European Perspective of Consumers' Food Choices. Dordrecht: Springer Science+Business Media B.V. doi:10.1007/978-3-662-04601-2

Friedlander, M., & Levy, I. (1995). Cultivation of Gracilaria in Outdoor Tanks and Ponds. *Journal of Applied Phycology*, *7*(3), 315–324. doi:10.1007/BF00004005

Gami, B., Naik, A., & Patel, B. (2011). Cultivation of Spirulina Species in Different Liquid Media. *Journal of Algal Biomass Utilization*, *2*(3), 15–26.

Government of Italy. (2012). *Testo del decreto-legge 18 ottobre 2012. Pubblicato nel supplemento ordinario n. 194/L alla 19 ottobre 2012, No. 245. Coordinato con la legge di conversione 17 dicembre 2012, No. 221. Recante: "Ulteriori misure urgenti per la crescita del Paese" (Publication No. 179).* Rome: Gazzetta Ufficiale.

Guiry, M. D., & Blunden, G. (Eds.). (1991). *Seaweed Resources in Europe: Uses and Potential*. Chichester, UK: John Wiley & Sons Ltd.

Habib, M. A. B., Parvin, M., Huntington, T. C., & Hasan, M. R. (2008). *A Review on Culture, Production and Use of Spirulina as Food for Humans and Feeds for Domestic Animals*. Rome: Food and Agriculutre Organization of the United Nations.

Hallström, E., Carlsson-Kanyama, A., & Börjesson, P. (2015). Environmental Impact of Dietary Change: A Systematic Review. *Journal of Cleaner Production*, *91*, 1–11. doi:10.1016/j.jclepro.2014.12.008

Hepburn, P., Howlett, J., Boeing, H., Cockburn, A., Constable, A., Davi, A., & Wal, J. M. et al. (2008). The Application of Post-Market Monitoring to Novel Foods. *Food and Chemical Toxicology*, *46*(1), 9–33. doi:10.1016/j.fct.2007.09.008 PMID:17950974

Hermann, M. (2009). The Impact of the European Novel Food Regulation on Trade and Food Innovation Based on Traditional Plant Foods from Developing Countries. *Food Policy*, *34*(6), 499–507. doi:10.1016/j.foodpol.2009.08.005

Hsieh, Y. P., Leong, F. M., & Rudloe, J. (2001). Jellyfish as Food. *Hydrobiologia*, *451*(1–3), 11–17. doi:10.1023/A:1011875720415

Ingram, M., Nabhan, G., & Buchmann, S. L. (1996). Our Forgotten Pollinators: Protecting the Birds and Bees. *Global Pesticide Campaigner*, *6*(4), 1–8.

Iversen, E. S., & Hale, K. K. (1992). Species for Human Food. In E. S. Iversen & K. K. Hale (Eds.), *Aquaculture Sourcebook* (pp. 35–204). New York, NY: Springer US. doi:10.1007/978-1-4684-1428-8_3

Jongema, Y. (2015). *List of Edible Insects of the World*. Retrieved November 26, 2016, from http://www.wur.nl/en/Expertise-Services/Chair-groups/Plant-Sciences/Laboratory-of-Entomology/Edible-insects/Worldwide-species-list.htm

Kapraun, D. F. (1999). Red Algal Polysaccharide Industry: Economics and Research Status at the Turn of the Century. In *Proceedings of the Sixteenth International Seaweed Symposium*. Cebu City: Springer-Science+Business Media, B.V. doi:10.1007/978-94-011-4449-0_2

Katayama, N., Ishikawa, Y., Takaoki, M., Yamashita, M., Nakayama, S., Kiguchi, K., & Force, S. A. T. et al. (2008). Entomophagy: A Key to Space Agriculture. *Advances in Space Research*, *41*(5), 701–705. doi:10.1016/j.asr.2007.01.027

Kawashima, S. (1993). Cultivation of the Brown Alga, Laminaria "Kombu". In M. Ohno & A. T. Critchley (Eds.), *Seaweed Cultivation and Marine Ranching* (pp. 25–40). Yokosuka: Japan International Cooperation Agency.

Kim, W. C., & Mauborgne, R. (2015). *Blue Ocean Strategy, Expanded Edition: How to Create Uncontested Market Space and Make the Competition Irrelevant*. Brighton, MA: Harvard Business Review Press.

Kitamura, M., & Omori, M. (2010). Synopsis of Edible Jellyfishes Collected from Southeast Asia, with Notes on Jellyfish Fisheries. *Plankton and Benthos Research*, *5*(3), 106–118. doi:10.3800/pbr.5.106

Lüning, K., Yarish, C., & Kirkman, H. (1990). *Seaweeds: Their Environment, Biogeography, and Ecophysiology*. Hoboken, NJ: John Wiley & Sons.

Marsden, T., & Smith, E. (2005). Ecological Entrepreneurship: Sustainable Development in Local Communities through Quality Food Production and Local Branding. *Geoforum*, *36*(4), 440–451. doi:10.1016/j.geoforum.2004.07.008

Morikawa, T. (1984). Jellyfish. *FAO Infofish Marketing Digest*, *1*, 37–39.

Neori, A., Chopin, T., Troell, M., Buschmann, A. H., Kraemer, G. P., Halling, C., & Yarish, C. et al. (2004). Integrated Aquaculture: Rationale, Evolution, and State of the Art Emphasizing Seaweed Biofiltration in Modern Mariculture. *Aquaculture (Amsterdam, Netherlands)*, *231*(1), 361–391. doi:10.1016/j.aquaculture.2003.11.015

Nowak, V., Persijn, D., Rittenschober, D., & Charrondiere, U. R. (2014). Review of Food Composition Data for Edible Insects. *Food Chemistry*, *193*, 36–46. PMID:26433285

Oliveira, E. C., Alveal, K., & Anderson, R. J. (2000). Mariculture of the Agar-Producing Gracilarioid Red Algae. *Reviews in Fisheries Science*, *8*(4), 345–377. doi:10.1080/10408340308951116

Ramos-Elorduy, J., Moreno, J. M. P., & Camacho, V. H. M. (2009). Edible Aquatic Coleoptera of the World with an Emphasis on Mexico. *Journal of Ethnobiology and Ethnomedicine*, *5*(11). Retrieved from https://ethnobiomed.biomedcentral.com/articles/10.1186/1746-4269-5-11 PMID:19379486

Rumpold, B. A., & Schlüter, O. K. (2013). Potential and Challenges of Insects as an Innovative Source for Food and Feed Production. *Innovative Food Science & Emerging Technologies*, *17*, 1–11. doi:10.1016/j.ifset.2012.11.005

Shelton, A. M., Zhao, J. Z., & Roush, R. T. (2002). Economic, Ecological, Food Safety, and Social Consequences of the Deployment of Bt Transgenic Plants. *Annual Review of Entomology*, *47*(1), 845–881. doi:10.1146/annurev.ento.47.091201.145309 PMID:11729093

Sherman, R. A., & Pechter, E. A. (1988). Maggot Therapy: A Review of the Therapeutic Applications of Fly Larvae in Human Medicine, Especially for Treating Osteomyelitis. *Medical and Veterinary Entomology*, *2*(3), 225–230. doi:10.1111/j.1365-2915.1988.tb00188.x PMID:2980178

Simon, J. Y. (2008). Detoxification Mechanisms in Insects. In J.L. Capinera (Ed.), Encyclopedia of Entomology (pp. 1187–1201). Dordrecht: Springer Science+Business Media B.V.

Snowden, A. L. (2010). Post-Harvest Diseases and Disorders of Fruits and Vegetables: Vol. 1. *General Introduction and Fruits*. Boca Raton, FL: CRC Press.

Subasinghe, S. (1992). Jellyfish Processing. *INFOFISH International*, *4*, 63–65.

Torzillo, G., Carlozzi, P., Pushparaj, B., Montaini, E., & Materassi, R. (1993). A Two-Plane Tubular Photobioreactor for Outdoor Culture of Spirulina. *Biotechnology and Bioengineering*, *42*(7), 891–898. doi:10.1002/bit.260420714 PMID:18613137

Uye, S. I. (2008). Blooms of the Giant Jellyfish Nemopilema Nomurai: A Threat to the Fisheries Sustainability of the East Asian Marginal Seas. *Plankton and Benthos Research*, *3*(Supplement), 125–131. doi:10.3800/pbr.3.125

Van Huis, A. (2003). Medical and Stimulating Properties Ascribed to Arthropods and Their Products in Sub-Saharan Africa. In E. Motte-Florac & J. M. C. Thomas (Eds.), *Insects in Oral Literature and Traditions* (pp. 367–382). Paris: Peeters.

Van Huis, A., van Itterbeeck, J., Klunder, H., Mertens, E., Halloran, A., Muir, G., & Vantomme, P. (2013). *Edible Insects: Future Prospects for Food and Feed Security*. Rome: Food and Agriculture Organization of the United Nations.

Vantomme, P. (2015). Way forward to Bring Insects in the Human Food Chain. *Journal of Insects as Food and Feed*, *1*(2), 121–129. doi:10.3920/JIFF2014.0014

Wollina, U., Karte, K., Herold, C., & Looks, A. (2000). Biosurgery in Wound Healing – The Renaissance of Maggot Therapy. *Journal of the European Academy of Dermatology and Venereology*, *14*(4), 285–289. doi:10.1046/j.1468-3083.2000.00105.x PMID:11204517

World Health Organization. (2014). *World Malaria Report 2013*. Geneva: World Health Organization.

Yamanaka, R., & Akiyama, K. (1993). Cultivation and Utilization of Undaria Pinnatifida (Wakame) as Food. *Journal of Applied Phycology*, *5*(2), 249–253. doi:10.1007/BF00004026

Yong-Woo, L. (1999). *Silk Reeling and Testing Manual*. Rome: Food and Agriculture Organization of the United Nations.

ADDITIONAL READING

Bixler, H. J. (1996). Recent Developments in Manufacturing and Marketing Carrageenan. *Hydrobiologia*, *326*(1), 35–57. doi:10.1007/BF00047785

Crouch, I. J., & Van Staden, J. (1993). Evidence for the Presence of Plant Growth Regulators in Commercial Seaweed Products. *Plant Growth Regulation*, *13*(1), 21–29. doi:10.1007/BF00207588

Jiang, L., Hoegg, J., Dahl, D. W., King, J., Metzger, M., Owen, S., & Kahle, L. R. (2015). Balanced Innovations in New Product Development: Sunny Faces, Sunny Technology. In *Proceedings of the 2010 Academy of Marketing Science (AMS) Annual Conference*. New York, NY: Springer International Publishing. doi:10.1007/978-3-319-11797-3_165

Jonker, J. (2015). The Future of Corporate Sustainability: Towards an Ecology of Organisations Focused on Sustainability. In L. O'Riordan, P. Zmuda, & S. Heinemann (Eds.), *New Perspectives on Corporate Social Responsibility* (pp. 23–47). Wiesbaden: Springer Fachmedien Wiesbaden. doi:10.1007/978-3-658-06794-6_2

Kenis, A., & Lievens, M. (2015). Greening the Economy or Economizing the Green Project? When Environmental Concerns are Turned into a Means to Save the Market. *Review of Radical Political Economics*. Retrieved November 26, 2016, from http://rrp.sagepub.com/content/early/2015/07/09/0486613415591803.full.pdf+html

Larsen, H. B., Lindkvist, K. B., Sánchez-Hernández, J. L., & Trondsen, T. (2015). Analysing Food Chain Development: A Theoretical Framework. In K. B. Lindkvist & T. Trondsen (Eds.), *Nordic-Iberian Cod Value Chains* (pp. 7–22). Basel: Springer International Publishing. doi:10.1007/978-3-319-16405-2_2

Leonidou, L.C., Christodoulides, P., Kyrgidou, L.P., & Palihawadana, D. (2015). Internal Drivers and Performance Consequences of Small Firm Green Business Strategy: The Moderating Role of External Forces. *Journal of Business Ethics*, 1–22. Retrieved November 26, 2016, from file:///C:/Users/Vasily/Downloads/10.1007%252Fs10551-015-2670-9.pdf.

Mepham, B. (2000). A Framework for the Ethical Analysis of Novel Foods: The Ethical Matrix. *Journal of Agricultural & Environmental Ethics*, *12*(2), 165–176. doi:10.1023/A:1009542714497

Rollin, F., Kennedy, J., & Wills, J. (2011). Consumers and New Food Technologies. *Trends in Food Science & Technology*, *22*(2), 99–111. doi:10.1016/j.tifs.2010.09.001

Salafsky, N., Salzer, D., Stattersfield, A. J., Hilton-Taylor, C., Neugarten, R., Butchart, S. H., & Wilkie, D. et al. (2008). A Standard Lexicon for Biodiversity Conservation: Unified Classifications of Threats and Actions. *Conservation Biology*, *22*(4), 897–911. doi:10.1111/j.1523-1739.2008.00937.x PMID:18544093

Schneider, U. A., Havlík, P., Schmid, E., Valin, H., Mosnier, A., Obersteiner, M., & Fritz, S. et al. (2011). Impacts of Population Growth, Economic Development, and Technical Change on Global Food Production and Consumption. *Agricultural Systems*, *104*(2), 204–215. doi:10.1016/j.agsy.2010.11.003

Shahidullah, A. K. M., & Haque, C. E. (2015). Green Microfinance Strategy for Entrepreneurial Transformation: Validating a Pattern towards Sustainability. *Enterprise Development and Microfinance*, *26*(4), 325–342. doi:10.3362/1755-1986.2015.027

Sili, C., Torzillo, G., & Vonshak, A. (2012). Arthrospira (Spirulina). In B. A. Whitton (Ed.), *Ecology of Cyanobacteria II: Their Diversity in Space and Time* (pp. 677–705). Heidelberg: Springer Verlag. doi:10.1007/978-94-007-3855-3_25

Taylor, J., Garat, J. P., Simreen, S., & Sarieddine, G. (2015). An Industry Perspective: A New Model of Food Safety Culture Excellence and the Impact of Audit on Food Safety Standards. *Worldwide Hospitality and Tourism Themes*, *7*(1), 78–89. doi:10.1108/WHATT-12-2014-0041

Verbeke, W. (2005). Consumer Acceptance of Functional Foods: Socio-Demographic, Cognitive and Attitudinal Determinants. *Food Quality and Preference*, *16*(1), 45–57. doi:10.1016/j.foodqual.2004.01.001

Wilcock, A., Pun, M., Khanona, J., & Aung, M. (2004). Consumer Attitudes, Knowledge and Behaviour: A Review of Food Safety Issues. *Trends in Food Science & Technology*, *15*(2), 56–66. doi:10.1016/j.tifs.2003.08.004

KEY TERMS AND DEFINITIONS

Alternative Sources: A food that is regarded as a healthier alternative to food that is prepared or produced by conventional methods. Alternative sources of food which mean that substitute food can replace the principal food can help to solve food problems. Many kinds of alternative food sources can be helpful to solve food problems such as starvation, lack of nutrition, and economy and business crisis.

Economic Impact: An effect that an event, policy change, or market trend will have on economic factors such as consumer confidence, stock market activity, and others. Events such as regulatory changes, supply shortages or introduction of new products can have a significant economic impact due to the way that they affect business activities.

Emerging Trend: A trend in which people are becoming more interested over a period of time.

Food: Anything serving for consumption or use.

Green Start-Ups: Used to describe someone who is not experienced, either in a particular area or in an overall sense. A business functioning in a capacity where no negative impact is made on the local or global environment, the community, or the economy.

Novel Foods: Newly developed, innovative food or food produced using new technologies and production processes as well as food traditionally eaten.

Safety Foods: Refers to the conditions and practices that preserve the quality of food. These practices prevent contamination and foodborne illnesses.

Chapter 5
Implications of Trade Liberalization for Food Security Under the ASEAN–India Strategic Partnership:
A Gravity Model Approach

Ishita Ghosh
Symbiosis International University, India

Ishita Ghoshal
Symbiosis International University, India

ABSTRACT

The objective of the chapter is to understand India's role in providing food security through trade with the ASEAN under the India-ASEAN Strategic Partnership. A dynamic panel has been employed to assess and estimate the said objective, given that India and most of the ASEAN countries are considered emerging markets. Econometric investigation points out that while almost more than half the variables in the augmented gravity model hold good, food imports from India, agricultural labour force, common maritime border, distance from India, food price index, food production index, GDP (size) of the importing country, per capita GDP of the importing economy, and trade openness have statistically significant effect on the food trade from India to the ASEAN. Improving maritime infrastructure and agri-logistics, investing in climate change and water management while augmenting the agri-labour productivity are of paramount importance in order to improve food trade between India and the ASEAN.

INTRODUCTION

The world population is predicted to cross nine billion by the year 2050, resulting in growing concerns to ensure food security (Food and Agriculture Organization of the United Nations [FAO], 2001). Food security has had evolving definitions since the mid-1970s. The latest definition given as per the State of

DOI: 10.4018/978-1-5225-2733-6.ch005

Food Insecurity, 2001, is: "Food security [is] a situation that exists when all people, at all times, have physical, social and economic access to sufficient, safe and nutritious food that meets their dietary needs and food preferences for an active and healthy life" (FAO, 2001).

In the modern context, it is not really a concern if the world is capable of meeting the growing demands to adequately cover calorific and nutritional demands; however, barriers to worldwide access such as political factors, ownership, institutions, poverty and inequality still pose a threat to global supply chains and distribution (Gillson & Fouad, 2016).

Improvements in food security have been noted in almost all the regions of the world and particularly in emerging markets (Emerging Market Economy, n.d.). An emerging market economy describes a nation's economy, that is, progressing toward becoming more advanced, usually by means of rapid growth and industrialization. These countries experience an expanding role both in the world economy and on the political frontier (Emerging Market Economy, n.d.). The Global Food Security Index [GFSI] (The Economist, 2016) mentions that barring Europe, the low-income and lower-middle-income countries have progressed towards ensuring better measures for food security, thereby reducing the global gap between the least and the most food secure nations. Although structural changes in economies have increased the access to a wide range of affordable and nutritious food, put in place more extensive food safety-net programs and expanded crop storage capacity and dietary diversity, food security still remains a concern and challenge in the modern day. Despite investment in infrastructure, most developing and emerging economies are still grossly short on road and port facilities that aid in accessing food. Also, absorbing urban migration is a gradual process – pressures of urbanization have been growing worldwide and agricultural labor is depleting, while there has been a downward trend in the world Gross Domestic Product (GDP) growth. Macroeconomic improvements and stability have been compromised by geopolitical turmoil and climate change (The Economist, 2016).

According to the Asian Development Bank [ADB] (2016), two-thirds of the world's one billion hungry reside in Asia and the Pacific. While food prices are rising, bringing about food shortages and malnutrition in the region's poor, both rural and urban dwellers are forced to spend more than half of their income on food and related goods. The persistent volatility in food price has become an impediment to reaching the Millennium Development Goals, especially the ones related to the mitigation of hunger and poverty (Asian Development Bank, 2016).

The projected population growth that will reside in urban areas of Asia by 2026 is close to 50% of the constantly growing population. This will need comprehensive planning and significant investments to cater to food distribution, storage and marketing facilities to ensure food security. There will also be a heavy requirement of improved land cultivation techniques and labor productivity (Briones, Chavez, Durand-Morat, & Wailes, 2012).

Climate change also plays a vital role in ensuring food security. The Asia-Pacific El Niño has, and the predicted La Niña will cause droughts and hence, reduction in agricultural output. This has affected and will continue to affect inflation rates, poverty, exports, etc., adversely. Prolonged effects have may affect macroeconomic and geopolitical stability (El Niño in Asia, 2016; Tabor, Ginting, & Aji, 2015).

Currently, Asia faces serious water supply crisis. The demand for water in expected to rise by 40% towards 2030. While 80% of the water supply is directed to agricultural production, the water crisis is going to compound the food security issue. Using less water to grow more food alongside sustainable water management will have to be the order of the day.

The ADB (2016) has also projected that domestic food prices will rise in the developing and emerging markets in Asia by 10% between 2020 and 2025. This is likely to push a population of 64 million Asians into poverty, over and above the current status. Extraordinary food price surges, globally increase food insecurity thereby posing great economic difficulties to the region, nations trading in food and agriculture, and governments (Huh & Park, 2013; Jha, Kubo, & Ramaswami, 2013).

As a consequence of the 2007-2008 food price crisis, the Association of Southeast Asian Nations [ASEAN] Summit of 2009 promised to ensure measures towards food security as an urgent and long-term priority. By 2010, the ASEAN Ministry of Agriculture and Forestry [AMAF] fostered economic ties between India, China and the Republic of Korea, as well as, within the regions of the Gulf Cooperation Council [GCC] and the ASEAN+3 (ASEAN plus China, Japan, and South Korea) (Choulamany, n.d.).

It should also be noted that according to Morgan Stanley Capital International (n.d.), India and several nations in ASEAN are emerging economies. The ASEAN being largely agrarian in nature, it is pertinent to examine this trading bloc from the viewpoint of food security and trade.

In this backdrop, the chapter focuses on the agricultural trade between India and ASEAN under their strategic partnership and regional cooperation. The background and problems pertaining to food security under the said alliance are discussed and a gravity equation of trade is employed to empirically identify the variables that may be significant in improving food security in the region through trade with India as well as, variables that impede the bilateral trade in agriculture. The authors hope to provide insights from the results of the analysis to augment the food trade relationship between India and the ASEAN economies.

BACKGROUND

Out of the seven aims and purposes of the ASEAN Declaration (1967), collaboration "for the better utilization of agriculture and industry to raise the living standards of the people", is a key tenet (Overview, n.d.). As mentioned in the introductory section, ASEAN has made ties with India to cope with and cater to food security issues. In 2010, the ASEAN – India Ministerial Meeting on Agriculture and Forestry [AIMAF] and the ASEAN – India Working Group on Agriculture and Forestry [AIWGAF] were put in place to help improve relations in bilateral and sustainable trade in food and agriculture (Choulamany, n.d.). For a while (up to 2011-2012), ASEAN has been India's fourth-largest trading partner after China, the EU, and the USA. Trade between the two regions touched nearly $80 billion in 2014-2015 while the expected potential has been set for $200 billion by 2022. In 2016, India has slid down to the ninth position from the former fourth. Possible reasons for this stagnation and decline could be a change in the government and policies which are yet to take effect. Moreover, the "Make in India" initiative since 2014 has been "inward looking" where the country has essentially concentrated on inward Foreign Direct Investments (FDI) for self-sufficiency and sustainability in manufacturing. The India-ASEAN regional partnership has spanned over two decades now (ASEAN-India Relations, 2016). However, since the execution of the India-ASEAN Free Trade Area (FTA) in 2010, India's exports of garlic, onions, turmeric, wheat, meslin, cane sugar, groundnuts, oilcake/ oilcake meal of soybean, oilcake/oilcake meal of rape/ colza seed, and millets (sorghum and bajra) to ASEAN have registered a sharp increase, year-on-year (Government of India, 2016).

MAIN FOCUS OF THE CHAPTER

Drivers of Food Security and Bilateral Trade in Agriculture: A Literature Review

According to Holt-Giménez (2008), the well-established causes of food crises have been poor weather and climate change, low grain reserves, high oil prices that push up the prices of fertilizers and food transportation, increase in meat consumption worldwide with the rise in global GDP, that has resulted in focusing on the farming of crops for cattle and livestock, thereby decreasing or replacing production of other crops. Also, the new-found need for agrofuels has increased grain prices. Deregulation of stock and commodity markets and the over-enthusiasm to make quick and large amounts of profits have also added to the speculative bubbles in the futures markets. In fact, right after the 2007-2008 subprime crisis, investors sought new avenues to put in money. Rising food prices showed the way to attractive commodity futures investments, which in turn further increased the food prices. When they saw food prices going up, they poured investments into commodities futures, pumping up the price of grains and worsening food price inflation (Holt-Giménez, 2008).

The four main concepts of food sufficiency are given differently by various researchers. According to Reutlinger and Knapp (1980) and Gouel and Jean (2013), food sufficiency is the minimal level of food consumption, while according to Siamwalla and Valdes (1980), food sufficiency is a target level. Barraclough and Utting (1987) have said that food sufficiency is the food adequate to meet nutritional needs. Another way to evaluate food sufficiency is to ensure that the population in question have "enough" food. This can be measured by estimating by how far below the consumption is from the threshold (Lipton & Heald, 1984; Clay, Shaw, & Singer, 1987), while Maxwell (1990) has tried measuring the "intensity" of food insecurity.

The second concern is access to food, which assesses if individuals are able to obtain adequate food. This is motivated by Sen's (1983) concept of "entitlements" of food – that is, "an individual's entitlement is rooted in his/her endowment and is then transformed via production and trade into food or commodities which can be exchanged for food". While the secure access to adequate food is the third concern, the concepts of risk and its avoidance are key to the idea of food security due to various reasons such as variability in crop production and food supply, market and price variability, risks in employment and wages, etc. (Applanaidu, Bakar, & Baharudin, 2014).

It is also important to mention that the temporal dimension in food security has become a vital concept in the latter day policymaking. The World Food Programme (2009) notes that there may be several ways of incorporating the time variable in case of food insecurity – chronic (long term, or persistent for six months), transitory (short-term, or temporary), and cyclical (seasonal variability).

In Malaysia, Mohamed Arshad and Hameed (2010) have inspected the effect of increasing food prices on food security. According to their study, it is the fundamentals like decreasing agricultural production growth leading to falling supply, decreasing stocks of global cereal and rising food demand from the emerging economies around the world.

Aker and Lemtouni (1999) looked into the national level factors and integrated domestic and global markets that affect food security. Macroeconomic variables considered in the study were domestic food production, average annual rainfall, world food prices, gross domestic product, Gini coefficient, and exports of goods and services.

The study of Applanaidu et al. (2014) on Malaysia analyzed the dynamic impact of macroeconomic factors like biodiesel production, exchange rate, government expenditure on rural development, GDP, food price index and population on food security with the vector autoregressive (VAR) method. Variance decomposition analysis shows that biodiesel production, exchange rate and government expenditure on rural development affect food security the most in the tenth year, GDP in the sixth year, and population and exchange rate in the fifth year.

Volatility and rise in food prices have received great attention from the policy makers due to the challenges posed by the dynamics of food price inflation and volatile prices (Lee & Park, 2013). They conducted a panel analysis for 72 countries for the span 2000 to 2011 assessing how food price inflation and volatilities are transmitted globally alongside checking for the impact of several internal and external variables on them. The findings suggest that lagged global food price inflation is significantly associated with domestic food price inflation in Asia. Increasing economic growth rates, shares of food imports and fall in growth rate of food imports lead to a rise in food prices. Higher economic growth rate lowers volatility of food prices. On the other hand, appreciation of the currency, higher political stability, and income of the economy lead to fall in food price inflation.

Among the various economic and econometric concepts to assess food security, the use of the gravity model is gaining pace amid modern day researchers. Very simply put, the model has its origin in the law of universal gravitation in physics, which was developed by Sir Isaac Newton in 1687. The law of universal gravitation describes the gravitational force between two masses in relation to the distance that lies between these two masses. The gravitational constant is an empirically determined value (equation 1)

$$F_{ij} = G * (M_i M_j) / d_{ij}^2 \tag{1}$$

The gravitational force F_{ij} proportionate to the product of the two masses M_i and M_j and inversely proportional to the square of the distance d_{ij}^2 that keeps the two masses apart from each other. The gravitational constant G is an empirically determined value. This relationship is applicable to any context where the modeling of flows or movements is demanded. The gravity model in international economics puts the law of universal gravity into an economic context. Assuming that the trade volume between countries i and j is X_{ij} and furthermore economic masses defined as E_i and E_j, trade flows can be characterized. Thus, the predicted movement of goods or labor between countries i and j would be (equation 2):

$$X_{ij} = (E_i E_j) / \varphi_{ij}^2 \tag{2}$$

In other words, a mass of goods, labor, or other factors of production at origin i (E_i) is attracted by a mass of demand for goods or labor at destination j (E_j), but the potential flow is reduced by the distance between them (φ_{ij}^2).

However, the problem with drawing parallels between trade and physical force of gravity is that there is no set of parameters for which equation (2) will hold exactly for an arbitrary set of observations. Departing from strict analogy, traditional gravity model allows the coefficients of (2) applied to the mass variables and of (3) applied to the bilateral distance to be generated by data to fit statistically inferred the relationship between data on flows and the mass variables and distance. Typically, the stochastic version of the gravity equation has the following form (equation 3):

$$X_{ij} = a_0 E_i^{\alpha 1} E_j^{\alpha 2} \varphi_{ij}^{\alpha 3} \varepsilon_{ij}, \tag{3}$$

where a_0, α_1, α_2, and α_3 are unknown parameters.

In the original version by Tinbergen (1962), as given in (2), the model is expressed in a log-log form, so that the parameters are the elasticity of the trade flow with respect to the explanatory variables. With respect to equation (3), adjacent countries are assumed to have a more intense trade than what distance alone would predict; the adjacency is indicated by the dummy variable N_{ij}, that took the value 1 if the two countries share a common border. Moreover, the equation is augmented with political factors: a dummy variable V_{ij} indicates that goods traded received a preferential treatment if they belonged to some unilateral trade or system of preferences. The strategy of considering the effect of Preferential Trade Agreements (PTA) through the use of dummy variable has been prominent in the literature. Only recently, the alternative strategy of explicitly including the preferential margin guaranteed by the agreement has been taken into account: the authors will come back to this issue in the following. As customary, an independent, identically distributed stochastic term, ε_{ij}, is included. Therefore, the expression would take the form (equation 4):

$$log\, X_{ij} = a_0 + \alpha_1 logE_i + \alpha_2 logE_j + \alpha_3 log\varphi_{ij} + \alpha_4 logN_{ij} + \alpha_5 logV_{ij} + \varepsilon_{ij}, \quad (4)$$

where:

a_0 = constant,
$\alpha_1 \log E_i + \alpha_2 \log E_j$ = economic indicators,
$\alpha_3 \log\varphi_{ij} + \alpha_4 \log N_{ij}$ = distance,
$\alpha_5 \log V_{ij}$ = policy,
ε_{ij} = error term.

Bilateral trade flows are determined by the variables included in the right-hand-side of the gravity equation. This implies a clear direction of causality that runs from income and distance to trade. This direction of causality is, however, theory-driven and based on the assumption that the gravity equation is derived from a microeconomic model where income and tastes for differentiated products are given. Three decades of theoretical work has shown that the gravity equation can be derived from many different – and sometimes competing – trade frameworks (Salvatici, 2013; Starck, 2012; Masudul Haq et al., 2014; Sarker & Jayasinghe, 2007; van Dijk, 2011; Natale, Borrello, & Molotova, 2015; Wang, 2016).

Material and Methods

Given the previous discussion, the following seventeen variables were considered for estimating the gravity model in the current study:

1. Food import from India into the ten ASEAN countries.
2. GDP of India (origin country) in current USD.
3. GDP of the importing country in current USD.
4. GDP per capita of importing country in current USD.
5. Trade as a percentage of GDP to measure trade openness of importing country.
6. Official exchange rate of the importing country (local currency unit (LCU) per USD, period average).
7. Agriculture, value added (percentage of GDP).
8. Depth of the food deficit (kilocalories per person per day).

9. Food production index (2004-2006 = 100).
10. Distance from India.
11. Per capita food production variability.
12. A dummy for the common maritime border.
13. A dummy for Bilateral Trade Agreement (BTA).
14. Food price index for importing country.
15. Quality of port infrastructure, WEF (1 = extremely underdeveloped to 7 = well developed and efficient by international standards).
16. Agricultural labor force as a share of total labor force.
17. Political stability and absence of violence/terrorism.

The gravity equation of trade in this context will help only to determine the variables that play a role in agricultural trade between India and the ASEAN countries. Customs and other import duties and corruption index were dropped from the analysis due to the lack of available data for all the countries in the study.

For the purposes of analysis, the authors have considered the following model using data for the ten ASEAN countries (variables 3-17) and India from 1995 to 2015. Given the nature of the data, the study has utilized panel data methods to estimate the augmented gravity model for deducing the impact of various macroeconomic and policy variables on food import from India to these countries.

$$\begin{aligned}\log(\text{Food Import}_{it}^{\text{India to } i\text{th country}}) &= a + b_1 \log(GDP_t^{\text{India}}) + b_2 \log(GDP_{it}^{\text{Importing Country}}) + b_3 D_{it} \\ &+ b_4 \log(GDP_{it}^{PC}) + b_5 TO_{it} + b_6 ER_{it} + b_7 GDP_{it}^{\text{Agriculture}} + b_8 \text{Food Deficit}_{it} + b_9 F\text{Prod}I_{it} \\ &+ b_{10} \text{Food Prod}n\text{Variability}_{it}^{PC} + b_{11} CMB_i + b_{12} BTA_i + b_{13} FPI_{it} + b_{14} QPI_{it} \\ &+ b_{15} \text{Agri Labour Force}_{it}^{\text{Share of total}} + b_{16} \text{Political Stability}_{it} + \varepsilon_{it}\end{aligned} \quad (5)$$

where:

$\log(\text{Food Import}_{it}^{\text{India to } i\text{th country}})$ is the log of food import from India into the ten ASEAN countries over the years;
$\log(GDP_t^{\text{India}})$ is the log of India's GDP which is the host or origin country;
$\log(GDP_{it}^{\text{Importing Country}})$ is the log of the importing country's GDP;
$\log(GDP_{it}^{PC})$ is the log of per capita GDP of the importing country (to capture the population factor in the model);
D_{it} is the distance of the importing country from India;
TO_{it} is the trade openness of importing country measured by total trade as a share of GDP;
ER_{it} is the exchange rate of the importing country;
$GDP_{it}^{\text{Agriculture}}$ is the share of agricultural production in total GDP of the importing country;
Food Deficit_{it} is the extent of the food deficit that the importing country suffers from;
$F\text{Prod}I_{it}$ is the food production index;
Food Prod*n* Variability $_{it}^{PC}$ is the per capita food production variability;

CMB_i is the dummy for common maritime borders where $CMB_i = 1$ if the importing country share a maritime border with India, 0 otherwise;
BTA_i is the dummy for bilateral trade agreement where $BTA_i = 1$ if the importing country has a separate bilateral trade agreement other than the India-ASEAN agreement, 0 otherwise;
FPI_{it} is the food price index;
QPI_{it} is the quality of port infrastructure for the importing country;
$\text{Agri Labour Force}_{it}^{\text{Share of total}}$ is the share of labor force involved in agricultural activities in total labor force of the importing country;
$\text{Political Stability}_{it}$ is the index for political stability and absence of violence/terrorism.

Data

Panel data for the ten ASEAN countries and India has been used for the period of 1995 to 2015. Data for food import from India into the ten ASEAN countries and share of labor force involved in agricultural activities in total labor force of the importing country have been collected from the United Nations Conference on Trade and Development [UNCTAD] reports; GDP of importing country, per capita GDP, trade openness, exchange rate, food production index, agricultural GDP, quality of port infrastructure have been collected from World Bank Indicators; depth of food deficit (kilocalories per person per day), per capita food production variability, food price index for importing country, political stability and absence of violence/terrorism have been collected from FAO Food Security Indicators. The dummy for BTA was constructed based on information available from Asian Regional Integration Centre of Asian Development Bank. The dummy for CMB has been constructed by the authors based on the current world map. The distances of the importing countries from India have been calculated based on the flying distances available from www.distancefromto.net.

Customs and other import duties data is available with the World Bank, and Corruption Index data is available from Transparency International – The Global Coalition against Corruption website.

Methodology

Panel data estimation techniques have been used to understand the extent and direction of the relationships of the above-mentioned variables with a value of food imported from India by the ASEAN countries. The authors have used panel vector autoregression analysis (VAR) to estimate the gravity model. The fixed effect or random effect methods of estimation were not relied upon since these are single equation analyses which assume explicit exogeneity of the regressor variables. The nature of the economic variables considered for the study does not allow the authors to strictly assume that the regressors are exogenous. VAR analysis allows the authors to treat all the variables as endogenous and then estimate a set of equations instead of a single equation. This gives the more reliable result as the inter-relationships among all the variables in the model are taken into consideration for the estimation process. The following elucidates the detailed analytical methods used for the study:

Panel Data Unit-Root Tests

The authors have used panel data unit root tests to check whether the variables are stationary or not. Several statistical methods (Levin, Lin, & Chu, 2002; Im, Pesaran, & Shin, 2003; Choi, 2001; Breitung, 2000; Hadri, 2000) are constructed to test for unit roots in panel data. Among those, the study has employed Levin-Lin-Chu test and Im-Pesaran-Shin tests to check for stationarity in the variables.

Panel Causality for a Pair of Variables

Granger causality is checked to analyze the causality hypotheses. In the case of a long-run relationship, panel Vector Error Correction Model (VECM) is used while panel VAR method is used in case of no cointegration. The conventional Granger causality tests raise two critical issues for a panel data case, both dealing with the potential heterogeneity of the individual cross sections. The first source of heterogeneity is cross-sectional variation due to the distinctive intercepts; such heterogeneity may be addressed with a fixed effects model (that is, it is controlled by introduction of individual effects in the model). The more crucial case is where heterogeneous slope coefficients should be considered (that is, it should be controlled by introducing individual dimension for regression slopes in the model). The second source of heterogeneity affects the causality relationships. For instance, for some individuals, the introduction of past values may improve the forecast, whereas for others there may be no improvement. Therefore, we should distinguish two subgroups of individuals according to the causality relationships. If this heterogeneity is not considered, the test of causality hypothesis may lead to a fallacious conclusion concerning the relative size of the two subgroups. In a nutshell, the Granger causality for panel data sets should consider the different sources of heterogeneity of the data-generating. A newly suggested theory of Granger Causality in Panel by Hurlin and Venet (2001) is used here to test causality which controls for both sources of heterogeneity. The procedure has three main steps, which are related to the homogeneous non-causality (HNC), homogeneous causality (HC) and heterogeneous non-causality (HENC) hypotheses.

Panel Vector Auto Regression (VAR)

To check for causality in the case of no long-run relationship, panel VAR is used. Following fixed effects model is estimated:

$$\Delta Y_{it} = \sum_{k=1}^{p} a_k \Delta Y_{it-k} + \sum_{k=1}^{p} b_k \Delta X_{it-k} + \lambda_i + \mu_{it}$$

$$\Delta X_{it} = \sum_{k=1}^{p} a_k \Delta X_{it-k} + \sum_{k=1}^{p} b_k \Delta Y_{it-k} + \lambda_i + \mu_{it}$$

where p = lag length, λ_i: country fixed effects – controls the potential heterogeneity of cross sections.

Panel ARDL Approach: Multivariate Analysis

Let us consider the following ARDL process for panel data:

$$y_{it} = \rho_i y_{i(t-1)} + X_{it}\delta_i + \varepsilon_{it}$$

where $i = 1,2,\ldots N$ cross-section units or series, that are observed over periods $t = 1,2,\ldots T$; X_{it} represents the exogenous variables in the model, including any fixed effects or individual trends; ρ_i are the autoregressive coefficients; ε_{it} are the errors which are assumed to be a mutually independent idiosyncratic disturbance.

If $\rho_i < 1$, y_i is said to be weakly (trend) stationary. On the other hand, if $\rho_i = 1$ then y_i contains a unit root. For purposes of testing, there are two assumptions that we can make about the ρ_i. First, one can assume that the persistence parameters are common across cross-sections so that $\rho_i = \rho$ for all *i*. The Levin-Lin-Chu (LLC), Breitung and Hadri tests all employ this assumption. Alternatively, one can allow ρ_i to vary freely across cross-sections. The Im-Pesaran-Shin (IPS), and Fisher-ADF and Fisher-PP tests are of this form.

After determining the level of integration of all variables, the study proceeds to the application of panel ARDL to estimate the long-run and short-run relationships between regression variables. Autoregressive Distributed Lag (ARDL) models are standard least squares regressions which include lags of both the dependent variable and independent variables as regressors (Greene, 2008). Although ARDL models have been used in econometrics for decades, they have gained popularity in recent years as a method of examining long-run and co-integrating relationships between variables (Pesaran, Shin, & Smith, 1999). In panel settings with individual effects, standard regression estimation of ARDL models is problematic due to bias caused by the correlation between the mean-differenced regressors and the error term. This bias only vanishes for large numbers of observations *T*, and cannot be corrected by increasing the number of cross-sections *N*. To address this problem, a number of small T–large *N*, dynamic panel data GMM estimators have been developed (for example, Arellano and Bond, 1991). In large datasets, these assumptions underlying dynamic GMM are often inappropriate, and the estimator breaks down. In these cases, a popular alternative is the Pooled Mean Group (PMG) estimator of Pesaran et al. (1999). This model takes the cointegration form of the simple ARDL model and adapts it for a panel setting by allowing the intercepts, short-run coefficients, and co-integrating terms to differ across cross-sections.

Analysis

The authors begin by checking for stationarity of the variables. The IPS test has been conducted including a trend for all the panel variables. LLC test requires panel data to be strongly balanced and hence the test could not be run for most of the variables. However, the authors have relied on the LLC test instead of IPS for the variable measuring depth of food deficit since all the observations of the variable had zero value for Singapore as the country does not suffer from food deficit at all and IPS test requires at least six non-zero observations per panel. The dummy variables, distance from India, India's GDP and quality of port infrastructure index (on a scale of 1-7) were excluded from the tests due to the non-panel nature of the variables. The results show that all the variables considered for the tests are stationary in

Table 1. Unit root tests

Variables	IPS with trend	LLC
	p-value	
	Level	Level
Agricultural lab force, total	0.0575*	
Agricultural GDP	0.0445**	
BTA	NA	
CMB	n.a.	
Food deficit	....	0.0002***
Distance from India	n.a.	
FPI	0.0595*	
FProdI	0.0012***	
log(ER)	0.0629*	
log(Food Import from India)	0.0006***	
log(GDP Importing Country)	0.0178**	
log(GDP India)	n.a.	
log(GDP Per Capita)	0.0155**	
Food production variability	0.0408**	
Political stability and absence of violence/terrorism	0.0028***	
QPI	n.a.	
TO	0.0038***	

Source: Authors' development

Note: *, ** and *** indicate statistically significant values at 10%, 5%, and 1% levels of significance, respectively

the level form (p-values are lesser than the level of significance) hence they are all integrated of order 0 [I(0)]. The results are presented in Table 1.

Given that neither food import from India nor any of the regressors for the model are integrated of order 1 [I(1)], we cannot establish a co-integrating relationship between the regressors and food import from India that would have shed light on the long run relationship among the said variables. Hence we move on to check for causality among the panel variables running from the factors that are crucial for food import, to food import from India.

The pairwise causality tests were run for various lag lengths. It was seen that given the sample size of the panel data, the maximum lag length that can be considered for testing causality is 9. It is found that the GDP of the importing country has a causal effect on food import from India at lag lengths 1, 2, 3, 6, and 9. This emphasizes the importance of the size of the economy in determining its capacity to import food. Food production index causes food import from India at the 1st lag. The dearth or abundance of domestically produced food products will promote or stunt food import from India. The index for political stability and absence of violence or terrorism is also seen to cause import of food from India at the 3rd lag. A more stable and peaceful economy may pave the way for sustainable policies towards food security of the economy, triggering a change in the food import pattern for the country. However, this process being a long drawn one, the effect of the stability is expected to be seen only after three years.

Food production variability is found to cause food import from India at the 8th lag implying again that a more (or less) dispersed food production will prompt (or curtail) food imports only when the volatility has been experienced over some time and not right away. Agricultural GDP, depth of food deficit and the food price index of the importing economies is found to cause food import from India at a lag length of 9 which leads us to believe that it is only the persistent behaviour of these variables that may take shape through effects on other variables which finally go on to facilitate or impede the food imports from India. The rest of the variables were not found to have any causal effect on food import from India. The results of the causality tests for various lag lengths are presented in Table 2. The statistically insignificant results have not been reported here.

Next, the authors estimate a VAR model to understand the extent and direction of the effects of the chosen variables on food import from India. The estimated model has been reported below:

$$\begin{aligned}&\log(\text{Food Import}_{it}^{\text{India to }i\text{th country}}) = 26.39 - 0.04\log(GDP_t^{\text{India}}) + 2.923log(GDP_{it}^{\text{Importing Country}})\\&-2.133\log(GDP_{it}^{PC}) - 0.005D_{it} + 0.006TO_{it} - 0.337ER_{it} - 0.061GDP_{it}^{\text{Agriculture}} + 0.004\text{Food Deficit}_{it}\\&+0.021F\text{Prod}I_{it} - 0.012\text{Food Prod}n\text{Variability}_{it}^{PC} - 9.058CMB_i + 1.025BTA_i + 0.215FPI_{it}\\&-0.109QPI_{it} - 18.835\text{Agri Labour Force}_{it}^{\text{Share of total}} - 0.019\text{Political Stability}_{it}\\&+0.068\log(\text{Food Import}_{i(t-1)}^{\text{India to }i\text{th country}}) - 0.396\log(\text{Food Import}_{i(t-2)}^{\text{India to }i\text{th country}})\end{aligned}$$

Table 3 reports the coefficients and the student's t-statistic associated with the variables which allow the authors to establish the direction and value of the impact of a unit change in the regressor variable on the endogenous variable, log of food import from India and the statistical significance of the variables in explaining the variability of the endogenous variable. The critical values of the t-statistic for 10%, 5%, and 1% levels of significance and relevant degrees of freedom were found to be 1.684, 2.021, and

Table 2. Panel causality

Lag Length	Causal Variables	F-statistic	p-value
1	Food production index (2004-2006 = 100)	3.51351	0.0627
	log(GDP Importing Country)	10.6838	0.0013
2	log(GDP Importing Country)	4.21413	0.0163
3	log(GDP Importing Country)	2.79414	0.0418
	Political stability and absence of violence/terrorism	6.1952	0.0007
4, 5, and 7	-	-	-
6	log(GDP Importing Country)	2.15896	0.0497
8	Per capita Food Production variability	1.96164	0.0691
9	Agriculture, value added (% of GDP)	1.89106	0.0608
	Depth of the food deficit (kilocalories per person per day)	1.74314	0.0847
	Domestic food price index	4.05604	0.0019
	log(GDP Importing Country)	1.92637	0.0528

Source: Authors' calculation

2.704, respectively. The authors reject the null hypothesis of no significant effect of the regressor on the regressand when the calculated *t* statistic is greater in absolute value than the critical/tabulated statistic. The results show that the 2nd lag of food imports from India, agricultural labor force, common maritime border, distance from India, food price index, food production index, GDP (size) of the importing country, per capita GDP of the importing economy, and trade openness have statistically significant effect on food imports from India. Lagged food import from India, agricultural labor force, common maritime border, distance from India, and per capita GDP of the importing country affect food import from India in an adverse manner. The rise in the values of these variables will lead to a decline in the flow of food imports to these economies from India. On the other hand, food prices, food production, GDP of the importing country, and trade openness have a positive effect on food imports.

The values of food import from India to the ASEAN countries in the past have a negative impact on the current value of the variable. It may be due to the fact that the country of origin prefers to diversify with respect to trading partners for food items exports and there is a cyclical pattern for inflows of food imports into the importing economy that recurs periodically. It can also be attributed to weather conditions of the importing country – periodic bad weather may prompt the need to import food from India rather than every year.

The rise in agricultural labor force leads to a dip in food imports. It may be because the additional labor units are adding to the food production of the economy making it more self-sufficient and at the same time reducing its dependence on foreign countries for food needs. This also goes on to suggest that the agricultural sector of the ASEAN economies suffer from insufficient labor inputs comparative to capital and land resources. Hence, increase in the agricultural labor force takes the economy closer to the production possibility frontier with greater and more efficient utilization of other factors of production which may have been left comparatively less productive due to lack of human capital.

The existence of a common maritime border reduces food imports from India due to the higher cost of transportation through waterways. The port facilities of the Indian sub-continent are still poor. Also, the high traffic in these very waterways which lead to the common maritime borders for India and the ASEAN countries add to the time and costs of transportation of the traded products.

The distance of the ASEAN economy from India is also a deterring factor for food imports. This result emphasizes the significance of the gravity model considered to estimate the model for the study. The farther apart the economies are, the lesser the chances of them entering into trade relations – this is the essence of the basic gravity model. The obvious reason for this is the increased per unit cost of transportation with the rise in the distance to be covered to trade the products.

Per capita GDP of the importing country adversely affects food imports from India since a rise in the population of the economy will bring along greater demand for food, which in turn will force the economy to rely more on imports than what it used to earlier. Rising population, at the same time, will lead to falling in the per capita GDP. Hence, it may be natural to expect a negative relationship between per capita GDP and food imports.

The rise in food prices (probably because of inefficient production processes with high costs of production) will put the economy in a comparatively disadvantageous position making it more cost efficient to import the necessary food products rather than producing them inside the country. Either absolute or comparative advantage will be at work and push the economy towards importing food at cheaper prices and maximize the welfare of the nation.

The increase in food production may lead to increased food imports if the products produced within the country and the ones imported are not substitutes but complements of each other.

Table 3. VAR results

Variables	Co-efficient	t-statistic
log(Food import from India) _ (t-1)	0.068	0.578
log(Food import from India) _ (t-2)	-0.396***	-5.559
Agriculture GDP	-0.061	-1.148
Agriculture lab force, total	-18.835***	-3.33
BTA	1.025	0.52
CMB	-9.058**	-2.134
D	-0.005***	-3.102
Food deficit	0.003	0.637
FPI	0.215*	1.724
FProdI	0.021***	3.501
log(ER)	-0.337	-0.801
log(GDP Importing Country)	2.924***	5.189
log(GDP India)	-0.041	-0.085
log(GDP Per Capita)	-2.133***	-3.935
Per capita Food Production variability	-0.012	-0.675
Political stability and absence of violence/terrorism	-0.019	-0.058
QPI	-0.109	-0.448
TO	0.006**	2.125

Source: Authors' development

Note: *, ** and *** indicate statistically significant values at 10%, 5%, and 1% levels of significance, respectively

GDP of the importing country reflects the size of the nation and the potential demand of the economy. With higher purchasing power, the economy may choose to opt for more variety of food products and/or better quality of the same food products, necessitating and facilitating food import.

Higher trade openness may be attributable to the overall temperament of the economy towards new, better and/or cheaper food products. It is a characteristic of a more globalized economy which will automatically lead to a rise in the food imports as well.

For the ARDL model, the optimal number of lags for the endogenous variable, that is, food imports from India was found to be 2. Hence, the results are exactly the same as the VAR results reported above. As a result, the authors decided not to report them separately.

SOLUTIONS AND RECOMMENDATIONS

As shown through the analysis above, the augmented gravity model establishes the factors that are significant in determining agricultural trade flows.

Literature indicates, that while India features in the list of top-ten countries trading in agricultural commodities with the ASEAN nations, it still remains at the bottom of that list (ASEAN Statistics, 2015). India has high potential in making the ASEAN nations food secure but probably not through agricultural trade, alone.

The ASEAN nations, by tradition, have been involved in wide-spread extensive intra- and extra-regional trade in agricultural products. Since the ASEAN Economic Community (AEC) propagates "free flow of goods" as its principal tenet, trade facilitation through tariff reduction has been the key focus in this bloc – not only among the ASEAN nations but also with the non-ASEAN trade alliances. While tariff reductions have been on a steady decline (as indicated by the increase in the proportion of agricultural trade), the removal of non-tariff barriers (NTBs) is still not a success story. The quantification of the effects of NTBs and non-tariff measures (NTMs) are not easy. The costs arising from such sources are directly or indirectly captured as "trade costs" under the distance variable in the gravity equation. Although, the above analysis undertaken does not consider tariff rates but it does throw light on some of the reasons why agricultural trade is sluggish between India and the ASEAN. Based on the significant variables, some of the following recommendations can be made.

Since most of India and ASEAN are oil-importing nations, keeping food prices under check is a challenge for both and will have to be achieved by primarily reducing transportation costs – a factor embodied within the distance variable of the gravity equation of trade.

Barring Lao PDR, ASEAN countries and India have large coastlines, with many ports. However, there are large structural differentials in the shipping and maritime infrastructure within ASEAN and also between India and ASEAN. Within the ASEAN economies, Cambodia, Laos, Myanmar, Vietnam (CLMV) fall far behind when compared to other ASEAN countries and India, in terms of cargo volume and shipping tonnage. Predominant ports in ASEAN are located in developing parts of ASEAN and have been growing faster than those located in CLMV or the Mekong sub-region. In the containerized cargo segment, out of world's top 25 ports, seven ports come from ASEAN and only one from India. Singapore Port ranks second in the world in terms of efficiency and infrastructure whereas India's key port, Jawaharlal Nehru Port, is only in the 21st position worldwide. Also, the Jawaharlal Nehru Port is on the western coast of India, which means that the cargo has to travel additionally from west to east. If the eastern ports are made as efficient, India will cut down on trading costs, by far.

Pan-Asian trade integration is possible through better maritime connectivity. Well-connected ports and feeder routes can substantially improve trade between ASEAN and India. While only a few ports are connected directly by shipping lines in this trading bloc, a lot of resources is being allocated to reach ports in Cambodia, Indonesia, Malaysia, Myanmar, and Thailand with whom India's trade has been growing steadily. The existing route through Malacca Strait is heavily congested and also relatively unsafe. As an alternative, economic corridor based multimodal connectivity such as Mekong-India Economic Corridor will connect the Indian coastline with Southeast Asian coastlines in a shorter time and at a lesser cost. Other numerous infrastructural projects also serve to tie India closer to East Asia. India is participating in the United Nations Economic and Social Commission for Asia and the Pacific and the Pacific initiatives for an Asian Highway Network and the Trans-Asian Railway network. Discussions are also proceeding on reopening the World War II-era Stilwell Road linking the India's state of Assam with China's Yunnan province through Myanmar, under the Kaladan Multi-modal Transit Transport project. India should also explore developing the Dawei port and SEZ project in Myanmar under a joint venture. Cooperation in terms of financial credit and infrastructural support would also go a long way in enhancing trade volumes and reducing costs.

Simplifying and standardizing customs and trade logistics will improve the trade competitiveness and integration. Customs Single Window and such like should be implemented urgently.

Poverty reduction and capacity building will also increase labor productivity. Proper healthcare and education augment food utilization and bring about stability in the labor force. India has competencies in research and development (R&D), education and healthcare, which can benefit several ASEAN countries. The nation can categorically help through technology transfers, R&D, augmented human capital by means of providing better healthcare and education and making investments focused specifically towards supporting sectors that will strengthen food security directly or indirectly. This kind of an arrangement will probably involve more of services (and some goods) trade initially, but that, in turn, will pave the way to increasing the pace for agricultural trade and food security between India and the ASEAN economies.

As for managing water crises, climate change and the onslaught of population increase and urbanization, joining hands with each other and finding solutions to integrate as a region, would go a long way for this trading union.

FUTURE RESEARCH DIRECTIONS

This chapter could not include the Corruption Index and import duties due to the lack of adequate data. The inclusion of these variables would have made a better impact on the analysis and findings. Instead of considering total food imports, if the import composition can be bifurcated, more robust results and better insights may have been obtained. Again, if the composition of bilateral investment between India and ASEAN could be disaggregated, then determining how much is apportioned to food and agriculture related infrastructure and development leading to better food security, may have been determined.

CONCLUSION

Emerging countries have played an important role in increasing global food availability in the past decade. While they have increased their agricultural production in order to meet growing demand, the fast increase in food supplies in these markets have raised concerns regarding social and environmental impacts. India and the ASEAN countries may be similar in many ways but as far as economic structure is concerned, these nations are not equals. Growth, infrastructure, and the determinants of the supply-side economy vary widely within this FTA. In order to move onwards sustained food security, measures to augment bilateral trade in agriculture through improved global value chains, port and road infrastructure, better storage and logistics for access to food, reduced tariffs, R&D and knowledge sharing to effectively manage the consequences of climate change will put India and ASEAN in good stead, in the future.

REFERENCES

Aker, J. C., & Lemtouni, A. (1999). A Framework for Assessing Food Security in the Face of Globalization: The Case of Morocco. *Agroalimentaria*, *8*, 13–26.

Applanaidu, S. D., Bakar, N. A., & Baharudin, A. H. (2014). An Econometric Analysis of Food Security and Related Macroeconomic Variables in Malaysia: A Vector Autoregressive Approach (VAR). *UMK Procedia*, *1*, 93–102. doi:10.1016/j.umkpro.2014.07.012

Arellano, M., & Bond, S. (1991). Some Tests of Specification for Panel Data: Monte Carlo Evidence and an Application to Employment Equations. *The Review of Economic Studies*, *58*(2), 277–297. doi:10.2307/2297968

ASEAN-India Relations. (2016). In *ASEAN India Progress & Prosperity*. Retrieved October 17, 2016, from http://www.mea.gov.in/aseanindia/20-years.htm

ASEAN Statistics. (2015). *Top Ten ASEAN Trade Partner Countries/Regions*. Retrieved December 13, 2016, from http://www.asean.org/storage/2016/01/statistic/table20_asof121Dec15.pdf

Asian Development Bank. (2016). *Agriculture and Food Security Issues in Asia and the Pacific*. Retrieved October 6, 2016, from https://www.adb.org/sectors/agriculture/issueshttps://www.adb.org/sectors/agriculture/issues

Barraclough, S., & Utting, P. (1987). *Food Security Trends and Prospects in Latin America*. Notre Dame, IN: The Hellen Kellogg Institute of International Studies.

Breitung, J. (2000). The Local Power of Some Unit Root Tests for Panel Data. In B. H. Baltagi (Ed.), *Nonstationary Panels, Panel Cointegration, and Dynamic Panels* (pp. 161–177). Amsterdam: Elsevier. doi:10.1016/S0731-9053(00)15006-6

Briones, R. M., Chavez, E. C., Durand-Morat, A., & Wailes, E. J. (2012). *Climate Change and Price Volatility: Can We Count on the ASEAN Plus Three Emergency Rice Reserve*. Manila: Asian Development Bank.

Choi, I. (2001). Unit Root Tests for Panel Data. *Journal of International Money and Finance*, *20*(2), 249–272. doi:10.1016/S0261-5606(00)00048-6

Choulamany, X. (n.d.). *Issues and Strategies to Assure Food Security in ASEAN: ASEAN Level Initiative on Food Security* [PowerPoint slides]. Retrieved September 12, 2016, from https://www.oecd.org/tad/events/Issues%20and%20Strategies%20to%20Assure%20Food%20Security%20in%20ASEAN%20-%20%20%20by%20Xaypradeth%20Update.pdf

Clay, E. J., Shaw, D. J., & Singer, H. W. (1987). *Poverty, Development, and Food: Essays in Honour of H.W. Singer on his 75th Birthday*. Houndmills, UK: Macmillan. doi:10.1007/978-1-349-09214-7

El Niño in Asia. (2016). In *The United Nations Office for the Coordination of Humanitarian Affairs*. Retrieved September 11, 2016, from http://www.unocha.org/el-nino-asia

Emerging Market Economy. (n.d.). In *InvestingAnswers*. Retrieved October 2, 2016, from http://www.investinganswers.com/financial-dictionary/world-markets/emerging-market-economy-1518

Food and Agriculture Organization of the United Nations. (2001). *The State of Food Insecurity in the World*. Rome: Food and Agriculture Organization of the United Nations.

Gillson, I., & Fouad, A. (2016). *Trade Policy and Food Security: Improving Access to Food in Developing Countries in the Wake of High World Prices*. Washington, DC: World Bank.

Gouel, C., & Jean, S. (2013). Optimal Food Price Stabilization in a Small Open Developing Country. *The World Bank Economic Review*, *29*(1), 72–101. doi:10.1093/wber/lht018

Government of India. (2016). *State of Indian Agriculture 2015-16.* Retrieved October 10, 2016, from http://agricoop.nic.in/imagedefault/state_agri_1516.pdf

Greene, W. H. (2008). *Econometric Analysis*. New York, NY: Prentice Hall.

Hadri, K. (2000). Testing for Stationarity in Heterogeneous Panel Data. *The Econometrics Journal*, *3*(2), 148–161. doi:10.1111/1368-423X.00043

Holt-Giménez, E. (2008). *The World Food Crisis: What Is behind It and What We Can Do?* Oakland, CA: Institute for Food and Development Policy.

Huh, H., & Park, C. (2013). *Examining the Determinants of Food Prices in Developing Asia*. Manila: Asian Development Bank.

Hurlin, C., & Dumitrescu, E. (2012). *Testing for Granger Non-causality in Heterogeneous Panels*. Retrieved November 17, 2016, from https://halshs.archives-ouvertes.fr/halshs-00224434/document

Hurlin, C., & Venet, B. (2001). *Granger Causality Tests in Panel Data Models with Fixed Coefficients*. Paris: EURIsCO.

Im, K. S., Pesaran, M. H., & Shin, Y. (2003). Testing for Unit Roots in Heterogenous Panels. *Journal of Econometrics*, *115*(1), 53–74. doi:10.1016/S0304-4076(03)00092-7

Jha, S., Kubo, K., & Ramaswami, B. (2013). *International Trade and Risk Sharing in the Global Rice Market: The Impact of Foreign and Domestic Supply Shocks*. Manila: Asian Development Bank.

Lee, H., & Park, C. (2013). *International Transmission of Food Prices and Volatilities: A Panel Analysis*. Manila: Asian Development Bank.

Levin, A., Lin, C. F., & Chu, C.-S. J. (2002). Unit Root Tests in Panel Data: Asymptotic and Finite Sample Properties. *Journal of Applied Econometrics*, *108*(1), 1–22. doi:10.1016/S0304-4076(01)00098-7

Lipton, M., & Heald, C. (1984). *African Food Strategies and the EEC's Role: An Interim Review* (1st ed.). Brussels: Centre for European Policy Studies.

Masudul Haq, M., Kaitibie, S., Huda, S., Vock, N., Coughlan, K., Abdella, K., … Moustafa, A. (2014). *Using a Gravity Approach to Explain Food Imports for Improved Food Security in Qatar.* Paper presented at Qatar Foundation Annual Research Conference, Doha.

Maxwell, S. (1990). Food Security in Developing Countries: Issues and Options for the 1990s. *IDS Bulletin*, *21*(3), 2–13. doi:10.1111/j.1759-5436.1990.mp21003002.x

Mohamed Arshad, F., & Hameed, A.A.A. (2010). Global Food Prices: Implications for Food Security in Malaysia. *Journal of the Consumer Research and Resource Centre*, 21–38.

Morgan Stanley Capital International. (n.d.). *Market Classification.* Retrieved September 6, 2016, from https://www.msci.com/market-classification

Natale, F., Borrello, A., & Molotova, A. (2015). Analysis of the Determinants of International Seafood Trade Using a Gravity Model. *Marine Policy*, *60*, 98–106. doi:10.1016/j.marpol.2015.05.016

Overview. (n.d.). In *Association of Southeast Asian Nations*. Retrieved November 2, 2016, from http://asean.org/asean/about-asean/overview/

Pesaran, M. H., Shin, Y., & Smith, R. J. (1999). *Bounds Testing Approaches to the Analysis of Long-run Relationships.* Cambridge Working Papers in Economics, 9907. Cambridgeshire, UK: University of Cambridge.

Reutlinger, S., & Knapp, K. (1980). *Food Security in Food Deficit Countries* (1st ed.). Washington, DC: World Bank.

Salvatici, L. (2013). *The Gravity Model in International Trade*. Washington, DC: African Growth and Development Policy Modelling Consortium.

Sarker, R., & Jayasinghe, S. (2007). Regional Trade Agreements and Trade in Agri-food Products: Evidence for the European Union from Gravity Modelling Using Disaggregated Data. *Agricultural Economics*, *37*(1), 93–104. doi:10.1111/j.1574-0862.2007.00227.x

Sen, A. (1983). *Poverty and Famines: An Essay on Entitlement and Deprivation*. Oxford, UK: Oxford University Press. doi:10.1093/0198284632.001.0001

Siamwalla, A., & Valdes, A. (1980). Food Insecurity in Developing Countries. *Food Policy*, *5*(4), 258–272. doi:10.1016/0306-9192(80)90055-X

Starck, S. (2012). *The Theoretical Foundation of Gravity Modeling: What are the Developments that have Brought Gravity Modeling into Mainstream Economics?* (Unpublished master dissertation). Copenhagen Business School, Copenhagen, Denmark.

Tabor, S., Ginting, E., & Aji, P. (2015). *Preparing for El Niño: Policy Options*. Manila: Asian Development Bank.

The Economist. (2016). *The Global Food Security Index.* Retrieved September 10, 2016, from http://foodsecurityindex.eiu.com/

Tinbergen, J. (1962). *Shaping the World Economy: Suggestions for an International Economic Policy* (1st ed.). New York, NY: Twentieth Century Fund.

Van Dijk, M. (2011). *African Regional Integration: Implications for Food Security*. Retrieved October 18, 2016, from http://edepot.wur.nl/168845

Wang, J. (2016). Analysis and Comparison of the Factors Influencing Worldwide Four Kinds of Vegetable Oil Trade: Based on Gravity Model. *Modern Economy*, *7*(2), 173–182. doi:10.4236/me.2016.72019

World Food Programme. (2009). *Emergency Food Security Assessments (EFSAs). Technical Guidance Sheet No. 5.* Retrieved September 11, 2016, from http://documents.wfp.org/stellent/groups/public/documents/manual_guide_proced/wfp197295.pdf?_ga=1.244370689.508075766.1479356271

ADDITIONAL READING

Bajpaee, C. (2007). *India Rediscovers East Asia.* Retrieved October 14, 2016, from http://www.atimes.com/atimes/South_Asia/IJ31Df01.html

Engle, R. F., & Granger, C. W. J. (1987). Cointegration and Error Correction: Representation, Estimation, and Testing. *Econometrica, 55*(2), 251–276. doi:10.2307/1913236

Henneberry, S. R., & Carrasco, C. D. (2015). Food Security Issues: Concepts and the Role of Emerging Markets. In A. Schmitz, P. L. Kennedy, & T. G. Schmitz (Eds.), *Food Security in an Uncertain World* (pp. 63–79). Bingley: Emerald Group Publishing Limited. doi:10.1108/S1574-871520150000015005

Johansen, S. (1988). Statistical Analysis of Cointegrating Vectors. *Journal of Economic Dynamics & Control, 12*(2-3), 231–254. doi:10.1016/0165-1889(88)90041-3

Kapila, S. (2016). *India's 'Look East' and 'Act East' Policies Reviewed 2016.* Retrieved October 9, 2016, from http://www.southasiaanalysis.org/node/1924

MacKinnon, J. G., Haug, A., & Michelis, L. (1999). Numerical Distribution Functions of Likelihood Ratio Tests for Cointegration. *Journal of Applied Econometrics, 14*(5), 563–577. doi:10.1002/(SICI)1099-1255(199909/10)14:5<563::AID-JAE530>3.0.CO;2-R

Maddala, G. S., & Wu, S. (1999). A Comparative Study of Unit Root Tests with Panel Data and a New Simple Test. *Oxford Bulletin of Economics and Statistics, 61*(s1), 631–652. doi:10.1111/1468-0084.61.s1.13

Pedroni, P. (1999). Critical Values for Cointegration Tests in Heterogeneous Panels with Multiple Regressors. *Oxford Bulletin of Economics and Statistics, 61*(s1), 653–670. doi:10.1111/1468-0084.61.s1.14

Pedroni, P. (2004). Panel Cointegration: Asymptotic and Finite Sample Properties of Pooled Time Series Tests with an Application to the PPP Hypothesis. *Econometric Theory, 20*(03), 597–625. doi:10.1017/S0266466604203073

Wailes, E., Clarete, R., Briones, R., & Pochara, F. (2012). *Food Security in Asia: The 2007-2008 Food Price Crisis*. Manila: Asian Development Bank.

World Health Organization. (n.d.). *Micronutrients Database.* Retrieved November 1, 2016, from http://www.who.int/vmnis/database/en/

KEY TERMS AND DEFINITIONS

Agriculture: A cultivation of animals, plants and fungi for food, fiber, biofuel, medicinal plants, and other products used to sustain and enhance human life.

Association of Southeast Asian Nations: A regional organization comprising ten Southeast Asian states which promotes intergovernmental cooperation and facilitates economic integration amongst its members. Since its formation on August 8, 1967, by Indonesia, Malaysia, the Philippines, Singapore, and Thailand, the organization's membership has expanded to include Brunei, Cambodia, Laos, Myanmar (Burma), and Vietnam. Its principal aims include accelerating economic growth, social progress, and sociocultural evolution among its members, alongside the protection of regional stability and the provision of a mechanism for member countries to resolve differences peacefully.

Causality: It indicates that one event is the result of the occurrence of the other event, that is, there is a causal relationship between the two events. This is also referred to as cause and effect.

Food Access: It exists when all people, at all times, have physical, social and economic access to sufficient, safe and nutritious food, which meets their dietary needs and food preferences for an active and healthy life.

Food Availability: It refers to the supply of food through production, distribution, and exchange.

Free Trade Area: The geographical expanse encompassing the member countries or a trade bloc who have signed a free-trade agreement. Such agreements involve cooperation between at least two countries to reduce trade barriers – import quotas and tariffs – and to increase trade of goods and services with each other.

Gravity Model: A model used to estimate the amount of interaction between two cities or countries. It is based on Newton's universal law of gravitation, which measured the attraction of two objects based on their mass and distance. The amount of interaction between two cities or countries is proportional to the size of the cities and inversely proportional to their distance squared.

Panel Data: This term is used in econometrics and statistics, for multi-dimensional data frequently involving measurements over time. Time series and cross-sectional data are special cases of panel data that are in one dimension only (one panel member or individual for the former and one-time point for the latter).

Stability: An absence of excessive fluctuations in the macroeconomy. An economy with fairly constant output growth and low and stable inflation would be considered economically stable. An economy with frequent large recessions, a pronounced business cycle, very high or variable inflation, or frequent financial crises would be considered economically unstable.

Trade Integration: A unification of economic policies between different states through the partial or full abolition of tariff and non-tariff restrictions on trade taking place among them prior to their integration. This is meant in turn to lead to lower prices for distributors and consumers with the goal of increasing the level of welfare while leading to an increase of economic productivity of the states.

Trade Liberalization: A removal or reduction of restrictions or barriers on the free exchange of goods between nations. This includes the removal or reduction of tariff obstacles, such as duties and surcharges, and nontariff obstacles, such as licensing rules, quotas, and other requirements. The easing or eradication of these restrictions is often referred to as promoting "free trade".

Chapter 6
Maritime Piracy in Southeast Asia:
Consequences on Economic and Political Integration

Giacomo Morabito
University of Messina, Italy

Davide Di Fatta
University of Messina, Italy

Vittorio D'Aleo
University of Messina, Italy

Roberto Musotto
University of Messina, Italy

Walter Vesperi
University of Messina, Italy

ABSTRACT

The costs of the Southeast Asia maritime piracy to the global economy are largely underestimated regarding the issue of the fight against piracy. In particular, the region suffers from severe negative socio-economic impacts due to over-fishing, and the reduction and collapse of the fisheries have led to a widespread loss of income and employment. The attacks on fishers take place regularly in most Southeast Asian countries. Also, in many cases, the pirates are the fishers themselves, and they attack fishers of other nationalities. All those attacks have a serious negative impact on food security in the region by damaging fishing industry and threatening artisanal fishers who risk losing their entire income. Given the fact that large parts of Southeast Asia's fish populations are overfished, fishers have less possibility to avoid areas prone to pirate attacks.

INTRODUCTION

Maritime piracy threat is difficult to define, understand, and counteract. There are different definitions. According to the United Nations, piracy cannot be considered as in international waters (piracy anchoring or territorial waters are considered to be of brigandage).[1] More comprehensive definition adopted by the International Maritime Bureau (IMB) qualifies piracy as any act against a ship collision with the intent

DOI: 10.4018/978-1-5225-2733-6.ch006

to commit theft or any other crime and with the ability to use force to accomplish the act, regardless of the nature or severity of the crime and without distinction of zone (International Chamber of Commerce [ICC], 2015). This definition thus circumvents the cumbersome legal definition of the United Nations Convention on the Law of the Sea (UNCLOS), less operational one. Finally, each of the actors, state or private, retains the definition of risk that suits him best at the media, diplomatic or from marine insurance.

This threat is difficult to understand in view of the extent of maritime areas in terms of their geographical characteristics, varying objectives of pirates, and geopolitical frameworks (presence of constitutional states, their ability to ensure their sovereignty missions, and their commitment to the powers of their own). In addition, national tactics specific to the interventions at sea once pirates are captured are very complex (proof to provide custody, the presence of police officers embarked on military buildings).

Piracy is also an evolving threat. Depending on the region, pirates in a greater of lesser extent may be related to some movements for independence, political actions, terrorist attacks, or simply a crime. Whatever threat is in question, the answer is ultimately important to the humankind, on the one hand, by helping establishing the rule of law in case of a failed state like Somalia and, on the other hand, by promoting safety of riparian countries bordering with "pirates areas" and developing, wherever possible, common regional responses. Countries for some "complacent" and corrupt often have different internal priorities and refuse to be deprived of their prerogatives.

The anti-piracy measures should be harmonized with the policies of the countries concerned, considering not offending national sensibilities. This is the case of Nigeria which refuses any foreign intervention in its territorial waters, and to a lesser extent, Indonesia and other states bordering the Straits of Malacca between the Malay Peninsula and the Indonesian island of Sumatra, in Southeast Asia, which also refuse any foreign interference into the area. Finally, the sea is a common good to the states, and the pirates play well encouraging or compelling states to work together to provide a regional and global response to a transnational threat. This phenomenon has always attracted the attention of defense authorities, as only the navies have offshore resources to intervene far in the sea and protect maritime communication channels (Sea Lanes of Communication, or SLOC) through which 90% of manufactured goods pass.

Among the most plagued areas to piracy are the mouth of the Persian Gulf, the Strait of Bab el Mandeb, and the Indian Ocean, where 30.000 tankers and container ship pass annually. This represents 30% of oil trade and 70% of the merchant traffic in the world. Three-quarters of maritime traffic between the Middle East and the Far East pass through the Malacca Strait, 300 ships per day, including 25 tankers. Each container may be targeted as an economic issue. The embedded value is huge, carrying the equivalent of 6.000 semi-trailer trucks. The annual turnover of maritime transportation in this area is about $500 billion, what stir cravings and temptations among certain groups of people. The strategic and economic issues are therefore essential in the event of overall destabilization of those maritime zones.

What is the state of piracy in the world? The IMB recorded 264 attacks in 2013, including 141 in Southeast Asia and 51 in the Gulf of Guinea. It should not be limited to numbers and statistics, which can be as contradictory as the agreed definitions of piracy.

In the region of Southeast Asia, coastal states are facing a common transnational threat with certain common developments and similarities in the regional political context. In the Strait of Malacca and in the area ranging from southwestern China Sea to the south of the Philippian and Indonesian waters, maritime piracy remains alive. The attacks usually take place at night, on targets of opportunity, ships at anchor or at berth (in 80% of cases), which are considered, according to the definition adopted by the UN, as simple banditry maritime with handguns and knives. Ships of small tonnages, barges and tugs are attacked, despite a recent recovery in attacks against larger vessels, as evidenced by the attack on

Taiwanese tanker Orapin in the south of the Straits of Malacca on May 28, 2014. Piracy moves being less present in the strait since the establishment of coordinated Malacca Straits Patrols (MSP patrols). They were deployed despite the reservations that may exist between Indonesia, Malaysia, and Singapore, in purpose to more effectively secure the strategic maritime route and avoid the interference of foreign powers. Nevertheless, the presence of the USA in the region and the US policy of Asian pivot, with the emphasis on the Pacific allows increasing the presence of vessels of the Seventh US Fleet in the Philippines and Singapore. US ships are involved, too, in the fight against piracy missions in rapid spreading their light frigates (Littoral Combat Ship, or LCS).

If the number of piracy attacks in the Malacca Straits decreased from 38 attacks in 2012 to only one in 2013, in Indonesia it increased from 15 in 2009 to 106 in 2013. The threat remains present and cannot be overlooked, especially because of the possibility of an alliance with some terrorist groups like Abu Sayyaf in the Philippines.

BACKGROUND

Since 2008, media paid attention to maritime piracy, bringing with them an interest and desire to understand this phenomenon among the population. The issue was to understand from a legal point of view why piracy, the phenomenon from adventure books and fiction films is back on the stage in the XXI century. Has maritime piracy not been eradicated?

The origin of piracy dates back to ancient times at least. The first written reports of acts of piracy appeared in the third century BC. Pliny the Elder in his Historia Naturalis noted that archers must take on board the seagoing merchant ships off the Tamil country, implying that the sea in the south of India was at that time infested with pirates. This criminal activity has never really ceased to exist.

At the end of the XX century, piracy was not in the focus of media, with an exception of few cases, such as the diversion of transatlantic Achille Lauro cruise ship in 1985 in the Mediterranean Sea. However, pirates remained active throughout the world, especially in the region of Southeast Asia. This area suffered the brunt of maritime piracy actions launched by the local octopus which extended its grip on the seas.

Given the increasing acts of maritime piracy in Somalia and the Gulf of Aden over the past two years, the authors subsequently reviewed the major consequences of piracy and developed certain solutions to this phenomenon. Before making the panorama of piracy in the contemporary world, the authors begin with defining the term "piracy operation" that may seem obvious at first glance, but not practical.

According to the IMB (ICC, 2015), acts of piracy are mainly concentrated in certain countries and regions. Indeed, with 92 acts in 2008 (against 13 in 2007), the Gulf of Aden is positioned in the first place having been the site of 31% of the acts recorded by IMB worldwide. The total amount of acts of piracy recorded by IMB in 2008 amounted to 293. The third place is occupied by Indonesia, which experienced 28 acts of piracy in 2008, 10% of the world total. In this country, the number of acts of piracy has been decreasing since 2003, when 121 cases were recorded. Until 2007, Indonesia had been the country the most affected by piracy. Improvement of the situation in the Strait of Malacca in general and in neighboring countries, in particular, is the result of the cumulative effect of increased vigilance and patrols operated by neighboring countries, along with the precautions taken by the ships passing through this area.

Pirates attack all types of ships. In 2008, attacks were targeted on chemical tankers, container ships, bulk carriers, cargo, and tankers. Taken together, they account for 75% of the acts of piracy reported in 2008. The authors also note a rising number of attacks against chemical and oil tankers in 2006-2008. The attacks against those two types of boats decreased substantially between 2003 and 2006 but rebound strongly between 2006 and 2007. The authors have no major explanation for such development; however, they assume that this may be the result of the importance given to such ships from the attackers. As a part of a diversion for obtaining a ransom, it is indeed more interesting for hackers to attack big ships carrying high-value commodities enabling the attackers to derive a good ransom. Bulk carriers and cargo are usually attacked by those hackers whose purpose is to steal cash and personal belongings of the crew or those seeking to recover easily salable commodities on the black market.

The costs of the Southeast Asia maritime piracy to the global economy are largely underestimated regarding the issue of the fight against piracy in that region. Reconstruction of the overall costs of piracy means, therefore, assessment of a diversity of different costs that fall on a variety of subjects and beneficiaries. Some of those costs are traceable and statistically detectable. Others can be estimated only based on forecast parameters. Every year billions of tons of cargo are transported by sea in Southeast Asia, and the vessels carrying the cargo have to pass a "choke point" between the Indian and Pacific Oceans. There are two major options: the Straits of Malacca and the Lombok/Makassar Strait. The Straits of Malacca are a narrow 805 km stretch of water between Peninsular Malaysia, Singapore, and Indonesia: about one-quarter of the world's traded goods and a quarter of all oil carried by sea pass through the straits annually. The Lombok/Makassar Strait is used by the vessels moving between Australia and East Asia, as well as by the fully-laden tankers of over 200,000 dead-weight tons traversing from the Indian Ocean to the Pacific Ocean since they cannot use the Straits of Malacca.

According to Biloslavo and Quercia (2013), there has to be a distinction between different types of costs related to the phenomenon of maritime piracy. The first distinction is between the real costs and macroeconomic costs. Real costs are all costs incurred during prevention, reaction to prevention, reducing, or managing the consequences of the acts of piracy. Macroeconomic costs are those caused by the effects which piracy produces on the countries of the region or their trading partners. For example, real costs associated with piracy are redemption payments, naval missions, insurance costs, while macroeconomic costs are represented by an impact of piracy on the economies of the countries in the region, especially those neighboring the coast.

In turn, Biloslavo and Quercia (2013) distinguish the real costs in direct costs and indirect costs. Direct costs are all of those costs which are incurred when a kidnapping takes place. They are directly related to meeting the economic demands of pirates in order to secure the release of the ship crew, cargo, and the ship itself. Such costs are distinguished in material costs and instrumental costs. Material costs are procured by pirates under the threat of violence and following a criminal act such as the seizure of a ship. Instrumental costs are all of those an owner has to bear in order to negotiate and pay the ransom with all the necessary legal guarantees and security. Indirect costs are incurred in advance in order to counter or reduce the risk of actions of piracy, increase the resilience of ships, and cover possible economic losses. Depending on whether those costs fall on the certain country, international organization, or private owners, they can be divided into costs of enforcement and deterrence and protection costs.

There are few contributions concerning maritime piracy with a quantitative approach. Martínez-Zarzoso and Bensassi (2013) tested the effect of modern piracy on maritime trade cost proposing a simple model of transport cost determination and deriving a transport costs equation augmented with maritime piracy as an additional explanatory variable. Their results showed a significant and positive impact of piracy on maritime transport cost; in addition, they showed that localized conflicts could selectively harm some international trade routes. Generally, the negative impacts of maritime piracy disrupt the global economy. In fact, it may have a significant impact on GDP of the trading countries through a drop in trade. According to the analyses carried out by Anderson and Marcouiller (2005) and Blomberg and Hess (2006), several types of violent acts, such as maritime piracy and terrorism, have a significant negative impact on bilateral trade. De Sousa, Mirza, and Verdier (2009) focused exclusively on terrorist activities. They showed that terrorist acts have an impact on the bilateral trade relations between the targeted nation and the perpetrator nation, as well as between the neighboring countries of the perpetrator nation and the targeted nation. Bensassi and Martínez-Zarzoso (2012) also estimated the impact of piracy on maritime trade between Europe and Asia using data on incidents of maritime piracy between 1999 and 2008. They applied the gravity model, widely used to investigate the role played by specific policy or geographical variables in bilateral trade flows.

The region of Southeast Asia suffers from severe negative socio-economic impacts due to over-fishing, and the reduction and collapse of the fisheries have led to a widespread loss of income and unemployment (United Nations Environment Programme, 2006). In most of the countries in Southeast Asia, pirate attacks on fishers take place regularly, as they often carry cash to buy supplies or pay-off officials in order to fish in certain areas. Pirates are often after this cash, fish, gear, or even the vessels. In order to intimidate the fishers, pirates that attack smaller fishing vessels are in most cases armed, carrying knives, small arms, or light weapons. However, in many cases, the pirates are fishers themselves, and they attack fishers from other countries. Throughout the Southeast Asia, pirates attack all types of fishing vessels. In most cases, they are after cash, personal belongings, and fish, but sometimes they also take gears or even vessels. There are also reports of crew members being kidnapped for ransom. All these attacks have a serious economic impact on the fishing industry of the region, especially on the artisanal fishers who risk losing their entire income when gear or engines are stolen. Given the fact that large parts of fish populations in the region of Southeast Asia are over-fished, fishers have less possibility to avoid areas prone to pirate attacks. They have to fish in certain areas in order to make their living and thus bear a risk of being attacked. However, they do take safety measures. Most fishers in the Southeast Asia carry knives. In certain high-risk areas, some of them may even have firearms onboard (Liss, 2007). In other cases, fishers do not carry firearms and the owners of vessels often have sufficient financial resources to pay substantial ransoms for the recovery of their vessels and crew. Pirates appear to accept substantially lower ransom payments than those they initially demand. In some cases, when the vessel owners cannot pay, pirates just take off with all valuable assets available onboard (Liss, 2007). Apart from the relatively large sums of money paid as ransom, negotiations between pirates and a ship-owner or the victim's family can take three-four months. For a ship-owner, a fisher, and his family it means a substantial loss of income. Not in all hijacking cases, pirates are after ransom.

Overall, it is clear that fishers in the region of Southeast Asia are strongly impacted by piracy in economic terms. However, the fishers do not only suffer losses due to piracy but are often the perpetrators themselves. For example, Malaysian fishers have been attacked by Indonesian or Thai fishers that are mainly interested in the catch of the targeted boat. In addition, owners of fishing vessels and fishers also buy stolen fishing gears. This saves them money compared to buying the new equipment.

MAIN FOCUS OF THE CHAPTER

Economic Consequences

The Asian system is in revealing meaning. Piracy historically originates from Europe, but Asian approach is significantly different. In Europe, piracy was widely regarded as a crime and not integrated into the economic functioning of any country, the exception being Barbarossa (or Khayr al-Din) who established the state based on piracy in Algiers in 1516 and was appointed the Governor General of the said city by the Ottoman Sultan. In Asia, piracy sometimes led to the foundation of states, their traditional economy of subsistence, and seat of power.

During the XX century, internationalization has opened the way for a more effective fight against piracy. Indeed, as the history shows, before the XX century, there had been no joint efforts taken by many countries to combat piracy. Multiple conflicts between countries prevented any real international efforts.

Piracy has always existed based on rapid economic development and increasing income. If people are turning to maritime crime, it is to escape poverty or unemployment which hit them. They also hope for easier income gain with fewer efforts from their side. They look for an attractive income. If in addition, the activity is less dangerous than it appeared, pirates end up winning. How people are turning more and more to this kind of activity, and how their claims increase?

The economic consequences of piracy primarily affect international shipping. The annual cost of piracy to the maritime industry is estimated at between $1 billion and $16 billion, mainly because of delays in delivery and cargo theft. Since the cost of insurance increases for every ship passing a seaport or an area prone to piracy, the economies of some countries could suffer serious losses. Piracy is not content with simple robbery. It requires a certain degree of courage, organization, and even certain unconsciousness. Next, to the survival, piracy has developed ambitious pirates who rubbed legal modes of operation. In some countries, piracy has become an honorable way of living, a regulated economic system that allowed survival of many people and development of power. Countries soon realized the benefits given by an army of hackers gained to their cause: power and economy.

Pirate activity has always been shrouded in an aura, which, though not always positive, never let people indifferent. The lure of easy income always attracts impoverished people. On the other hand, the "profession" of a pirate is not ostracized by the society. Instead, in some cultures, it is valued and even promoted as an artisan profession that is exerted from a father to a son. In addition, the existing lax protection of vessels has made attacks much simpler than they had been before. The diversity and complexity of pirates' organizations also made the practice safer for pirates.

In Southeast Asia, repressive measures against piracy have been developed on the national and regional levels. The problem of piracy raised in the region of Southeast Asia during the 1990s. Traditionally, Southeast Asia is recognized as a hotbed of piracy. This believed myth was maintained and became more aggressive with the development of technology. Despite many risks to ships in the region, the countries of Southeast Asia have been slow to respond, demonstrating the usual reluctance of states to surrender some of their sovereignty in the fight against a common evil. However, by necessity, they had to work together to build the legal enforcement system adapted to their problems. Today, this system is exemplary in the fight against piracy.

Nowadays, piracy in Southeast Asia declines. The efforts implemented, both by states and by agencies, have borne fruit. Nevertheless, because of its particular geographical situation, the region of Southeast Asia remains an area to keep under surveillance. Historically, Southeast Asia with its multiple islands outside any control of national authorities, ambient superstition, and intensive maritime traffic is a region where pirate acted. The autonomous state actions brought no improvement because of their weakness, specific geographical situation, particularly in the Malacca Strait, but also because of the dent in sovereignty.

Attempts agreements have finally ended up producing a result as the number of attacks has been declining, according to the IMB (ICC, 2015). The IMB data are to be taken lightly because they identify crimes that the only definition of Montego Bay would recognize. The IMB stands for piracy acts committed at anchor in territorial waters and as well as they committed in international waters. Nevertheless, regarding Southeast Asia and its particular configuration, it seems logical to consider the acts committed in territorial waters or archipelagic as well as those committed in international waters if one seeks to obtain a homogeneous reaction effectively.

According to the IMB (ICC, 2015), maritime piracy in Southeast Asia has been declining steadily since the early 2000s. In 2003, the IMB recorded 445 acts of piracy including 151 in the region of Southeast Asia. In 2004, there were 329 acts recorded, among which 140 in Southeastern Asia. However, since 2005, the trend has been reversing–among 276 attacks recorded worldwide, 98 happened in Southeast Asia. The year of 2004 is the year of developing the Regional Cooperation Agreement on Combating Piracy and Armed Robbery against Ships in Asia (ReCAAP) agreement. The authors assume that its influence, despite the faults that they could point out, pushed the states in an effective response. It was in 2008 when the trend finally reversed since, among 293 acts recorded worldwide, 111 were attributed to Somali pirates and the Gulf of Aden, and only 36 to the pirates in Asia. In 2009, only two attacks reported happening in the region.

By dint of effort, the region has been pacified. It would be more accurate to say that piracy has moved from one region to another. Therefore, it is necessary for the countries not to relax their efforts, both in the actual fight, but also in the search for triggers of piracy in order to avoid a new flare.

The decrease in piracy does not mean that it has disappeared or has become obsolete, as it was anticipated. The danger of piracy alone is sufficient to justify the continuation of military and legal actions. It is also a political necessity: countries are even more inclined to continue fighting against piracy and maintaining a significant military presence in the region. In addition, the need for cooperation offers them greater legitimacy internationally. This military cooperation also involves the assistance of other countries, concerned about the terrorist aspect the piracy may take. Faced with this little help if desired, the riparian states have chosen to maintain joint patrols at sea and in air. Facilities in border prosecution are also implemented.

After many attempts, countries of Southeast Asia have formed an alliance strong enough to curb piracy, in which they all found their advantages. It took unilateral, bilateral and multilateral attempts to find the right balance. In addition to patrols and cooperative military measures that maintain fear in the minds of criminals, there are other measures, closer to the home of piracy that maintain the attention of governments. The authors observed that poverty, political instability, corruption are at the origin of many vocations. Governments have therefore tried to establish authority in the fight against corruption, maintaining an extensive presence even in the most remote areas, where hackers could easily find refuge and assistance.

These measures have produced satisfactory results, driving piracy away from Asian shores. It has even reappeared on other seas, such as Somalia and in the Gulf of Aden. Can Asian solutions be applicable to other regions?

Fleets of Southeast Asia

For marine called modern, it is to cope with traditional threats: ensuring sovereignty and integrity of its territorial waters and be able to fight against a hostile navy. As for post-modern marine, they target non-traditional threats, which are in the context of contemporary issues of maritimization of world trade. Depending on the case, war marines face a range of missions (naval diplomacy, deterrence, protection of SLOC and transnational interests, protection of nationals of the exclusive economic zones and the environment) as well as a variety of threats (state conflict, piracy, drugs, independence movements, trafficking, and, increasingly, natural disasters).

Fleets must be able to defend a maritime system, consisting of SLOC and strategic passages, requiring information management and complex logistical constraints of external operations.

Considering this analytical framework, yet few Southeast Asian navies are able to carry out such missions, except Singapore (which took the command of CTF 151several times) and to a lesser extent Malaysia which sent several patrol vessels in the Indian Ocean. Conversely, riparian marine Malacca Strait has striven to reduce the number of pirate attacks through coordinated Malacca Straits sea patrols(MSSP) and Malsindo (Malaysia, Singapore, and Indonesia) in 2004, followed its air component, EIS (Eyes in the Sky) and intelligence component, IFC (Information Fusion Center) in 2005.

Various regional coordinated missions which soon showed their operational limits should encourage the states of Southeast Asia to further push their cooperation in order to respond more effectively to transnational threats. Specifically, these modernizing marines still lack doctrinal continuity and strategic coherence. The shipping industry is more responsive to economic and political opportunities as operational requirements.

Recently, the use of naval tool during natural disasters (typhoon Haiyan in the Philippines) or human catastrophes (disappearance of flight MH 370) demonstrated operational deficiencies and lack of coordination and mutual trust in the region. Reasonable exploitation of maritime resources is also a part of the challenges of the XXI century which riparian countries are not yet sufficiently aware.

The disparity of forces and targets in Southeast Asia is not conducive to strategic or doctrinal convergence because of the following reasons:

- Indonesia expands fleet (65,000 men and 54,200 tons), being aware that its current naval forces are not able to protect the territory of 5.8 million square kilometers and 17,508 islands that make up the archipelago. The military and navy serve as a link between the islands, provide supplies, combat illegal maritime activities, paying special attention to illegal fishing, fight against an insurgency (or West Papua to Aceh), and participates in recent humanitarian missions (since the tsunami of 2004). On May 20, 2014, admiral Marsetio, head of the Indonesian navy (Angkatan Laut Tentara Nasional Indonesia, or TNI AL), explicitly requested that the next government would pay more attention to maritime issues. Jakarta says increasingly concerned about China's ambitions south of the South China Sea, not hesitating to talk about race in intra-regional armament. The government encourages the continuation of the military efforts, with the acquisition of 124 aircraft and 12 submarines by 2024. TNI AL has also conducted amphibious exercises recovery by force (as Japan did in September 2016 with the expeditionary force of US Marines) on the outskirts of Natunas islands. Virulent diplomatic tensions have divided Jakarta and Canberra after forays into Australian territorial waters in the end of 2013.

- Malaysia aligns a strong fleet of 14,000 men (27,900 tones). The Royal Malaysian Navy (RMN) develops its submarine capabilities with the acquisition of two submarines (SM) French Scorpènes (two other desired) and new amphibious capabilities to be able to intervene in the South China Sea. The missions are to ensure sovereignty in territorial waters, particularly between Peninsular Malaysia and East Malaysia enclave on the island of Borneo, the fight against illegal activities at sea and piracy.
- The Philippines are investing heavily in modernizing their fleets, whose main tasks are committed to the anti-insurrectional fight. National budget of the Philippines increased by 20% in one year. With the help of the USA and Japan, the Armed Forces of the Philippines develop capacity coast-guard (Philippine ships are former US Coast Gard Hamilton class and Dallas), to patrol around the disputed Scarborough Shoal reef. Tensions with China are high, with the recent arrest of Chinese fishermen in the disputed area and the dispatch of troops by General Emmanuel Bautista in February 2014. In general, relations with the USA are good. The base of Subic Bay will re-open in the US Navy. The two states have conducted joint exercises in May 2014, involving 5,500 men and the territory of 220 km from Scarborough Shoal.
- Singapore has the most modern fleet in the region, including strong Trident frigates, six Swedish SM and two German SM 218SG. The Royal Singapore Navy (RSN) provides missions of sovereignty and protection of SLOC in the region, but also in the Indian Ocean. It is one of the few modern fleets framed by a sustainable strategy and interoperable with the US Navy.
- Thailand (74,000 men and 50,160 tons) essentially ensures sovereignty missions in offshore operating areas and the fight against illegal fishing in the disputed waters.
- Vietnam has undertaken an access denial strategy towards China by an ambitious modernization of its underwater fleet program by acquiring six Russian Kilo-636 in photo-cons, nicknamed black hole (through their excellent acoustic signature). They will ensure a presence of Vietnam to continue Wed. The first two buildings were delivered in 2012; the last will be delivered in 2018. At the recent escalation of tensions with China, Prime Minister Nguyen Tan Dung has accused China of "serious violations dangerous".

The Role of ASEAN

The deployment of coast guard which is stronger and less lethal compared to warships is a symbolic message. These paramilitary forces often are not armed but more exposed. The risk of escalation is important, and collisions are frequent. For example, Malaysia created its body coastguard in 2005 (Malaysian Maritime Enforcement Agency, or MMEA) with support from the USA and Japan.

Association of Southeast Asian Nations (ASEAN) should serve as a platform for discussion with China, which seeks to focus on bilateral negotiations. Since 2011, ASEAN innovates on maritime cooperation in order to improve the dialogue between China and other riparian states. The establishment of the ASEAN Maritime Security Cooperation shall promote the exchange of information between member states, while the ASEAN Defence Ministers' Meeting (ADMM) + Expert Working Group meetings try to highlight regional interests before the sovereign concerns, including environmental safety.

Other initiatives can improve communication and cooperation between the fleets, as in the Western Pacific Naval Symposium. Current divisions within the ASEAN member states are too pronounced to be able to talk about a front door to China. Littoral states (Vietnam, Philippines, Indonesia, Malaysia, and

Thailand) should initiate discussions behind the scenes and with China and finally with extra-regional powers, before treating the problem at the level of ASEAN.

Shared between competition and cooperation, regional navies have trouble making body against Chinese ambitions in the area. The risk of escalation of tensions between the member states of ASEAN persists. The conflict between Malaysia and the Philippines in Sabah is the most recent example.

ReCAAP: The Key-Solution for Multilateral Security Cooperation

The Indo-Pacific high seas do not immediately bring international cooperation to mind. Indeed, most of what is newsworthy along Asia's maritime frontier has to do with impressive new submarines or disputes over where one state's exclusive economic zone ends and another's begins. Despite the many sources of friction in these waters, broad multilateral security cooperation does exist in the form of the Regional Cooperation Agreement on Combating Piracy and Armed Robbery against Ships in Asia (ReCAAP).

ReCAAP does not often find its way into the mainstream news, but those in the commercial shipping industry know it well. Its prime mandate is to fight back against pirates and maintain open access to Asian sea lines of communication (SLOCs). In its own words, it strives to "serve as a platform for information exchange," "facilitate communications and information exchange among participating governments," "analyze and prove accurate statistics of the piracy and armed robbery incidents to foster better understanding of the situation in Asia," "facilitate capacity building efforts that help improve the capability of member countries in combating piracy and armed robbery," and "cooperate with organizations and like-minded parties".

The ReCAAP Agreement was finalized in November 2004. Currently, it has 20 signatories, including China, India, Japan, South Korea, and Australia among others. Extra-regional countries are also participants in the regime, including the United Kingdom, the USA, the Netherlands, Denmark, and Norway. The organization is headquartered in Singapore. The impetus for its founding was an acute surge in maritime piracy in the wake of the 1997 Asian financial crisis. Most recently, ReCAAP was lauded for decreasing the total number of pirate incidents on the Asian high seas in 2013.

ReCAAP's primary value-add in combating non-state threats to intercontinental sea-based commerce is through its information-sharing functions. ReCAAP's website pitches itself as an "Information Sharing Centre". The organization also maintains a great deal of transparency in how it presents its information: it maintains an excellent resource called the "Consolidated Incident Reports" which quantifies several aspects of a pirate or criminal incident.

ReCAAP's response to pirate incidents is to immediately alert proximal maritime authorities. In the example of the Danai 4, ReCAAP alerted Malaysia's Maritime Enforcement Agency to aid in recovery operations. Incidentally, Malaysia and Indonesia remain the two significant states that have yet to accede to the agreement. ReCAAP de facto involves Indonesia and Malaysia to a disproportionate degree given that the Strait of Malacca is a hot flashpoint for maritime piracy in Asia. The two states have nuanced reasons for not joining ReCAAP. Malaysia feels slighted by its headquarters being in Singapore and views it as a redundant addition to Asia's security apparatus in light of the IMB Piracy Reporting Centre. As for Indonesia, it has refused to join on concerns for its sovereignty.

Awareness of ReCAAP and its activities is essential to understanding maritime security and broad multilateralism in Asia. ReCAAP is not an ambitious competitor to the North Atlantic Treaty Organization (NATO) or even the Shanghai Cooperation Organization (SCO) – far from it. It represents a coordinated attempt by Asia's sea-faring states to preserve freedom of navigation and maritime security for the prosperity of the region as a whole.

SOLUTIONS AND RECOMMENDATIONS

At present, maritime trade is developing at high speed in Southeast Asia, given the region's significant role in international commerce and the general progress of regional integration. With the growing importance of seaborne trade, the maritime sector has increasingly played a significant role in the development of port cities and coastal communities in the region, contributing to their wellbeing and prosperity. In addition, the fishery is an integral part of the life of people in many countries of Southeast Asia. The most illustrative example is Indonesia where fish accounted for only 5.2% of GDP but provided 72% of animal protein consumption in the country in 2011. Fishing is also the primary source of livelihood for coastal communities that have few other alternatives for employment. Indonesia faces many serious threats to its fishery industry, such as climate change. However, fisheries and commercial ships in this region have always been particularly vulnerable to maritime criminal activities due to the narrow waterways and countless small islands that define the region's geography. Southeast Asian countries have recently increased their defense expenditure and started to modernize their military capabilities, particularly in the air and naval domains, but the regional cooperation to counter maritime attacks has been insufficient. This is due to unresolved conflicts between international and domestic laws concerning any coastal state's obligation and jurisdiction to combat pirates. Consequently, these conditions have encouraged the increase of maritime attacks in Southeast Asia and limited success of the international cooperation on combating maritime attacks in the region, where the related rates of economic growth have been among the highest in the world in the past twenty years.

FUTURE RESEARCH DIRECTIONS

The increasing development of maritime trade in Southeast Asia requires a stable maritime security. Achieving maritime security cooperation in this region requires that the relevant countries work hard to reach consensus, build up mutual confidence, and eliminate those concerns that affect the claim of sovereign right in the sphere of maritime cooperation. In addition, they should adopt effective measures to promote economic and social development, with the aim of eliminating threats to maritime security, such as maritime piracy. Notably, these efforts could affect maritime piracy in different ways depending on the contemporaneous fishing conditions in the region. Indeed, in areas with poor fishing conditions, the patrols could be much more successful at reducing piracy. This is mainly due to lack income opportunities contributed to the larger number of potential pirates in those locations. No direct connection exists between organized crime and maritime piracy. However, carrying out a research on this nexus and examining the dynamics of maritime piracy in the region could both help analyzing the risk of collaboration between pirates and criminals. In conclusion, to eradicate maritime piracy, it would be appropriate to adopt a preventive strategy by strengthening the domestic economic conditions in the countries affected by maritime piracy. Achieving maritime security cooperation would have an advantage of creating an area of strong economic stability, which is also a strategic region for international trade.

CONCLUSION

In September 2014, the United States became the latest nation to join the Singapore-based ReCAAP. When originally agreed in 2004, ReCAAP was the first government-to-government partnership to promote this type of enhanced anti-piracy collaboration. The agreement also established the Information Sharing Center (ISC) for information exchange and incident reporting among contracting parties. The addition of the USA adds important strength to an often under-recognized institution.

Following the 1997 Asian Financial Crisis, a surge in maritime piracy threatened the security of trade routes in Southeastern Asia, which is a problem that persists today. During 2013, 161 attacks were reported around the Strait of Malacca, according to Deutsche Welle (Diogo Mateus, 2014, July 11). While 106 of those attacks occurred in Indonesian waters, and another 9 near Malaysia, neither of those nations has signed the ReCAAP.

In October 2014, a Vietnamese oil tanker, Sunrise 689, disappeared from the radar on its way from Singapore to Vietnam. The ship reappeared days later near the Singapore Straight with a third of its cargo – some 2,000 tons of oil – missing. No crew members were killed, and the identity and location of the attackers are still unknown.

The Strait of Malacca, one of the most vital sea-lanes in the world, carries around 40% of global trade through its waters each year. The USA has a great interest in the area, particularly in its position as a crucial oil transit chokepoint. Enormous resources of the USA will assist ReCAAP in becoming an information hub for combating maritime piracy in the region. The International Maritime Organization sees the ReCAAP as an example of regional cooperation that can be hopefully replicated elsewhere.

ReCAAP reports that there has been 18% increase in a number of incidents of piracy and armed robbery against ships reported in Asia during January-June 2015 compared to the same period in 2014.

The bulk of the increase was due to the increase in the number of petty theft cases onboard ships while underway in the Straits of Malacca and Singapore (SOMS).

During January-June 2015, 106 incidents were reported in Asia, comprising 100 actual incidents and 6 attempted incidents. Among 106 incidents, 11 were piracy incidents occurred in the South China Sea and 95 were incidents of armed robbery against ships. Amongst these, the number of petty theft incidents was highest during January-June 2015. Out of 62 incidents of petty theft, 35 (56%) occurred onboard ships while underway in the eastbound lane of the Traffic Separation Scheme of SOMS and 12 (19%) onboard ships at ports and anchorages in Vietnam. A key area of concern is incidents involving siphoning of ship fuel/oil from product/oil tankers, most of them below 5,000 GT.

Eight incidents of siphoning were reported during January-June 2015 compared to five incidents reported during the same period in 2014. However, the collective efforts by enforcement agencies, relevant international organizations, and shipping industry had produced results, as demonstrated from the arrest of the perpetrators involved in the hijacking of a chemical tanker, Sun Birdie on January 29, 2015, and product tanker, Orkim Harmony on June 11, 2015.

These outcomes were encouraging and demonstrated the effectiveness of the cooperative mechanism in information sharing and operational cooperation among ReCAAP Focal Points, regional maritime authorities from Malaysia, Indonesia, and Australia, INTERPOL, and the ReCAAP ISC. Except for petty theft and siphoning incidents, the other type of incidents reported during January-June 2015 had decreased compared to the same period in the past four years.

The majority of the incidents involved perpetrators operating in groups of one to six men. This accounts for 63% of the incidents reported during January-June 2015. Two incidents involved more than nine men, and both were incidents involving siphoning of ship fuel/oil that occurred in the South China Sea; and they were Dongfang Glory which was boarded by 25 perpetrators on April 1, 2015, and Orkim Harmony boarded by 13 perpetrators on June 11, 2015.

The report for the period of 1 January – 30 June 2015 from the International Chamber of Commerce (ICC) and IMB (ICC, 2015) highlights a continuing trend in South East Asia in the hijacking of small coastal tankers by maritime pirates, averaging one attack every two weeks. According to the report, five small tankers were hijacked in Southeast Asian waters in the second quarter of 2015 alone, bringing the total number of vessels hijacked globally in 2015 to 13.

However, IMB has stressed that enhanced cooperation between regional authorities is paying off and that early detection of approaching skiffs has resulted in attacks being aborted.

IMB highlighted the tracking and arrest off Vietnam of an eight-man Indonesian gang responsible for the hijacking of a Malaysian tanker. IMB also praised the Malaysian authorities for the conviction of nine Indonesian pirates, apprehended after the hijacking of an anchored product tanker off Johor.

Pottengal Mukundan, IMB Director declared: "Information sharing and coordinated action between concerned coastal states are crucial in responding to this threat. We commend the effort that caught one gang and also the hefty custodial sentences imposed on another which will help deter further incidents" (ICC, 2015).

The latest global report reveals that a total of 134 incidents of piracy and armed robbery against ships were reported to the IMB Piracy Reporting Centre (PRC) in the first six months of 2015; an increase in the 116 reports for the corresponding period in 2014.

Pirates managed to board 106 vessels and were responsible for 13 hijackings and 15 attempted attacks worldwide. So far, in 2015, 250 crewmembers have been taken a hostage, 14 assaulted, ten kidnapped, nine injured, and one killed.

According to the latest statistics, while more than a third of incidents reported to the IMB took place off the coast of Indonesia, the majority of these related to low-level, opportunistic thefts from vessels. Littoral states need to do more, including expanding and strengthening existing coordinated counter-piracy patrols in SOMS and exploring the possibilities of enhancing their involvements in the strait by strengthening its respective patrolling/presence. In this regard, the engagements of littoral countries' navies are highly welcome.

Whilst the overall increase in piracy and robbery at sea is noted by both authorities, and as confirmed by the authors' data, it is clear that enhanced cooperation and information sharing is vital to enable timely response to any incident, as demonstrated in the Sun Birdie and Orkim Harmony cases in Asia.

However, it should be emphasized that greater collaboration, between not only regional navies, but also reporting centers, authorities, and coast guard authorities in order to provide comprehensive statistics in support of law enforcement activities.

REFERENCES

Anderson, J. E., & Marcouiller, D. S. J. (2005). Autarky and Anarchy: Endogenous Predation and Trade. *International Economic Review*, *46*(1), 189–213. doi:10.1111/j.0020-6598.2005.00315.x

Bensassi, S., & Martínez-Zarzoso, I. (2012). How Costly is Modern Maritime Piracy for the International Community? *Review of International Economics, 20*(5), 869–883. doi:10.1111/roie.12000

Biloslavo, F., & Quercia, P. (2013). *Il tesoro dei pirati. Sequestri, riscatti, riciclaggio*. Rome: Ministero della Difesa.

Blomberg, S., & Hess, G. (2006). How Much Does Violence Tax Trade? *The Review of Economics and Statistics, 88*(4), 599–612. doi:10.1162/rest.88.4.599

De Sousa, J., Mirza, D., & Verdier, T. (2009). Trade and the Spillovers of Transnational Terrorism. *Swiss Journal of Economics and Statistics, 145*(IV), 453–461.

Diogo Mateus, S. (2014, July 11). 'Worrying' Rise in Piracy Attacks Around Malacca Strait. *Deutsche Welle*. Retrieved from www.dw.com/en/worrying-rise-in-piracy-attacks-around-malacca-strait/a-17780275

International Chamber of Commerce. (2015). *Piracy and Armed Robbery against Ships. Report for the Period of 1 January – 30 June 2015*. Retrieved February 11, 2017, from www.hellenicshippingnews.com/wp-content/uploads/2015/07/2015-Q2-IMB-Piracy-Report-ABRIDGED.pdf

Liss, C. (2007). *The Roots of Piracy in Southeast Asia*. Retrieved February 11, 2017, from http://nautilus.org/apsnet/the-roots-of-piracy-in-southeast-asia/

Martínez-Zarzoso, I., & Bensassi, S. (2013). The Price of Modern Maritime Piracy. *Defence and Peace Economics, 24*(5), 397–418. doi:10.1080/10242694.2012.723156

United Nations Environment Programme. (2006). *Challenges to International Waters – Regional Assessments in a Global Perspective*. Nairobi: UNEP.

ADDITIONAL READING

Anderson, J. E., & Bandiera, O. (2006). Traders, Cops and Robbers. *Journal of International Economics, 70*(1), 197–215. doi:10.1016/j.jinteco.2005.07.009

Anderson, J. E., & Marcouiller, D. S. J. (2002). Insecurity and the Pattern of Trade: An Empirical Investigation. *The Review of Economics and Statistics, 84*(2), 342–352. doi:10.1162/003465302317411587

Anderson, J. E., & Van Wincoop, E. (2004). Trade Costs. *Journal of Economic Literature, 42*(3), 691–751. doi:10.1257/0022051042177649

Bateman, S. (2007). Building Good Order at Sea in Southeast Asia: The Promise of International Regimes. In K. C. Guan & J. K. Skogan (Eds.), *Maritime Security in Southeast Asia* (pp. 97–116). New York, NY: Routledge.

Bradford, J. F. (2005). The Growing Prospects for Maritime Security Cooperation in Southeast Asia. *Naval War College Review, 58*(3), 63–86.

Geise, T. (2007). Maritime Terrorism in Southeast Asia. *Journal of Current Southeast Asian Affairs, 26*(5), 8–43.

Helpman, E., Melitz, M., & Rubinstein, Y. (2008). Estimating Trade Flows: Trading Partners and Trading Volumes. *The Quarterly Journal of Economics*, *123*(2), 441–487. doi:10.1162/qjec.2008.123.2.441

Huang, V. (2009). Building Maritime Security in Southeast Asia: Outsiders not Welcome? *Naval War College Review*, *61*(1), 87–105.

International Chamber of Commerce. (2015). *Maritime Piracy Report Reveals Continued SE Asian Attacks against Small Tankers*. Retrieved February 11, 2017, from https://icc-ccs.org/index.php/news/1111-maritime-piracy-report-reveals-continued-se-asian-attacks-against-small-tankers

Johnson, D., & Valencia, M. (Eds.). (2005). *Piracy in Southeast Asia: Status, Issues, and Responses*. Singapore: Institute of Southeast Asian Studies.

Marcouiller, D. S. J. (2000). *Hijacking, Hold-Up, and International Trade*. Boston, MA: Boston College.

Ong-Webb, G. G. (Ed.). (2006). *Piracy, Maritime Terrorism and Securing the Malacca Straits*. Singapore: Institute of Southeast Asian Studies.

Raymond, C. Z. (2009). Piracy and Armed Robbery in the Malacca Strait: A Problem Solved? *War College Review*, *62*(3), 31–42.

Storey, I. (2009). Maritime Security in Southeast Asia: Two Cheers for Regional Cooperation. *Southeast Asian Affairs*, 36–58.

Teitler, G. (2002). Piracy in Southeast Asia: A Historical Comparison. *Maritime Studies*, *1*, 67–83.

KEY TERMS AND DEFINITION

Fishing Industry: Any industry or activity concerned with taking, culturing, processing, preserving, storing, transporting, marketing or selling fish or fish products. The Food and Agriculture Organization defines it as a sphere which includes recreational, subsistence and commercial fishing, and harvesting, processing, and marketing sectors.

Food Security: A condition related to the supply of food, and individuals' access to it. Concerns over food security have existed throughout history. The Food and Agriculture Organization identified the four pillars of food security such as availability, access, utilization, and stability.

International Trade: An exchange of capital, goods, and services across international borders or territories. It is an exchange of goods and services among nations of the world. In most countries, such trade represents a significant share of gross domestic product (GDP).

International Maritime Bureau: A specialized department of the International Chamber of Commerce. The IMB's responsibilities lie in fighting crimes related to maritime trade and transportation, particularly piracy and commercial fraud, and in protecting the crews of ocean-going vessels.

Maritime Piracy: An act of robbery or criminal violence by ship- or boat-borne attackers upon another ship or a coastal area, typically with the goal of stealing cargo and other valuable items or properties.

Maritime Terrorism: A use or threat of violence against a ship (civilian or military), its passengers or sailors, cargo, a port facility, or if the purpose is solely a platform for political ends.

Regional Cooperation Agreement on Combating Piracy and Armed Robbery against Ships in Asia: A multilateral agreement between 16 countries in Asia, concluded in November 2004. It includes the RECAAP Information Sharing Centre, an initiative for facilitating the dissemination of piracy-related information.

Southeast Asia: The subregion of Asia, consisting of two geographic regions: Mainland Southeast Asia (also known historically as Indochina), which comprises Vietnam, Laos, Cambodia, Thailand, Myanmar, and West Malaysia; Maritime Southeast Asia, which comprises Indonesia, East Malaysia, Singapore, Philippines, East Timor, Brunei, Cocos Islands, and Christmas Island.

United Nations Convention on the Law of the Sea: The international agreement that resulted from the third United Nations Conference on the Law of the Sea, which took place between 1973 and 1982. The Law of the Sea Convention defines the rights and responsibilities of nations with respect to their use of the world's oceans, establishing guidelines for businesses, the environment, and the management of marine natural resources. The Convention concluded in 1982, replaced four 1958 treaties. UNCLOS came into force in 1994, a year after Guyana became the 60th nation to ratify the treaty.

ENDNOTE

1 The definition of the crime of piracy is contained in article 101 of UNCLOS, which reads as follows: "Piracy consists of any of the following acts: (a) any illegal acts of violence or detention, or any act of depredation, committed for private ends by the crew or the passengers of a private ship or a private aircraft, and directed: (i) on the high seas, against another ship or aircraft, or against persons or property on board such ship or aircraft; (ii) against a ship, aircraft, persons or property in a place outside the jurisdiction of any State; (b) any act of voluntary participation in the operation of a ship or of an aircraft with knowledge of facts making it a pirate ship or aircraft; (c) any act of inciting or of intentionally facilitating an act described in subparagraph (a) or (b).

Chapter 7
Value-Added Agriculture for Central Asian Countries

Khabibullo Pirmatov
Slovak University of Agriculture in Nitra, Slovakia

Jana Galova
Slovak University of Agriculture in Nitra, Slovakia

Elena Horska
Slovak University of Agriculture in Nitra, Slovakia

ABSTRACT

The goal of this chapter is to analyze the socio-economic role of value-added agriculture (VAA) for Central Asian (CA) countries. The agricultural sector of the region provides raw materials for the food, textile, and leather industry. Cotton, wheat, rice, and fruit (fresh and dried) play an important role in the foreign trade of each CA country. These countries have unrealized potential for storing, freezing, processing, and packaging of the wide nomenclature of fruit, food production, and drinks with the organization of their further exports to perspective markets. Adding value to agricultural products lead to increasing the share of finished goods in export, supplying import-substituting products, improving infrastructure in rural areas, providing new jobs, and growing farmers' income. Based on the analysis, the authors recommend using value-added agriculture for the CA countries by attracting domestic and foreign investments to rural areas, establishing tax incentives, and allocating preferential credits for agribusiness.

INTRODUCTION

The chapter describes the introduction of Central Asia (CA) region, the theoretical aspects of value-added agriculture (VAA), foreign trade analysis of cotton and wheat, as well as the volume of main fruit production in Central Asian countries (CACs). Assessing the high competitive advantage and great potential of CACs in the agricultural sector, the implementation of VAA in the region is recommended to improve socio-economic conditions in rural areas. The database of the Asian Development Bank Institute

DOI: 10.4018/978-1-5225-2733-6.ch007

(ADBI), the Food and Agriculture Organization of the United Nations (FAO), the International Trade Center (ITC), the United Nations Industrial Development Organization (UNIDO), the United States Department of Agriculture (USDA), World Factbook – Central Intelligence Agency (CIA Factbook), and the World Bank (WB) are represented as the sources of numerical data.

There are several publications on value-added agriculture for developing countries. Prominent studies in the related literature include those from Anderson and Hanselka (2009), Dani (2015), Goletti and Samman (2002), Jaffee and Gordon (1993). The studies were conducted on value-added processes, food supply chain from management and social perspectives, exporting high-value food commodities, as well as post-harvest systems in world agriculture in general. Focus on the value-added agriculture, innovations and integration perspectives on the case of Central Asian countries can be found in scientific works of Bobokulov (2006), Bloch (2002), Pomfret (2010), Rakhimov (2010), Shtaltovna (2016), and Turaeva and Hornidge (2013). Babu and Tashmatov (1999) as well as Lioubimtseva and Henerbry (2009) have mentioned that agricultural challenges lead to food insecurity in the region. The intensification of VAA drives to provide more processed food products within the region that have longer food durability while protecting raw agricultural products from the risk of early deterioration. The goal of the chapter is to examine the potential and development of VAA for CACs and give recommendations for its wide implementation. Firstly, the major terms of the topic are described. Next, the focus is made on Central Asian countries as well as the socio-economic role and prospective development of value-added agriculture in this region. In the conclusion part, recommendations are provided.

BACKGROUND

Value added (value-added, value add, added value) can be defined within marketing and economics studies. From the marketing point of view, value added specifies the increase in the value of a product or/and service while passing through the stages of being developed, processed or produced. Value added is oriented for goods, services or with the combination of goods and services (Figure 1).

All activities during preparing the products/services are divided into value add and non-value add. Both of them require cost and time from producers, but customers pay only for value-added activities. The main non-value activities include overproduction, transportation, over-processing, inventory, defects, motion, and creativity (Ohno, 1988). In order to create value-added action, it is essential to pay attention to three criteria. Firstly, customers intend to pay for the activity. Secondly, it should be done right for the first time. Thirdly, the action should change the product/service to some extent. The studies of the US enterprises revealed that up to 95% of the activities related to a given process do not add value to

Figure 1. Value added divisions in the categories of production and offer
Source: Authors' own development

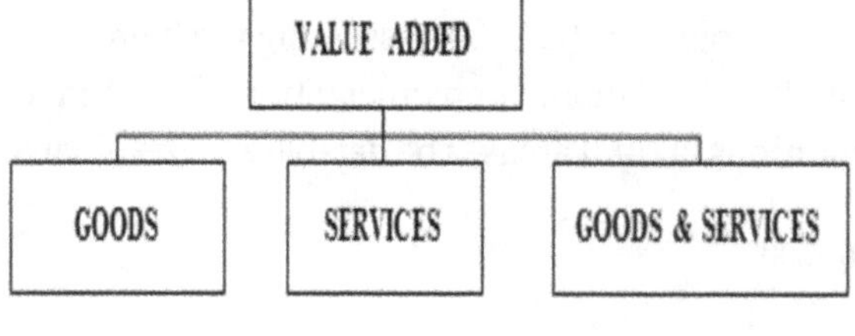

the product or service provided by the organization, as viewed by their customers (Termini, 1996). The value-added process is the difference between input costs (land, labor, and capital) and output prices (goods and services). Therefore, producers try to minimize non-value added activities by operating resources effectively and efficiently to satisfy customer demands. The value of a product or service can also be added with the combination of other products.

In the digital era, when consumers can have access to any product they want and have it delivered in record time, companies are struggling for obtaining competitive advantages. Companies are constantly challenged to find a way to add value to justify their pricing to a more discerning market. Companies are learning that consumers are less concentrated on the product, and more focused on what the product do for them. Finding out what the customer truly values is critical to how the company produces, packages, markets, and delivers its products (Investopedia, 2016).

In economics, according to the System of National Accounts determination (European Commission, International Monetary Fund, Organization for Economic Co-operation and Development, United Nations, & World Bank, 2009), gross value added (GVA) is the difference between output and intermediate consumption. Value added represents the contribution of labor and capital to the production process. GVA is important, as it has a close relation with the Gross Domestic Product (GDP). It is used in the calculation of GDP, which is a key indicator of the state of an entire economy of a state. GDP is the sum of gross value added of all resident producer units plus that part (possible the total) of taxes on products, fewer subsidies on products, that is not included in the valuation of output (European Commission, International Monetary Fund, Organization for Economic Co-operation and Development, United Nations, & World Bank, 2009).

VAA creates more advantage for both rural and country development as a whole. It increases the share of finished goods in the country's export, as well as supplies import-substituting products. Adding value to agricultural products leads to providing new jobs, improving the infrastructure, and increasing of farmers' income. VAA is to transform the primary agricultural commodities from their original form to a more valuable one. It is not only increasing values of the agricultural products, but also reducing deterioration risks, diversifying agricultural commodities, and providing food security.

Value is usually created by focusing on the benefits associated with the agribusiness product or service that arise from quality, functionality, form, place, time, and the ease of possession. A product should have one or more of these qualities to generate additional value (Anderson & Hanselka, 2009). Most of the consumers will pay more for those products, which have been already processed and therefore ready for consumption than buying raw agricultural products and then spending time on cooking.

By passing different processes and activities (mainly labor), primary agricultural commodities become converted to semi-products or finished products. Each step of the process should add value to a product. The final step focuses on the product's uniqueness (Table 1).

Table 1. Value added agriculture for cotton clothes and bakery products

Primary products	Semi-product	Finished goods
Cotton	Cotton lint	Clothes
Wheat	Flour	Bakery products

Source: Authors' own development

The primary value added starts with postharvest operations. The postharvest operations can include cleaning, grading, sorting, storing, etc. After the postharvest operation, there are two options: either sell/consume the products as semi-products/fresh food or use them as ingredients for processed food products. Each kind of activity generates an additional income.

Cotton is not only fiber but also a food crop. Its value chain covers a wide range of sectors such as textile production, food, and pharmaceutical industries, and chemical and construction manufacture. There is a positive correlation between developing VAA and small and medium enterprises in the rural area. At the same time, it gives an impulse to the creation of new jobs (Table 2).

Table 2. Cotton properties and products

Fact Sheet: Properties and Cotton Products
General information
• Cotton is not only fiber but also a food crop. • Bolls are the fruit of the cotton plant. • After picking the cotton crop, the cotton fiber is separated from the seed (ginning). • Approximately 85% of the total income from a cotton crop is cotton lint, which accounts for around 42% of the picked cotton by weight. The rest of the income (about 15%) is from cotton seed. • Almost all parts of the cotton plant including the lint, cottonseed, linters, stalks, and seed hulls are used in various industries.
Natural cotton fiber properties
• Cotton is a perfect fiber for clothing and undergarments because of its softness, absorbedness, and breathable nature. • Cotton is a good heat conductor, it keeps cold in summer and warmth in winter. • Cotton fiber is a natural product considered as non-allergenic. • Compared with oil-based synthetic fabrics, cotton is more comfortable and breathable as it has a unique fiber structure. • Due to its natural whiteness and high rate of absorbency, cotton is one of the easiest fabrics to dye. • Australian cotton, in particular, is perfectly suited to color application processes as it offers spinners crisp white lint, low breakages and stoppages, good throughput efficiency and uniformity of yarn. • When cotton is wet, it becomes stronger. It holds up to 27 times its own weight in water. • Eliminating static cling, cotton cannot hold an electric charge.
Products made from cotton
• 215 pairs of denim jeans, 250 single bed sheets, 750 shirts, 1,200 T-shirts, 3,000 nappies, 4,300 pairs of socks, 680,000 cotton balls, or 2,100 pairs of boxer shorts can be produced from one 227 kg bale of cotton lint. • As cotton is strong and absorbent, it is a perfect fabric for a production of many household goods, such as sheets and towels, tarpaulins and tents, pharmaceuticals and medical supplies. Even astronauts' in-flight space suits are made from cotton. • Only about 40% of the global cotton harvest is used for home furnishings and industrial products, as for the highest share (60%) is used to make apparel. • Firstly, cotton lint is spun, then it is woven or knitted into fabrics such as velvet, corduroy, chambray, velour, jersey, and flannel. • Other natural fibers like wool and synthetic fibers like polyester can be mixed with cotton and thus create dozens of a different kind of fabric. • Tents, car tire cord, fishnets, and book binding also contain cotton fiber.
Products made from cotton linters
• Cotton linters are curly fibers typically less than 3 millimeters long, which remain on the cottonseed after ginning. • Linters are also used in the manufacturing of paper and as a raw material in the manufacturing of cellulose plastics. • Medical supplies such as bandages, cotton buds, and cotton balls are generally made from cotton linters.
Products made from cottonseed
• Cottonseed weighs approximately a half of the picked cotton and is mostly used to make oil. • Around 200 kg of oil, 500 kg of cottonseed meal, and 300 kg of hulls can be produced from one ton of cotton seed. • Cottonseed has a high content of anti-oxidants, which ensures its longer durability. Cottonseed oil is cholesterol free and high in polyunsaturated fats. It can be used for deep frying, and for some margarine and salad dressings. • Cottonseed oil is also an essential substance to make products such as soap, emulsifiers, cosmetics, pharmaceuticals, rubber, paint, waterproofing agents, candles, etc. • Cottonseed meal and hulls are also valuable feed sources for livestock. • World cottonseed production can potentially meet protein requirements for half a billion people and many billions of animals as well.

Source: Cotton Australia, 2017

The common practice of an agriculture supply chain in developing countries shows that farmers harvest crops and sell them to collectors. In turn, they add hired labor, taxes, financial and fixed costs to the farmers' price and resell to processors for preparing finished products. The next step is to market ready-to-use products to retailers, who also add their value and finally present a product to the consumers (Figure 2).

As a result, farmers gain minimal benefits from these processes. Moreover, consumers also pay for products more than they should. It is recommended to shorten these procedures by involving farmers in agribusiness, i.e. small and medium enterprises (SMEs) which should be established close to farms so that farmers have an opportunity to market their products directly to consumers. It creates jobs, extra income for farmers, and suitable economic infrastructure in rural areas, as well as provides reasonable prices for consumers.

To improve socio-economic development in rural areas, countries with competitive advantages in agricultural production should pay more attention to VAA and create conditions (organizational, legal, and regulatory) for involving farmers in agribusiness. Farming is very risky, as it depends on weather conditions, land productivity, and market instability.

By 2050, the urbanization rate is expected to reach 85.9% for developed and 64.1% for developing countries (Gilder, 2012, October 27). Generally, consumption is higher in urban areas than in rural ones. Moreover, the total world population will have reached 9.7 billion people, as well as more than 54% of the world population will have lived in Asia by this time (United Nations Department of Economic and Social Affairs, 2015). Fortunately, food sector is able to make innovations in food products, business models, packaging, technology and so forth, which add value to supply chains (Dani, 2015).

The global food industry, with a value of $2.2 trillion annually, is one of the most important industries in the world economy. The agroindustry is expected to be worth $10 trillion by 2028. The most of this growth will come from the developing world as processed agricultural products provide a crucial source of export revenue to developing countries, particularly in the light of the long-term trend towards declining prices for traditional agricultural commodity exports (Goletti & Samman, 2002; Shengming, 2000). When analyzing food consumption and buying behavior in developed and developing countries, it can be seen that people in developed countries buy more processed foods than people in developing countries. There is a positive correlation: the higher the level of development, the more processed food is consumed.

Figure 2. Agriculture supply chain
Source: Authors' own development

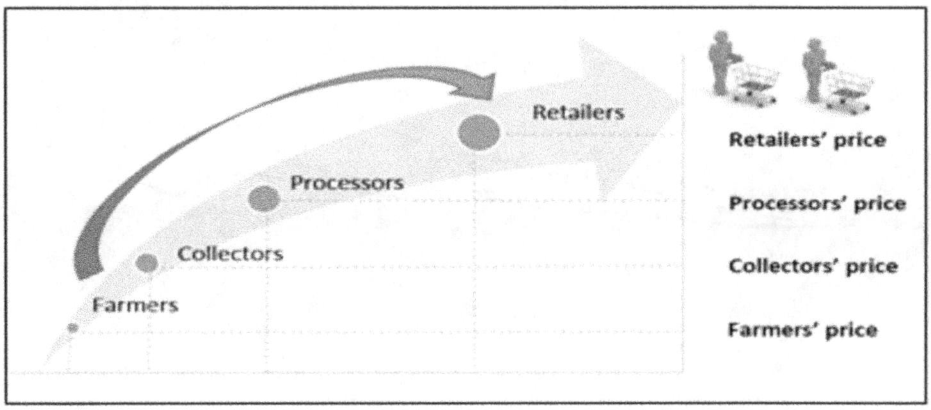

While international trade has been present throughout much of history, its economic, social, and political importance has been on the rise in the recent centuries (Berčík, 2013; Kleinová & Ürgeová, 2011). As stated by the Industrial Development Report 2016 of the United Nations Industrial Development Organization (2016), manufacturing remains the main driving force of economic growth, largely attributable to its higher productivity and scope for innovation as well as the fact that over the past few decades, the majority of global manufacturing has steadily shifted from the West to the East and from the North to the South. The report also informs that since the beginning of the century, rapid growth in the manufacturing value added (MVA) has been a major source of poverty reduction in many developing and emerging industrial economies (DEIEs) through employment and generation of income (Figure 3).

Based on statistic calculation, the United Nations Industrial Development Organization (2015) has forecasted positive scenario in which DEIEs still have considerable capacity for manufacturing growth and technological progress in the coming decades. Currently, such a rapid growth of DEEIs is the expense of certain countries such as China, India, Mexico, Brazil, and Turkey (Figure 4).

China was the second-biggest manufacturing producer in 2014 after the United States, which contributed more than all non-China emerging countries combined. China's impact on the MVA growth rate in DEIEs is very important, as the country's share in DEIEs' MVA increased from 15.8% in 1990 to 51.3% in 2014. China's manufacturing industry has become the largest sector in the economy and accounted for one-third of GDP (United Nations Industrial Development Organization, 2015).

In 2014, the share of value-added agriculture in the global GDP reached 3.1%. In comparison to other sectors of the world economy (industry and services), the share was very low (Rovný, 2016). Analyzing the sectoral composition of world manufacturing value added in 2000, 2005, and 2013, food and beverages have the highest share. It confirms that food market has been and continues to be the most important sector. The share of tobacco products decreased from 1.1% (2000) to 0.7% (2013) in industrialized countries and from 3.5% (2000) to 2.5% (2013) in DEIEs. The share of textiles and wearing apparel with fur together in DEIEs accounted for 7.4%, as for these sectors' share among industrialized countries

Figure 3. World manufacturing value added, by country group and worldwide, 1990-2014
Source: United Nations Industrial Development Organization, 2015

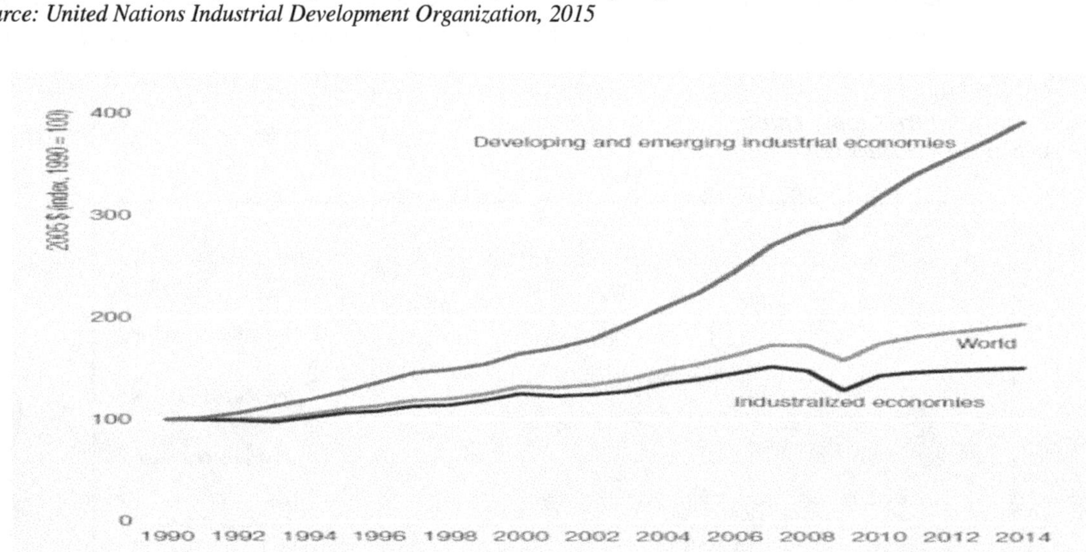

Figure 4. Manufacturing value added share of the five largest countries in developing and emerging industrial economies' group total, 1990, 2000, and 2014
Source: United Nations Industrial Development Organization, 2015

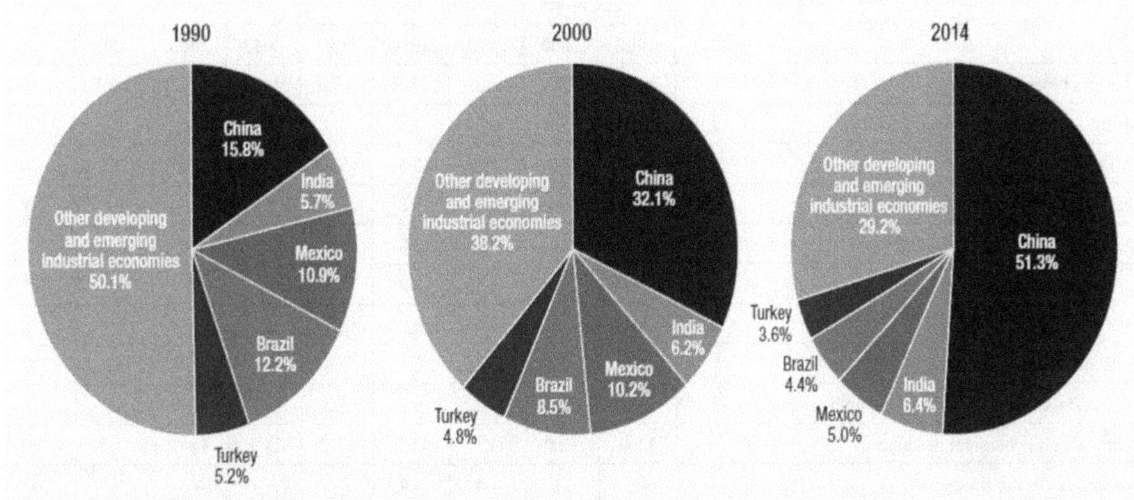

amounted to 1.8% in 2013 (Table 3). This is due to the fact that most of the leading cotton producers are developing countries. Their huge low-cost labor resources make foreign companies invest their capital and enter developing markets with their technologies. In turn, the governments of emerging countries create conditions by presenting tax incentives in order to stimulate investment.

Chemicals and machinery equipment also have substantial shares in the global manufacturing equaling 11.7% and 8.5% in 2013, respectively. In the long term, the machinery equipment market may become saturated, and its share may drop. The same situation is with refined petroleum products, inasmuch as petroleum is a limited natural resource. Nowadays, several countries (China, the USA, Germany, Japan, Italy, etc.) have started to use renewable energy sources (e.g., hydropower, geothermal, biomass, wind, and solar). Conversely, the demand for food and beverage products is growing due to increase in the number of the world population each year as well as these products being the main source for life.

Taking into account global challenges and trends, CA countries should update development strategies putting forward right goals to be winners among emerging and developed economies. Continuing to produce more high value-added agriculture products and encouraging investment in this sector may improve other related branches: people welfare, rural development, infrastructure, etc. For this purpose, all resources and opportunities are present in these countries.

MAIN FOCUS OF THE CHAPTER

In the modern context, Central Asia (CA) includes Kazakhstan, Kyrgyzstan, Tajikistan, Turkmenistan, and Uzbekistan. All five republics declared their independence from the Soviet Union in 1991. Central Asia is a landlocked region bordering the Caspian Sea in the west, China in the east, Russia in the north, and China, Afghanistan, and Iran in the south. Central Asia's landscape can be divided into the

Table 3. Share of manufacturing value added, by industry group within country groups and worldwide, 2000, 2005, and 2013, percentage

International Standard Industrial Classification (ISIC) description	Industrialized countries			DEIEs			World		
	2000	2005	2013	2000	2005	2013	2000	2005	2013
Food and beverages	11.0	11.4	11.2	16.6	14.9	13.3	12.1	12.2	12.0
Tobacco products	1.1	0.9	0.7	3.5	2.9	2.5	1.6	1.4	1.4
Textiles	2.3	1.8	1.1	5.9	5.3	4.5	3.0	2.7	2.5
Wearing apparel, fur	1.7	1.0	0.7	3.8	3.4	2.9	2.1	1.6	1.6
Leather, leather products and footwear	0.7	0.4	0.3	2.0	1.4	1.3	0.9	0.7	0.7
Wood products (excluding furniture)	2.0	2.0	1.5	1.7	1.4	1.4	1.9	1.8	1.5
Paper and paper products	3.0	2.8	2.4	3.0	2.9	2.7	3.0	2.9	2.5
Printing and publishing	4.6	4.2	3.4	2.1	1.8	1.4	4.1	3.6	2.6
Coke, refined petroleum products, nuclear fuel	3.1	3.5	3.2	6.2	5.2	3.7	3.7	3.9	3.4
Chemicals and chemical products	11.0	12.0	12.0	11.2	11.2	11.1	11.0	11.8	11.7
Rubber and plastic products	4.7	4.6	4.4	3.4	3.5	3.3	4.5	4.3	3.9
Non-metallic mineral products	4.0	3.8	3.1	5.5	5.4	5.8	4.3	4.2	4.2
Basic metals	5.0	5.0	4.5	7.8	10.0	11.2	5.5	6.2	7.1
Fabricated metal products	8.0	7.5	7.1	3.9	4.0	4.6	7.2	6.6	6.1
Machinery and equipment (not elsewhere classified)	9.7	9.7	9.2	4.8	5.9	7.4	8.8	8.7	8.5
Office, accounting and computing machinery	1.5	1.4	2.0	1.2	1.5	1.5	1.5	1.5	1.8
Electrical machinery and apparatus	4.0	3.8	3.9	2.8	3.3	4.5	3.8	3.6	4.1
Radio, television and communication equipment	5.2	6.2	9.7	3.7	4.6	5.1	4.9	5.8	7.9
Medical, precision and optical instruments	3.5	3.9	4.7	0.7	0.9	1.1	3.0	3.1	3.3
Motor vehicles, trailers, semi-trailers	7.7	8.3	8.3	6.2	6.4	6.7	7.4	7.8	7.7
Other transport equipment	2.9	3.0	3.8	1.5	1.7	1.9	2.6	2.7	3.1
Furniture, manufacturing (not elsewhere classified)	3.3	3.0	2.7	2.3	2.3	2.1	3.1	2.8	2.5
Total	100.0	100.0	100.0	100.0	100.0	100.0	100.0	100.0	100.0

Source: United Nations Industrial Development Organization, 2015

vast grassy steppes of Kazakhstan in the north and the Aral Sea drainage basin in the south. Along the margins of the Amu Darya and Syr Darya river systems are suitable for agriculture. The rivers flow northwestward through Kyrgyzstan, Tajikistan, and eastern Uzbekistan and Turkmenistan after rising in mountain ranges to the south and east. Those two major rivers drain into the Aral Sea and provide most of the region's water resources, though northern Kazakhstan is drained by rivers flowing north to Russia. On the east and south, Central Asia is bounded by the western Altai and other high mountain ranges extending into Iran, Afghanistan, and western China.

The region as a whole experiences hot summers and cool winters, with plenty of sunshine and very little precipitation (Encyclopedia Britannica Online, 2016). Therefore, the region's agriculture mostly depends on irrigation. The Intergovernmental Panel on Climate Change (IPCC) in its Fifth Assessment Report stated that Central Asia is expected to become warmer in the coming decades and increasingly arid, especially in the western parts of Turkmenistan, Uzbekistan, and Kazakhstan (Hijioka et. al., 2014; Lioubimtseva & Henerbry, 2009). However, Central Asia's glaciers are already melting, which decreases long-term water availability in Uzbekistan and Turkmenistan, and causes changes to land cover, cropping systems and livestock grazing distribution (Sedik & Lerman, in press; United Nations Office for Disaster Risk Reduction & World Bank, 2009; World Bank, 2009).

The annual cost of land degradation due to land use change is about $6 billion, most of which is due to rangeland degradation ($4.6 billion), followed by desertification ($0.8 billion), deforestation ($0.3 billion), and abandonment of croplands ($0.1 billion). The costs of action against land degradation are lower than the costs of inaction in Central Asia by more than 5 times over a 30-year horizon, meaning that each dollar spent on addressing land degradation is likely to have about $5 of return. This is a very strong economic justification favoring action vs. inaction against land degradation. Thus, the costs of action were found to equal about $53 billion over a 30-year horizon, whereas if nothing is done, the resulting losses may equal almost $288 billion during the same period (Mirzabaev et al., 2016). In the future, land degradation can lead to low agricultural productivity and food deficiency in Central Asia. Implementation of sustainable agricultural intensification, process, and marketing innovations, value-added approaches at different stages of the food chain, provision of sufficient, accessible, and nutritious food can all be solutions to food insecurity issue in the region while enabling economic and social development in rural areas. Such soil protective technology as conservation agriculture, which becomes a popular farming practice in South America, North America, and Australia, also has a potential to secure sustained productivity in Central Asia.

This region is abundant with human and natural resources, which give the advantage of cooperation with other economic centers. In the report of the World Bank (2016), there are totally 68.5 million people and the annual growth accounted for 1.7% in 2015 (Table 4). By country area, Kazakhstan is on the 9th position with 2,724,900 sq. km. in the world. With 30 billion in barrels, Kazakhstan takes the 12th position in the world by the proved reserves of crude oil. There are 17.5 tons cubic meters of proved reserves of natural gas in Turkmenistan (4th place in the world), Kazakhstan with 2.4 tons cubic meters is on the 15th place, and Uzbekistan is in the 20th place with 1.8 tons cubic meters (World Factbook, 2016).

Central Asian countries are considered as multinational and multi-confessional ones, as there are many ethnic groups, tribes, languages, and religions in the region. Islam is the dominant religion, with the most adherents belonging to the Sunni branch. The five largest ethnic groups in Central Asia are, in descending order, the Uzbeks, the Kazakhs, the Tajiks, the Turkmens, and the Kyrgyz. All these groups have the common background in terms of tradition, culture, and history.

Table 4. Population and area of CACs in 2015

Central Asian Countries	Population, people	Territory, square kilometers
Kazakhstan	17,544,126	2,724,900
Kyrgyzstan	5,957,000	199,951
Tajikistan	8,481,855	144,100
Turkmenistan	5,373,502	488,100
Uzbekistan	31,299,500	447,400
Total	68,655,983	4,004,451

Source: Authors' development based on World Bank, 2016; World Factbook, 2016

After the collapse of the Soviet Union, the gross domestic product (GDP) of the CA countries was negative results until the late 1990s. The main reason of the economic downturn was the interrelation between the former republics of the Soviet Union. CA rural areas recovered the economic shock of the transition by their agricultural production capacity. Annual GDP rose significantly in all CA countries in the period of 1997-2014 (Figure 5).

All five countries of the region are transition economies changing from centrally planned economies to the market (Feige, 1994). The transition process is usually characterized by changes and development of institutions, particularly private enterprises; changes in the role of a state, thereby, the creation of fundamentally different governmental institutions and promotion of private-owned enterprises, markets, and independent financial institutions (Falke, 2002).

Uzbekistan has been following its own path of development ("the Uzbek Model") from the beginning of its independence. It has transitioned to a socially oriented market economy based on the five key principles: priority of economics over politics; state as the main reformer; superiority of law in all aspects of life and business; strong social policy; step-by-step transition to market relations (Karimov, 1993). Due to the successful implementation of this model, Uzbekistan has achieved macroeconomic stability, equilibrium of domestic and foreign sectors of the economy, growth of currency reserves, as well as maintenance of steady high rates of annual economic growth (8% at average) (The Korean Post, 2013). Furthermore, these main five principles mitigated a negative influence of the global financial-economic crisis on the economy of Uzbekistan in 2008-2009.

Kazakhstan presented its development strategy ("Kazakhstan-2030") in October 1997, where seven long-term priorities were outlined: national security; domestic political stability and consolidation of the society; economic growth based on the developed market economy with a high level of foreign investments; health, education and welfare of the citizens of Kazakhstan; power engineering resources; infrastructure, more particularly transport and communication; professional state (Nazarbayev, 1997). Based on the Global Competitiveness Report 2014-2015 (Schwab, 2015), Kazakhstan has achieved to

Figure 5. GDP growth in CACs (annual percentage)
Source: Authors' development based on World Bank, 2016

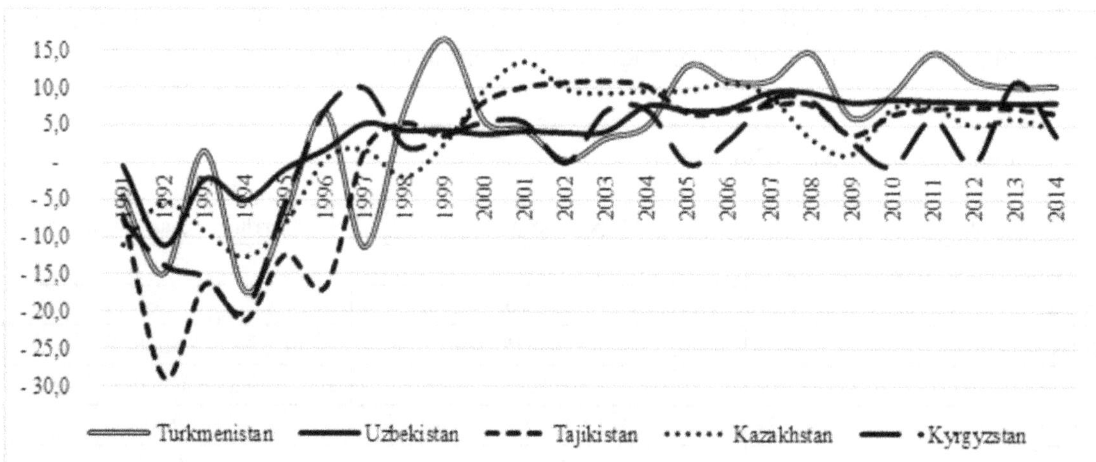

be among the 50 most competitive economies in the world. The strategy "Kazakhstan-2050" has been introduced as the continuation of the previous one and put forward the goal to join the top 30 world economies by 2050. In order to reach this goal, "100 specific steps" have been developed.

Kazakhstan, Turkmenistan, and Uzbekistan got the advantage by trading and attracting investment by virtue of the high prices for oil, gas, and cotton on the global market. There are prospects that China's increasing economic ties with the Commonwealth of Independent States (CIS) and the European Union (EU) along with the development of transit routes through the region of Central Asia will improve linkages with outside markets. In turn, China with its world-biggest population can be the oriented-market for the export of the high value-added food products from the CA countries. Furthermore, the CA region borders with Russia, which offers the high possibility to increase the demand for food due to the sanction imposed by the USA and the EU and Russia's countersanctions, including an embargo on food import from a number of developed countries of the world.

The economic performance of Kyrgyzstan is mainly due to precious metal (gold), re-exports and tourism services (Asian Development Bank Institute, 2016). Agriculture sector plays an important role in all CA economies, including Tajikistan. The most common characteristic of the CA countries is a low level of product diversification (Erokhin, 2016). Therefore, there is a need to pay attention to value-added agriculture in the CA countries and diversify finished products from raw agriculture commodities. For the implementation of high value-added products, these countries have sufficient natural and human resources.

SOLUTIONS AND RECOMMENDATIONS

The agricultural sector of CACs provides the raw materials for the food, textile, and leather industries. CA countries' average agriculture value added accounted for 16.5% in 2014. During the 1992-2014 period, the agriculture sector decreased in Kazakhstan, Uzbekistan, and Kyrgyzstan. As for Tajikistan, this sector was quite stable ranging between 27.1% and 27.5%. There was a fluctuation rising from 10.6% (1992) to 20.3% (2003) and falling down to 14.5% in 2014. The industry sector shrank in the CACs, except Turkmenistan, as this country invested in fuel and mining industries with those sectors rapidly growing from 11.9% in 1992 to 48.4% in 2014 (Figure 6).

In spite of declining, which was the overall tendency for the agricultural sector in the CACs, the volume of agricultural production increases in Kyrgyzstan, Tajikistan, and Uzbekistan. In Turkmenistan, food and livestock production indexes rose, but crop production descended from 84.2 in 2003 to 76.7 in 2013. Crop, food, and livestock production indexes in Kazakhstan in 2003 went down and then went up again in 2013 (Table 5).

There is a gradual growth in the number of rural population in the CACs, except Turkmenistan. The main kind of employment of rural population in the active working age is agriculture. The average rural population in the CACs was 58.2% in 1991 and there was a slight increase up to 59.8% in 2014 (Figure 7). In order to improve the socio-economic conditions of the people in rural areas, VAA should be implemented widely, which creates more advantages for both people living in rural areas and the country as a whole.

Uzbekistan is the sixth largest cotton producer and the fifth largest cotton exporter. Turkmenistan is also among the top ten countries for cotton production and export in the world. Uzbekistan, Turkmenistan, and Tajikistan are among the top 20 countries in the production of cotton in the world (National

Figure 6. GDP contribution by sectors (percentage)
Source: Authors' development based on World Bank, 2016

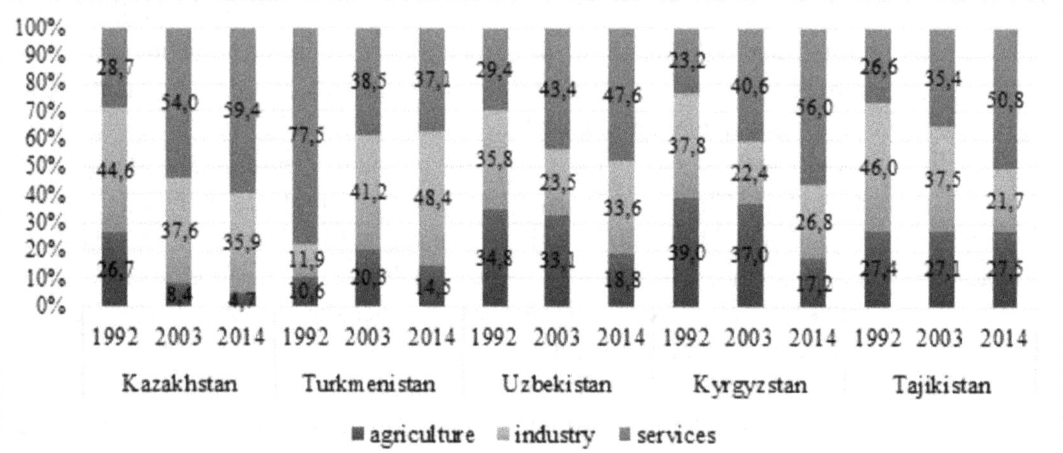

Table 5. Agricultural production of CA countries

Central Asian countries	Crop production index (2004-2006=100)			Food production index (2004-2006=100)			Livestock production index (2004-2006=100)		
	1993	2003	2013	1993	2003	2013	1993	2003	2013
Kazakhstan	101.8	98.1	145.9	127.5	93.8	125.4	154.7	90.6	111.0
Kyrgyzstan	54.9	96.1	111.9	75.1	100.9	112.2	106.0	101.4	112.6
Tajikistan	79.8	89.3	151.1	81.0	88.7	155.7	86.6	86.6	163.9
Turkmenistan	74.6	84.2	76.7	47.6	89.8	124.4	45.9	92.5	136.3
Uzbekistan	80.7	83.2	144.4	71.8	88.2	163.3	80.6	88.9	165.8

Source: Authors' development based on The Global Economy, 2016; Food and Agriculture Organization of the United Nations [FAO], 2016

Cotton Council of America, 2016). By analyzing the foreign trade of each CAC, it is seen that cotton, wheat, rice, and fruits (fresh and dried) play important role in agricultural trade. For instance, the cotton export of four CACs has high export share compared with other agricultural commodities. Based on the International Trade Center (2016), 15% of Uzbekistan's export in 2015 was cotton, which accounted for $847.018 million. In the same period, the value of cotton export and its share in the country's export was $486.683 million (5.25%) in Turkmenistan, $96.774 million (12.83%) in Tajikistan, $55.425 million (0.13%) in Kazakhstan, and $21.173 (2.45%) in Kyrgyzstan. In spite of the positive balance of trade in cotton, the import of articles and apparel and clothing accessories is by 78% more than export. Instead of exporting such primary agricultural commodities as cotton, there is need to export products desired by customers and acquire more benefits in term of their cost (Table 6).

Kazakhstan is among the top ten largest wheat and flour producers in the world. Wheat and flour are marketed domestically and also exported to Uzbekistan, Tajikistan, and Kyrgyzstan. About 80% of Kazakhstan's wheat is produced in the three northern provinces of Akmola, Kostanay, and North Kazakh-

Figure 7. Rural and urban population in the CACs
Source: Authors' development based on World Bank, 2016

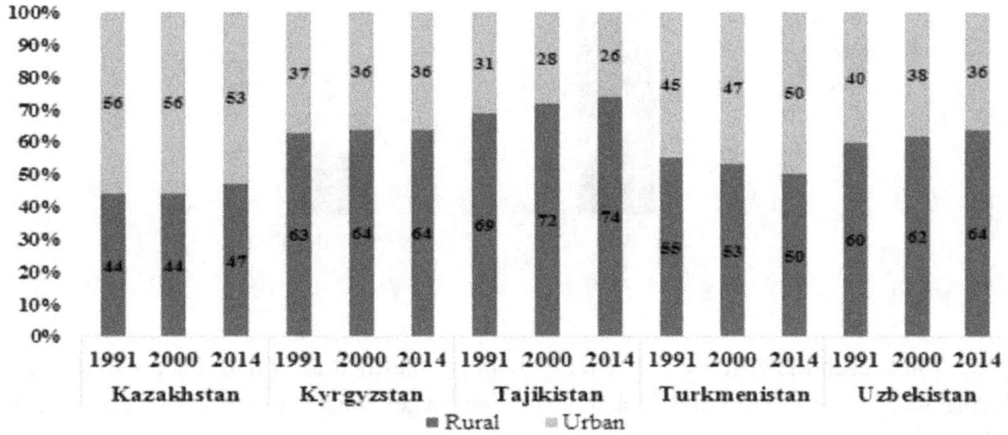

Table 6. Foreign trade of cotton and import of articles of apparel and clothing accessories in the CACs in 2015 ($ million)

Central Asian countries	Export of cotton	Import of cotton	Export of articles of apparel and clothing accessories	Import of articles of apparel and clothing accessories
Kazakhstan	55.425	12.246	22.150	564.164
Kyrgyzstan	21.173	321.471	40.529	131.847
Tajikistan	96.774	24.496	12.537	345.446
Turkmenistan	486.683	0.276	23.392	46.317
Uzbekistan	847.018	2.307	158.658	75.635
Total	1,507.073	360.796	257.266	1,163.409

Source: Authors' development based on International Trade Center, 2016

stan bordering with Russia. From these regions, the country's grain exports origin (Petrick, Oshakbaev, & Wandel, 2014). As a result of an active grain production policy in the end of 1990s, the area sown to cereals and grain legumes reached 22.7 million hectares including 13.5 million hectares of wheat. This represents 2.9% and 13.5% of world area under these crops, respectively (Morgounov, 1998).

The total value of wheat export accounts for $1,244.415 million only by Kazakhstan among the CACs in 2015. Wheat is the most imported agricultural commodity for Uzbekistan, Tajikistan, and Kyrgyzstan. The total value of wheat import equals $434.104 million. The shares of Kyrgyzstan, Tajikistan, and Uzbekistan are 11%, 42%, and 47%, respectively (Figure 8).

Annual growth of the fruit production excluding grapes is 6% in Central Asia, with Uzbekistan racing 7.2% and Turkmenistan at 6.1%. Uzbekistan's output represents 70% of Central Asia's entire fruit (excluding grapes) production (FAO, 2014). However, according to the estimates, about 25-30% (more than 1 million tons annually) of the total annual grown fruits and vegetables in Uzbekistan is lost to

Figure 8. Total value of wheat export and import of the CACs in 2015 ($ million)
Source: Author's development based on the International Trade Center, 2016

waste and does not reach the customer. The long chain of intermediaries and dealers to the end user, reducing profit margins of farmers and dehkan farms, and in general, their cost-effectiveness (Malikov, Qineti, & Pulatov, 2016).

In the case of pome fruits (not elsewhere specified) the other pome fruit is not separately identified; in some countries apples, pears and quinces are reported under this general category. Furthermore, cereals (not elsewhere specified) include the other cereal crops that are not identified separately because of their minor relevance at the international level; because of their limited local importance, some countries report cereals under this commodity heading that are classified individually by the FAO. The category stone fruits (not elsewhere specified) covers also other stone fruit not separately identified; in some countries, apricots, cherries, peaches, nectarines and plums are reported under this general category.

Uzbekistan was leading in the world in the production of pome fruits (apples, pears, and quince) in 2013. The total volume was 120,000 tons. It also takes the second place with producing 1,641,882 tons of carrots and turnips. Kazakhstan is ranked on the first place in the global production of safflower seed (Table 7). There are future prospects for decreasing more the share of raw cotton, fruit, and vegetables in the export by converting them to semi (cotton lint, fabric, etc.) or finished products (clothes), processed food (jam, juice, etc.).

FUTURE RESEARCH DIRECTIONS

In the future, the demand for petroleum can be declining due to two factors: oil resource limitation and massive use of alternative energy sources. Conversely, demand for food and beverage products is growing due to the increasing number of the world population.

Agriculture sector plays an important role in all CA economies, however, production of processed food with high value added is sufficient for domestic, but not for potential foreign markets. In order to export more high VAA products and to be winners in the foreign trade, these countries have enough natural and human resources. China increases economic cooperation with the CIS and the EU countries and develops transit routes through the CA region. Countries located in this region should take advantage of China's huge population and focus on this market in terms of export of processed food products. Due

Table 7. Kazakhstan and Uzbekistan by the following food and agricultural commodities in the top 10 world rankings in 2013

World Ranking	Country	Commodity	Quantity (tons)	Processed products
1	Uzbekistan	Fruit, pome (not elsewhere specified)	120,000	Jam, juice, alcohol, fruit chips, baby food, vinegar, fruit tea, compote, dried fruits, canned fruits
1	Kazakhstan	Safflower seed	174,900	Oil, coloring and flavoring foods, margarine, medicines, cosmetics, dye, bird food
2	Uzbekistan	Carrots and turnips	1,641,882	Juice, baby food, compote, turnip chips, cosmetics, carrot roundels, pickled turnips
3	Uzbekistan	Apricots	430,000	Jam, juice, alcohol, baby food, compote, dried apricots, canned apricots, apricot pits and seeds, apricot kernel oil, apricot seed extracts, cosmetics, medicines
4	Kazakhstan	Cereals (not elsewhere specified)	276,840	Muesli, sweets, flours, alcohol, porridge, powder, baby food, bread
4	Uzbekistan	Fruit, stone (not elsewhere specified)	40,000	Jam, compote, juice, fruit tea, alcohol, dried fruits, canned fruits
4	Uzbekistan	Jute	21,000	Bags, shoes, fiber
5	Uzbekistan	Cherries	100,000	Jam, juice, compote, canned cherries
6	Kazakhstan	Buckwheat	83,500	Flour, pasta, cake, porridge, bread, bird food
7	Uzbekistan	Persimmons	50,000	Baby food, sauce, dried persimmons
7	Uzbekistan	Watermelon	1,558,301	Juice, ice-cream
8	Kazakhstan	Melon, other (including cantaloupes)	774,190	Jam, dried melon
8	Uzbekistan	Garlic	203,585	Garlic oil, spice, powder, flakes, pickled, vinegar, jelly, sauce, insecticide
9	Uzbekistan	Cherries, sour	40,000	Jam, juice, dried cherries, compote
10	Uzbekistan	Cucumber and gherkins	607,397	Pickled, cosmetics

Source: Authors' development based on FAO, 2016

to the sanctions imposed by the USA and the EU countries on Russia and Russia's countersanctions in the form of food import embargo, Russia's import of food and agricultural products from the CA countries is increasing. All these contemporary trends and challenges have to be studied in order to develop solutions for ensuring food security and to increase the share of VAA in agricultural production.

Moreover, land degradation in the region should not be eliminated from the top of the agenda, as this factor can decrease a share of arable land and become the major threat to food security. Therefore, there is a need to a wider use of sustainable agricultural practices, which maximize production in an environmentally friendly manner without increasing production costs. Conservation agriculture is a technology able to reduce land degradation and increase food security in a more sustainable way in the CA countries.

CONCLUSION

In the context of globalization, VAA plays an important socio-economic role in the economy of the CA countries. It positively influences such spheres as rural development, employment, and foreign trade. CA countries have a capacity for storing, freezing, processing, and packaging of a wide nomenclature of fruit, food products, and drinks, as well as for their further exports to the perspective markets.

By supporting VAA, new food processing companies create job opportunities in rural areas while reducing unemployment in the CA countries. Local and foreign investments need to be attracted not only to urban centers but also to rural areas. As a consequence, incomes of people grow, while the rural infrastructure improves. In addition, the benefits from the export of VAA products are bigger than of raw agricultural commodities. The increase of the share of finished goods in export and foreign capital flows in the economy likely bring the positive balance in foreign trade.

There is a probability in the CA countries of facing land degradation, water scarcity or natural disasters such as earthquakes, floods, or seasonal extreme weather, which negatively impact agricultural productivity and food security of the CA countries. VAA provide more processed food products within the region that have longer food durability compared to raw agricultural products. In the case of difficult periods caused by natural phenomenon, food products with higher value added are very important to moderate negative influences.

Furthermore, VAA requires updating farmers' skills and training in order to get higher yields and run agribusiness in a more efficient manner. Currently, there is a lack of cooperation between farmers and research institutions or universities specialized on agriculture. Farmers face agricultural challenges, which can be solved only through coordinated and collaborative approaches. Due to the high intellectual potential of the CA scholars, many researches on agricultural issues are carried out in life science institutes, but target groups such as agricultural consultants in rural areas are insufficiently informed about the results of such researches. With the assistance of information and communications technologies, close cooperation should be established between agricultural consultants, life science universities, and research institutes.

To sum up, the governments of the CA countries should take measure to attract domestic and foreign investments to rural areas, establish tax incentives and allocate preferential credits for agribusiness. With the extensive practice of VAA, existing agricultural challenges in the region can have positive solutions, the exports of value added agriculture products are increasing while foreign exchange earnings are flowing, as well as it can be meeting the domestic demands providing food security within the region.

REFERENCES

Anderson, D. P., & Hanselka, D. (2009). *Adding Value to Agricultural Products.* Texas AgriLife Extension Service Publication No. E-573. Retrieved February 7, 2017, from https://oaktrust.library.tamu.edu/bitstream/handle/1969.1/86940/pdf_1302.pdf?sequence=1&isAllowed=y

Asian Development Bank Institute. (2016). *Connecting Central Asia with Economic Centers.* Tokyo: Brookings Institution Press.

Babu, S. C., & Tashmatov, A. (1999). Attaining Food Security in Central Asia – Emerging Issues and Challenges for Policy Research. *Food Policy*, *24*(4), 357–362. doi:10.1016/S0306-9192(99)00052-4

Berčík, J. (2013). The Modern Forms of Marketing Research in Selected Agri-Food Businesses. In *Proceedings of the International Scientific Conference "Business Management – Practice and Theory in the 21st Century"*. Nitra: Slovak University of Agriculture.

Bloch, P. C. (2002). *Agrarian Reform in Uzbekistan and Other Central Asian Countries*. Madison, WI: University of Wisconsin-Madison, Land Tenure Center.

Bobokulov, I. (2006). Central Asia: Is There an Alternative to Regional Integration? *Central Asian Survey, 25*(1–2), 75–91. doi:10.1080/02634930600903056

Cotton Australia. (2017). *Sydney, Australia: Properties and Cotton Products.* Retrieved February 7, 2017, from http://cottonaustralia.com.au/cotton-library/fact-sheets/cotton-fact-file-cotton-properties-and-products

Dani, S. (2015). *Food Supply Chain Management and Logistics: From Farm to Fork.* London: Kogan Page Publishers.

Encyclopedia Britannica Online. (2016). *Central Asia.* Retrieved February 7, 2017, from https://www.britannica.com/place/Central-Asia

Erokhin, V. (2016). Rising Economic Powers among Economies in Transition: Is There Enough Space on the Global Market? In V. Erokhin (Ed.), *Global Perspectives on Trade Integration and Economies in Transition* (pp. 1–25). Hershey, PA: IGI Global. doi:10.4018/978-1-5225-0451-1.ch001

European Commission. International Monetary Fund, Organization for Economic Co-operation and Development, United Nations, & World Bank. (2009). System of National Accounts 2008. New York, NY: The United Nations.

Falke, M. (2002). *Community Interests: An Insolvency Objective in Transition Economies?* F.I.T. Discussion Paper No. 01/02. Frankfurt: Frankfurt Institute for Transformation Studies, European University Viadrina.

Feige, E. L. (1994). The Transition to a Market Economy in Russia: Property Rights, Mass Privatization and Stabilization. In G. S. Alexander & G. Skąpska (Eds.), *A Fourth Way?: Privatization, Property, and the Emergence of New Market Economics* (pp. 57–78). London: Routledge.

Food and Agriculture Organization of the United Nations. (2014). *Europe and Central Asia Food and Agriculture*. Budapest: Food and Agriculture Organization of the United Nations, Regional Office for Europe and Central Asia.

Food and Agriculture Organization of the United Nations. (2016). *FAOSTAT.* Retrieved February 7, 2017, from http://www.faostat.org

Gilder, G. (2012, October 27). Urban Life: Open-Air Computers. *The Economist.* Retrieved March 5, 2017, from http://www.economist.com/news/special-report/21564998-cities-are-turning-vast-data-factories-open-air-computers

Goletti, F., & Samman, E. (2002). Post-Harvest Systems in World Agriculture. In P. Golob, G. Farrell, & J.E. Orchard (Eds.), Crop Post-Harvest: Science and Technology: Vol. 1. Principles and Practice (pp. 1–34). Oxford, UK: Blackwell Science. doi:10.1002/9780470751015.ch1

Hijioka, Y., Lin, E., Pereira, J. J., Corlett, R. T., Cui, X., Insarov, G. E., … Surjan, A. (2014). Asia. In V.R. Barros, C.B. Field, D.J. Dokken, M.D. Mastrandrea, K.J. Mach, T.E. Bilir, … L.L. White (Eds.), Climate Change 2014: Impacts, Adaptation, and Vulnerability. Part B: Regional Aspects. Contribution of Working Group II to the Fifth Assessment Report of the Intergovernmental Panel on Climate Change (pp. 1327–1370). Cambridge, UK: Cambridge University Press.

International Trade Center. (2016). *Trade Map.* Retrieved February 7, 2017, from http://www.trademap.org

Investopedia. (2016). *Value Added Definition.* Retrieved February 6, 2017, from http://www.investopedia.com/terms/v/valueadded.asp

Jaffee, S., & Gordon, P. (1993). *Exporting High-Value Food Commodities: Success Stories from Developing Countries*. Washington, DC: World Bank Publications.

Karimov, I. A. (1993). *Building the Future: Uzbekistan-Its Own Model for Transition to a Market Economy*. Tashkent: Uzbekistan Publishers.

Kleinová, K., & Ürgeová, J. (2011). Country Image and Branding of Slovakia. *Polityki europejskie, 54*(5), 50–58.

Lioubimtseva, E., & Henerbry, G. (2009). Climate and Environmental Change in Arid Central Asia: Impacts, Vulnerability, and Adoptions. *Journal of Arid Environments, 73*(11), 963–977. doi:10.1016/j.jaridenv.2009.04.022

Malikov, N., Qineti, A., & Pulatov, A. (2016). *Agriculture and Economic Development in Uzbekistan.* Paper presented at the International Scientific Days, Nitra.

Mirzabaev, A., Goedecke, J., Dubovyk, O., Djanibekov, U., Le, Q. B., & Aw-Hassan, A. (2016). Economics of Land Degradation in Central Asia. In E. Nkonya, A. Mirzabaev, & J. von Braun (Eds.), *Economics of Land Degradation and Improvement – A Global Assessment for Sustainable Development* (pp. 261–290). Cham: Springer International Publishing. doi:10.1007/978-3-319-19168-3_10

Morgounov, A. (1998). *Spring Wheat in Kazakstan: Current Status and Future Directions.* CIMMYT.

National Cotton Council of America. (2016). *Rankings.* Retrieved February 7, 2017, from http://www.cotton.org/econ/cropinfo/cropdata/rankings.cfm

Nazarbayev, N. A. (1997). *Kazakhstan-2030*. Almaty: Bilim.

Ohno, T. (1988). *Toyota Production System: Beyond Large-Scale Production*. Boca Raton, FL: CRC Press.

Petrick, M., Oshakbaev, D., & Wandel, J. (2014). *Kazakhstan's Wheat, Beef and Dairy Sectors: An Assessment of Their Development Constraints and Recent Policy Responses*. Halle: Leibniz Institute of Agricultural Development in Central and Eastern Europe.

Pomfret, R. (2010). Trade and Transport in Central Asia. *Global Journal of Emerging Market Economies, 2*(3), 237–256. doi:10.1177/097491011000200302

Rakhimov, M. (2010). Internal and External Dynamics of Regional Cooperation in Central Asia. *Journal of Eurasian Studies, 1*(2), 95–101. doi:10.1016/j.euras.2010.04.002

Rovný, P. (2016). The Analysis of Farm Population with Respect to Young Farmers in the European Union. *Procedia: Social and Behavioral Sciences, 220*, 391–398. doi:10.1016/j.sbspro.2016.05.513

Schwab, K. (2015). *World Economic Forum's Global Competitiveness Report, 2014-2015*. Retrieved February 7, 2017, from http://www3.weforum.org/docs/WEF_GlobalCompetitivenessReport_2014-15.pdf

Sedik, D., & Lerman, Z. (in press). *Food Security – Prospects of the Farming Sector and Rural Development across the Eurasian Plain: The Case of Wheat Production Potential in the Belarus, Uzbek and Turkmen Republic*. Brussels: Joint Research Centre.

Shengming, H. (2000). Present Situation, Policies and Outlook for China's Food Industry Development. In *The Agro-food Processing Sector in China: Developments and Policy Challenges* (pp. 75–82). Paris: OECD Publishing.

Shtaltovna, A. (2016). Knowledge Gaps and Rural Development in Tajikistan: Agricultural Advisory Services as a Panacea? *Journal of Agricultural Education and Extension, 22*(1), 25–41. doi:10.1080/1389224X.2014.997257

Termini, M. J. (1996). *The New Manufacturing Engineer: Coming of Age in an Agile Environment*. Dearborn, MI: Society of Manufacturing Engineers.

The Global Economy. (2016). *Countries*. Retrieved February 7, 2017, from http://www.theglobaleconomy.com

The Korean Post. (2013). *Uzbekistan Uses Uzbek Model of Transition to Socially Oriented Market Economy*. Retrieved February 7, 2017, from http://www.koreapost.com/news/articleView.html?idxno=377

Turaeva, R., & Hornidge, A. K. (2013). From Knowledge Ecology to Innovation Systems: Agricultural Innovations and Their Diffusion in Uzbekistan. *Innovation (Abingdon), 15*(2), 183–193.

United Nations Department of Economic and Social Affairs. (2015). *World Population Prospects: The 2015 Revision*. New York, NY: The United Nations.

United Nations Industrial Development Organization. (2015). *Industrial Development Report 2016: The Role of Technology and Innovation in Inclusive and Sustainable Industrial Development*. Vienna: United Nations Industrial Development Organization.

United Nations Office for Disaster Risk Reduction, & World Bank. (2009). *Central Asia and Caucasus Disaster Risk Management Initiative: Risk Assessment for Central Asia and Caucasus*. Geneva: ISDR.

World Bank. (2009). *Adapting to Climate Change in Europe and Central Asia*. Washington, DC: World Bank.

World Bank. (2016). *World Development Indicators*. Retrieved February 7, 2017, from http://data.worldbank.org

World Factbook. (2016). *Central Intelligence Agency*. Retrieved February 8, 2017, from https://www.cia.gov/library/publications/the-world-factbook

ADDITIONAL READING

Alpas, H., Smith, M., & Kulmyrzaev, A. (Eds.). (2011). *Strategies for Achieving Food Security in Central Asia. NATO Science for Peace and Security Series C: Environmental Security*. Amsterdam: Springer Netherlands.

Dukhovny, V. A., & Galina, S. (2011). Water and Food Security in Central Asia. In C. Madramootoo & V. A. Dukhovny (Eds.), *Water and Food Security in Central Asia* (pp. 1–23). Amsterdam: Springer Netherlands. doi:10.1007/978-90-481-9974-7_1

Friedrich, T., Derpsch, R., & Kassam, A. (2012). Overview of the Global Spread of Conservation Agriculture. *Field Actions Science Reports. The Journal of Field Actions. Special Issue 6.*

Government of the Republic of Uzbekistan. (1998). *The Law on Dehkan Farms (No. 604-I)*. Tashkent: Government of the Republic of Uzbekistan.

Schroeder, K. G., & Meyers, W. H. (2016). The Status and Challenges of Food Security in Europe and Central Asia. In A. Schmitz, P. L. Kennedy, & T. G. Schmitz (Eds.), *Food Security in a Food Abundant World: An Individual Country Perspective* (pp. 85–103). Bingley: Emerald Group Publishing. doi:10.1108/S1574-871520150000016003

United States Department of Agriculture. (2016). *USDA*. Retrieved February 7, 2017, from http://www.usda.gov/

KEY TERMS AND DEFINITIONS

Agricultural Business: Various commercial operations in the sector of agriculture, aimed at receiving profit.

Dehkan Farm: A small-scale family farm carrying out production and sale of agricultural products based on family members' labor in the household plots provided to the head of the family for a lifelong inherited possession.

Export Activities: Carrying out of operations on the import of goods and services from one country to another.

Foreign-Trade Balance: A difference between the volumes of export and import of a country, expressed in monetary terms and calculated for a certain period of time.

Import Activities: Carrying out of operations on the export of goods and services to one country from another.

International Trade: The system of trade and economic relations between countries, which is connected with export and import activities regarding goods, services, and capital.

Transitional Economies: Economies which are changing from a centrally planned economy to a market economy due to economic and administrative reforms.

Chapter 8
Food Security and Rural Development on Emerging Markets of Northeast Asia:
Cases of Chinese North and Russian Far East

Gao Tianming
Harbin Engineering University, China

ABSTRACT

The impact of economic globalization on sustainable development has been more and more perceptible since the 21st century. Regions of Russian Far East and Northeast China are the important regions affecting the sustainable development of the world because of their rich natural resources, good environment condition, and large potential for economic development. Due to a number of historical and geopolitical reasons, the cooperation between Russia and China in the region of Northeast Asia is progressing, but at a quite slow speed and in an unsustainable manner. The chapter addresses some problems and challenges of sustainable rural development in the Heilongjiang Province, P.R. China and neighbor cross-border regions of Russia (Far East Federal District and Siberian Federal District), investigates similarities and differences of rural way of life in those regions, and discusses measures to promote the cooperation between the two countries and ensure sustainable rural development and food security in the macro-region of Northeast Asia.

INTRODUCTION

In developing economies, demand for food and agricultural products is rising. China, India, Brazil, Russia, and other emerging economies which altogether unite over 40% of global population, have achieved certain success in increasing of living standards of their citizens. People increased consumption and changed their dietary habits. Demand surpassed supply which led to the problems related to food security.

DOI: 10.4018/978-1-5225-2733-6.ch008

Over the 1980-2010s, rapid economic growth and urbanization in China caused a substantial shift in the structure of food consumption. Zhou (2010) records growth of consumption of food products of higher quality, nutrient value and price in China (meat and meat products, milk and dairy products, and seafood) – to the prejudice of cheaper and fewer nutrient crops. Consequently, Chinese meat and dairy producers demand more crops as fodder for agricultural animals, which increases the load on agricultural lands and accelerates their degradation (Liu & Diamond, 2005). According to Nath et al. (2015), by 2030, China will need to increase agricultural lands under crops by 21% to be able to meet the increasing internal demand for food. Obviously, in the years ahead, China will face new challenges in the sphere of food security.

Russia, abundant with agricultural land, experience food security problems of different origin. In 2014-2015, because of Western sanctions, collapsing oil prices, and ensuing ruble depreciation Russian agricultural producers were hurt by the rise in prices for imported inputs. While retaining sanctions and countersanctions, high-interest rates and low state support in the longer term, Russian agricultural production will suffer from the drop in investment and, accordingly, in production. Consumers are hurt as well since the import ban along with a potential decrease in domestic agricultural production increases food prices and fuels food inflation. On the case of Russia, trade protection measures do not protect the market, but reduce food availability and quality of food products, and threaten the overall food security of the country.

Against the backdrop of different sources of food security problems, both countries have to search for effective solutions. Regardless of the concrete mechanisms selected in China and Russia, the basis for ensurance of food security is the development of domestic agricultural production, sustainable to economic shocks and international market fluctuations and responsive to shifts in consumption patterns. The author believes that development of rural territories may become the solution of food security problems in view of the new global challenges since rural territories have strong natural, demographic, economic, and cultural potential. Its rational usage may ensure sustainable development of agricultural production, the supply of agricultural products and food on the domestic markets, and high standards of life of rural dwellers in any country.

BACKGROUND

In some of the emerging economies, including Russia, understanding of food security differs from the internationally accepted one of the Food and Agricultural Organization of the United Nations (FAO). Thus, in some governmental documents and the works by Nechaev (2008), Ushachev (2013), and Shagaida and Uzun (2015) food security is treated as increasing of domestic food production up to the certain threshold set by the government. Spoor and Robbins (2012) point out the high political background of food security issue in Russia and countries of the Eurasian Economic Union (EAEU), which is confirmed by recent governmental decisions on food import bans taken in Russia in purpose to support domestic food producers.

A similar point of view is found in the Chinese sources. Many researchers, including Luan, Cui, and Ferrat (2013), Ghose (2014), and Jianhua (2011) food security treated as food independence or food self-sufficiency. More attention is paid to such dimensions of food security as ensurance of physical and economic access to food products of high quality and sustainability of such access in the long-term. While FAO considers food security as access by people to enough food on domestic market by means of both domestic production and import, some Chinese experts argue that increasing scale of international trade threatens food security of China (Zhu, 2016; Yu, Elleby, & Zobbe, 2015; Yu, Feng, Hubacek, & Sun, 2016; Zhou, 2010).

However, in the conditions of rising demand for food among Chinese people consumption in a greater extend shifts to animal production (Zhou, 2010; Zhou, Liu, Cao, Tian, & Wang, 2014). Even in the mid-term perspective, China will run out of available agricultural land resources to ensure the sustainable food security of the country in accordance to the four FAO pillars: availability, access, utilization, and stability of food (Liu & Diamond, 2005).

One of the key solutions to food security problem is sustainable development of those territories which provide a supply of agricultural products and food. Rural territories have strong natural, demographic, economic and cultural potential which rational usage may ensure sustainable development and high standards of life of rural dwellers in any country. Over the transition period of 1990-2000s, there were certain reforms implemented in agricultural production and land relations not only in China but in the most of the economies in transition. According to Ivolga (2014), in Russia, those reforms let to stabilize the situation in rural areas during the transition period. However, the globalized and urbanized world has brought social and economic sustainability of rural areas into increasingly challenged surroundings (Chen, Xiangzheng, Jianing, & Peiying, 2016; Kneller, Morgan, & Kanchanahatakij, 2008). In the case of China (as in many regions of Russia, especially those in the eastern part of the country), rural areas lag behind urban ones in terms of living standards and quality of life because of their low level of industrial development and urbanization (Long, Liu, Li, & Chen, 2010). As of Heilig (2003), the economic boom in China during 200-2010s was mainly generated by urbanized coastal provinces in southern and eastern parts of China. Rural areas in the northern parts of the country, on the contrary, experienced little economic growth after the first wave of development in the early 1980s, when family farming was re-introduced after decades of centrally planned agricultural communes (Chen & Fleisher, 1996).

In Russia, too, gaps in infrastructural development between rural and urban (even suburban) areas are continuing to grow. A number of rural settlements in the eastern parts of the country, especially in the Far East, goes down because of huge migration outflow from rural areas to cities. Migration brings together related negatives: ageing of population, lack of labour of high qualification, degradation of population, growing social tensions, abandonment of rural settlements and agricultural lands, lowering effectiveness of agricultural production, and growing environmental load because of outdated machinery and low culture of farming (Christiaensen, Demery, & Kuhl, 2011).

Two countries, China and Russia, have much in common in the sphere of rural development, especially in those districts which are not very much industrialized and lay outside the economically developed southern and southeastern coastal provinces (in the case of China) and central regions of Russia. Contemporary rural development policies are much broader than they used to be in the early 2000s. They shifted from agriculture itself to a broader spectrum, which included the social and economic situation in rural territories, development of rural infrastructure, employment and involvement of rural households into economic activities, rural tourism and other alternative sources of income, environmental and recreational issues, etc. (Erokhin, 2014). The effectiveness of such policies directly affects living standards of rural people, the social and demographic situation in rural areas, national food security, social and economic control over rural territories, and development of traditional cultures and rural way of life (Ivolga, 2014).

MAIN FOCUS OF THE CHAPTER

Northern China can be divided into northeast, north, and northwest region. They are also considered as province-level administration areas and consist of fourteen provinces, including Heilongjiang, Jilin, Liaoning, Beijing, Tianjin, Hebe, Inner Mongolia, Shanxi, Shandong, Ningxia, Shaanxi, Gansu, Qinghai, and Xinjiang (Ma, 2011).

Russia's Far Eastern Federal District is composed of nine regions, which are Republic of Sakha (Yakutia), Kamchatsky Krai, Primorsky Krai, Khabarovsky Krai, Amurskaya Oblast, Magadanskaya Oblast, Sakhalinskaya Oblast, Jewish Autonomous Oblast, and Chukotsky Autonomous District.

For the purpose of this research, we will study the Heilongjiang Province located in the northeast part of China, since the province directly borders with Primorsky Krai, Khabarovsky Krai, Amurskaya Oblast, Jewish Autonomous Oblast, and Zabaykalsky Krai of the Siberian Federal District of Russia (Figure 1).

For the purposes of the current research, we have primarily addressed the works related to the analysis of local specifics of rural development and unique economic, social and environmental features of certain regions (Merzlov, Ovchintseva, & Popova, 2012; Zhou, Wu, Chen, & Chen, 2007). Regional specifics of sustainable development of rural areas in the northern and eastern parts of China are obtained from the works of Ma (2011), Chen and Fleisher (1996), Chen et al (2016), and Liu, Heijman, Zhu, Dries, and Huang (2016), while overview of contemporary issues of sustainable rural development in Russia in terms of existing differences between regions is made based on the works of Ivolga (2014) and Zhuplei (2015). Impacts of international markets, international trade, and international cooperation on sustainable development of agricultural production, rural areas, and ensurance of food security are addressed in the works of Stiglitz and Greenwald (2006), Taranov and Panasyuk (2010), Taranov and Panasyuk (2012), and Zhu (2016).

Figure 1. Map of the Far East of Russia and the Northeast China
Note: 1 is for Jewish Autonomous Oblast, Russia
Source: Author's development based on https://upload.wikimedia.org/wikipedia/commons/8/82/Heilongjiang_in_China.svg.

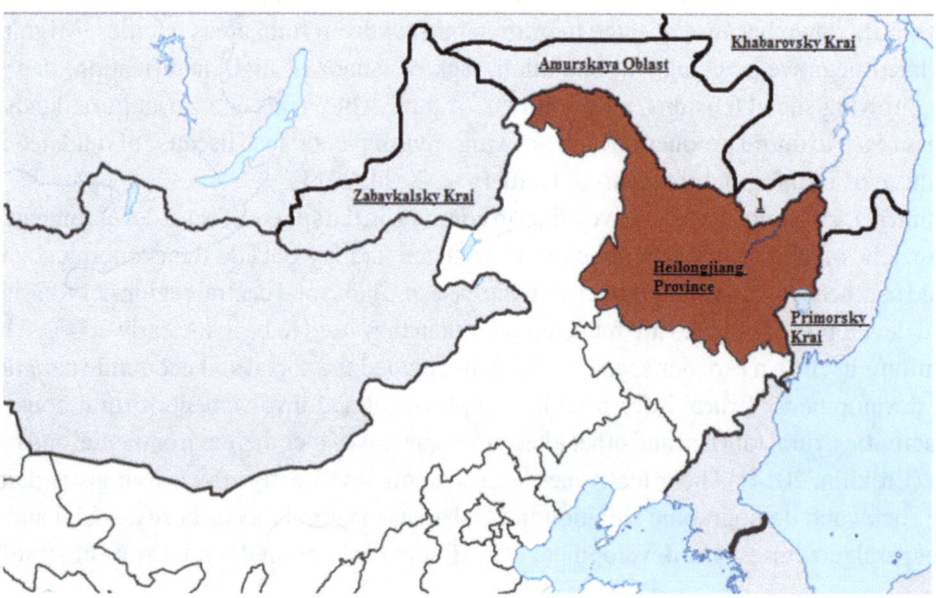

Development of agricultural production is crucially important for the stabilization of social and economic situation and the establishment of comfortable living and working environment in the rural areas of China's North and Russia's Far East (hereinafter, CNRFE region). The revival of the agricultural sector is a key to retention of the workforce in rural areas and sustainable development of rural areas in this region of the world (Erokhin & Ivolga, 2012).

The major strategic goal of the agri-industrial complex is the provision of people with agricultural products and food. However, for CNRFE region, this overall goal has specific implications: sustainable development of rural territories which are vast, sparsely inhabited, and underdeveloped; increase of competitiveness of agricultural products; ensurance of financial viability and modernization of agricultural production; accelerated development of traditional agricultural industries (animal husbandry, hunting, etc.; preservation and recovery of agricultural lands and other natural resources involved in agricultural production and rural life (Li, Ali, & Zhang, 2007).

Both the overall macroeconomic conditions and specific features of the region condition development of rural territories of the CNRFE region. Objective specific features include natural and climatic conditions for agricultural production; existing infrastructure in the agricultural sector, including production, storage, transportation, and distribution of agricultural products; specific dietary habits of rural people in the region, including indigenous communities.

In Northern China and Eastern Russia, average air and land temperatures are rather low for agricultural production. Almost 90% of land lay in permafrost zones. Continuous soil freezing is peculiar even for the southern parts of the Russia's Far East. Predominant soils are boggy, peaty, and grey-peaty. Russia's Far East lay in the zone unfavorable for vegetable growing, while in the China's North conditions for vegetable growing are rather favorable.

Despite unfavorable temperature regime and low fertility of soils, the land is still the most important resource for agricultural producers of the CNRFE region. Over 44% of the territory are occupied by forests, another 22% by deer pastures (northern parts of the Russia's Far East). Average share of agricultural lands in the overall territory of the region is only 1.3%, however, it varies from one administrative region to another (10-14% in Primorsky Krai, Amurskaya Oblast, and Jewish Autonomous Oblast; 0.5-2.1% in the Republic of Sakha (Yakutia), Khabarovsky Krai, Kamchatsky Krai, and Sakhalinskaya Oblast; 0.3% in Magadanskaya Oblast; only 0.01% in Chukotsky Autonomous District).

The share of arable land in the overall territory of the region is even lower – only 0.4% (0.4 hectares per capita) in the Russia's Far East in average (1.6 hectares per capita in Amurskaya Oblast and only 0.07 hectares per capita in Khabarovsky Krai and Chukotsky Autonomous District). Most of the fertile arable lands are concentrated in the southern part of the Russia's Far East (Amurskaya Oblast, Primorsky Krai, Khabarovsky Krai, and Jewish Autonomous Oblast) and in the Heilongjiang Province.

Subjective features which affect the agricultural sector and rural territories in the CNRFE region include continuing migration of people from rural areas to cities or to other regions of Russia and China; worsening social and demographic conditions in rural areas; lack of qualified labour force; degradation of rural territories in the light of growing expansion of China; relatively weak entrepreneurial behaviour of rural people; low effectiveness of state support programs in the sphere of small and medium entrepreneurship; low competitiveness of Russian agricultural and food products in comparison with Chinese ones; low effective demand in the region, especially in Russia.

The agricultural sector of the CNRFE region has always been functioning in the difficult conditions, but in the course of the previous two decades, existing disproportions have deepened. In the Russia's Far East, the share of agricultural production in Gross Regional Product (GRP) fell down to 2.7% as of the beginning of the 2000s (Russia's average was 3.5%). Some positive changes started in 2006 when Russian government launched several development projects in the sphere of agricultural production and rural development (Bokusheva, Hockmann, & Kumbhakar, 2012). Since 2006, agricultural output in the Russia's Far East has increased by 22.3% (Russia's average is 29.5%). The productivity of milking herd increased; production of eggs raised by 17.6%; pig stock grew. In the Republic of Sakha (Yakutia) population of meat herd horses increased by 32%, deer – by 40%, in comparison with 2000. However, livestock cattle herd is decreasing – by 13% in 2015 compared to 2005. This negative tendency is peculiar for the majority of the administrative regions of the Russia's Far East, except Sakhalinskaya Oblast and Amurskaya Oblast.

The growth of agricultural output is provided by crop production. Cultivated areas under crops increased by 24.9% during 2005-2015. The structure of lands under crops changed significantly: shares of vegetables and potato decreased by 31.0% and 28.1%, respectively, while the share of soya in the total land under cultivation reached 54%. Redistribution of lands and other resources in favor of soya may be fatal for Russian agricultural producers in the light of the Russia's accession to the World Trade Organization (WTO) (Taranov, 2015). According to the WTO obligations, Russia has to decrease import duties for soya from 20% down to 0%. Neighboring China is the world biggest producer and exporter of soya. When Russia's import duties are decreased, the domestic market will be flooded with less expensive soya from China, and Russian soya producers will become non-competitive. Soya cluster with deep soya processing has to be established in order to support the competitiveness of Russian producers in the conditions of liberalizing trade environment.

Another important element of the agricultural sector of the Russia's Far East used to be the hydro land reclaiming system which supported the cultivation of rice. During the 1970-1980s, there were 60 thousand hectares of rice hydro land reclaiming system constructed in the Russia's Far East. The productivity of rice was 3.0-3.2 tons/hectare which provided 8 kg of rice per capita (medical recommendation of healthy nutrition). During the 1990s, cultivated area under rice decreased from 60 thousand hectares down to only 2 thousand hectares. Agricultural producers in the Russia's Far East almost abandoned rice cultivation.

The agricultural sector of the Russia's Far East lacks investments. Agricultural production is not attractive for domestic investors. In 2015, a portion of investments in agriculture is below 1% of the total investments in the region (Russian average is 3.3%). Investments in the social sphere are 15% of the total fixed capital expenditures. The average share of expenditures on agriculture among the consolidated budgets of the administrative regions in the Russia's Far East is 3.3% (the highest for the Republic of Sakha (Yakutia) and Amurskaya Oblast, and the lowest for Magadanskaya Oblast and Khabarovsky Krai).

The total capacity of the food market of the Russia's Far East equals $4.3 billion. Administrative regions differ one from another in terms of this parameter. Thus, the food market in the Republic of Sakha (Yakutia) is $1.2 billion, in Primorsky Krai and Khabarovsky Krai – $1.1 billion and $1.0 billion, respectively, while in the remaining administrative regions – below $0.08 billion. Despite the announced in Russia policy of import substitution in agriculture and embargo on the import of some kinds of food products from a number of countries, an essential part of the Russian food market, including the regions of the Far East, is provided by means of import. Food security problem is worsened by the fact that in the light of decreasing agricultural production consumption of many products is below the normative level of healthy nutrition. Consumption goes down, and quality of consumed products decreases. In the Far East, consumption of meat is below the average national level by 23% (except the Republic of Sakha

(Yakutia), Sakhalinskaya Oblast, and Khabarovsky Krai). The similar situation is for the consumption of milk and dairy products.

By now, the potential of crop production in the southern part of the Far East is exhausted. Due to the more severe climatic conditions and lower fertility of soils, the central part of the region is more difficult for the development of agricultural production compared to the southern areas. In the meantime, the economic crisis in the central part of the region expands to an extent when it can not be overcome with the existing amount of state support. In agricultural organizations, land areas under cultivation are decreasing. In some of the districts, all agricultural organizations have disappeared completely. In rural areas, they have been replaced by small and medium enterprises. However, the majority of them bankrupted within a short period of time. The situation in the less inhabited northern territories is even more difficult.

One of the most influential factors constraining the effective development of agricultural production and sustainable development of rural territories in the CNRFE region is low labor potential of rural areas. Rural population in both Russia and China is decreasing. During the decade of 2004-2014, the number of rural dwellers in Russia decreased by 1.7% (from 38.62 mln people in 2004 down to 37.98 mln people in 2014), in China – by 18.3% (from 757.05 mln people in 2004 down to 618.66 mln people in 2014) (Table 1).

Unemployment along with poor standards of living caused a rapid increase of migration outflow from rural areas. During 2002-2010 (2010 is the year of the most recent census in Russia), rural population of the Russia's Far East decreased by 14.3%. The decline of the rural population happens in parallel with the liquidation of rural settlements, particularly in the northern areas of the Far East. During 2002-2010, 104 rural settlements were liquidated, including 91 in the northern part of the region.

Just as Russia, China is now experiencing the "rural flight", when fewer people are needed to bring the same amount of agricultural output to market, and related agricultural services and industries are consolidated. Rural flight is exacerbated when the population decline leads to the loss of rural services (such as business enterprises and schools), which leads to greater loss of population as people leave to seek those features. Surplus labor in rural areas due to higher average fertility rates, and improved living conditions all play a role in contributing to the flows of migrants from rural to urban areas.

Until 1983, the Chinese government, through the hukou system, restricted the ability of the citizens to internally migrate. A hukou is a record in a government system of household registration required by law in mainland China and Taiwan and determines where citizens are allowed to live. Since 1983, the Chinese government has slowly lifted the restrictions on internal migration leading to a great increase in the number of internal migrants, especially from rural areas towards the urbanized regions. Currently, a huge surplus of the labor force in rural areas of China form the labor market for crop production. In the very first years of economic reforms, a number of rural migrants reached 100 million people. The problem of excessive labor supply is quite pressing for China. 100-200 million people migrate within

Table 1. Rural population in China and Russia in 2004-2014, mln people

Country	2004	2005	2006	2007	2008	2009	2010	2011	2012	2013	2014
China	757.05	745.44	731.60	714.96	703.99	689.38	671.13	656.56	642.22	629.61	618.66
Russia	38.62	38.42	38.13	37.88	37.82	37.78	37.44	37.31	37.23	37.12	37.98

Source: State Council of the Russian Federation, 2015

the country seeking for income opportunities. Their major field of employment is agricultural sector, so-called rural settlement enterprises. State regulation in China is targeted on redistribution of migration flows, determination of the hosting districts, and facilitation of easy migration of rural people from northern provinces of China to neighboring regions of Russia (primarily, Primorsky Krai, Khabarovsky Krai, and Amurskaya Oblast) for employment in agricultural production, farming, entrepreneurial, intermediary, and other types of business activities.

Despite the easier regime, even today, the hukou system limits the ability of rural migrants to receive full access to urban social services at the urban subsidized costs. The share of rural population in total population in China is much higher in comparison to Russia (Table 2). Over 42% of the population of Heilongjiang Province lives in rural areas, while in neighbor regions of Russia – only 24.64% (Far Eastern Federal District in total, average).

Amid high birth rate which ensures reproduction of rural population on a progressively increasing scale, both Heilongjiang Province in China and regions of Far Eastern Federal District of Russia experience common problems related to low life expectancy in rural areas and outflow of employable population to more urbanized and developed territories (Table 3). In Russia, despite the growing crude birth rates, a number of regions with the natural increase of rural population are only 34, which is about 41%.

Depopulation of rural areas is partly caused by unemployment. In China, about 40% of total employment is in rural areas. The poorest rural households tend to derive a large share of their income from agricultural activities, which often show low levels of productivity and net profits. In Russia, employment in the agri-industrial sector had been decreasing during 2004-2014, while employment in non-agricultural industries had been increasing. The share of agriculture-related industries in rural employment is very

Table 2. Share of rural population in total population in the selected regions of China and Russia in 2014, %

Country / region	2014
China, total	45.23
Heilongjiang Province	42.60
Russia, total	26.03
Far Eastern Federal District of Russia, total	24.64
Republic of Sakha (Yakutia)	34.71
Kamchatsky Krai	22.45
Primorsky Krai	23.11
Khabarovsky Krai	18.17
Amurskaya Oblast	32.70
Magadanskaya Oblast	4.58
Sakhalinskaya Oblast	18.63
Jewish Autonomous Oblast	31.73
Chukotsky Autonomous District	31.45
Siberian Federal District of Russia, total	27.39
Zabaykalsky Krai	32.61

Source: State Council of the Russian Federation, 2014; National Bureau of Statistics of the People's Republic of China, n.d.

Table 3. Grouping of Russia's Far Eastern regions on natural increase (decline) of rural population in 2014

Groups on natural increase/decline of population	Regions
Natural increase	Republic of Sakha (Yakutia); Zabaykalsky Krai; Khabarovsky Krai; Jewish Autonomous Oblast; Chukotsky Autonomous District
Natural decline	
Including mortality/birth rate:	
below 120%	Altaysky Krai; Amurskaya Oblast; Primorsky Krai
120-150%	Kamchatsky Krai; Magadanskaya Oblast; Sakhalinskaya Oblast

Source: Author's development based on State Council of the Russian Federation, 2014; Ivolga, 2014

much affected by natural and environmental factors. As environmental conditions for agricultural production are worsened, the share of agriculture in rural employment decreases (Ivolga & Erokhin, 2011).

Lack of employment opportunities within the traditional places of habitation gave rise to leaving for seasonal works outside the rural areas. In some of the settlements of the Russia's Far East, almost all menfolk are employed outside the rural areas. In terms of promotion of employment opportunities in rural settlements, the situation is particularly difficult in the regions with high investment activity. Implementation of big federal programs, such as the construction of hydropower station, spaceport, highways, and oil pipelines wash labor force out of rural settlements in favor of construction sites. Such construction super-projects also increase demand for labor in the cities, where production and service facilities are located, and where people from rural areas find jobs.

Considering the backwardness of agriculture compared to other industries in terms of income, lack of full-time employment opportunities, and existing deficit of skilled human resources in the region, the task of provision of agricultural sector with sufficient labor force of relevant quality is becoming more and more challenging.

One on the most recent measures to fight depopulation in rural areas and develop agricultural production of the Russia's Far East is the program of free-of-charge distribution of one hectare of land per person to everyone interested. The initiative suggests a solution of several problems: involvement of underutilized land into the turnover, establishment of new employment (and self-employment) opportunities, and attraction (or return) of the employable population to the rural areas of the Russia's Far East. Land plots are assumed to lease for a period of five years, upon the expiration of which the decision on future use of the plot should be taken. If the land is used in a proper manner, the plot is transferred to the property of the land user. In the case of improper use, or no use, the land plot is confiscated.

The author argues that the initiative will not cause a mass migration of Russian people to the Far East. In such a vast region, where access to land is not a problem, hence the distribution of land is not an incentive. People in rural areas in the Far East have excessive access to land – it is doubtful that new land plots may retain them in rural areas. During the 1990-2000s, cultivated agricultural lands in the Far East decreased by 40%. These are abandoned lands in the areas where agricultural production is not economically viable due to climatic conditions. According to Table 2, such regions are predominantly urban, except Amurskaya Oblast, where the share of rural population in the total population of the region is relatively high.

In China, on the contrary, much attention is paid to the problem of decreasing areas under cultivation. The country loses agricultural lands because of their removal from turnover in favor of the construction of industrial enterprises, roads, mining sites, and other objects. Land policy of China assumes stabilization of arable lands and increase of areas under cultivation by means of tilling of wild land in the northern border regions. In recent years, Chinese government concentrates on the protection of agricultural lands from erosion and bogging, their melioration and drainage.

In Russia, areas of abandoned lands are increasing, and rural territories are becoming desolated. Those who want to start farming may purchase land in the central regions of Russia, close to available infrastructure. In the Far East, land plots are allotted in the areas far away from cities, without access to any infrastructure, including roads. Therefore, it is supposed to be problematic to store agricultural products and deliver them to the final customer. Even in the case of reasonable access to infrastructure, utilization value of its use due to its scarcity is higher than in other regions of Russia and in the neighbor Heilongjiang province. Hence, rural households and farmers are constrained to target on the domestic market of their rural settlement, municipal district, or administrative region, at best. However, the disposable income of population in Russia is shrinking, including in the Far East, which means a decrease of effective demand.

Because of the critical situation in terms of the development of infrastructure in the rural territories, agricultural production in the Far East is developing in two formats. The first format is the development of not purely rural, but suburban agricultural enterprises (collective farms and farmers) located in the rural areas nearby cities or linked with urban markets with appropriate transport infrastructure. Among such suburban regions are rural areas in Amurskaya Oblast – grain and soya production belt of the Far East. Even in those regions, distribution of agricultural products is the major problem. The second format is about the usage of agricultural lands in the Far East by incoming workers from Northern China, who compete with Russian labor force on the market. Chinese farmers concentrate on the best available lands, get land plots for a period of one to three years, utilize them with maximum productivity, often against land-use recommendations. Because of low costs and abandon of any environmental protection measures, Chinese farmers are able to deliver cheap vegetables and fruit on the market and ensure the existing demand in the region.

The author believes that the problem of depopulation of rural territories in the Far East can not be solved by the provision of land plots only. Such an idea can be productive only in the case when the people moving to the Far East are provided with substantial support from the state, including financial (interest-free loan, resettlement benefits, etc.) and infrastructural (access of allotted land plots to infrastructure in terms of power, transportation, and water supply) measures. When asking for guarantees of rational land use from a farmer, the state in its turn has to provide some guarantees of support.

Summarising the results of analysis of agricultural production and development of rural territories in the Far East, the author emphasises the following negative impacting factors: low rates of economic growth; low effectiveness of reproduction processes in the entire economy and in particular sectors and regions, including agri-industrial complex of the Far East; absence of strategic vision of sustainable development of the region; uncertainty of strategic and tactic tasks of agricultural development; low effectiveness of existing system of state regulation of agricultural production and food market; irregularity of relations between the federal government and regional administrations; lack of financial support for research and development in the sphere of agricultural production.

There are some conclusions based on the analysis of agricultural production and food market of the Russia's Far East:

1. The region of the Far East is not the one with developed agricultural production. Only 9% of the employable population of the region are involved in agricultural production. Among the administrative regions of the Far East where agricultural sector is relatively important in terms of GRP are Jewish Autonomous Oblast and Amurskaya Oblast. In these two regions, the share of the employable population involved into agricultural production is 15%. As for the northern regions of the Far East, the agricultural sector is developed in the Republic of Sakha (Yakutia) with 11% of the employable population.
2. The agricultural sector of the Far East is more sensitive to various changes compared to other regions of Russia in terms of its response to economic reforms and macroeconomic fluctuations. The decline of agricultural production during the 1990-2010s has been rather severe.
3. Agricultural production experienced essential shifts in terms of its structure. Some crops, primarily wheat, and soya, dominate in Amurskaya Oblast (50% of total grain production of the entire region), Primorsky Krai, Khabarovsky Krai, and the Republic of Sakha (Yakutia). Vegetable production and animal husbandry are dispersed equally. On the one hand, domestic agricultural producers utilize existing competitive advantages and focus on the cultivation of the most marketed crops in pursuance of maximum commercial effect. On the other hand, they decrease the degree of diversification and promote dependence on the narrow range of products. In the case of any price fluctuations on the global market and in the light of the expansion of soya producers coming from Northern China, such reliance on grain and soya decreases food security in the region.

Economic reforms of the 1990-2010s in Russia resulted in the essential decline of agricultural production and degradation of rural territories. Regions became differentiated in terms of the level of development of production and social infrastructure, demographics, the income of the rural population, and standards of living in rural areas. Unemployment rates in rural areas in Russia's regions of Far East vary (Table 4). Some regions had 10% in 2014, while others ended up with 20%. Average unemployment rate in Russia in 2014 was 6%.

The labor market in the rural areas of the Russia's Far East is characterized by two divergent tendencies: population decreases, but the share of people in active working age increases. The growth of population in active working age is faster than economic expansion rates, which drives up unemployment in rural areas. Despite the slowly growing employment level, rates of unemployment in rural areas of the Far East are still very high (above 30% in 2015). Moreover, growth rates for employment are slower than those for the economically active population, which forces people to migrate from rural settlement to urban areas within the administrative region, to the more developed regions of Siberia and Central Russia, and even to neighbor China in search of employment opportunities.

Table 4. Grouping of Russia's Far Eastern regions on unemployment in rural areas in 2014

Level of unemployment in rural areas, %	Regions
Below 10	Kamchatsky Krai; Chukotsky Autonomous District; Magadanskaya Oblast; Sakhalinskaya Oblast
10-20	Republic of Sakha (Yakutia); Altaysky Krai; Zabaykalsky Krai; Khabarovsky Krai; Primorsky Krai; Amurskaya Oblast; Jewish Autonomous Oblast

Source: Author's development based on State Council of the Russian Federation, 2014; Ivolga, 2014

There are four main reasons for the high rate of unemployment among rural people and low sustainability of rural areas in the Far East. The first reason is a structural imbalance between the demand for labor and its supply in rural areas, caused by an inconsistency between qualification of rural employees and needs of agricultural employers. The share of the population with higher and secondary professional education increases gradually but still remains lower compared to the national average. Employers seek certain qualifications and skills, which current labor market is not able to meet. This has a serious negative impact on the technological and innovative development of agricultural production and represents a threat to sustainable rural development.

The second reason is the exceptionally low level of income in rural areas compared to cities and central regions of Russia.

The third reason is demographic ageing. The natural decline in population is not compensated by its natural replacement. Recovery of labor potential is provided by means of migration gain, primarily, from China, but incoming people lack qualifications needed for high-paid jobs and therefore compete with for low-skilled positions. It drives unemployment even higher and fuels social tensions in rural areas.

The fourth reason is the underdevelopment of non-agricultural activities in rural areas, which means a lack of alternative sources of income for rural households. The transition to new technological patterns and a growing workforce productivity in agriculture inevitably bring about a reduction in employment. In order to secure rural areas as social and production subsystems, it is necessary to diversify the rural economy through the promotion of non-agricultural activities, as the most of the developed countries do. Programs for the development of non-agricultural activities in rural territories have been successfully implemented in the EU, the USA, China, and other countries (Bondarenko, 2011).

Merzlov et al. (2012) developed the classification of Russia's regions, which included four types and nine subtypes of regions depending on the character of rural development, utilization of available agricultural and environmental resources, the level of social and demographic development, and threats to the sustainable development of rural areas. Far Eastern regions of Russia are included into the Subtype 3c ("poor social and natural conditions") and are characterized by a combination of unfavorable social and environmental conditions. The density of rural population is very low (two inhabitants per square km in average). Agricultural production is not a predominant type of rural activities since many people are involved in forestry and mining.

Republic of Sakha (Yakutia), Kamchatsky Krai; Chukotsky Autonomous District; Magadanskaya Oblast; Sakhalinskaya Oblast are related to Type 4 ("low level of territorial development"). Those are north-eastern parts of Russia with an area of 62% of Russia's territory and a population of only 6% of total population. The regions are not heavily involved in agricultural production, and their contribution to the national gross agricultural production is very small. Rural people in those regions are employed in forestry and mining. Agricultural production is supported by regional and federal budgets. The bankruptcy of forest industry enterprises, high unemployment, and migration outflow create serious threats to the development of rural areas.

According to this classification, northern-eastern provinces of China, namely Heilongjiang Province, can also be referred to as Type 3 ("relatively unfavorable conditions"), as Russia's regions of Far East. According to Zhou et al (2007) and Ma (2011), the land infertility is a severe issue and a plenty of land in the northern parts of China is in arid, semiarid or desert regions. Moreover, the environment in northern China is vulnerable due to serious soil erosion, soil desertification, shortages of agricultural water resources and lowering of groundwater levels. Based on this kind of natural environment, agricultural productivity is comparatively low in comparison to than that of southern China (Ma, 2011).

Another problem in China's rural areas, pointed out by Heilig (2003), is the structure of agriculture. He indicates that the average farm size in China is extremely small, which limits market opportunities for farmers. Consequently, rural people have very little monetary income and have to find nonagricultural labor in the urban areas to gain enough money. The problem is that a large part of China's agriculture is traditional subsistence farming, which contributes little to a modern economy (Heilig, 2003).

In China, compared to the majority of developed countries, the level of state support of agriculture is lower. During the reforms in the agricultural sector (1980-2010s), output increased four times, while export of agricultural products fourteen times. The government of China put in place some measures related to the development of agricultural production. The strategy of economic development of less developed regions of the country, including northern and western provinces, has been accepted. According to this strategy, designated regions of the country received increased amounts of financial support and benefited from preferential tax treatment. Agricultural producers in those regions got access to concessional credits and subsidies for purchasing of advanced resource-saving equipment. Promotion of favorable investment climate for the attraction of foreign capital to the agricultural sector is among the priorities for the Chinese government.

The major goals of state support of agricultural production are:

1. Exemption of farmers from various taxes which decreases production costs and increases indirect support in agriculture.
2. Professional development programs offered for rural people and farmers throughout the major agricultural regions of the country.
3. Development of agricultural insurance.
4. Development of infrastructure for the purposes of agricultural production, the introduction of innovations, expansion of construction of irrigation facilities, and economic reforms in rural areas.

The Chinese government pays attention to the development of market mechanisms in rural areas and their control. In China, an increase of farmers' income by 1% adds 0.51% to GDP growth (Beloglazov, 2007). That is why the increase of prosperity in rural areas is so important for the government.

The positive experience of development of rural territories in China should be adopted by Russia. Implementation of those measures which have been effective in China may cause essential positive influence on the development of neighboring regions of the Russia's Far East. The most important lesson to be studied from Chinese experience is the model of gradual transition from administrative economy to market environment with effective state regulation, without declines in production and standards of living in rural areas. Due to the unfavorable climatic conditions and related difficulties in agricultural production in the majority of administrative regions of the Russia's Far East, it is necessary to provide support for domestic agricultural organizations (by means of subsidies) and rural households (through agricultural consumer cooperatives) which focus on production of the most competitive and value-added products, or ensure sustenance of rural population.

Support has to be provided to the common implementation of innovative agri-technical decisions, the introduction of high-productive breeds of agricultural animals and high-yielding crops, the involvement of new resources into agricultural production in the purpose of output increase.

Development of agricultural production in the eastern parts of Russia and northern parts of China should be facilitated by the measures focused on improvement of soil fertility, an increase of cultivated lands under high-quality varieties of crops, development of forage resources for animal husbandry. It is necessary to create economic conditions for equipping crop production organizations with modern resource-saving machinery, modernization of animal husbandry farms, development of the genetic potential of animals in meat and dairy production, the attraction of investments into the agricultural sector, research and development of innovation technologies.

Rural territories require a substantial breakthrough in the development of social infrastructure and engineering facilities, housing construction and improvement of the living environment in rural areas, the creation of new employment opportunities in rural areas, and education of rural people in the agricultural colleges and universities of the region.

An important task is the elimination of disparities in the agricultural sector by means of support of those industries which have potential advantages on both domestic and international markets but are not able to unlock this potential without state support and regulation (Anderson, Jha, Nelgen, & Strutt, 2013). In particular, state support has to be focused on stabilization of population of agricultural animals, including the sectors of animal husbandry which are traditional for indigenous communities (primarily, deer, horses, and sheep). Development of these sectors will help to increase production of meat and ensure the preservation of a traditional way of living and employment in rural areas of the CNRFE region.

Development of crop production and an increase of productivity of grain crops have strategic importance for the fulfillment of potential advantages of the region on domestic and international markets. In some subsectors of agricultural production, state policy should be focused on support of small entities: rural households and farmers. They have to be provided with equal rights on the distribution of agricultural products and access to production and transport infrastructure.

Nowadays, due to degradation of traditional large-scale agricultural production of Soviet-type rural households remain the only form of employment in many rural settlements in the Russia's Far East. In the light of ensurance of sustainable development of rural territories and food security of the region, support for rural households and retention of people in their traditional places of habitation have to be among the top priorities for the government. In Russia, the effectiveness of support of small entrepreneurship (farmers, rural households, and agricultural consumer cooperatives) is low. During 2000-2015, the share of rural households in the gross production of agricultural products declined (from 75% in 2000 down to 45% in 2015). The increase in productivity of small economic entities can not be achieved without substantial changes in state regulations. The major portion of rural households produces agricultural products for self-consumption only (Trukhachev, Ivolga, & Lescheva, 2015). Rural people are not ready to become entrepreneurs and do not feel strong support on the part of the state. One of the perspectives ways to increase the involvement of rural households to commercial production of agricultural products is the promotion of cooperatives in rural areas. The most productive ones may become those cooperatives which unite households on the basis of the family contract.

Russia has adopted many of the practices of agricultural and rural transformation and ensurance of food security from China. For example, Federal Law of the Russian Federation "About Development of Agriculture" has many intersections with clauses of agricultural and agrarian policy legislation of China (particularly, in relation to the articles "State Agrarian Policy", "Measures to Implement State Agrarian Policy", "Major Directions of State Support in the Sphere of Development of Agriculture", "State Program for Regulation of Markets of Agricultural Products, Raw Materials, and Food", etc.).

China's experience in the promotion of cooperation in rural areas also has potential to be adopted in Russia. Big integrated agricultural holdings approved themselves as effective forms of agricultural production. Establishment and development of similar economic entities in the regions of the Russia's Far East may be facilitated on the basis of public-private partnerships. Cross-border cooperation has certain potential too since it allows launching radical transformations in the agricultural sector. Such cooperation includes the mass use of seasonal workforce from northern China, which ensures the development of high-productive production of vegetables in the region.

Taking into account low food self-sufficiency of the Russia's Far East, especially in the northern parts of the region, it is reasonable to promote interregional cooperation. Such problems as ensurance of food security, supply of population of the Far East with food products, land reclamation measures, and development of fodder reserves for animal husbandry have to be developed in the form of joint actions, including the mechanisms of state procurement, interregional co-financing of construction and infrastructure projects, establishment of interregional food complex in the southern part of the Russia's Far East.

In China, investment in the agricultural complex is one of the major tools of state regulation. The government pays attention both to big investment projects (like in Russia) and small projects in rural areas. The consolidated budget of the agricultural sector is composed of four sources: direct state financing; credits to agricultural producers through state banks; resources of collective enterprises in rural areas; tax revenues obtained from farmers and rural households (Beloglazov, 2007). In the flow of market reforms in agriculture, Chinese government implements the concept of development of individual entrepreneurial initiative in rural areas. Aside from preferences obtained from state and regional sources, enterprises in rural areas also have a right to attract foreign investments. It allows the creation of effective enterprises with a high level of development of technology, the large scale of production, and high commercial results (Zhou, 2010).

For the rural areas of the Russia's Far East, such an approach would have been more effective than the free-of-charge lease of abandoned agricultural lands. The state capital is to come to the agricultural complex in the form of public-private partnership. This may help to increase production facilities in most of the sectors, including the production of milk and dairy products, meat and meat products, vegetables, animal husbandry, seed production, and development of infrastructure of agricultural and food markets. Such policies will drive agricultural production and ensure the provision of the population with domestically produced agricultural products and food.

SOLUTIONS AND RECOMMENDATIONS

Diversification of rural economics and expansion of income opportunities for rural inhabitants are the key tasks on the way to increasing the sustainability of rural development. Agricultural production in both Russia and China operates in rather complicated social, economic and environmental conditions. Elaboration and implementation of targeted development programs and regionalized state policies are required for ensurance of sustainable rural development of the Far Eastern regions of Russia of Subtype 3c ("poor social and natural conditions") and Type 4 ("low level of territorial development"), as well as Heilongjiang Province of China of Type 3 ("relatively unfavorable conditions").

Aims of such policies are:

- Creation of favorable social and economic conditions for performance of national and territorial development functions by rural areas;
- Stabilization of rural population size and creation of conditions for its sustainable growth by means of lower mortality, increasing life expectancy, and decreasing of migration outflow;
- Development of employment opportunities, increase of social standards of living and welfare with due consideration of internationally recognized standards;
- Performance increase of agricultural production and bigger impact of rural areas into the social and economic development of far east of russia and northern east of china.

In our opinion, the relevant state policies in the sphere of ensurance of sustainable rural development in the considered regions of russia and china have to be based on the following principles:

- Approach to the development of rural areas as entire territorial complexes which perform important social functions and impact to the social and economic development of russia and china;
- Assurance of rights of those people living in rural areas, including ensurance of access to state and municipal services of adequate quality;
- Usage of various forms of state support for ensurance of favorable conditions of social and economic development of rural areas and utilization of existing economic, demographic and natural potential;
- Partnership between the state, local authorities in the regions, business, and rural dwellers;
- Differentiated approach to the development of rural areas in various regions of russia's far east and chinese northeast, consideration of local peculiarities and their effects on agricultural production and rural areas in general. The goal here is to minimize differences between regions in terms of living standards of rural dwellers, employment opportunities, development of infrastructure, etc.;
- Development of links between rural areas and cities, integration of rural areas into a common economic system on a basis of agricultural and industrial cooperation, development of modern effective forms of agribusiness;
- Investment projects in agricultural production coupled with the development of social and engineering infrastructures in rural settlements;
- Rational use of natural resources, preservation, and development of agricultural landscapes, environmental protection of rural areas;
- Creation of better living conditions in rural settlements.

FUTURE RESEARCH DIRECTIONS

Both Russia and China will definitely benefit from studying practices of each other in the sphere of ensurance of food self-sufficiency and food security through the sustainable development of rural areas. This sphere of knowledge is underinvestigated. In the works by Chinese and Russian authors, the author was able to find the very limited amount of information related to food security issues in the light of Russian-Chinese collaboration between the regions of the Russia's Far East and China's North and North-East. Analysis of Chinese reforms in the sphere of agricultural production and ensurance of food security has to be made for the purpose of adaptation of best practices to specific conditions of the Russia's Far East.

CONCLUSION

Experiencing unfavorable environmental conditions of agricultural production, suffering from a rural flight and high unemployment rates, facing a lot of internal structural problems, Russia and China have to make every possible effort to develop diversification of the rural economy, support farmers and alternative forms of employment and self-employment, including local crafts and rural tourism. The set of measures has to include implementation of rural development issues into the national and regional development strategies; consideration of tasks of sustainable rural development in the rural area planning schemes; improvement of rural infrastructure, including transport and communications; analysis of environmental problems and existing threats to sustainable environmental development; elaboration of measures to secure biodiversity; expand special support measures, such as for young people and families, in order to retain them in rural areas; increase of investment attractiveness of rural areas in general and rural settlements in particular as local centers of rural development.

Sustainable regional and rural development in both Russia's Far East and Chinese Northern East is only possible if the regional diversity is taken into account in development concepts (Heilig, 2003). Like Russia, China has all possible extremes in climate, population, and infrastructure availability. It is, therefore, essential, that detailed regionalized analyses are undertaken for understanding the specific development constraints and options of various regions, their similarities, and differences, and those social and economic conditions that determine a development potential of rural areas in those regions.

REFERENCES

Anderson, K., Jha, S., Nelgen, S., & Strutt, A. (2013). Re-examining Policies for Food Security in Asia. *Food Security*, *5*(2), 195–215. doi:10.1007/s12571-012-0237-5

Beloglazov, G. (2007). Food Security of the People's Republic of China and its Russian Vector. *Russia and Pacific RIM*, *3*, 75–83.

Bokusheva, R., Hockmann, H., & Kumbhakar, S. (2012). Dynamics of Productivity and Technical Efficiency in Russian Agriculture. *European Review of Agriculture Economics*, *39*(4), 611–637. doi:10.1093/erae/jbr059

Bondarenko, L. (2011). Employment in Rural Areas and Diversification of Rural Economics. *Economics of Agriculture of Russia*, *1*, 71–76.

Chen, J., & Fleisher, B. M. (1996). Regional Income Inequality and Economic Growth in China. *Journal of Comparative Economics*, *22*(2), 141–164. doi:10.1006/jcec.1996.0015

Chen, Z., Xiangzheng, D., Jianing, D., & Peiying, H. (2016). Urbanization and Sustainability: Comparison of the Processes in "BIC" Countries. *Sustainability*, *400*(8), 1–18.

Christiaensen, L., Demery, L., & Kuhl, J. (2011). The (Evolving) Role of Agriculture in Poverty Reduction. *Journal of Development Economics*, *96*(2), 239–254. doi:10.1016/j.jdeveco.2010.10.006

Erokhin, V. (2014). Approaches to Sustainable Rural Development in a Predominantly Non-Rural Region. *Economics of Agriculture*, *2*, 291–306.

Erokhin, V., & Ivolga, A. (2012). How to Ensure Sustainable Development of Agribusiness in the Conditions of Trade Integration: Russian Approach. *International Journal of Sustainable Economies Management*, *2*(1), 12–23. doi:10.4018/ijsem.2012040102

Ghose, B. (2014). Food Security and Food Self-Sufficiency in China: From Past to 2050. *Food and Energy Security*, *3*(2), 86–95. doi:10.1002/fes3.48

Heilig, G. K. (2003). *Sustainable Regional and Rural Development in China: Where Do We Stand?* Laxenburg: IIASA.

Ivolga, A. (2014). Overview of Contemporary Issues of Sustainable Rural Development in Russia in Terms of Existing Differences between Regions. *Economics of Agriculture*, *2*, 331–345.

Ivolga, A., & Erokhin, V. (2011). Rural Development and Agrarian Reform: Russian Experience during the Transition Period. In *Proceedings of the International Conference "Rural Development Policies from the EU Enlargement Perspective"*. Ecka: Institute of Agricultural Economics.

Jianhua, Z. (2011). China's Success in Increasing Per Capita Food Production. *Journal of Experimental Botany*, *32*, 1–5.

Kneller, R., Morgan, C. W., & Kanchanahatakij, S. (2008). Trade Liberalisation and Economic Growth. *World Economy*, *31*(6), 701–719. doi:10.1111/j.1467-9701.2008.01101.x

Li, W. J., Ali, S. H., & Zhang, Q. (2007). Property Rights and Grassland Degradation: A Study of the Xilingol Pasture, Inner Mongolia, China. *Journal of Environmental Management*, *85*(2), 461–470. doi:10.1016/j.jenvman.2006.10.010 PMID:17129663

Liu, J., & Diamond, J. (2005). Chinas Environment in a Globalizing World. *Nature*, *435*(7046), 1179–1186. doi:10.1038/4351179a PMID:15988514

Liu, M., Heijman, W., Zhu, X., Dries, L., & Huang, J. (2016). Individual and Social Optima of Rural Land Allocation by Stakeholders: A Case Study on Eco-Fragile Areas of Northern China. *The Rangeland Journal*, *38*, 57–72.

Long, H., Liu, Y., Li, X., & Chen, Y. (2010). Building New Countryside in China: A Geographical Perspective. *Land Use Policy*, *27*(2), 457–470. doi:10.1016/j.landusepol.2009.06.006

Luan, Y., Cui, X., & Ferrat, M. (2013). Historical Trends of Food Self-Sufficiency in Africa. *Food Security*, *5*(3), 393–405. doi:10.1007/s12571-013-0260-1

Ma, L. (2011). Sustainable Development of Rural Household Energy in Northern China. *Journal of Sustainable Development*, *5*(4), 115–124.

Merzlov, A., Ovchintseva, L., & Popova, O. (2012). *Regional Experience of Elaboration of Programs of Sustainable Rural Development*. Moscow: Rosinformagrotekh.

Nath, R., Luan, Y., Yang, W., Yang, C., Chen, W., Li, Q., & Cui, X. (2015). Changes in Arable Land Demand for Food in India and China: A Potential Threat to Food Security. *Sustainability*, *7*(5), 5371–5397. doi:10.3390/su7055371

National Bureau of Statistics of the People's Republic of China. (n.d.). *Database.* Retrieved December 3, 2016, from http://www.stats.gov.cn/

Nechaev, V. (2008). *Conceptual Foundations of Ensurance of Food Security in Russia*. Moscow: All-Russian Research Institute of Agricultural Economics.

Shagaida, N., & Uzun, V. (2015). *Food Security in Russia: Monitoring, Tendencies, and Threats*. Moscow: Russian Academy of National Economy and Public Administration under the President of the Russian Federation.

Spoor, M., & Robbins, M. J. (Eds.). (2012). *Agriculture, Food Security, and Inclusive Growth.* The Hague: Institute of Social Sciences.

State Council of the Russian Federation. (2014). *Report on Sustainable Development of Rural Territories of the Russian Federation.* Moscow: Council under the President of the Russian Federation on Implementation of Priority National Projects and Demographic Policy.

Stiglitz, J., & Greenwald, B. (2006). Helping Infant Economies Grow: Foundations of Trade Policies for Developing Countries. *The American Economic Review, 96*(2), 141–146. doi:10.1257/000282806777212206

Taranov, P. (2015). The Russian Economy in the Context of Trends in International Trade and Economic Policy. *Economics and Entrepreneurship, 10*(63), 94–98.

Taranov, P., & Panasyuk, A. (2010). Food Dependence of Russia in the Light of Joining the World Trade Organization (WTO). *Bulletin of Orenburg State Agrarian University, 26*, 106–109.

Taranov, P., & Panasyuk, A. (2012). Competitiveness of Agri-food System of Russia in the Light of Eurasian Economic Integration. *National Interests: Priorities and Security, 40*(181), 56–60.

Trukhachev, V., Ivolga, A., & Lescheva, M. (2015). Enhancement of Land Tenure Relations as a Factor of Sustainable Agricultural Development: Case of Stavropol Krai, Russia. *Sustainability, 7*(1), 164–179. doi:10.3390/su7010164

Ushachev, I. (2013). *Food Security of Russia in View of the Global Partnership*. Moscow: Russian Academy of Agricultural Sciences, All-Russian Research Institute of Agricultural Economics.

Yu, W., Elleby, C., & Zobbe, H. (2015). Food Security Policies in India and China: Implications for National and Global Food Security. *Food Security, 7*(2), 405–414. doi:10.1007/s12571-015-0432-2

Yu, Y., Feng, K., Hubacek, K., & Sun, L. (2016). Global Implications of Chinas Future Food Consumption. *Journal of Industrial Ecology, 20*(3), 593–602. doi:10.1111/jiec.12392

Zhou, Z. (2010). Achieving Food Security in China: Past Three Decades and Beyond. *China Agricultural Economic Review, 2*(3), 251–275. doi:10.1108/17561371011078417

Zhou, Z., Liu, H., Cao, L., Tian, W., & Wang, J. (2014). *Food Consumption in China: The Revolution Continues*. Cheltenham, UK: Edward Elgar Publishing. doi:10.4337/9781782549208

Zhou, Z., Wu, W., Chen, Q., & Chen, F. (2007). Study on Sustainable Development of Rural Household Energy in Northern China. *Renewable & Sustainable Energy Reviews, 12*(8), 2237–2239.

Zhu, Y. (2016). International Trade and Food Security: Conceptual Discussion, WTO and the Case of China. *China Agricultural Economic Review*, *8*(3), 399–411. doi:10.1108/CAER-09-2015-0127

Zhuplei, I. (2015). Sustainable Development of Agriculture: Regional and Applied Aspect. *Almanac of Modern Science and Education*, *99*(9), 64–67.

ADDITIONAL READING

Arayama, Y., & Miyoshi, K. (2004). Regional Diversity and Sources of Economic Growth in China. *World Economy*, *27*(10), 1583–1607. doi:10.1111/j.1467-9701.2004.00668.x

Bay, B. J. (2011). The World Trade Organization and the Millennium Development Goals: The Role of Multilateral Trade Negotiation in Achieving Food Security for the World's Most Vulnerable Populations. *Georgetown Journal of International Law*, *43*(3), 165–169.

Birchfield, L., & Corsi, J. (2010). Between Starvation and Globalization: Realizing the Right to Food in India. *Michigan Journal of International Law*, *31*(2), 735–740.

Chen, Y. F., & Tang, H. P. (2005). Desertification in North China: Background, Anthropogenic Impacts and Failures in Combating. *Land Degradation & Development*, *16*(4), 367–376. doi:10.1002/ldr.667

Du Puis, E. M., & Goodman, D. (2005). Should We Go Home to Eat? Towards a Reflexive Politics of Localism. *Journal of Rural Studies*, *21*(3), 359–371. doi:10.1016/j.jrurstud.2005.05.011

Erokhin, V. (2015). Russian Trade in Agricultural Products: Current State and Influences of Trade Integration. In A. Schmitz & W. H. Meyers (Eds.), *Transition to Agricultural Market Economies. The Future of Kazakhstan, Russia and Ukraine* (pp. 164–171). Boston, Oxfordshire: CABI. doi:10.1079/9781780645353.0164

Erokhin, V., Heijman, W., & Ivolga, A. (2014). Sustainable Rural Development in Russia through Diversification: The Case of the Stavropol Region. *Visegrad Journal on Bioeconomy and Sustainable Development*, *3*(1), 20–25. doi:10.2478/vjbsd-2014-0004

Farsund, A. A., Daugbjerg, C., & Langhelle, O. (2015). Food Security and Trade: Reconciling Discourses in the Food and Agriculture Organization and the World Trade Organization. *Food Security*, *7*(2), 383–391. doi:10.1007/s12571-015-0428-y

Gandhi, V. P., & Zhou, Z. (2014). Food Demand and the Food Security Challenge with Rapid Economic Growth in the Emerging Economies of India and China. *Food Research International*, *63*(1), 108–124. doi:10.1016/j.foodres.2014.03.015

Gazuda, S. (2013). Regional Prerequisites of Rural Development. *Economics and Management*, *4*, 180–186.

Haggblade, S., Hazell, P. B. R., & Reardon, T. A. (2007). *Transforming the Rural Nonfarm Economy: Opportunities and Threats in the Developing World*. Baltimore, MD: Johns Hopkins University Press.

Haggblade, S., Hazell, P. B. R., & Reardon, T. A. (2010). The Rural Non-farm Economy: Prospects for Growth and Poverty Reduction. *World Development*, *38*(10), 1429–1441. doi:10.1016/j.worlddev.2009.06.008

Hoggart, K., & Paniagua, A. (2001). What Rural Restructuring? *Journal of Rural Studies*, *17*(1), 41–62. doi:10.1016/S0743-0167(00)00036-X

Huang, J. K., & Rozelle, S. (1995). Environmental Stress and Grain Yields in China. *American Journal of Agricultural Economics*, *77*(4), 853–864. doi:10.2307/1243808

Kiseleva, N., & Orlyanskaya, A. (2012). Assessment of Level of Social and Economic Rural Development (Case of Stavropol Region). *Fundamental Researches*, *5*(11), 1266–1270.

Kovalenko, E. (2012). Mechanism of Sustainable Development of Regional Rural Territories. *Contemporary Problems of Science and Education, 2*. Retrieved December 3, 2016, from www.science-education.ru/102-5823

Kruseman, G., & Bade, J. (1998). Agrarian Policies for Sustainable Land Use: Bio-Economic Modelling to Assess the Effectiveness of Policy Instruments. *Agricultural Systems*, *58*(3), 465–481. doi:10.1016/S0308-521X(98)00041-9

Li, R., & Lin, H. (2014). Developing the Agro-Grassland System to Insure Food Security of China. *Journal of Agricultural Chemistry and Environment*, *3*(3), 9–15. doi:10.4236/jacen.2014.33B002

Li, X. D. (2003). Rethinking the Peasant Burden: Evidence from a Chinese Village. *The Journal of Peasant Studies*, *30*(3-4), 45–74. doi:10.1080/03066150412331311229

Lin, J. Y. (1992). Rural Reforms and Agricultural Growth in China. *The American Economic Review*, *82*, 34–51.

Matsuyama, K. (1992). Agricultural Productivity, Comparative Advantages, and Economic Growth. *Journal of Economic Theory*, *58*(2), 317–334. doi:10.1016/0022-0531(92)90057-O

Otimana, P. I., Toderoiua, F., Alexandria, C., Floriana, V., Gavrilescua, C., Ionela, J., & Tudo, M. M. et al. (2014). Sustainable Development Strategy for the Agri-food Sector and Rural Area – Horizon 2030. *Procedia Economics and Finance*, *8*, 510–517. doi:10.1016/S2212-5671(14)00121-X

Robinson, G. M. (2004). *Geographies of Agriculture: Globalisation, Restructuring and Sustainability*. Harlow: Pearson.

Rozelle, S., Huang, J. K., & Zhang, L. X. (1997). Poverty, Population and Environmental Degradation in China. *Food Policy*, *22*(3), 229–251. doi:10.1016/S0306-9192(97)00011-0 PMID:12348536

Sun, S. (2010). Heterogeneity of FDI Export Spillovers and Its Policy Implications: The Experience of China. *Asian Economic Journal*, *24*(4), 289–303. doi:10.1111/j.1467-8381.2010.02042.x

Topalova, P. (2007). Trade Liberalization, Poverty, and Inequality: Evidence from Indian Districts. In A. Harrison (Ed.), *Globalization and Poverty* (pp. 291–335). Chicago, IL: University of Chicago Press. doi:10.7208/chicago/9780226318004.003.0008

Tutwiler, M. A., & Straub, M. (2005). *Making Agricultural Trade Reform Work for the Poor*. Washington, DC: International Food and Agriculture Trade Policy Council.

Van Keulen, H., Kuyvenhoven, A., & Ruben, R. (1998). Sustainable Land Use and Food Security in Developing Countries: DLVs Approach to Policy Support. *Agricultural Systems*, *58*(3), 285–307. doi:10.1016/S0308-521X(98)00032-8

Vuković, P., Kljajić, N., & Arsić, S. (2012). Multifunctional Agriculture as an Assumption and a Condition for Rural Development in Serbia-Special Turn to Rural Tourism. *International Journal of Sustainable Economies Management*, *2*(1), 24–32. doi:10.4018/ijsem.2012040103

Woods, M. (1998). Advocating Rurality? The Repositioning of Rural Local Government. *Journal of Rural Studies*, *14*(1), 13–26. doi:10.1016/S0743-0167(97)00044-2

Yang, H., & Li, X. B. (2000). Cultivated Land and Food Supply in China. *Land Use Policy*, *17*(2), 73–88. doi:10.1016/S0264-8377(00)00008-9

Zhao, S. X. B., & Wong, K. K. K. (2002). The Sustainability Dilemma of Chinas Township and Village Enterprises: An Analysis from Spatial and Functional Perspectives. *Journal of Rural Studies*, *18*(3), 257–273. doi:10.1016/S0743-0167(02)00005-0

Zykova, N., Ikonnikova, O., & Kononov, O. (2011). Diversification of Rural Economics: Problems and Perspectives. *Russian Entrepreneurship*, *11*(2), 151–155.

KEY TERMS AND DEFINITIONS

Far East: The regions of East Asia, including Northeast Asia and the Russian Far East, the extreme east parts of Russia, between Lake Baikal in Eastern Siberia and the Pacific Ocean.

Food Security: Available access to sufficient, safe, and nutritious food to maintain a healthy and active life.

Krai: A type of geographical administrative division in Russia. Historically, krais comprised vast territories located along the periphery of Russia. There are 9 krais in Russia.

Oblast: A type of administrative division in Russia and some other states of the former Soviet Union. There are 46 oblasts on Russia.

Province: A type of the highest-level administrative division in Chine. There are 23 provinces in China.

Rural Population: People living in rural areas as defined by national statistical offices.

Rural Area: A geographic area that is located outside towns and cities, with low population density and small settlements.

Sustainable Development: A combination of economic development, including productive sectors providing employment, desired consumption, and wealth; human development, including increased life expectancy, education, and equity; society development, including the values of security and well-being of national states, regions, and institutions as well as the social capital of relationships and community ties.

Chapter 9
Retrospective of Economic and Trade Policies Focused on Agricultural Development: Case of Azerbaijan

Khatai Aliyev
Baku Engineering University, Azerbaijan

Ilkin Gasimov
Institute for Scientific Research on Economic Reforms, Azerbaijan

ABSTRACT

The importance of the agricultural sector in the economy of Azerbaijan is high. This sector has always been at the center of economic reforms by the government. This chapter overviews economic and trade policies of the government focused on the development of agricultural production since 1991. Authors carry out analysis of policy changes during separate development stages. The research output presents agricultural policy before the oil boom as mainly devoted to achieving structural transformation from centrally planned economy to the market environment. Within the oil boom period, the government provided substantial fiscal and technical support to the agricultural sector as well as applied tax incentives to farmers but did not pay attention to the transformation from family farming to medium and large-scale production. However, fiscal and macroeconomic challenges of post-oil boom period forced the government to focus on increasing efficiency of the subsidies and incentives and implementation of further agricultural reforms.

INTRODUCTION

Azerbaijan is a country situated in the South Caucasus region. After regaining its independence in October 1991, the country declared its intention to transform economic system from centrally planned to market economy. Before 1991, Azerbaijan's economic system was totally owned and governed by the central government. Public property dominated in agricultural sector. That is why the primary focus

DOI: 10.4018/978-1-5225-2733-6.ch009

of agricultural policy after regaining the independence was privatization of public property. However, efficiency of this transformation is still open to question.

Agriculture is a crucial sector for Azerbaijan economy due to different reasons such as maintaining food security of the country and share of population employed in this sector. That is why economic policies addressed to stimulate agricultural production have been the priority for the government. For this purpose, the government has provided subsidies and applied tax freedom for domestic agricultural producers. Moreover, preferential loans have been offered for agricultural enterprises with favorable conditions through Azerbaijan National Fund for Entrepreneurship Support (ANFES).

Several studies analyze the tendencies of agricultural production performance and efficiency of financial and technical support to this sector, i.e. Food and Agriculture Organization of the United Nations (2012); Welton, Astaryan, & Jijeleva (2013); Danilowska, Ismayilov, and Aliyev (2014); Huseyn (2014); Center for Economic and Social Development (2015); Karimov (2015); Aliyev and Suleymanov (2015); Khalilov (2015). The novelty of the current study is analyzing the tools of the economic policy of the Azerbaijan government oriented on agricultural development in retrospective context, including the recent policy implementation behavior. Especially, agricultural policy changes in post oil boom period are new for the literature.

The core purpose of this chapter is to outline the agricultural policy of Azerbaijan in the first years of independence, as well as before and after the oil boom by using descriptive tools. Within this framework, background section was devoted to present a detailed description of Azerbaijan's agriculture sector since regaining the independence. Meanwhile, this section also represents the role of agriculture sector in Azerbaijan economy. Then, we review agricultural transformation process along with the governmental financial and technical support to this sector. Furthermore, problems, solutions, and recommendations are discussed.

BACKGROUND

Overview of Agricultural Production in Azerbaijan

Azerbaijan is a country with 86.6 thousand km^2 area and more than 9.5 million population. It is situated on the shore of Caspian Sea and borders with Russia in the North, the Islamic Republic of Iran in the South, Georgia in the North–West and West, and Armenia in the West. Azerbaijan is surrounded by the Greater and Lesser Caucasus Mountains and Kura-Araz lowland (Khalilov, Shalbuzov, & Huseyn, 2015, p. 4).

Depending on geographical features, there are various climate patterns in Azerbaijan, such as arid and humid subtropical. The annual average temperature in lowlands in summer is 25-30°C, while not above 5°C in highlands. In winter, the temperature is 3-6°C and (-3°C)–(-5°C), respectively (Khalilov et al., 2015, p. 4).

The share of arable soil in the total area of Azerbaijan is only 40%. Except this, 5% of its area is useful for perennial planting, 57% for haymaking and pasture. Salinization problem is specific for Azerbaijan lands. The total share of salinized areas is above 60%, which causes a decrease of agricultural lands suitable for farming. Total land area per capita declined to 0.97 hectares in 2010, where it was 2.26 in 1960. The decrease is also observed in arable lands per capita. It declined from 1.09 hectares to 0.5 hectares in 1960-2010. Cultivated lands per capita were 0.38 and 0.17 hectares within the given period of time (Huseyn, 2014, p. 10).

The history of agriculture in Azerbaijan goes to early ages. Azerbaijan as a country situated in convenient geographical position and having rich flora attracted the first humans to the region. According to writings by ancient historians such as Herodotus, Theophrastus, Polybius and others, wheat was cultivated from the early ages in this region. Archeological discoveries in different parts of Azerbaijan also prove this fact. Except for wheat, humans of this region cultivate garlic, pea, watermelon, vegetables, etc. (Abbasov, 2013, pp. 58–59).

Agriculture sector of Azerbaijan started to develop during the Soviet period. The growth especially increased after the World War II and continued until the collapse of the Soviet Union. During the 1960's, agriculture sector of Azerbaijan could end the recession and enter to the development stage. In 1969, there were produced 574 thousand tons of wheat, 299 thousand tons of cotton, 372 thousand tons of vegetables, 272 thousand tons of grapes, 52 thousand tons of fruit, 113 thousand tons of potatoes, and 24.5 thousand tons of tobacco in Azerbaijan. In that year, productivity of grain was 9.5 quintals per ha, cotton – 15.1 quintals per ha, vegetables – 120 quintals per ha, grape – 43.6 quintals per ha, fruit – 10.4 quintals per ha, tobacco – 18 quintals per ha, and potato – 79 quintals per ha (Abbasov, 2013, pp. 2–3).

Cattle breeding also grew within the given period. In 1969, the productivity of calf born per 100 cows was 49, where this figure was 43 per 100 sheep. In average, 677 kg milk output per cow was 677 kg, and wool output per sheep was 1.6 kg. The development tendency continued during the 1970-1975s. As a result, production in agriculture sector increased by 34%, where it is expected to be 30% (Abbasov, 2013, p. 3).

In Azerbaijan, a number of decrees were approved for development of agriculture sector during the 1970's. One of such decrees was the Decree on Actions on Development of Agriculture of Azerbaijan Soviet Socialist Republic (ASSR) approved by the Central Committee of the Soviet Communist Party and the Soviet of Ministers of the Union of Soviet Socialist Republics (USSR) in July 1970. It was followed by the Decree on Intensification of Agricultural Production in ASSR in 1975 and the Decree on Specialization of Agriculture Sector and Development of Vine-growing and Winemaking in ASSR in 1979. As a result, total productivity of agriculture sector increased 2.5 times during 1981-1985 in comparison to 1966-1970. The upward trend is also observed in separate fields of agricultural production. For example, output of wheat increased 1.7 times and reached 1,240 thousand tons, vegetables – increased 2.4 times up to 891 thousand tons, cotton – increased 2.2 times up to 707 thousand tons, tobacco – 2.7 times up to 57 thousand tons, tea – increased 3.2 times up to 29 thousand tons, fruit – increased 3.8 times up to 317.5 thousand tons, vine – increased 7.9 times up to 1,829.2 thousand tons (Abbasov, 2013, p. 3).

With the collapse of the Soviet Union in 1991, Azerbaijan regained its independence. The first years of independence were very dramatic for Azerbaijan. Economic problems occurred along with the collapse of the Soviet Union, political difficulties, and the war with Armenia negatively impacted the economy of the country. Agriculture was also influenced by these processes. As a result, agricultural production decreased by 58% during 1990-1995 (Aras, Suleymanov, & Mammadov, 2016, p. 70). Data for the production of various agricultural products in 1990-1995 is presented in Table 1.

Cattle breeding was also negatively influenced by the events occurred at the beginning of the 1990s. Consequently, almost all branches of cattle breeding witnessed downward trend till 1995 (Table 2).

After 1995, the country entered the recovery period, and a number of actions were taken for development of agricultural sector. Firstly, privatization was launched in 1996. As a result, 200 farms were privatized, and approximately 36,100 new farms were established (State Statistical Committee of the Republic of Azerbaijan, 2016). In other words, 3.5 million people got private lands; the share of private lands in the total land area of the country was approximately 20%.

Table 1. Output of selected agricultural product in Azerbaijan in 1990-1995, thousand tons

Years	Grain	Cotton	Tobacco	Grape	Tea	Sugar beet	Potato	Vegetable	Fruit
1990	1,413.6	542.9	52.9	1,196.4	30.7	n.a.	185.2	856.2	367.4
1991	1,346.4	539.7	57.3	1,125.6	26.6	n.a.	179.9	805.3	498.3
1992	1,337.2	336.3	52.3	607.0	22.6	19.8	156.0	555.1	400.9
1993	1,147.9	284.5	44.9	411.3	24.0	11.6	152.2	487.8	346.4
1994	1,039.2	283.7	20.8	313.8	19.4	17.8	150.3	482.9	323.5
1995	921.4	274.1	11.7	308.7	9.4	28.1	155.5	424.1	324.4

Source: State Statistical Committee of the Republic of Azerbaijan, 2016

Table 2. Output of selected animal products in Azerbaijan in 1990-1995

Years	Meat (thousand tons)	Milk (thousand tons)	Egg (million)
1990	175.5	970.4	985.3
1991	153.5	947.7	958.2
1992	112.7	850.4	812.2
1993	92.5	798.5	584.5
1994	84.4	783.7	494.0
1995	82.0	826.5	455.8

Source: State Statistical Committee of the Republic of Azerbaijan, 2016

Before the privatization, 99.8% of cattle stock and 98% of small animals belonged to the state. With the launch of privatization in the agricultural sector in 1996, the share of private sector in cattle breeding increased sharply (Ibadoglu, 2014, p. 3).

Except for privatization, a number of other decrees related to agricultural production were approved, i.e. the Decree on Agrarian Reforms (February 18, 1995), the Decree on Distribution of Property in Agricultural Enterprises and Transfer of Some Parts to Municipalities (January 10, 1997), the Decree on Land Leasing (March 12, 1999), the Decree on Land Market (July 11, 1999), the Land Code of the Republic of Azerbaijan (June 25, 1999). Decree by the President of the Republic of Azerbaijan on Acceleration of Reforms in Agrarian Sector on March 22, 1999, also positively affected the development of agricultural production (Abbasov, 2013, p. 6).

In consequence of the measures undertaken by the government, development in the agricultural sector was observed. Agricultural statistics, particularly, statistics of cattle breeding shows the increase of agricultural production (Table 3).

Role of Agricultural Sector in National Economy

During recent years, agricultural sector played an important role in the economy of Azerbaijan. This fact is especially true for the period before the oil boom, which started after 2005. The share of agricultural production in total Gross Domestic Product (GDP) can be a good evidence for this fact. In 2000, the

Table 3. Output of selected agricultural products in Azerbaijan in 1996-2014, thousand tons

Years	Grain	Cotton	Tobacco	Grape	Sugar beet	Potato	Vegetable	Fruit	Meat	Milk	Egg (million)
1996	1,018.3	274.4	11.2	275.0	40.1	214.6	570.0	3212	85.7	843.3	477.3
1997	1,127.1	124.6	15.1	145.3	33.7	223.4	495.4	330.9	90.5	881.5	492.4
1998	950.3	112.9	14.6	144.2	41.4	312.5	502.3	390.6	99.9	946.5	509.0
1999	1,098.3	96.8	8.6	112.5	42.2	394.1	670.8	436.5	104.6	993.4	526.3
2000	1,540.2	91.5	17.3	76.9	46.7	469.0	780.8	477.0	108.7	1,031.1	542.6
2001	2,016.1	83.6	12.7	68.1	41.3	605.8	916.4	497.5	114.1	1,073.7	555.5
2002	2,195.7	80.4	3.3	62.1	115.8	694.9	974.6	516.8	124.6	1,119.9	561.6
2003	2,057.8	99.6	4.7	65.0	84.6	769.0	1,046.3	572.1	134.4	1,167.8	670.6
2004	2,158.2	135.7	6.5	54.9	56.8	930.4	1,076.2	424.6	196.8	1,213.7	829.4
2005	2,254.3	191.2	7.2	79.7	36.6	1,083.1	1,127.3	625.7	198.0	1,251.8	874.6
2006	2,564.2	125.9	4.8	94.1	167.2	999.3	1,186.4	662.4	204.7	1,299.3	760.9
2007	2,725.5	95.5	2.9	103.4	141.9	1,037.3	1,227.3	677.7	218.7	1,341.0	953.6
2008	2,848.3	55.4	2.5	115.8	190.7	1,077.1	1,228.3	712.8	228.0	1,381.3	1,101.2
2009	2,988.3	31.9	2.6	129.2	188.7	983.0	1,178.6	718.2	231.5	1,432.8	1,209.4
2010	2,000.5	38.2	3.2	129.5	251.9	953.7	1,189.5	729.5	244.9	1,535.8	1,178.6
2011	2,458.4	66.4	3.6	137.0	252.9	938.5	1,214.8	765.8	254.9	1,597.5	1,011.0
2012	2,802.2	57.0	4.3	151.0	173.8	968.5	1,216.2	810.0	276.0	1,605.6	1,226.7
2013	2,955.3	45.2	3.5	148.5	187.9	992.8	1,236.3	853.8	286.9	1,796.7	1,401.5
2014	2,383.3	41.0	2.9	147.7	177.3	819.3	1,187.7	850.8	291.2	1,855.8	1,562.7

Source: State Statistical Committee of the Republic of Azerbaijan, 2016

contribution of agricultural sector to the GDP of Azerbaijan was above 16%. However, after that year, it started to decline down to 9.0% in 2005 and 5.5% in 2010. Negative trend continued till 2014. At the end of that year, oil prices started to decline in the global market. As a result, the share of the oil sector in total GDP of Azerbaijan also decreased. In such condition, the share of agricultural sector increased after a long period up to 6.2% in 2015 (State Statistical Committee of the Republic of Azerbaijan, 2016).

Despite the decrease in percentage, the upward trend was observed in the overall volume of agricultural output, including forestry and fishing. So that, in 2000 these sectors provided an output equivalent to \$972,948, where this number reached \$3,005,897 in 2010 and \$3,224,095 in 2015 (State Statistical Committee of the Republic of Azerbaijan, 2016). More detailed data is presented in Table 4 and Figure 1.

The agricultural sector of Azerbaijan plays a significant role in employment, being the biggest employer in the country. Statistics data show that the share of agricultural employment in overall employment was 38.7% in 2005. In 2010, this figure decreased down to 38.2%, in 2015 – down to 36.4%.

Export of agricultural products has always been one of the major problems of Azerbaijan economy. In the countries rich in natural resources, like Azerbaijan, export of such resources naturally prevails over agricultural products. Azerbaijan is a big exporter of crude oil and natural gas. Statistics shows that before the 2000s, the share of agricultural export in total export in Azerbaijan was below 10%. For example, in 1996 – 4.5%, in 1997 – 7.0%, and in 1998 – 7.7%. (Table 5)

Table 4. Output in agriculture, forestry and fishing production in Azerbaijan in 2000-2015

Years	Agriculture, forestry, and fishing, $ million	Share of agricultural sector in GDP, %
2000	972,948	16.1
2001	1,011,154	14.8
2002	1,084,615	14.0
2003	1,138,461	12.4
2004	1,201,667	11.0
2005	1,458,846	9.1
2006	1,704,358	7.1
2007	2,437,179	6.7
2008	2,879,487	5.6
2009	2,794,230	6.1
2010	3,005,897	5.5
2011	3,389,102	5.1
2012	3,607,307	5.1
2013	4,002,820	5.4
2014	4,024,615	5.3
2015	3,224,095	6.2

Source: State Statistical Committee of the Republic of Azerbaijan, 2016

Figure 1. Dynamics of output in agriculture, forestry and fishing production in Azerbaijan in 2000-2015
Source: State Statistical Committee of the Republic of Azerbaijan, 2016

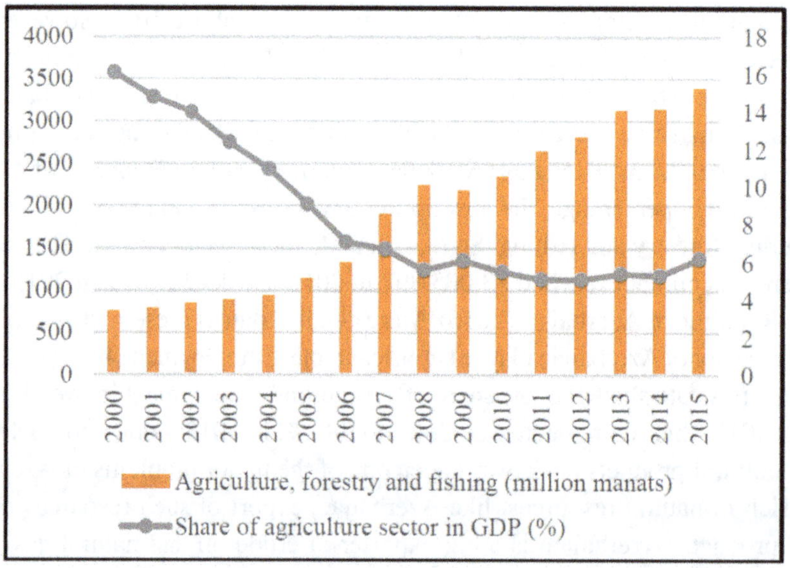

Table 5. Agricultural export and its share in total export of Azerbaijan in 1995-2015

Year	Agricultural export, $ thousand	Share in total export of the country, %
1995	38,531.8	6.04
1996	28,557.8	4.52
1997	55,343.5	7.00
1998	46,733.9	7.70
1999	60,180.7	6.47
2000	56,694.1	3.25
2001	55,535.0	2.40
2002	67,666.4	3.12
2003	84,269.8	3.25
2004	156,014.8	4.31
2005	326,915.9	7.52
2006	165,330.5	2.59
2007	507,554.6	8.37
2008	532,988.8	1.11
2009	526,152.4	3.58
2010	593,342.9	2.78
2011	717,079.1	2.70
2012	832,207.3	3.48
2013	878,256.8	3.66
2014	840,797.9	3.85
2015	770,461.3	6.74

Source: State Statistical Committee of the Republic of Azerbaijan, 2016

During 2000-2005, the share of agriculture in total export decreased down to 4.32% (in 2004). By the beginning of the oil boom after 2005, share agriculture in total export started to decrease and fell down to 2.78% in 2010. However, after 2014, the decline in oil prices positively impacted agricultural exports. As a result, the share of agriculture in total export increased up to 6.74% in 2015 (Figure 2).

The agricultural sector is important not only for its contribution to various spheres of the economy but also for food security. Despite all shortcomings in agriculture, Azerbaijan government succeeded in achieving food security and improving the self-sufficiency of the country during the recent years. In order to solve the food security problem, the Food Security Programme of Azerbaijan for 2001-2010 was adopted in 2001. In addition, in 2008, the President of Azerbaijan approved the State Programme on Reliable Food Supply of Population of Azerbaijan for 2008-2015 (Huseyn, 2014, p. 8).

Despite the efforts of the government on ensuring food security in the country, there are still certain problems in this area. The most important one is that Azerbaijan is dependent on food imports (Table 6).

Figure 2. Dynamics of agricultural export and its share in total export of Azerbaijan in 1995-2015
Source: State Statistical Committee of the Republic of Azerbaijan, 2016

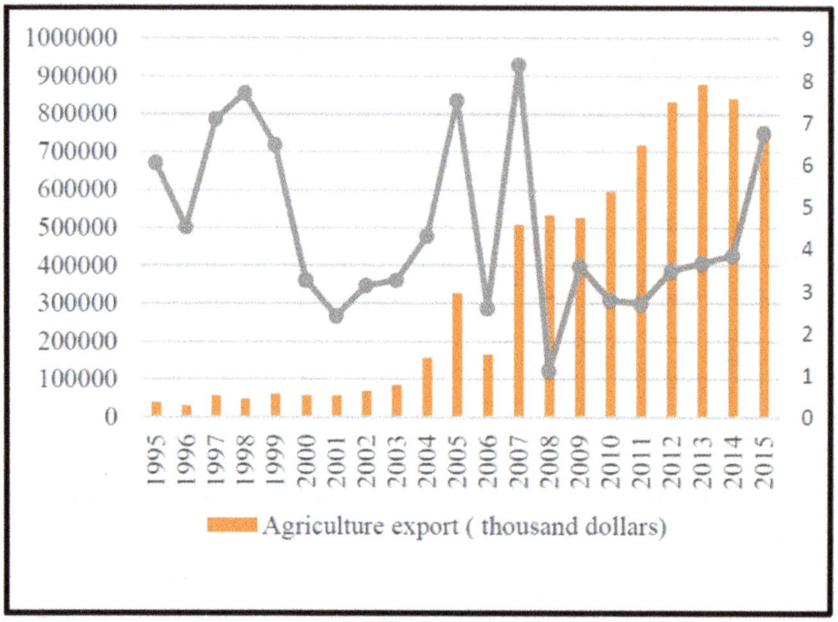

Table 6. Self-sufficiency in some crop products in Azerbaijan in 2007-2013, percentage

Product	2007	2008	2009	2010	2011	2012	2013
Crops, total	57.0	63.3	74.0	56.5	64.8	64.3	64.0
Wheat	48.0	55.7	69.2	48.9	57.7	56.8	56.0
Barley	99.2	97.2	98.7	87.7	93.7	95.1	97.8
Corn	76.4	68.1	67.2	64.5	68.0	67.3	60.0
Oats	68.2	88.6	94.5	80.6	82.8	84.4	89.3
Other crops	10.0	4.4	0.4	1.0	4.0	8.7	2.3
Legume	76.4	77.7	68.9	65.8	70.4	71.6	76.6
Potato	98.0	103.6	104.7	100.5	101.6	98.2	97.6
All types of vegetables	98.8	104.6	101.2	97.6	95.7	98.9	102.3
Plantation goods	100.1	100.2	100.2	100.0	100.2	100.1	100.0
Fruits and berries	138.1	161.4	136.0	107.9	116.8	125.7	121.8
Grape	93.7	93.3	90.7	90.4	89.9	94.3	95.0

Source: State Statistical Committee of the Republic of Azerbaijan, 2016

MAIN FOCUS OF THE CHAPTER

Considering the essential role of agricultural sector in the economy, Azerbaijan government has been continuously supporting this sector in the forms of subsidies, concessional credits, and technical support. In this part of the chapter, the authors discuss transformation from centrally planned to market economy in relation to agriculture, as well as the financial system in Azerbaijan agriculture and direct financial support to producers of agricultural products. In addition, this part of the chapter covers other types of incentives and tax concessions.

Agricultural Transformation: Privatization

As mentioned above, Azerbaijan quit centrally planned economic system immediately after regaining its independence and declared its intention to build a new economic system based on market principles. Before that, economic system of the country was addressed to strategic needs of the former Soviet Union (Soyak & Nesirova, 2003). Farming used to be held under the control of central administration within massive production cooperatives and state farms (Lerman, 2001). One of the major features of that time was collective farming in the forms of the so-called "sovkhozes" and "kolkhozes".

Due to the military conflict with Armenia and occupation of 20% of the territory, as well as severe economic crises and instability, the launch of economic reforms was delayed for a number of years. On July 16, 1996, Azerbaijan introduced the Law about Land Reform. This law entered into force on August 2, 1996, and was targeted on agricultural transformation. For this purpose, the State Commission on Agrarian Reform was created along with its regional and local bodies. According to the Law, lands were classified on the type of ownership, i.e. public, municipal, and private.

Most of the arable lands were transformed into private property for the use of citizens born before August 2, 1996. In other words, the government launched reforms in the agricultural sector, distributed land among private owners, and implemented privatization of the property of "kolkhozes" and "sovkhozes" (Aliyev, 2014; Aliyev & Suleymanov, 2015). As a result of agricultural reforms, 56.9% of total area of the country (4.92 million hectares) was left in state property, 23.5% (2.03 million hectares) was left to municipalities, and 19.6% (1.70 million hectares) was allocated as private property (Ahadov, 2009). Meanwhile, land of former state farms was allocated as following: 38.1% (1.37 million hectares) as private property, 56.4% (2.03 million hectares) to municipalities, and only 5.5% (0.20 million hectares) was left to the state (Ahadov, 2009). According to Ahadov (2009), 3.5 million citizens in rural areas of the country (872 thousand families) received land plots as their private property as result of the land reform. The size of land plot per person changed depending on the land stock availability and land quality across the country. Resulting from the land reform, 2,028 collective farms were liquidated, while 41 state farms were established in order to maintain people in rural areas with seeds and agricultural animals. Land reform within the internationally recognized territory of Azerbaijan was not implemented in the areas of Nagorny Karabakh and surrounding districts, which are still under the occupation of Armenia.

Land privatization led to a structural transformation in the agricultural sector of Azerbaijan. It also determined the structure of production by various types of ownership. Family farming started to dominate in the structure of agricultural production. According to the official statistics, in 2014-2015, 92-93% of total output in agriculture was provided by private landowners, family peasant farms, and households (Figure 3).

In the course of the land reform, the share of agricultural production provided by private owners, family peasants, and households have been increasing continuously, while total output by agricultural enterprises and organizations remained stable. In 1998-2006, the output of agricultural enterprises and organizations did not grow. Only after 2007, it started to increase slightly.

Figure 3. Structure of agricultural production in Azerbaijan, million AZN
Source: State Statistical Committee of the Republic of Azerbaijan, 2016

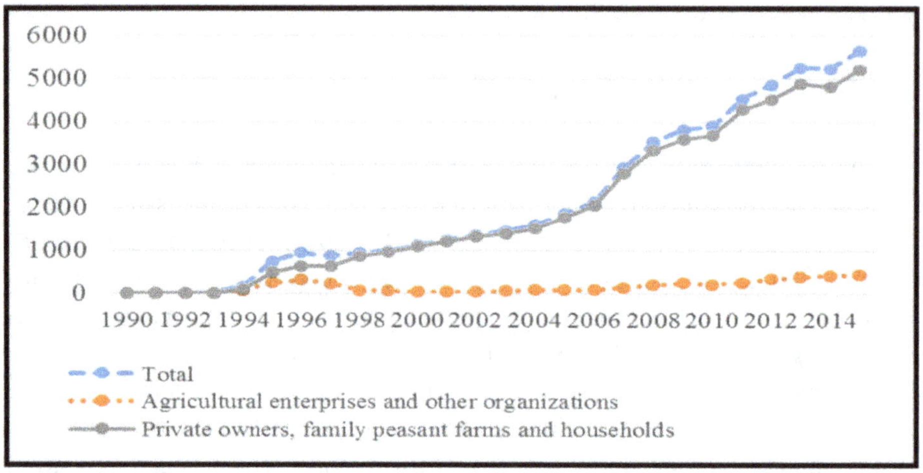

The effects of agricultural reforms and structural transformation embody themselves in the statistical data related to the number of agricultural enterprises and private ownerships (Table 7). Private ownerships are those which carry out activities without forming a legal entity.

The number of agricultural enterprises and service organizations had decreased till 2003. During 2003-2006, there was some growth. Hence, the number of agricultural enterprises and service organizations in 2006 was 26.9%, which was more than in 2003. However, during the oil boom period of 2005-2014, the agricultural sector lost its investment attractiveness, which led to a permanent decrease in a number of enterprises in this sector. By 2015, the downward trend still exists.

A similar trend is observed in relation to the number of different categories of agricultural organizations. Resulting from the agrarian reform, only two collective farms decided to keep their previous legal status voluntarily. Others were liquidated, and their property was allocated. The transition from collective farming to family farming affected production in a negative manner. From this point of view, the efficiency of the reforms in the context of agricultural development is doubtful. The main problem is that family farming is a tradition-bound way of agricultural production. While discussing agricultural development policies and government support to farmers, production performance should be considered.

Agricultural Financial System and Preferential Loans

During various stages of economic development, the agricultural sector of Azerbaijan was challenged by a shortage of financial resources. During the periods of recession (1991-1994) and restructuring (1995-2005), such a shortage was natural due to the poor economic situation in the country. However, this might be changed during the oil boom (2005-2015) as the country gained huge revenues from the export of crude oil. Despite its crucial role in maintaining employment and social welfare, the financial system in Azerbaijan agriculture has not been improved.

The agricultural financial system in Azerbaijan should be considered within the overall financial system of the country where both public and private organizations take their roles in establishing a fi-

Table 7. Agricultural enterprises, service organizations, and private ownerships in Azerbaijan in 2000-2015

Year	Agricultural enterprises, service organizations	including							Private ownerships
		Agricultural enterprises	including				Service organizations	
			State enterprises	Collective farms	Cooperatives	Other private organizations		
2000	2,910	2,653	408	2	250	1,993	257	3,248
2001	2,420	2,191	247	2	182	1,760	229	2,604
2002	1,984	1,759	319	2	168	1,270	225	2,589
2003	2,143	1,918	302	2	157	1,457	225	2,607
2004	2,241	1,988	302	2	165	1,519	253	2,661
2005	2,506	2,182	303	2	164	1,713	324	2,681
2006	2,518	2,208	274	2	156	1,776	310	2,456
2007	2,282	1,925	264	2	162	1,497	357	2,488
2008	2,308	1,854	257	2	88	1,507	454	2,613
2009	2,409	1,913	249	2	83	1,579	496	2,571
2010	2,336	1,825	217	2	73	1,533	511	2,618
2011	2,343	1,917	254	2	69	1,592	426	2,593
2012	1,864	1,360	241	2	72	1,045	504	2,451
2013	1,860	1,337	238	2	72	1,025	523	2,334
2014	1,869	1,334	240	2	64	1,028	535	1,624
2015	1,837	1,298	235	2	58	1,003	539	1,534

Source: State Statistical Committee of the Republic of Azerbaijan, 2016

nancial framework for farmers. The system is composed of financial intermediaries and state agencies. Mechanism of the agricultural financial system in Azerbaijan is outlined in Figure 4.

Financial intermediaries are commercial banks and non-bank credit organizations which offer credits for agricultural enterprises and other organizations. By the end of 2015, there are 43 banks (2 public and 41 private) and 157 non-bank credit organizations in Azerbaijan. On the one hand, such big number of players is beneficial for the agricultural sector. However, agricultural production is considered as high-risk for investing, and therefore, credit interest rates are high. On the other hand, financial intermediaries require tangible guarantees when providing requested a number of financial resources. Considering the share of family farms in total agricultural production, most of which are of a small size and located in the regions of Azerbaijan with insufficient ability to provide any guarantees to credits organizations, financial resources of commercial banks and non-bank credit organizations are almost not available and economically reasonable for agricultural producers.

Therefore, the share of agricultural loans in total credit portfolio of banks and non-bank credit organizations is rather small. Changes in amount and shares of agricultural loans during 2005-2015 are presented in Table 8.

A number of loans provided by financial intermediaries to agricultural producers are not sufficient to ensure development in this sector. That is why the government provides financial support to farmers through different channels, such as subsidies or preferential credits. In this context, Azerbaijan National

Figure 4. Financial system of Azerbaijan agriculture
Source: Danilowska et al., 2014

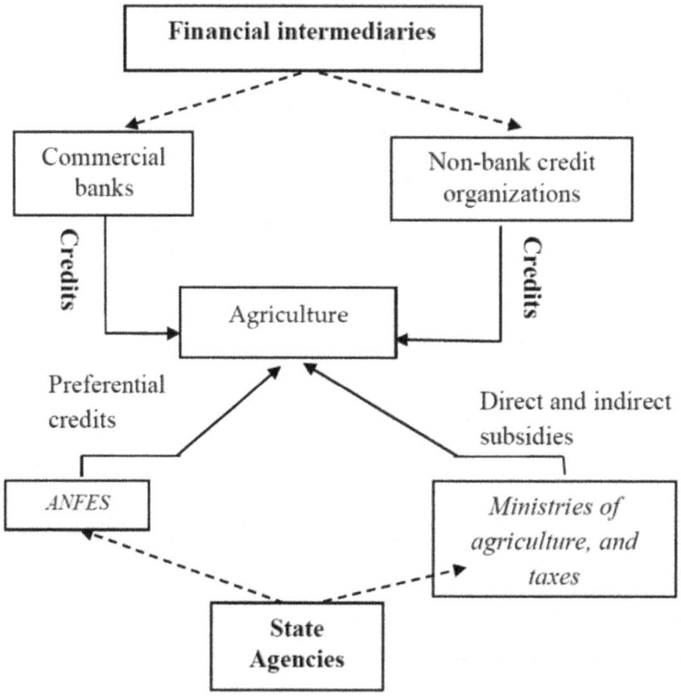

Table 8. Loans to agricultural producers in Azerbaijan in 2005-2015, $ million

	2005	2007	2009	2010	2011	2012	2013	2014	2015
Loans, total	1,847.4	6,002.3	10,778.8	11,747.9	12,628.6	15,697.0	19,772.9	23,772.6	21,303.9
Loans to agricultural producers	125.1	252.8	506.1	565.8	598.3	700.3	940.1	1086.3	498.14
% in total	6.8	4.2	4.7	4.8	4.7	4.5	4.8	4.6	2.3

Source: Central Bank of the Republic of Azerbaijan, 2016
Note: 1 USD = 0.78 AZN for 2005-2014, and 1.02 AZN for 2015

Fund for Entrepreneurship Support (ANFES) plays the dominant role in the distribution of preferential loans. The Fund has been established after the declaration of independence with the purpose to provide tax credits from the state budget for entrepreneurs, but actually started its active operations only after 2002 (Azerbaijan National Fund for Entrepreneurship Support, 2016). The agricultural sector is among the priorities of the ANFES (Table 9).

The Fund unites various financial institutions and collaborates with them. The mechanism is outlined in the Procedures on the Use of Funds for the National Fund for Entrepreneurship Support of the Republic of Azerbaijan, approved by the President of the Republic of Azerbaijan on October 7, 2009.

An entrepreneur who intends obtaining preferential credit from the Fund may firstly apply to one of the participating financial institutions. Assessment of submitted investment proposals is carried out by the corresponding participating institution, and the Fund implements project expertise. Once the

Table 9. ANFES loans to agricultural sector

Year	Sector	Number of projects	Share in total number of projects, %	Credit amount, $ thousand	Share in total credit amount, %
2002	Processing of agricultural products	8	16.3	444.61	23.5
	Agricultural production	19	38.8	566.66	29.9
2003	Processing of agricultural products	30	9.7	2,536.02	18.4
	Agricultural production	177	57.1	4,886.28	35.4
2004	Processing of agricultural products	70	5.2	3,522.94	14.9
	Agricultural production	1,055	78.8	9,716.66	41.0
2005	Processing of agricultural products	44	2.0	9,042.82	19.4
	Agricultural production	1,767	80.2	16,651.02	35.8
2006	Processing of agricultural products	54	3.0	16,123.71	14.0
	Agricultural production	1,256	70.9	30,107.69	26.1
2007	Processing of agricultural products	29	3.3	17,872.56	15.4
	Agricultural production	572	64.9	30,936.41	26.6
2008	Processing of agricultural products	25	3.1	10,823.71	9.6
	Agricultural production	547	68.6	36,871.41	32.8
2009	Processing of agricultural products	19	0.9	25,183.97	15.2
	Agricultural production	1,939	92.3	28,940.12	17.4
2010	Processing of agricultural products	16	1.1	13,269.23	9.0
	Agricultural production	1,303	91.1	71,284.48	48.4
2011	Processing of agricultural products	13	0.8	21,606.41	12.2
	Agricultural production	1,503	91.5	90,203.20	51.0
2012	Processing of agricultural products	10	0.4	35,717.94	12.8
	Agricultural production	2,234	92.3	128,164.74	45.8
2013	Processing of agricultural products	10	0.2	41,896.15	11.9
	Agricultural production	4,276	95.7	191,680.76	54.4
2014	Processing of agricultural products	11	0.2	24,867.31	7.5
	Agricultural production	5,344	96.1	210,501.3	55.7
2015	Processing of agricultural products	9	0.2	12,050.0	5.0
	Agricultural production	5,080	96.9	155,044.3	63.6
	Processing of agricultural products	348	n.a.	262,029.38	n.a.
	Agricultural production	27,072	n.a.	1,005,555.03	n.a.

Source: Azerbaijan National Fund for Entrepreneurship Support, 2016
Note: 1 USD = 0.78 AZN for 2005-2014, and 1.02 AZN for 2015

expertise is completed, the Fund informs the corresponding financial institution about its decision on acceptance or rejection.

According to ANFES statistical records, 27,420 projects were accepted for financing during 2005-2015, out of which 348 projects were related to the processing of agricultural products and 27,072 projects to the development of agricultural production. In total, approximately $1.26 billion were spent for stimulation of agricultural activities. This amount is quite big for such a small economy as Azerbaijan. However, the effectiveness of those investments is under the question. The problem is if these resources are really absorbed by agriculture, not other sectors.

Note that ANFES loans are comparatively cheap for entrepreneurs, especially in agriculture. Average interest rate is below 8%, but for the projects of the first priority for the government, the interest rate is around 4% only.

Tax Incentives and Direct Subsidies

Azerbaijan government declared agricultural sector as a priority for national economic policy. That is why financial support of agricultural producers is not limited to concessional credits only. Beyond this, tax incentives and direct subsidies also embody huge support for domestic farmers.

Tax incentives cover all types of taxes except land tax. According to the Law No. 824 approved on November 22, 2013, which introduced changes to the Tax Code of the Republic of Azerbaijan, agricultural producers, both private persons and legal entities, were freed of income tax, value added tax (VAT), and property tax (Ministry of Agriculture of the Republic of Azerbaijan, 2016). This is a strong incentive as a number of tax concessions in the total amount of tax until August 2013 was 1.4 billion AZN or $1.79 billion ($1 = 0.78 AZN) (Danilowska et al., 2014). More recent statistic data are not available.

Meanwhile, according to corresponding decrees by the Cabinet of Ministries of the Republic of Azerbaijan, imported machines and other type of technical vehicles, which are subjects to leasing contracts and are implemented in production and manufacturing of agricultural products, are freed of custom tariffs and VAT (Ministry of Agriculture of the Republic of Azerbaijan, 2016). In addition, import of all kinds of machines and their parts as well as breeds and various agricultural products intended to be used in production, manufacturing, and sorting of domestic agricultural products are also freed of customs tariffs and VAT.

Moreover, Azerbaijan government covers 50% of costs of fuel and fertilizers used by agricultural producers in the form of subsidies as well as finances purchasing of seeds that are more productive. Another type of subsidy is the amount of money paid to farmers per hectare of grain crops. The amount of subsidy per hectare is 40 AZN.

When considering the total amount of subsidies, one may notice the prevalence of official support. Until August 2013, total amount of direct subsidies was 644.89 million AZN ($826.78 million), out of which fertilizers: 87.29 million AZN ($111.91 million), fuel and motor oils: 378.25 million AZN ($484.93 million), seeds: 33.05 million AZN ($42.37 million), grain crops: 146.3 million AZN ($187.56 million) (Danilowska et al., 2014).

Overall assessment of the economic policy of Azerbaijan government in the context of support of agricultural production shows that the support is rather substantial in physical terms, but not very effective. As the authors noted in the Background, the situation in the agricultural sector has not changed significantly during the recent years. However, serious economic challenges related to the sharp decrease of oil price since the end of 2014 force the government to implement certain reforms in order to reduce dependence on oil production and stimulate non-oil exports. In this context, agriculture as a traditional sector of Azerbaijan economy with the high level of employment should play its role. Next section of the chapter discusses latest attempts and policy changes to overcome the challenges and develop agricultural production.

SOLUTIONS AND RECOMMENDATIONS

One of the major problems related to the agricultural sector in Azerbaijan is that the majority of agricultural production is provided by small family farms. It is also a serious issue to be solved in order to expand export of agricultural products. There are only a few legal entities professionally engaged in export of agricultural products in Azerbaijan. Therefore, agricultural sector must be restructured and product diversification and quality control should be developed.

With a purpose to increase the amount of support to agricultural producers and increase the effectiveness of investments, Azerbaijan government made significant attempts during 2015-2016. There are several major policy changes implemented within the last two years. One of such attempts is approval of the Decree by the President Ilham Aliyev on January 2015 on declaring 2015 the Year of Agriculture with the purpose to trigger sectoral production and modernization, employ public finance support and comprehensively promote the agricultural potential of the country (President of Azerbaijan, 2015). According to the Decree, the number of employees under the Ministry of Agriculture increased to 1,215 in order to reach defined targets, Also, the regional branches of the Ministry were established.

On April 11, 2016, President signed the Decree on Establishment of the Open Joint Stock Company on Food Products Supply and Procurement under the Subordination of the Ministry of Agriculture of the Republic of Azerbaijan (President of Azerbaijan, 2016b). Establishment of such organization is very beneficial for small producers which constitute the major part of agricultural producers in Azerbaijan. The organization is supposed to lead the development of small and medium enterprises in agriculture and diversification of financial sources. Especially, need for direct financing of agricultural production from the state budget is supposed to be decreased through enhancing participation of agricultural enterprises in Baku Stock Exchange operations. Moreover, the company will determine certain standards for purchased goods which are very important for trade operations in terms of quality improvement.

Another institutional policy change, which also covers stimulation of agricultural production, is the Presidential Decree on Additional Actions on Promotion of Non-oil Products Export, approved on March 1, 2016 (President of Azerbaijan, 2016a). It is the first of such policy attempts in the history of Azerbaijan. According to the Decree, export subsidies are paid from the state budget to the exporters of domestically produced goods in the amount of 3% of their customs value.

On June 14, 2016, a new law which has been discussed in the society for a long time has been accepted. The Law About Agricultural Cooperatives should be considered as a milestone in state economic policy focused on agricultural development. According to the Law, agricultural cooperatives should be established by agricultural producers and their unions on a voluntary basis (President of Azerbaijan, 2016c). The core goal of this law is to establish new big organizations of agricultural enterprises. It is crucial for the development of agricultural sector in Azerbaijan. As stated above, agricultural production Azerbaijan is mostly realized by family farms, which makes maintaining high quality, sufficient productivity and application of modern agricultural technologies quite challenging. In addition, the small size of agricultural producers makes restricts most of them from an access to international markets. It is expected that cooperatives will solve this issue at least partially. Moreover, the government could have an opportunity to control the use of subsidies and increase the efficiency of financial support and other types of incentives. Despite the expected positive outcomes, the transformation will take long. The institutional side of establishment of cooperatives should be carefully planned, and cooperatives should be attractive for family farms to stimulate unification.

However, current transformation of agricultural policy is accompanied by fiscal challenges. State budget revenues from oil are decreasing, that is why the government employs contractionary fiscal policy after long lasting expansionary tendency within the last decade (Aliyev & Gasimov, in press). Thus, budget expenditures were cut in 2016 budget proposal, and are expected to be cut even more in the near future. Besides, the government expanded the struggle against tax evasions and focused on increasing tax revenues from the non-oil sector. In the context of current fiscal policy challenges, tax incentives and subsidization could be unsustainable to maintain. Therefore, it is recommended to provide more technical and institutional support to agricultural producers, especially medium and large-scale enterprises, and increase the efficiency of state subsidies.

There is a big shortcoming in the implementation of economic policies focused on agricultural development in Azerbaijan. It is not the establishment of a mechanism which stimulates innovations, not the development of new innovative products or process approaches, but their application in the production process. Although there is Azerbaijan State Agricultural University, its scientific potential is neither sufficient not employed effectively. In order to address this issue, there is Broker State Mechanism approach elaborated by Ismayilov and Aliyev (2013) for the non-oil sector and Ismayilov and Aliyev (2016) for the agricultural sector. The basic approach of this mechanism is that the state links innovation owners and agricultural producers, purchases innovative ideas and transfers them to producers without any charge. The only thing producers have to do is to provide guarantees for their applications.

FUTURE RESEARCH DIRECTIONS

This chapter disputed previous trends and the current situation in the agricultural sector of Azerbaijan. However, there are still some significant factors that have to be taken into account in future investigations.

Firstly, the effectiveness of governmental subsidies to agricultural sector should be researched in more details. Until now, all researchers of this issue mostly have been concentrating on subsidies and policies provided by Azerbaijan government to the agricultural sector. Nevertheless, the effectiveness of those policies was out of research focus in most of the previous investigations.

Though there are different researches investigating agricultural sector from different perspectives, to our best knowledge, there are almost no empirical papers on the agricultural sector of Azerbaijan. In order to measure different links between various factors in the agricultural sector, interdependence should be assessed in future studies in this field.

CONCLUSION

This study analyses economic policy of Azerbaijan, which is addressed to the development of the agricultural sector of the country. Considering the importance of this sector for the country, its development has always been one of the priorities for the Azerbaijan government within the past years. Agriculture is a traditional sector for Azerbaijan since nearly 37-39% of labor force is employed in agriculture. Moreover, this sector also plays a significant role in non-oil exports of the country and is considered as having a competitive advantage on the international market. Definitely, food security issue also keeps the attention of the government to this sector.

In retrospective context, it is observed that before 2000 economic policies oriented on agricultural development focused on the transformation from centrally planned to market economy, in other words, privatization. Ownership of agricultural facilities and land was transferred from public to private property. Such policy implementation led to the devastation of existing large-scale production infrastructures and brought family farms to the fore. This structural transformation was followed by sharp fall in agricultural output and serious policy challenges. Although the government attempted to provide financial support in different ways, the efficient level of production was not achieved by either family farms or the government.

During 2000-2005, agricultural privatization was almost completed, and restructuring of production facilities was launched. However, both government and private institutions expected upcoming oil boom period. Most probably, due to this reason, the agricultural sector did not get as much attention as needed. The government acknowledged that family farms could not bring sustainability to the agricultural sector and reversed on stimulation of establishment of medium and large-scale enterprises. Nevertheless, the number of agricultural enterprises decreased.

During the period of oil boom (after 2005), financial support to agricultural producers was quite sufficient to achieve a satisfactory level of progress. However, concessional loans from ANFES as well as tax incentives and subsidies did not bring along appropriate qualitative transformation in farming, production, and marketing activities. Some statistical parameters improved, but still insufficiently. To comprehend the issue better, three key characteristics of oil boom period should be considered. Firstly, huge revenues from the export of crude oil relatively appreciated the national currency and made imported agricultural products more competitive on the domestic market. Entrepreneurs found import and distribution of agricultural products more profitable compared to investing in domestic agricultural production. Secondly, factors of production moved from agriculture to other sectors, particularly oil industry, service, and public sectors. Family-based agricultural production could not avoid such resource flow. Thirdly, the government did not provide a satisfactory level of institutional support to the agricultural sector in terms of establishment of medium and large-scale enterprises, as oil revenues were flowing to the country.

This tendency changed only by the sharp fall in oil prices at the end of 2014 (the end of the oil boom). The government looked for remedies to stimulate non-oil exports and minimize output fall in the non-oil sector as well as support employment. The first solution seems to be considered as the development of traditional agricultural production. This is a new milestone in economic policies of the Azerbaijan government focused on the development of agricultural sector.

REFERENCES

Abbasov, I.D. (2013). *Azərbaycanın və Dünya Ölkələrinin Kənd Təsərrüfatı.* Baku: Şərq-Qərb Publishing House.

Ahadov, D. (2009). *Land Reform in Azerbaijan.* Paper presented at the FIG Working Week 2009, Surveyors Key Role in Accelerated Development, Eilat.

Aliyev, K. (2014). *WTO Membership and South Caucasus Economies.* Saarbrucken: LAP Lambert Academic Publishing.

Aliyev, K., & Gasimov, I. (in press). Fiscal Policy Implementation in Azerbaijan Before-Within-and-After the Oil Boom. *Contemporary Economics, 11.*

Aliyev, K., & Suleymanov, E. (2015). Macroeconomic Analysis and Graphical Interpretation of Azerbaijan Economy in 1991-2012. *Expert Journal of Economics*, *3*(1), 40–49.

Aras, N., Suleymanov, E., & Mammadov, K. (2016). *Economy of Azerbaijan 25 Years of Independence*. Baku: Sharg-Garb Publishing House.

Azerbaijan National Fund for Entrepreneurship Support. (2016). *Annual Reports*. Retrieved October 19, 2016, from http://anfes.gov.az/en/show.page.php?guid=d2f7fbbc-269c-11e0-8e86-0022190362dd

Center for Economic and Social Development. (2015). *Evaluation and Modeling of Comparative Advantages in Agriculture in Azerbaijan*. Baku: CESD.

Central Bank of the Republic of Azerbaijan. (2016). *Statistical Bulletins*. Retrieved October 19, 2016, from http://en.cbar.az/pages/publications-researches

Danilowska, A., Ismayilov, A., & Aliyev, K. (2014). The Comparative Analysis of Agricultural Financial Systems in Poland and Azerbaijan. *Acta Scientiarum Polonorum Oeconomia*, *13*(3), 41–53.

Food and Agriculture Organization of the United Nations. (2012). Assessment of the Agriculture and Rural Development Sectors in the Eastern Partnership Countries. The Republic of Azerbaijan. Budapest: FAO Regional Office for Europe and Central Asia (REU).

Huseyn, R. (2014). An Analysis of the Agricultural Subsidy Policy in Azerbaijan. *Azerbaijan Journal of Economics and Social Studies*, *1*(1), 6–18.

Ibadoglu, G. (2014). *Azerbaijan's Economic Model and Its Development since Independence*. Retrieved November 16, 2016, from http://azerireport.com/index.php?option=com_content&task=view&id=2981&Itemid=55

Ismayilov, A., & Aliyev, K. (2013). Innovation Based Economic Development and the Importance of Government Promotion. *Journal of Qafqaz University – Economics and Administration, 1*(2), 231–241.

Ismayilov, A., & Aliyev, K. (2016). Government Support to the Agriculture for Promotion and Application of Innovations in Azerbaijan: Broker-State Framework. In *Proceedings of the Scientific Conference "Expanding Innovation Activities in Azerbaijan's Non-oil Sector"*. Baku: Azerbaijan Technical University.

Karimov, R. (2015). Development of Non-Oil Sector in Azerbaijan: Tendencies and Opportunities. *Journal of Business & Economic Policy*, *2*(2), 39–52.

Khalilov, H. (2015). Sustainable Agricultural Production in Azerbaijan: Current Situation and Main Tasks. *Azerbaijan Journal of Economics and Social Studies*, *3*, 6–23.

Khalilov, H., Shalbuzov, N., & Huseyn, R. (2015). *Country Report: Azerbaijan*. Baku: Research Institute of Agricultural Economics.

Lerman, Z. (2001). Agriculture in Transition Economies: From Common Heritage to Divergence. *Agricultural Economics*, *26*(2), 96–114. doi:10.1111/j.1574-0862.2001.tb00057.x

Ministry of Agriculture of the Republic of Azerbaijan. (2016). *Tax Incentives*. Retrieved October 19, 2016, from http://www.agro.gov.az/vergi-gztlri

President of Azerbaijan. (2015). *Decree on Declaring 2015 the Year of Agriculture in the Republic of Azerbaijan*. Baku: Administration of the President of the Republic of Azerbaijan.

President of Azerbaijan. (2016a). *Decree on Additional Actions on Promotion of Non-oil Products Export*. Baku: Administration of the President of the Republic of Azerbaijan.

President of Azerbaijan. (2016b). *Decree on Establishment of the Open Joint Stock Company on Food Products Supply and Procurement*. Baku: Administration of the President of the Republic of Azerbaijan.

President of Azerbaijan. (2016c). *Law of the Republic of Azerbaijan on Agricultural Cooperatives*. Baku: Administration of the President of the Republic of Azerbaijan.

Soyak, A., & Nesirova, Z. (2003). The Consequences of IMF's Policies in the Globalization Process: The Case of Azerbaijan. In *Proceedings of the Caucasus and Central Asia in the Globalization Process International Conference*. Baku: Qafqaz University.

State Statistical Committee of the Republic of Azerbaijan. (2016). *Statistical Database*. Retrieved October 19, 2016, from http://www.stat.gov.az/menu/13/indexen.php

Welton, G., Astaryan, A., & Jijeleva, D. (2013). *Comparative Analysis of Agriculture in the South Caucasus*. Tbilisi: United Nations Development Programme.

ADDITIONAL READING

Center for Economic and Social Development. (2014). *Evaluation and Modeling of Comparative Advantages in Agriculture in Azerbaijan*. Baku: CESD.

The International Fund for Agricultural Development. (2010). *Republic of Azerbaijan. Integrated Rural Development Project (IRDP). Project Design Report*. Baku: International Fund for Agricultural Development.

World Bank. (2011). *Promoting Azerbaijan's Agricultural Productivity (1997–2010)*. Washington, DC: The World Bank Independent Evaluation Group.

KEY TERMS AND DEFINITIONS

ANFES: The National Fund for Support of Entrepreneurs, Ministry of Economy, Republic of Azerbaijan.

AZN: International code for manat – national currency of the Republic of Azerbaijan.

Food Security: An access for all people at all times to enough food for a healthy, active life.

Food Self-Sufficiency: An ability of a state to meet consumption needs by means of domestic production rather than buying or importing.

Privatization: A process which intends to transfer enterprises or industries from public to private sector. This process was launched in Azerbaijan after the collapse of the Soviet Union.

Chapter 10
Effects of International Trade Agreements on the Economy and Society of Africa:
Special Focus on Nigeria

Henrietta Nagy
Szent István University, Hungary

József Káposzta
Szent István University, Hungary

György Iván Neszmélyi
Budapest Business School, Hungary

Omokheka Gregory Obozuwa
Szent István University, Hungary

ABSTRACT

The aim of this chapter is to give an overview of the trade agreements existing in Nigeria to find out their contribution to the development of the economy. There are several such agreements, however, the authors investigate how much they really operate and serve the wellbeing of people as well as food security and safety. There have been various initiatives in the countries of Africa to develop the agricultural sector, which has employed millions of people and has had a significant role in keeping the population in the rural areas. However, the authors claim that less progress has been achieved than expected. In this chapter, the authors intend to analyze the situation in Africa, with a special focus on Nigeria, to summarize the achievements, and to list up some recommendations on future measures.

INTRODUCTION

Trade agreements are undoubtedly veritable instruments through which a country can open its economy and market to commercial opportunities present in the ever dynamic and global market. Nigeria recognizes this fact, and, over the years, has entered into several of such agreements in a bid to strengthen and expand its economic frontiers. The thrust of these agreements has come in the form of reduction or removal of tariffs, quota/trade restrictions or concessions. It could also relate to exchange of goods and services between countries, all of which come mostly either as bilateral or multilateral trade agreements. Nigeria has been a member of the World Trade Organization (WTO) since January 1, 1995, and used

DOI: 10.4018/978-1-5225-2733-6.ch010

to be a member of its predecessor organization, the General Agreement on Tariff and Trade (GATT) from November 18, 1960.

Nigeria's GDP value represented 0.78% of the world economy in 2015 that might be considered low. Nigeria's international trade more than doubled in the 2003-2009 period, with exports rising to nearly $50 billion and imports to nearly $34 billion (World Trade Organization [WTO], 2011). Nigeria is the 38th largest export economy in the world and the 125th most complex economy according to the Economic Complexity Index, ECI: -1.048 (The Observatory of Economic Complexity, n.d.). Nigeria exports 40 products with revealed comparative advantage (meaning that its share of global exports is larger than what would be expected from the size of its export economy and from the size of a product's global market). In 2014, Nigeria exported $99 billion and imported $52.3 billion, resulting in a positive trade balance of $47.4 billion. Although crude oil accounts for nearly all the value of exports, trade continues to play an important role, with total trade (imports and exports) accounting for over 53% of GDP (Central Bank of Nigeria, 2010) and tariffs being the main trade policy instrument. To encourage trade and investment, Nigeria provides a broad range of incentives most of which are tax or import-tariff related and apply to enterprises producing for export and domestic markets. Also to facilitate international trade, the Nigerian Export Processing Zone Authority superintends over eleven export processing zones (with additional ones under development).

The biggest trade partner of Nigeria is China. A range of bilateral agreements including the Agreement on Economic and Technical Cooperation, Trade Agreement, Agreement on Mutual Investment Promotion and Protection, and Double Taxation Agreement have been signed between China and Nigeria. According to the Chinese Embassy in Nigeria (Economic and Commercial Counsellor's Office in Nigeria, 2014), bilateral trade with Nigeria stood at $10.8 billion, exceeding $10.0 billion for the first time. In 2012, bilateral trade was $10.6 billion, accounting for 5.3% of China's total trade with Africa, while in 2013 it was $13.6 billion and $180.0 million of direct investment, including assistance and technical cooperation. Nigeria's trade agreements have either been at bilateral, multilateral (usually negotiated through the WTO), or regional level. This chapter will provide an overview of the agreements and their impacts on the economy and society of Nigeria.

BACKGROUND

Since the early 1990s, many countries in Africa have made significant progress in opening up their economies to external competition through trade and exchange rate liberalization, often in the context of the International Monetary Fund (IMF) and World Bank's support programs. At the same time, with the creation or expansion of a number of regional trading arrangements in other parts of the world, several African nations have also worked towards this, resulting in the establishment or renewal of such trading arrangement in Africa too. The continent is now home to several regional trade agreements (RTAs) or trade blocs, many of which are part of deeper regional integration schemes. While some RTAs have been revived, some other have been broadened and deepened. Whether regional/bilateral agreements are becoming a major feature of the global trading system (as opposed to a major preoccupation of trade negotiators) is not obvious (Pomfret, 2001).

Such trading arrangements are envisaged to foster trade and investment relation amongst member countries by removal of tariffs and other impediments to intra-regional trade flows. In some cases, the arrangement also aims at fostering common economic and monetary union amongst member states, as also a common currency. The success of these arrangements in fostering intra-regional trade has been diverse, with Southern African Community (SADC), Economic Community of West African States (ECOWAS), Common Market for Eastern and Southern Africa (COMESA), Cross Border Initiative, and West African Economic and Monetary Union (UEMOA) being the more successful ones (African Countries Trade Agreements, n.d.).

International trade by its very nature requires careful planning and substantial investments, which can be recouped only over long periods of time. All long-term investments are highly sensitive to uncertainty, and foreign-trade-related investments doubly so for their outcomes may be affected by policy changes in several countries. The trade part of the international economic order can thus be understood as a set of policy commitments exchanged between and among countries in order to minimize policy-generated uncertainty and so to maximize the gains from trade. Orumgbe (n.d.) also says that international trade system encourages specialization which leads to greater gains in productivity and efficiency. It allows countries to concentrate their resources on producing goods they make best and importing goods which are more efficiently produced elsewhere. However, in the case of Nigeria, it cannot be stated. What can be observed is that even if Nigeria has favorable conditions for agricultural and food production, the sector has not developed as it would have been necessary – considering the economic and social needs.

Africa's place in the multilateral trading system has often received special attention, even though it has mostly focused on the contextual and factual analysis of the weakness of the continent's contribution to global commercial transactions or the vagaries of the participation of African states in trade negotiations. There has been more than enough criticism suggesting that Africa is not making sufficient efforts to take part in international trade. On the contrary, African countries merit a spotlight on their significant progress to open up to trade.

However, an important question is how Africa can benefit from the opportunities created by international trade while minimizing the negative effects that go hand in hand with liberalization. Africa's inability to benefit from opening up to transactions can be explained by its integral position in international trade that offers little in the way of returns and produces little added value and wealth. Its status is that of a supplier of basic commodities and raw materials in very limited quantities, which restricts it to the bottom of the international value chains. In addition, due to the rushed liberalization policies that African countries have experienced in the past, their efforts towards industrialization, valorization, and transformation of raw materials and diversification were thwarted by the sudden, forceful competition of imported goods. Many countries continue to suffer from the narrowing of their political space as well as the loss of their sovereignty and control of their own economic and trade policy instruments created during this period (Dieye, 2015).

In a case of a country rich in oil and natural gas, it could be assumed that it has well-operating economy and trade, by utilizing its richness in natural endowments, it plays important role in the global economy. However, it is not true for many African countries, including Nigeria. It needs external assistance. By specializing in the production and by trading with other countries, it is possible for countries to increase their incomes. Even though countries as a whole benefit from specialization and international trade, some groups in society, providers of labor or capital, might lose. If international trade leads a country to specialize in producing goods that require lots of workers and little capital, such a specialization increases wages (which benefits the workers) but decreases the income of the capital owners. The country, as a whole, benefits because the gain of the workers is bigger than the loss of the capital owners.

In addition to trade agreements in Nigeria, the chapter focuses on food security issues as well. Thus, it should be noted that availability of food alone does not seem sufficient to explain the attainment of food security in a country. Food can be available in a country because of effective agricultural policy; good harvest in a particular year or massive importation of food; or food handout (aid). Massive food import, particularly by developing countries, usually has negative effect on foreign reserves and causes budgetary hemorrhage (Davies, 2009), while food and which is sometimes used as an economic instrument in the service of political goal of the donor countries (Ikoku, 1980), may even discourage food production activities in the recipient countries; any country that needs massive food input or food aid before its citizens could feed would have only a short-term solution to its food crisis but would not be food-secure for all times because the feeding of the people in that country will be dependent on the willingness and sometimes the ability of the external suppliers to supply. This is not to suggest that every country that has a reason(s) to import food lacks food supply. On the contrary, some countries may and do import food to offset production shocks and cover the shortfall in domestic food supplies (Lavy, 1992), encourage consumption of some food items or even assist the export trade of a particular target state with which they have bilateral trade agreements. Import of food by such countries may not necessarily be undertaken to solve any severe food shortage problem. To that extent, these countries are not food-insecure. The aim of this chapter is to focus on the relationship between the trade agreements and the food security in Nigeria – providing there is one.

MAIN FOCUS OF THE CHAPTER

About Nigeria

Nigeria is located on the Gulf of Guinea between Cameroon, Chad, Niger, and Benin. Its territory extends to 923 thousand km^2, and the huge economic potential makes it worthwhile to examine and discuss. Being the federal republic, Nigeria comprises 36 federal states and the Federal Capital Territory (FCT) of Abuja (Figure 1).

The Federal Republic of Nigeria has the highest number of population in Africa totaled to 188 million inhabitants. The population has been rapidly growing. It was estimated that between 1911 and 1931 the Nigerian population grew from 17 million to 20 million – only 3 million in 20 years. By 1964 (four years after independence), the population was around 35.6 million people (Perkins & Stembridge, 1966). It means that in the course of a half decade Nigerian population duplicated, but from 1964 until now – during another half century – the population increased fivefold, and the rapid growth of population has still been continuing

After all, it can be stated that Nigeria is the most significant country in West Africa. Its economy has been developing dynamically for several years. In the past ten years, annual economic growth was about 7%, which exceeded the average growth rate for Africa (Engelberth, 2015). In 2014, Nigeria's GDP increased by 7.1% in spite of the decline in the oil prices. However, the decrease of the oil prices may worsen the conditions of the country in the short run, since the oil market prices are almost below the production costs. In 2014, the GDP in 2014 reached all-time high $568 billion, while in 2015 it was $481.07 billion, and by this, the economy of Nigeria became the biggest in Africa. Tarrósy (2014) also pointed out that in 2014 Nigeria overtook South Africa by its economic performance, and in the meantime, the author refers to the idea that Nigeria in the future might even take over the position of South Africa in the G20 group (Tarrósy, 2014; Neszmélyi, 2014).

Figure 1. Map of the Federal Republic of Nigeria
Source: Geoatlas, 2010

The economy of Nigeria is the largest in West Africa and the second largest in sub-Saharan Africa, predominantly oriented to the production of agricultural products and crude oil. Nigeria's economy – compared to the size of population – is not considered large. Before the country became independent in 1960, its economy was dominated by agriculture, and agricultural products overwhelmed in exports. However, in the early 1970s, the discovery of the oil fields fundamentally changed the structure of Nigeria's economy. In 2014, agriculture accounted for about 24% of the GDP and 70% of employment but contributed only about 2.5% of export earnings. Crude oil and natural gas accounted for about 11.0% of GDP, 95.0% of export earnings, and 70.0% of government revenue. Thus, the share of agricultural products in the total exports decreased significantly. The country has become oil exporting, which opposed it to the volatility of the oil prices. Due to the huge oil reserves and rich natural endowments, Nigeria was supposed to be a flourishing economy, but in reality, it is among the least developed countries of the world. Oil has brought several negative impacts on the country: it neither has resulted in any sustainable development nor started any progress in infrastructure and social services. Moreover, corruption spread dramatically, and most of the population did not profit from the oil income. The government intended to spend some of the income on important investments, including industrial infrastructure development, however, due to the lack of competition, such actions resulted in the decrease of industrial production. Nowadays, the speed up in the industrial development can be observed through major political programs and private investments. However, poverty has not been moderated and still causes big problems. The crisis caused by the dramatic decrease in the oil prices forced the politicians to review their former foreign trade policy and to diversify exports.

In spite of the fact that most of the area is arable, the majority of food the Nigerian people consume comes from imports. Beyond the rapid growth of the population, one of the major reasons of such import-oriented consumption is the rich oil and natural gas reserves, the exploitation and export of which provided the country with "easy cash" for the recent decades. Another reason is that the agricultural holdings are small and scattered, and farming is carried out with outdated tools and techniques. Modern and large-scale farms are not common. Mefor (2007) in his work – a portrait book on senator Ken Nnamani – pointed out that the world spends around $3 billion per day on weapons and armies. The effect this huge amount of money could have if it is to be channeled to the improvement of human welfare can hardly be overestimated. However, defense budgets of almost all nations remain on the rise even at the risks of starvation of their citizens (Mefor, 2007).

Political leaders and economic decision makers of Nigeria already recognized the necessity of development of food production and agricultural sector, which – contrary to the oil industry – would exercise a deep and positive impact on the rural society as well. Nigerian agriculture is being transformed towards commercialization at small, medium, and large-scale enterprise levels (Neszmélyi, 2014).

The abovementioned economic growth has not been translated into the creation of new workplaces or poverty alleviation. In 2016, the unemployment rate was 13.3% because the sectors driving the economic growth were not those with high job-creating potential. Oil and gas sector, for example, is a capital-intensive "enclave" with low employment-generating potential. The major policy issues are a generation of employment opportunities, particularly among the young people, and inclusive growth. Thus, one of the major reasons of current economic and social problems in Nigeria is the lack of diversification. As the economy lacks diversification, agricultural sector lacks modernization. To address this, the government is encouraging the diversification of the Nigerian economy apart from the oil-and-gas sector. It addresses underdeveloped infrastructure in the country and development of agricultural sector through modernization and establishment of staple crop processing zones, with the value chain model to provide linkages to manufacturing sector (African Economic Outlook, n.d.).

Tarrósy (2009) pointed out that the principle of the modernist theory is wrong in the cases of African countries and societies. For the development of these newly established independent states, the critical issue is whether it is possible to implement industrialization in classical agricultural regions, and if so, in which extent. Models suggesting increasing rates of industrialization might not be relevant for Africa (Tarrósy, 2009). Many governments failed to diversify the economy away from its overdependence on oil export. In parallel with the increasing importance of the oil-and-gas sector, there has been a general decrease in agricultural output. In Nigeria, this fact is observed in worsening foreign trade balance of agro-food sector. Because of this disproportional development, there is an increasing social tension between urban and rural regions of the country. If there seems no possibility to increase the living standard of the rural population concentrated in less-favoured regions of Nigeria, the authors have to forecast a rapid growth of overburdened cities (e.g., the population of Lagos is over 10 million). This fact further aggravates criminal and environmental problems of big cities (Zsarnóczai, Lakner, & Mughram, 2011).

Boosting agricultural production involves targeted interventions and reforms, including technological innovation, productivity improvement, infrastructure development in agricultural production zones, commercialization, input supply, and distribution systems. Specific interventions should include an increase of land under farming, an increase of use of improved seeds and fertilizers, enhanced cultural practices, mechanization of agricultural production, and adoption of a value chain approach to boost agricultural production. All those measures should be complemented with improvements in infrastructure, particularly road transport, energy, irrigation, storage, and processing. Moreover, partnerships between private sector operators and associations of farmers should be developed, and long-term financing should be provided at a reduced cost to small- and medium-sized agricultural enterprises. Such measures are incorporated in the Agricultural Transformation Agenda of the Nigerian government.

The Transformation Agenda of the Nigerian economy for 2011-2015 (National Planning Commission, 2016) declared agriculture and food security among the seven growth drivers of the economy, which are agriculture, water resources, solid minerals, manufacturing, oil and gas, trade and commerce, and culture and tourism. The main task of the sector is to secure food and feed needs of the nation, by enhancing:

- Generation of national and social wealth through greater export and import substitution;
- Capacity for a value-added approach to industrialization and creation of employment opportunities;
- Efficient exploitation and utilization of available agricultural resources;
- Development and dissemination of appropriate and efficient technologies for their rapid adoption in the agricultural sector (national planning commission, 2016).

In November 2012, the Agricultural Transformation Agenda was adopted in which the government's vision was set to boost agricultural sector in order to make Nigeria an "agriculturally industrialized economy" (Adesina, 2012).

As it can be observed, there have been significant restructuring processes in Nigeria's economy recently. In addition, one of the major priorities of the Nigerian government is to make the country attractive for domestic and foreign investors. Without foreign capital, economic growth cannot be achieved. Non-oil activities and investments are welcomed the most. In order to decrease the dominance of export, such sectors as tourism, telecommunication, trade, and industrial production have come into the frontline. Apart from oil, other products are not competitive on the market, and their export is limited. Consequently, Nigeria is now very open to regional integration and cooperation.

Bilateral Trade Agreements

Nigeria has bilateral investment treaties (BITs) with Egypt, Finland, France, Germany, Italy, Korea, the Netherlands, Romania, South Africa, Spain, Switzerland, Turkey, the United Kingdom, and the United Arab Emirates (the most recent) (Table 1). Agreements with Belgium, the Russian Federation, and the United States of America are being negotiated now (United Nations Conference on Trade and Development [UNCTAD], 2013).

According to the Economist Intelligence Unit (2010), Nigeria has comprehensive double taxation agreements with Belgium (1991), Canada (2000), China (2010), France (1992), the Netherlands (1993), Pakistan (1991), Romania (1994), South Africa (2009), and the United Kingdom (1989). These treaties apply to personal income, corporate income, capital gains, and petroleum profits. As of 2011, treaties with the Czech Republic, Norway, the Philippines, Poland, the Slovak Republic, Korea, and Sweden were still being considered for finalization and ratification. Agreement negotiations with Denmark, Gambia, Germany, New Zealand, Sierra Leone, and Switzerland are worthy of mention too. In the case of Hungary, the first bilateral inter-governmental agreement on trade relations was signed with Nigeria in 1982, but in order to enlarge economic relations, a new agreement on economic and technical cooperation was signed in 2016.

Table 1. Bilateral investment treaties of Nigeria

Countries	Date of the agreement
Algeria – Nigeria	January 14, 2002
Austria – Nigeria	April 8, 2013
Bulgaria – Nigeria	December 21, 1998
Canada – Nigeria	May 6, 2014
China – Nigeria	May 12, 1997
China – Nigeria	August 27,2001
Egypt – Nigeria	June 20, 2000
Ethiopia – Nigeria	January 19, 2004
Finland – Nigeria	June 22, 2005
France – Nigeria	February 27, 1990
Germany – Nigeria	June 25, 1979
Germany – Nigeria	March 28, 2000
Italy – Nigeria	September 27, 2000
Jamaica – Nigeria	August 5, 2002
Korea, Republic of – Nigeria	March 27, 1998
Kuwait – Nigeria	March 23, 2011
Netherlands – Nigeria	November 2, 1992
Nigeria – Romania	December 18, 1998
Nigeria – Russian Federation	June 24, 2009
Nigeria – Serbia	June 1, 2002
Nigeria – South Africa	April 9, 2000
Nigeria – Spain	July 9, 2002
Nigeria – Sweden	April 18, 2002
Nigeria – Switzerland	November 30, 2000
Nigeria – Taiwan Province of China	April 7, 1994
Nigeria – Turkey	October 8, 1996
Nigeria – Turkey	February 2, 2011
Nigeria – Uganda	January 15, 2003
Nigeria – United Kingdom	December 11, 1990

Source: UNCTAD, 2013

Other Trade Agreements

World Trade Organization

Nigeria is an original member of the WTO having ratified the WTO Agreement on December 6, 1994 (GATT document Let/1957, December 7, 1994). The Marrakesh Agreement establishing the WTO has been ratified, but in cases where its provisions have not been incorporated into Nigerian legislation, traders and investors are unable to invoke WTO provisions in domestic courts. However, appropriate legislations are being sorted out. Under the agreement, Nigeria has been active in the negotiations under the Doha Development Agenda thanks to its membership of several negotiating groups. The country is a member of the African Group, Africa, the Caribbean and the Pacific (ACP) Group, and the G-90 Group. In the agriculture negotiations, it is in the G-20 and the G-33 groups of developing countries, where it plays a prominent role. In the negotiations on agriculture under this agreement, Nigeria sought for improved market access, particularly into the markets of developed countries, along with reductions in domestic support and the elimination of export subsidies. As part of the G-33 group of countries, it also asked for flexibility in terms of market access and reductions in tariff escalations (WTO, 2007). Under this agreement, it has enjoyed special and differential treatment, capacity building, and technical assistance.

Regional Agreements

African Union (AU) and African Economic Community (AEC)

Nigeria is a founding member of the African Union (AU). With the signing of the treaty (Abuja Treaty) by fifty-one head of states in June 1991, it gave birth to the African Economic Community (AEC), which came into force after required ratifications in May 1994. The aim of the AEC is to promote economic, social and cultural integration in order to increase self-sufficiency with human resource mobilization. This is with a view to ensure economic stability and the peaceful relationship among member states. The mechanism to achieve this was to build strong Regional Economic Communities (RECs) (of which Nigeria belongs to the ECOWAS) and strengthen intra-regional integration. The roadmap which includes the establishment of free trade areas in each REC, creation of a continental customs union and African common market, and the establishment of the African economic monetary union and parliament, is planned to be achieved by 2028 (Amadou, 2014).

Economic Community of West African States (ECOWAS)

Nigeria has been a member of ECOWAS since it was created by the Treaty of Lagos in May 1975 with the objective of promoting economic integration among its member countries. In the quest to ensure speedy economic integration, the Treaty was revised in 1993 and set objectives for the creation of the common market. Single currency (to be called ECO) was intended to be launched by the West African Monetary Zone (of which Nigeria is a member) by 2015. Unfortunately, the introduction of a single currency was postponed until January 1, 2020, due to what it termed "non-fulfillment of financial responsibilities" by member countries. However, Amadou (2014) noted that Nigeria potentially portends a great impediment to achieving a single monetary union in West Africa because of the size of its economy compared to other countries in ECOWAS. According to Amadou (2014), for instance, the entire West African Economy had the GDP of approximately \$75 billion in 2013, while Nigeria alone had the GDP of \$260 billion in 2013 (\$510 billion, according to the US Department of State (2016)). Amadou (2014) also sees the fact that Nigeria is a major oil exporter as opposed to other West African countries.

Non-Reciprocal Preferential Trade Agreements

Some countries provide non-reciprocal preferential treatment for imports from Nigeria, of which are: Belarus, Canada, Japan, New Zealand, the Russian Federation, Switzerland, Turkey, and the EU. The

United States also provide preferential access under the African Growth and Opportunities Act (AGOA) and the Global System of Trade Preferences (GSTP), of which Nigeria has been part of since its inception in 1988. The United States and Nigeria have signed a bilateral trade and investment framework agreement. Also, in a bid to strengthen trade ties, Penny Pritzker, the U.S. Secretary of Commerce, visited Nigeria on a fact-finding mission in January 2016 with senior U.S. business executives, who make up President Obama's Advisory Council on Doing Business in Africa (US Department of State, 2016).

ACP-EU Partnership Agreement (Cotonou Agreement)

The ACP-EU Partnership Agreement, signed in Cotonou on June 23, 2000, was concluded for a 20-year period, from 2000 till 2020. It was the most comprehensive partnership agreement between developing countries and the EU. Since 2000, it had been the framework for the EU's relations with 79 countries in the ACP group. In 2010, ACP-EU cooperation was adapted to new challenges such as climate change, food security, regional integration, state fragility, and aid effectiveness.

Before the Cotonou Agreement, Lomé Conventions (Lomé I – Lomé IV) were applied. However, important developments on the international stage and socio-economic and political changes in the ACP group of countries highlighted the need for a re-thinking of ACP-EU cooperation.

In accordance with the revision clause to re-examine the Agreement every five years, negotiations to modify the Agreement were launched in May 2004 and concluded in February 2005. The objective was to enhance the effectiveness and quality of the EU-ACP partnership and to reflect the recent major changes in international and ACP-EU relations.

In March 2010, the European Commission and the ACP group concluded the second revision of the Cotonou Agreement. The second revision adapted the partnership to the changes which have taken place over the last decade, in particular:

- The growing importance of regional integration among ACP countries and in ACP-EU cooperation was reflected. Its role in fostering cooperation and peace and security, in promoting growth and in tackling cross-border challenges was emphasized. In Africa, the continental dimension was also recognized, and the African Union became a partner of the EU-ACP relationship.
- Security and fragility: no development could take place without a secure environment. The new agreement highlighted the interdependence between security and development and tackled security threats jointly. Attention was paid to peace building and conflict prevention. A comprehensive approach combining diplomacy, security, and development cooperation was developed for situations of state fragility.
- ACP partners faced major challenges if they were to meet the Millennium Development Goals, food security, HIV-AIDS, and sustainability of fisheries. The importance of each of these areas for sustainable development, growth, and poverty reduction was underlined, and joint approaches for the cooperation were then agreed.
- The EU and the ACP claimed climate change as a global challenge and considered it as the major subject for their partnership. The parties committed to raise the profile of climate change in their development cooperation and to support ACP efforts in mitigation and adaption to the effects of climate change.
- The trade chapter of the Agreement reflected the new trade relationship and the expiry of preferences at the end of 2007. It reaffirmed the role of the Economic Partnership Agreements to

boost economic development and integration into the world economy. The revised Agreement highlights the challenges ACP countries are facing to integrate better into the world economy, in particular, the effects of preference erosion. It, therefore, underlines the importance of trade adaptation strategies and aid for trade.

- More actors in the partnership: the EU had been promoting a broad and inclusive partnership with ACP partners. The new agreement clearly recognized the role of national parliaments, local authorities, civil society, and the private sector.
- More impact, more value for money: this second revision was instrumental in putting in practice the internationally agreed aid effectiveness principles, in particular, donor coordination. It would also untie the EU aid to the ACP countries to reduce transaction costs. For the first time, the role of other EU policies for the development of the ACP countries was recognized, and the EU committed to enhancing the coherence of the policies to this end (European Commission, n.d.).

Nigeria and the Economic Partnership Agreement (EPA)

In July 2014, ECOWAS Heads of State endorsed the Economic Partnership Agreement (EPA) between the African Union and the European Union. This trade agreement was expected to give rise to the institution of a Common Free Trade Area (CFTA) by 2015, thereby extending the African Growth and Opportunity Act by fifteen more years and serve as a strategic response of Africa to the WTO negotiations.

At the beginning, most of the African countries received the idea with high antipathy and even refused to sign it. It happened at the second EU-Africa summit in Lisbon, where the main objective was to encourage the African countries to sign new trade agreements by December 31, 2007, in accordance with the Cotonou Convention of 2000 winding up the 1975 Lomé accords. They claimed the EPAs as a straitjacket and refused the idea of the complete liberalization of trade. They did it despite the fact that the WTO had insisted that these preferential arrangements should have been dismantled or replaced by trade agreements based on reciprocity, claiming that this was the only way African countries could have continued to enjoy different treatment (Ramonet, 2008).

Abdoulaye Wade, the President of Senegal in the interview published in Le Monde in 2007 expressed the African opinion in sharp words, like "Europe expects Africa to voluntarily bow its head under the axe of executioner" (Ramonet, 2008). At the same time, other African politicians accused the EU that it wanted to divide the developing world to the Least Developed Countries (LDC) and non-LDC countries by the EPA. After all, the EPAs aroused wide public concern. Social movements and trade union organizations in Africa were mobilized against them. Such a revolt bore fruit: the summit ended in failure. The president of the European Commission (that time: José Manuel Barroso) was forced to back down and accept the African countries' call for further discussions (Ramonet, 2008).

Since the very beginning, Nigeria refused to be a signatory of the Agreement. Olusegun Aganga, the Minister of Industry, Trade, and Investment of Nigeria, speaking at an extraordinary session of the Conference of African Union Ministers of Trade in Addis Ababa in 2014, said the trade liberalization deal with the EU would have a long-term negative impact on the continent's efforts towards industrialization and creation of jobs (Emeka, 2015). According to Egunsola (2014), there is a serious concern on the part of Nigeria because of the fear that it will replace the existing non-reciprocal Cotonou agreement between the EU and Africa, which allowed African countries access to the EU market without any reciprocal action from Africa. Under the new regime, the EPA will only allow trade preferences for African goods if Africa also provides the EU unbounded access to its markets. Under the agreement, Africa will remove 80% of an import tariff on goods coming from the EU. Furthermore, the EU also expects Africa to remove restrictions on the export of unprocessed raw materials and promote liberalization in other areas such as investment and services (Egunsola, 2014).

According to Emeka (2015), Aganga argues that rather than enter into the Agreement, African countries should efficiently take advantage of their abundant natural resources and large market to develop industries and create jobs for their people, increase intra-African trade and achieve regional integration. Furthermore, Egunsola (2014) says that liberalization of 80% of import suggests that African-based industries will compete with the EU industries for African markets. Given the infrastructural deficit in Africa and high costs of doing business, it will be impossible for domestic industries to compete favorably (at least in the short run). In his view, Africa will likely become a dumping ground for the EU products, and the gains the EPA is supposed to provide for African economies will be eroded eventually. Local industries will be closed down, and there will be consequent job losses. Some time ago, a similar fate befell Nigerian textile industries, most of which have been closed down due to inability to compete with cheap imports from China and Europe. What is good is that this argument questions the level of government's commitment over the years to create a favorable environment for local industry to thrive. On the other hand, there are different opinions on the prospects and the role of the globally emerging economic powers in Africa. Taylor (2014) examined the contradictions of growth without development and he expressed his hope that the presence of the newly emerging economies, like Brazil, Russia, China, and India (BRIC countries) could offer a viable alternative to the centuries of exploitation by the West.

With the effective and rigorous implementation of strategic blueprints for domestic manufacturers coupled with drastic infrastructural development, it will not only ensure increased production but also alternatives to the most populous nation in Africa and improved quality occasioned by the healthy competition that will emanate therein. The perfect case experienced in the telecommunication industry. Gambia, Nigeria, and Mauritania are the only three countries who are still to sign the EPA. It will be interesting to see how the EPA between the EU and West Africa influences the region.

Future: Vision 20:2020

Nigeria's Vision 20:2020 provides the framework for the national development strategy, which focuses on the diversification of the economy and poverty reduction, to be driven vigorously by the private sector by creating incentives and conditions to attract trade and investment. The objective of Vision 20:2020 is to place Nigeria among the top-20 economies in the world by 2020. The core elements of the strategy are a modernization of agriculture, rehabilitation of infrastructure, and development of the power sector (National Planning Commission, 2009).

According to the National Planning Commission (2009), the strategy is being implemented through three medium-term development plans, which would require economic growth by 14% per year from 2009 till 2020. The projection is at this rate to attain minimum GDP of $900 billion and annual per capita income of above $4,000, estimating that the population would have reached 193 million by 2020 and 289 million by 2050. One of the four thematic areas identified in the first plan is the productive sector that revolves around seven sub-sectors: agriculture and food security; oil and gas; manufacturing; small and medium enterprises; solid mineral and steel development; culture and tourism; and trade and commerce. The overall goal is to change to a service-based economy by 2018. Hence, the critical approaches are support of non-oil exports, an increase of utilization of preferential trade arrangements, and development of Nigeria's participation in international trade negotiations. Specific sectoral priorities and targets, including donor support, are identified in the National Implementation Plan (WTO, 2011).

The government recognises that for trade and investment to thrive provision of the legal framework and removal of institutional and practical barriers are required (as provided for by the Foreign Exchange (Monitoring and Miscellaneous Provisions) Act, 1995 and the Nigerian Investment Promotion Commission (NIPC) Act, 1995, including tackling corruption, social and economic instability, and applying the rule of law). Also, Nigeria recognizes the fact that its products need to be of high quality to be able to compete favorably on the international market. Hence, one of the country's major regulatory trade agencies, the Standard Organizations of Nigeria (SON) in partnership with several donors, has been receiving technical assistance and capacity building in the area of sanitary and phytosanitary (SPS) standards (International Cocoa Organization, 2010) and capacity building for private and public-sector organizations in the area of food export (Nigerian Export Promotion Council, 2011).

Furthermore, in a bid to boost the gains of a rapidly growing economy with respect to trade, Nigeria presented its trade and productive-capacity-related needs at the August 2010 Session on Aid for Trade of the WTO Committee on Trade and Development (WTO, 2010). Seven areas for support were identified, namely:

1. Trade policy advocacy,
2. Trade policy development,
3. Enhancing the competitiveness of supply capacity,
4. Compliance support infrastructure and services,
5. Physical infrastructure,
6. Legal and regulatory framework,
7. Trade facilitation and trade-related financial services.

However, after rigorous and regular consultations with development partners, the Government of Nigeria decided to streamline the seven categories of activities into three broad areas of support for trade and productive capacity. These are:

1. Mainstreaming trade into national development strategies,
2. Enhancing trade competitiveness,
3. Institutional and human resource capacity development (wto, 2011).

This made the focus and strategies of Vision 2020 in terms of trade more concise and precise to pursue.

Tripartite Free Trade Agreement (TFTA): An Example of Efficient Initiative

Even if Nigeria is not the member of this Agreement, it is important to mention the creation of a free trade area from Cape Town to Cairo in June 2015, which is possibly the most significant event in Africa since the establishment of the Organization of African Unity in 1963. It is a grand move to merge existing principal regional organizations into a single African Economic Community. The Tripartite Free Trade Area (TFTA) includes 26 countries that are members of the Common Market for Eastern and Southern Africa (COMESA), East African Community (EAC), and Southern African Community (SADC) (Figure 2). The TFTA covers a population of 632 million and a combined GDP of $1.3 trillion. The area spans 17.3 million square kilometers, which is near twice the size of China or the United States.

Figure 2. Map of the TFTA
Source: UNCTAD, 2013

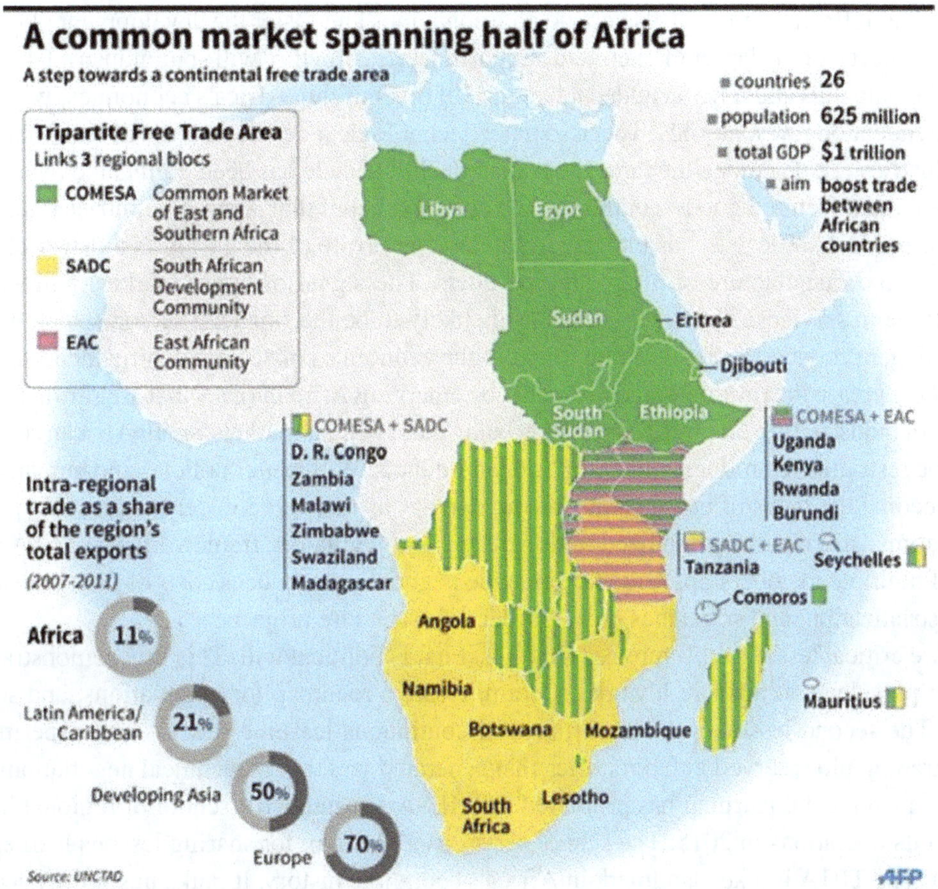

Critics argue a single trading bloc will not work where individual sub-regional ones have failed. On the contrary, consolidation of three trading blocs will be founded on previous trade gains. Between 2004 and 2014, trade within the COMESA region grew from $8 billion to $22 billion, respectively. Over the same period, trade within SADC grew from $20 billion to $72 billion, and for the EAC from $2.6 to $8.6 billion. The total trade between the three areas rose from $30.6 billion in 2004 to $102.6 billion in 2014. Despite the growth, only about 12% of Africa's trade is intra-regional, while it is 22% for South America, 40% for North America, 50% for Asia, and 70% for Western Europe. Tariff liberation of 60-85% will have a significant impact on the cross-border flow of goods and services. The TFTA will bring benefits to Africa in at least six mutually reinforcing ways. First, the conclusion of the Agreement will generate an impetus for the creation of similar arrangements in Western Africa, bringing economic powerhouses such as Nigeria into a continental free trade area. In addition, it is a bigger market, which free flow of goods and services will help to maintain economic growth at 6-7% per year. At this rate, the combined GDP of Africa is projected to reach $29 trillion by 2050, which would be equal to the current combined GDP of the EU and the US. With additional policies, such growth will contribute significantly to spreading prosperity and reducing poverty. The TFTA will serve as an impetus for investment in Africa's cross-border infrastructure. It is estimated that Africa needs to invest nearly $100 billion annually in infrastructure over the next decade. By now, less than half of this target is met. One of the reasons for the low level of investment has been poor coordination across the different trading blocs. Building infrastructure will also create additional jobs and foster the development of engineering services. Prospects for the larger markets and supporting infrastructure will spur industrial development. It both will create jobs and have an added advantage of diversifying Africa's economies, which are now largely dependent on raw materials. The associated technological development will lead to the creation of new industries. Trade among the three blocs over the last decade has been dominated by intermediate products and manufactured goods, contrary to the common belief that African countries trade in similar agricultural products. These trends underscore the potential role of the TFTA as a driver of industrial development and manufacture of high-value products. The signal of larger markets will also help to stimulate trade in services. The first beneficiary is likely to be the financial sector, which will be able to lend to bigger producers seeking to benefit from the economies of scale. Such financial services will reinforce the increase in cross-border investments by emerging African firms that are serving as regional champions of industrial development. Finally, by being part of larger markets, small African countries will no longer be restricted to producing their traditional products. With better policies and human resources, they can become the locus of new manufacturing operations that serve wider markets. By providing a single economic space with harmonized trade policies and regulatory framework, the TFTA solves the problem of multiple memberships, rationalizes trade negotiations, reduces costs of doing business, supports industrialization, and stimulates cross-border infrastructure projects.

There are critical lessons for future negotiations. First is political will. This was demonstrated by the decision of presidents to approve a work program, create a roadmap for negotiations, and stick to the timetable. The second lesson is the importance of continuous learning process and experimentation. The three trading blocs served as laboratories that generated lessons for technical negotiations. The importance of incremental learning has prompted COMESA to establish a school of regional integration that started its operations in 2015. The school serves as a platform for sharing lessons learned through integration. The TFTA is a key landmark in Africa's economic history. It ranks in significance with the independence of Ghana in 1957, the creation of the Organization of African Unity in 1963, and its reinvention as the African Union in 2002. To paraphrase Kwame Nkrumah, Ghana's first president, the best way to learn to be a continental free trade area is to be a continental free trade area (Juma & Mangeni, 2015). During recent years, there has been much talk of the potential benefits of creating an integrated African commercial bloc; the TFTA may be seen as the first tangible sign of action being taken. Following signature of the TFTA, negotiations to broaden its territorial ambit to West African nations opened on June 15, 2015, at the African Union summit in Johannesburg. According to the AU, the ultimate aim is to establish an intra-continental market benefiting from a youthful, fast-growing population of about 1 billion and a combined GDP of $3 trillion (Cannon & Major, 2015).

The TFTA includes a number of key African economies, such as South Africa, Egypt, Angola, and Kenya. Indeed, the nations constituting the bloc account for approximately 60% of Africa's economic output. The creation of the TFTA may serve to simplify Africa's complex overlapping trade zones, which in turn, will assist in attracting investors to the region. The TFTA does not currently include key West African economies, most notably, Nigeria. Although, negotiations on an extension of the TFTA area are underway. Estimated timeline for the conclusion of such negotiations has been set at two years. The prospect of obtaining a positive result is highly dependent on negotiators having learned valuable lessons from the seven-year period of negotiations, which culminated in the recent signature of the TFTA. The conclusion of such an agreement requires strong political will from the nations concerned in order to fuel progress and manage the process, for example, by way of approving a work program, creating a roadmap for negotiations, sticking to a clear and achievable timetable and utilizing technical groups. The tariff-focused nature of the TFTA has also come under scrutiny. It is widely recognized that there is a vital need for the improvement of regional infrastructure in Africa if the intra-continental trade is to increase. Although, as mentioned above, it is argued that the TFTA may trigger investment in infrastructure. Others argue that for intra-regional trade to be strengthened, infrastructure development must be addressed as a priority, before any removal of tariff barriers. The potentially positive effects of the TFTA may be stifled in reality by the continuing existence of infrastructure-related barriers to trade.

Major Effects of Trade Agreements on Food Production and Security in West-Africa

A study carried out by Busse and Großmann (2004) where the impact of trade agreements between the EU and some selected West African countries (Nigeria inclusive) was analysed, revealed that food production and income generated therein were threatened by trade liberalization as local producers cannot cope with the lower prices of food products occasioned by mass importation from trade partners from such trade partners. Most of the agricultural products consumed by African countries are imported and this goes a long way in determining the production strength and subsequent revenue of indigenous producers, hence, food safety.

Interestingly, the EU is West Africa's biggest trading partner. It accounted for 30% of the region's exports and 40% of its imports between 2002 and 2004 (United Nations Comtrade, 2006). The major exports include minerals (crude oil – 43%, iron – 3%, aluminum – 2%, and gold – 1%), agricultural products (cocoa – 19%, fresh fruit – 3%), fishery products (5%), and forest products (timber – 2%, rubber – 2%). The share of agricultural and processed food products was 18% of the total West African imports from the EU, of which dairy and milk products contributed the largest percentage of agricultural and processed food products (17%), and closely followed by preparation of cereals, tobacco, fish, preparations of vegetables, and fruits. However, it is worthy to point out that West Africa only commands less than 0.5% of total EU exports (United Nations Comtrade, 2006).

Milk and dairy products contribute the largest share of West African agricultural imports from the EU (17%). Over the last decade, import of milk and dairy products has increased significantly (Eurostat, 2006). The trade effects of partnership agreements for milk have varying peculiarities across various West African countries. Ghana provides the biggest significant effect with 24% increase in milk imports from the EU with only marginal increase in Guinea and 9.5% increase experienced in the region in general (Pannhausen, 2006).

Poultry exports to West Africa from the EU have experience significant increase in the 1990s but decreased towards the early 2000s (Eurostat, 2006). Apart from Guinea, most of the West African countries in 2004 imported at least 12% more from the EU than before the preferential trade agreement with Ghana's import increasing by 32%. Moreover, total trade diversion is 21% of non-EU imports, which is highly significant. Imported poultry products are less expensive compared to those produced domestically. This factor cannot be unconnected to the high cost of production connected with production inputs. This poses a threat to the indigenous farmers who in most cases are not given enough incentive to compete with international suppliers. It stifles the local economy and exposes local producers to unfavorable conditions. It also reduces enterprise and ingenuity which is a big problem for self-sufficiency in not only African countries but developing countries who rely heavily on imported food products which have implications for employment, production for the domestic and regional market, and food security.

Furthermore, poultry industry in Ghana obtains feed component from grains through local grain producers, who supply poultry feed to the agricultural industry. These producers are also at a risk in terms of their production supplies if most of these grains are also imported. Increased grain supplies from trade agreements will definitely reduce the price of local producers thereby lowering their income and put their continuous operation on the edge. Hence, these agreements have a ripple effect on the entire food and agriculture sector of the countries with Nigeria not being an exception.

Trade in wheat and wheat flour has increased steadily over the last ten years. However, recently the export value of the EU products has declined (Eurostat, 2006). The effect of trade agreements has seen these products' importation increase by 12.5% in West Africa (Pannhausen, 2006). He further explained that Ghana experienced the largest shift in their import structure, where the EU exports increased by 36.5%, whereas Mauritania recorded 3.5% with the EU-African Caribbean Pacific Agreement. However, this led to countries like Benin and Niger diverting 18.0% and 16.5% of their previously non-EU imports to the EU, respectively. The result of this is that grain producers have to compete with cheap, subsidized EU imported grains whose destination in most cases are the markets in the urban. With more negotiations on reduced tariff getting successful, it is more definite that imports to the countries will increase and become even cheaper, a serious concern for local producers. This is a situation that has really set the Nigerian agricultural sector aback as regards producing enough grains for production and consumption. In response to this, the government once banned the importation of rice and in some cases increased tariff to boost local production. However, given the current economic realities in the country and increase in population which is related to food demand, one will still have to see how long it will last. This is a key to trigger the development of local economy.

Though there was a significant drop in the EU exports to West Africa in 2004, nevertheless processed tomato products from the EU have increasingly entered the regional market, and Nigeria experienced about 25.5% increase in imports due to the EU trade agreements (Eurostat, 2006). In Nigeria, the challenge of tomato processing industries is that most of the domestic producers cannot compete with imports due to the high cost of production occasioned by poor infrastructural facilities and lack of economies of scale (UNCTAD, 2002).

SOLUTIONS AND RECOMMENDATIONS

Based on the abovementioned, it can be stated that the international trade agreements for Nigeria have been important rather from the diplomatic point of view so far. Because of corruption and political interest, they have not been able to contribute to the economic wellbeing of the entire society. From a sectoral point of view, Nigeria's economy is still not balanced. There are dominant sectors, like oil for export purposes, but other sectors, such as agriculture, food industry, and services have not experienced significant development in the past few decades. Empirical evidence from the analysis in Ijirshar (2015) shows that agricultural export can be as lucrative and profitable as any of the other sectors in Nigeria with respect to returns on investment. Since recent shock in oil prices could render Nigeria in economic shambles, much attention is needed in the agricultural sector to overcome such subsequent challenges. According to Olajide, Akinlabi, and Tijani (2012), in Nigeria, agriculture is an example of one key sector which role is, and would remain crucial to development fortunes. Economic history is replete with ample evidence that agricultural revolution is a fundamental pre-condition for economic growth, especially in developing countries (Woolf & Jones, 1969; Oluwasanmi, 1966; Eicher & Witt, 1964).

It can be stated that the solution for the expected development would be the elaboration of an integrated, multi-sectoral development policy, considering the availability of resources, labor force, and capital. The regions of Nigeria should contribute to the national economic development and the development of the society according to their endogenous resources and potentials. According to Edoumiekumo and Audu (2009), economic co-operation among regions and state should be encouraged. In long-terms, the one-focused export should be expanded by alternative products, which requires the development of some economic sectors. Import of food and agricultural product should be moderated by domestic production, which requires the use of abandoned lands and technological development of the sector. The role of international trade agreements should be to provide benefits for all the stakeholders – not exclusively in material terms, but they should create possibilities for participation for as many people as they can. People should use the benefits of such agreements as much as possible. Thus, the benefits of international trade agreements should have an impact on the everyday life of people.

FUTURE RESEARCH DIRECTIONS

Based on the information available, we should state that even if Nigeria and the countries of Africa have had international trade agreements for long, they could not result in significant improvement in the standard of living, in food security, or safety. In order to see the real impacts of trade agreements, the improvement in the global and national macro-indicators could be observed, despite the fact that they have not had a spectacular positive impact on the society and the economy as a whole yet. There are specific sectors which have played increasingly important role in the export and import relations, but agriculture still has unexploited capacities. Because of the increasing population, food production, food processing and securing safe food should be the core areas of future researches, with a special focus on the trends of food imports and their relation to the food safety and security, especially in the rural areas. It would be important to analyze how domestic agricultural and food production could be developed and how the unexploited capacities could be used to secure food and reduce the food-import dependence. Authors believe that the TFTA could provide a good basis for a more diversified development for Nigeria – after the decision on accession is made. Regarding the EU-Nigeria relations, more researches could be carried out on the experience with member states to define the basis for mutually beneficial fields of cooperation and to define the steps to be made to have more balanced trade relations – which is not the situation at this moment.

CONCLUSION

Nigeria is expected to be on track to becoming one of the twenty largest economies in the world by 2020, therefore structural reforms in the economy and society, including international trade are inevitable. Despite the fact that since 2008 the government has begun to implement the market-oriented reforms urged by the IMF, such as modernization of the banking system, removal of subsidies, resolving regional disputes over the distribution of earnings from the oil industry, and development of stronger public-private partnerships for roads, agriculture, and power, the improvement is not equalized.

The importance of trade and agreements between countries and/or organizations cannot be overemphasized because no country can stay totally independent of the others. However, developed countries must see less-developed ones as trade partners and not ones in need of aid or relief due to the impoverishment. Most countries in Africa have not come close to getting the best out of their trade agreements. The best they get is trade barrier reduction. In most cases, domestic goods cannot compete with the foreign ones that come in the reverse due to low capacity and technical deficiencies. Going forward, Nigeria and other countries of the same status must painstakingly look inwards to harness the untapped resources within (country and region), build capacity, and ensure political and economic stability, so that the local industries can benefit mutually and comprehensively from the gains occasioned by such agreements. Nigeria as a matter of necessity must go beyond trade agreements hinged primarily on oil and gas but focus more on diversification and service sector especially in the face of dwindling oil prices.

Nigerian economy needs growth in order to reduce the financial burden of imports, create jobs to absorb the growing unemployment, increase incomes, reduce poverty, and increase prosperity. Social development is vitally important for a country, which is the most populous in Africa, and where harmony and peaceful co-existence among various religions and ethnic communities is really fragile. Creating more jobs in rural areas may diminish tensions and popularity of radical political groups, especially among young generations. Application of new technologies and methods may rapidly increase the efficiency of agricultural production. Those lands which have been abandoned due to the scarcity of water could be still kept in cultivation.

After all, agricultural reforms should serve not only the development of the rural society but indirectly be a valuable contribution to peace and security of the West-African region.

According to the authors' judgment, multilateral agreements on economic co-operations have not yet contributed to the economic development of Nigeria to a significant extent. Even though in terms of GDP, the economy grew, but the real development, which would bring about the metamorphosis of the Nigerian economic structure still has not come about. From many aspects, ECOWAS members are competitors to each other rather than promising partners for future co-operation. In terms of trade, ECOWAS and other intra-African regional agreements are not success stories, however, it has a distinct and positive role in political dimensions.

The same can be claimed about the labored EPA agreements. The EU countries still have been following their own economic interests in the African region, and as a rule, the former colonizers have incomparable advantages. Not surprising that the overwhelming weight of the European-Nigerian trade and investment is represented by the United Kingdom, France, and other West European countries, while the new members – including Hungary – have lower trade volumes. In 2012, Spain was the 5th, the Netherlands was the 6th, while Portugal was the 8th in the top-10 fastest growing exporters to Nigeria. The fact that Nigeria's imports from the UK rose by 99.17% in 2012 also shows its strong trade relations with this country. Even if there is no free trade agreement with the UK, it is one of the largest investors in Nigeria. In 2011, the UK and Nigeria agreed on a joint mandate in a pursuit to double trade from £4 billion to £8 billion by 2014. The target has almost been achieved, with a bilateral trade of £7.2 billion in 2012. Regarding France, in 2014, export to France from Nigeria was $5.8 billion, while in 2011 it reached $7.4 billion. On the case of Nigeria and many other African countries, it looks that bilateral cooperation works much better than multilateral. At least the intensive and successful economic expansion of China in Africa seems to justify it.

Concisely, when one looks at the implications of trade agreements between Nigeria and international bodies, trade organizations and other countries, it is clear that since the country is a net importer of agricultural products, the local industry will have no other option than face serious competition from international producers. One of such challenges is a resultant decrease in prices for some of their products thereby reducing their incentives to produce for the market. This is worrisome as about 70% of Nigerians are into agriculture, hence their source of income and employment is threatened. Moreover, the majority of the population is in the rural area. In rural area being already bedeviled by high poverty rate and poor infrastructure, food insecurity is greatly threatened especially in the view of the perpetually growing population of the country. Not only smallholders will be affected, but also industries. However, this effect will not be felt in a bigger extent in the urban areas because they will benefit from the lower prices, as they are consumption driven (net-consumers). Therefore, there is not just a political angle to trade effects on food production and security, but also a socio-economic angle as well. It is important as a matter of necessity and urgency to monitor the sensitivity and consequences of this process on the local markets, households, and the entire economy.

REFERENCES

Adesina, A. (2012). *Agricultural Transformation Agenda: Repositioning Agriculture to Drive Nigeria's Economy*. Abuja: Ministry of Agriculture and Rural Development of the Federal Republic of Nigeria.

African Countries Trade Agreements. (n.d.). *Focus Africa*. Retrieved November 18, 2016, from http://focusafrica.gov.in/Trade_Agreements_Africa.html

African Economic Outlook. (n.d.). *African Economic Outlook – Nigeria 2013*. Retrieved November 17, 2016, from http://www.africaneconomicoutlook.org/fileadmin/uploads/aeo/2013/PDF/Nigeria%20-%20African%20Economic%20Outlook.pdf

Amadou, S. (2014). *Will There Be an African Economic Community?* Retrieved November 17, 2016, from https://www.brookings.edu/testimonies/hearing-on-will-there-be-an-african-economic-community

Busse, M., & Großmann, H. (2004). Assessing the Impact of ACP/EU Economic Partnership Agreements on West African Countries. Hamburg: Hamburgisches Welt-Wirtschafts-Archiv (HWWA), Hamburg Institute of International Economics.

Cannon, A., & Major, R. (2015). *Signature of the Tripartite Free Trade Agreement (TFTA) on 10 June 2015: The First Step Towards a United African Free Trade Bloc?* Retrieved November 17, 2016, from http://hsfnotes.com/publicinternationallaw/2015/07/07/signature-of-the-tripartite-free-trade-agreement-tfta-on-10-june-2015-the-first-step-towards-a-united-african-free-trade-bloc/

Central Bank of Nigeria. (2010). *Annual Report and Financial Statement for the Year Ended 31 December 2009*. Abuja: Central Bank of Nigeria.

Davies, A. E. (2009). *Food Security Initiatives in Nigeria: Prospects and Challenges*. University of Ilorin.

Dieye, C. T. (2015). *What Is Africa Worth in the International Trading System?* Retrieved November 18, 2016, from http://www.ictsd.org/bridges-news/bridges-africa/news/what-is-africa-worth-in-the-international-trading-system

Economic and Commercial Counsellor's Office in Nigeria. (2014). *Brief Introduction of Bilateral Relations*. Retrieved November 17, 2016, from http://nigeria2.mofcom.gov.cn/article/bilateralcooperation/inbrief/201407/20140700669579.shtml

Economist Intelligence Unit. (2010). *Country Commerce Nigeria*. New York, NY: Economist Intelligence Unit.

Edoumiekumo, S. G., & Audu, N. P. (2009). The Impact of Agriculture and Agro-based Industries on Economic Development in Nigeria: An Econometric Assessment. *Journal of Research in National Development*, *7*(1), 1–8.

Egunsola, O. (2014). *Should Nigeria Sign the Economic Partnership Agreement?* Retrieved November 18, 2016, from http://nigeriannet.com/should-nigeria-sign-the-economic-partnership-agreement/

Eicher, C., & Witt, L. (1964). *Agriculture in Economic Development*. New York, NY: McGraw Hill.

Emeka, A. (2015). *Experts Debate Proposed EU-ECOWAS Partnership Agreement*. Retrieved from 30 September 2016, from https://www.tralac.org/news/article/7793-experts-debate-proposed-eu-ecowas-partnership-agreement.html

Engelberth, I. (2015). Változó kínai gazdasági és politikai jelenlét Afrikában (The Changing Chinese Economic and Political Presence in Africa). In K. Solt (Ed.), *Alkalmazott tudományok II. Fóruma* (pp. 121–145). Budapest: Budapest Business School.

European Commission. (n.d.). *ACP – The Cotonou Agreement*. Retrieved November 17, 2016, from http://ec.europa.eu/europeaid/regions/african-caribbean-and-pacific-acp-region/cotonou-agreement_en

Eurostat. (2006). *Database*. Retrieved January 5, 2006, from http://epp.eurostat.cec.eu.int/portal/page?_pageid=1090,30070682,1090_33076576&_dad=portal&_schema=PORTAL

Geoatlas. (2010). *Map of the Federal Republic of Nigeria*. Retrieved December 12, 2016, from http://www.geoatlas.com/en/maps/countries-4/nigeria-1090

Ijirshar, V. U. (2015). The Empirical Analysis of Agricultural Exports and Economic Growth in Nigeria. *Journal of Development and Agricultural Economics*, *7*(3), 113–122. doi:10.5897/JDAE2014.0615

Ikoku, E. U. (1980). *Self-Reliance: African Survival*. Enugu: Fourth Dimension Publishers.

International Cocoa Organization. (2010). *SPS Capacity Building in Africa to Mitigate the Harmful Effect of Pesticides Residues in Cocoa and to Maintain Market Access*. Retrieved November 18, 2016, from http://www.icco.org/projects/10-projects/193-sps-capacity-building-in-africa-to-mitigate-the-harmful-effects-of-pesticides%20residues-in-cocoa-and-to-maintain-market-access.html

Juma, C., & Mangeni, F. (2015). *The Benefits of Africa's New Free Trade Area*. Retrieved November 18, 2016, from http://belfercenter.ksg.harvard.edu/publication/25444/benefits_of_africas_new_free_trade_area.html

Lavy, V. (1992). Alleviating Transitory Food Crisis in Sub-Saharan Africa: International Altruism and Trade. *The World Bank Economic Review*, *6*(1), 125–138. doi:10.1093/wber/6.1.125

Mefor, L. (2007). *Senator Ken Nnamani Portrait and the New Nigerian Senate*. Abuja: Psykom Publishing Ltd.

National Planning Commission. (2009). *Nigeria Vision 20:2020*. Abuja: Government of the Federal Republic of Nigeria.

National Planning Commission. (2016). *The Transformation Agenda – Summary of Federal Government's Key. Priority Policies, Programmes and Projects 2011–2015*. Abuja: Government of the Federal Republic of Nigeria.

Neszmélyi, G. (2014). The Issues of Sustainability and Self-sufficiency in the Nigerian Food Economy. In G. Neszmélyi (Ed.), *Socio-economic and Regional Processes in the Developing Countries* (pp. 98–116). Gödöllő: Szent István University Publishing.

Nigerian Export Promotion Council. (2011). *Expanding Nigeria's Exports of Sesame Seeds and Sheanut/Butter through Improved SPS Capacity Building for Private and Public Sector*. Retrieved November 18, 2016, from http://www.aflasafe.com/c/document_library/get_file?uuid=9adeb84f-67c2-4de8-9082-110a27b94489&groupId=524500

Olajide, O. T., Akinlabi, B. H., & Tijani, A. A. (2012). Agriculture Resource and Economic Growth in Nigeria. *European Scientific Journal*, *22*(8), 103–115.

Oluwasanmi, H. A. (1966). *Agriculture and Nigeria's Economic Development*. Ibadan: Oxford University Press.

Orumgbe, R. (n.d.). *The Effect of World Trade Organisation on Nigeria Maritime Industry*. Retrieved November 16, 2016, from http://www.academia.edu/11962449/THE_EFFECT_OF_WORLD_TRADE_ORGANISATION_ON_NIGERIA_MARITIME_INDUSTRY

Pannhausen, C. (2006). Economic Partnership Agreements and Food Security: What Is at Stake for West Africa? Bonn: German Development Institute (DIE).

Perkins, W. A., & Stembridge, J. H. (1966). *Nigeria – A Descriptive Geography* (3rd ed.). Ibadan: Oxford University Press.

Pomfret, R. (2001). *The Economics of Regional Trading Arrangements*. Oxford, UK: Oxford University Press. doi:10.1093/0199248877.001.0001

Ramonet, I. (2008). *Africa Says No – And Means It*. Retrieved November 18, 2016, from https://mondediplo.com/2008/01/01africa

Tarrósy, I. (2009). Politikai elméletek és módszertanok a mai afrikai problémák megértéséhez. In S. Csizmadia & I. Tarrósy (Eds.), Afrika Ma – Tradíció, átalakulás, fejlődés (pp. 13–17). Pecs: IDResearch Kft. & Publikon Kiadó.

Tarrósy, I. (2014). Nigéria választ – Újrázik-e Goodluck Jonathan? (Nigeria will Elect – Will Goodluck Jonathan Come Again). *Afrika Tanulmányok*, *4*, 5–12.

Taylor, I. (2014). *Africa Rising? BRICS – Diversifying Dependency*. Oxford, UK: James Currey.

The Observatory of Economic Complexity. (n.d.). *Nigeria*. Retrieved October 11, 2016, from http://atlas.media.mit.edu/en/profile/country/nga/

United Nations Comtrade. (2006). *UN Comtrade Database*. Retrieved December 6, 2016, from https://comtrade.un.org/

United Nations Conference on Trade and Development. (2002). *The Least Developed Countries Report*. Geneva: United Nations Conference on Trade and Development.

United Nations Conference on Trade and Development. (2013). *Nigeria Bilateral Investment Treaties (BITs)*. Retrieved November 17, 2016, from http://investmentpolicyhub.unctad.org/IIA/CountryBits/153#iiaInnerMenu

US Department of State. (2016). *U.S. Relations with Nigeria*. Retrieved November 18, 2016, from http://www.state.gov/r/pa/ei/bgn/2836.htm

Woolf, S. S., & Jones, E. I. (1969). *Agrarian Change and Economic Development: The Historical Problem*. London: Methuen.

World Trade Organization. (2007). *Council for Trade in Services – Special Session – Nigeria – Initial Offer – Document TN/S/O/NGA*. Retrieved November 18, 2016, from https://www.wto.org/english/docs_e/docs_e.htm

World Trade Organization. (2010). *Committee on Trade and Development – Sixteenth Session on Aid for Trade – Nigeria's Trade and Productive Capacity Reform – Document WT/COMTD/AFT/W/23*. Retrieved November 18, 2016, from https://www.wto.org/english/docs_e/docs_e.htm

World Trade Organization. (2011). *Trade Policy Review Body – Trade Policy Review – Report by the Secretariat – Nigeria – Document for Nigeria WT/TPR/S/247/Rev.1*. Retrieved November 18, 2016, from https://www.wto.org/english/docs_e/docs_e.htm

Zsarnóczai, J. S., Lakner, Z., & Mughram, Y. A. (2011). Hungarian Participation in Modernisation of the Third World. In I. Szűcs & J. S. Zsarnóczai (Eds.), *Economics of Sustainable Agriculture* (pp. 87–108). Gödöllő: Szent István University Publishing.

ADDITIONAL READING

Abayomi, O. (1997). The Agricultural Sector in Nigeria: The Way Forward. *CBN Bullion*, *21*, 14–25.

Búr, G., & Tarrósy, I. (2011). Az afrikai vándorlás hátteréről és jellemzőiről. In I. Tarrósy, V. Glied, & D. Keserű (Eds.), Új népvándorlás. Migráció a 21. században Afrika és Európa között (pp. 99–114). Pecs: IDResearch Kft. & Publikon Kiadó.

Emeka, O. M. (2007). Improving the Agricultural Sector toward Economic Development and Poverty Reduction in Nigeria. *CBN Bullion*, *4*, 23–56.

Iganiga, B. O., & Unemhilin, D. O. (2011). The Impact of Federal Government Agricultural Expenditure on Agricultural Output in Nigeria. *Journal of Economics*, *2*(2), 81–88.

KPMG. (2014). *2014 Quarter 1*. Retrieved November 19, 2016, from http://www.kpmg.com/Africa/en/KPMG-in-Africa/Documents/2014%20Q1%20snapshots/KPMG_Nigeria%202014Q1.pdf

Oji-Okoro, I. (2011). Analysis of the Contribution of Agricultural Sector on the Nigerian Economic Development. *World Review of Business Research*, *1*(1), 191–200.

Ojo, E. O., & Adebayo, P. F. (2012). Food Security in Nigeria: An Overview. *European Journal of Sustainable Development*, *2*(1), 199–222.

Olayide, S. O., & Essang, S. M. (1976). The Role of Commodity Exports in Nigeria Economic Growth. *Nigerian Journal of Rural Economic Development*, *10*, 105–122.

Oruruo, K. I. (2014). *Agriculture in Nigeria.* Retrieved November 19, 2016, from http://logbaby.com/news/agriculture-in-nigeria---by-king-i--oruro_15856.html

United Nations Development Programme. (2004). *Annual Report 2004. Mobilizing Global Partnership.* Retrieved November 17, 2016, from http://www.undp.org/content/undp/en/home/librarypage/corporate/undp_in_action_2004.html

KEY TERMS AND DEFINITION

Balanced Economy: It has two meanings: on the one hand, it means a condition of finances in a country or nation in which both its imports and exports are of an equal proportion. On the other hand, it refers to an ideal combination of economic sectors in which many of them generate export incomes and sources for further development.

Comprehensive Double Taxation Agreement (or Comprehensive Agreement for Avoidance of Double Taxation): Double taxation is generally defined as the imposition of comparable taxes in two or more places on the same taxpayer in respect of the same subject matter and for identical periods. The international community generally recognizes that double taxation hinders the exchange of goods and services, movements of capital, technology, and human resources, and poses an obstacle to the development of economic relations between economies. A comprehensive avoidance of double taxation agreement helps to minimize double taxation.

Endogenous Resources: Such resources, which are related to the specific area and can provide a basis for development.

Food Security: Food secure countries are those where food import is necessary to offset the food shortages and because the countries try to broaden the supply of food only.

International Trade Agreement: A treaty between two or among more countries with the view of intensifying their cooperation and gaining mutual benefits in the field of trade, taxation, tariffs, and investment. The most common trade agreements are of the preferential and free trade types. They are concluded in order to reduce (or eliminate) tariffs, quotas, and other trade restrictions on items traded between the signatories. By number and type of signatories a trade agreement is classified as bilateral (BTA), when signed between two sides, where each side could be a country (or other customs territory), a trade bloc or an informal group of countries (or other customs territories). A trade agreement signed between more than two sides (typically neighboring or located in the same region) is classified as multilateral.

Regional Integration: A process in which neighboring or geographically close countries enter into an agreement in order to upgrade cooperation through common institutions and rules. These agreements serve economic and political objectives.

Valorization: Establishment and maintenance of a price by governmental action.

Chapter 11
The Potential of Traditional Leafy Vegetables for Improving Food Security in Africa

Praxedis Dube
Wageningen University, The Netherlands

Rico Ihle
Wageningen University, The Netherlands

Wim J. M. Heijman
Wageningen University, The Netherlands

Justus Ochieng
World Vegetable Center, Eastern and Southern Africa, Tanzania

ABSTRACT

Feeding the quickly growing population in Africa remains a global challenge. As the demand for food increases, climate change, on the other hand, poses more challenges to agricultural productivity, implying that the provision of sufficient quantities and qualities of food is threatened. Traditional leafy vegetables (TLVs) in Africa are resilient to adverse weather conditions and are naturally rich in nutrients including vitamins A & C, iron, protein and other micronutrients. The objective of this chapter is to assess the potential of TLVs improving food security in Africa. TLVs represent a robust local source of food, their consumption is crucial in complementing Africa's diet with micronutrients. TLVs have therefore the potential to play a major role in improving food security and facilitating food sovereignty. Research on seeds, developing seed systems, coupled by preparing and processing methods to TLV products like cakes or flour is needed to increase their consumption, particularly among the young, elite and urban dwellers.

INTRODUCTION

Feeding the quickly growing population in Africa remains a worldwide challenge. Globally, close to a billion people are hunger stricken (Food and Agriculture Organization of the United Nations [FAO], 2015) and approximately one-third of them live in Africa (Sasson, 2012). As the demand for food increases, climate change, on the other hand, poses challenges to domestic food production in Africa, implying that the provision of sufficient quantities of food of the necessary quality is threatened. A lack of sufficient quantities of sufficiently diversified food compromises human health, resulting in an increase

DOI: 10.4018/978-1-5225-2733-6.ch011

in the risk of contracting diseases related to malnutrition, which affect one in every four people (FAO, 2015). Figure 1 illustrates the prevalence of hunger and the distribution of the undernourished population across Africa. The map reveals substantial differences between African regions. Sub-Saharan Africa is the most severely affected by the prevalence of hunger and undernourishment.

In contrast, Northern Africa accounts for 1% of the undernourished population, as depicted in Figure 2.

The differences in undernourishment in Africa are attributed to price volatility, an exploding population, droughts (FAO, 2015), poor governance (Ihle, von Cramon-Taubadel, & Zorya, 2009) and political instability (Veninga & Ihle, in press). For instance, highly volatile food prices affect mainly the poor people in urban and rural areas, since they spend a higher proportion of their limited incomes on their food needs (Luckmann, Ihle, Kleinwechter, & Grethe, 2014). They tend to consume smaller quantities less frequently as well as cheaper and less nutritious foods that lack vitamins, minerals, and other essential micronutrients, e.g. zinc or iron, both of which are required by the human body for healthy growth. This leads to widespread undernourishment. Vegetables, particularly traditional leafy vegetables

Figure 1. Hunger and undernourishment persistence in Africa in 2015
Source: FAO, 2015

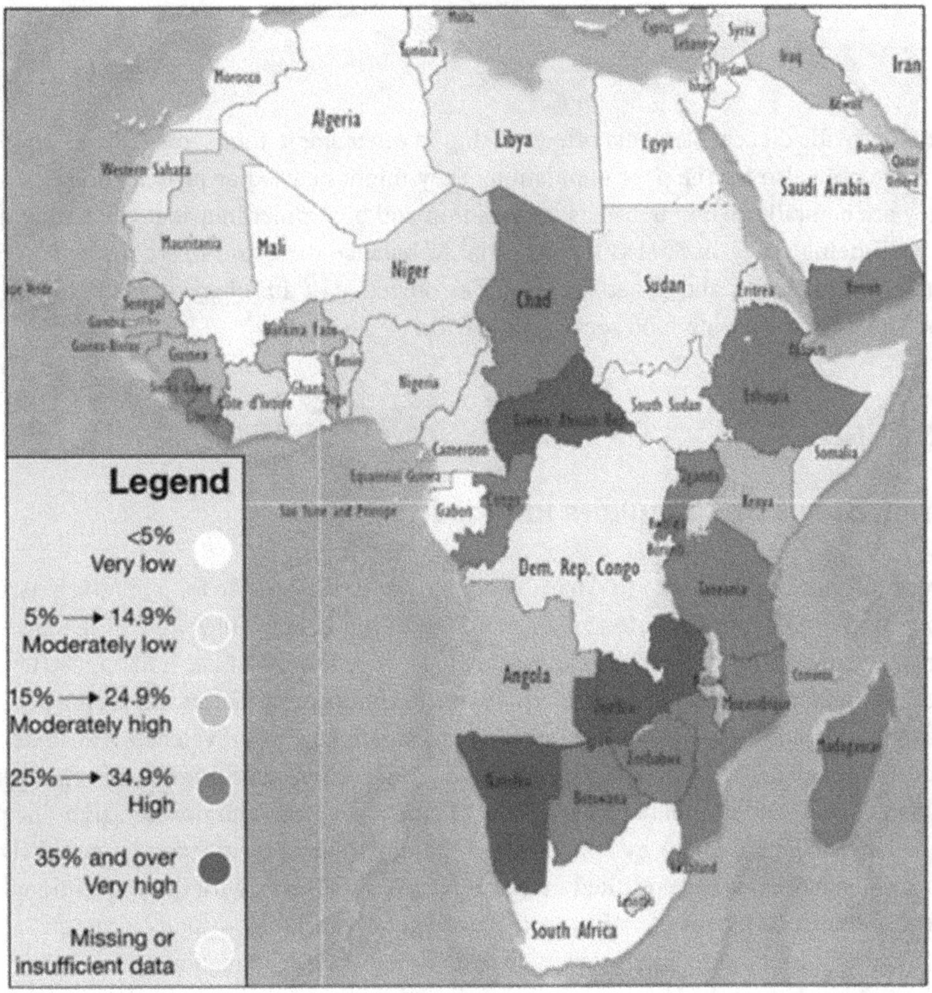

Figure 2. Prevalence percentage of undernourishment in Africa in 2014-2016
Source: Authors based on FAO, 2015

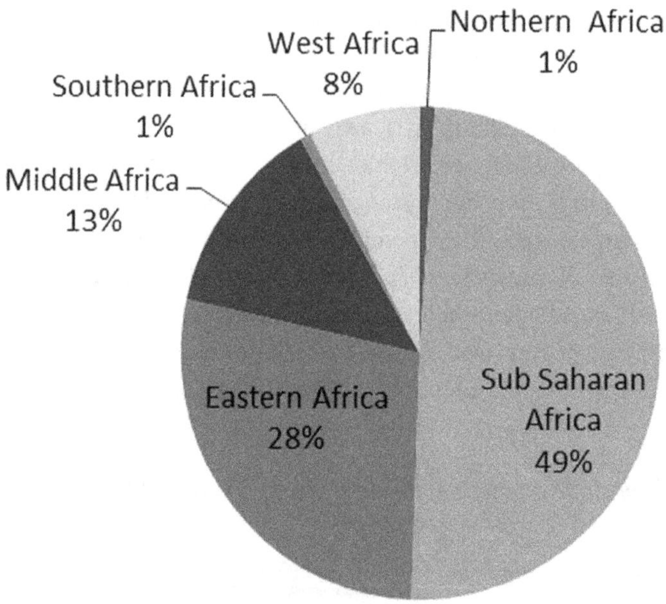

(TLVs) might provide dietary elements often missing in other staple foods, e.g. cereals and root foods, which are affordable also for the poor population. They might be used as primary sources of minerals because they are naturally rich in nutrients such as iron and other micronutrients that are important for a human's well-being. They have also the potential to balance diets and make them more healthy as they also include vitamins A and C, and proteins. The objective of this chapter is therefore to assess the potential of TLVs for improving food security in Africa.

BACKGROUND

The Current Status of Agriculture in Africa

Agriculture contributes significantly to Africa's economy. It is responsible for a relatively large share of the economy in terms of labor force. It accounts for 70% of jobs, contributes 33% of the gross domestic product, and generates 40% of the total export earnings in Africa (Otsuka, Larson, & Hazell, 2013). Africa faces the challenge of low food productivity due to a number of factors, including poor uptake of modern technologies, and high incidences of poverty culminating in the prevalence of an undernourished population among others. At the United Nations Sustainable Development Summit on September 25, 2015, world leaders adopted the Sustainable Development Goals (SDGs) to end poverty, fight inequality and injustice, and tackle climate change by 2030 (United Nations Development Programme [UNDP], 2016).

The modern technologies have resulted in the replacement of TLVs with the conventional vegetable varieties such as rape, cabbages etc. However, these exotic vegetables usually grow in high-altitude areas with a cool climate and require modernized technologies, like irrigation which is far beyond the reach of many smallholder farmers in Africa. In the humid lowland areas of West Africa, the arid areas of Namibia and Botswana, in the blistering heart of Kalahari, in hot dry regions in other parts of the continent, the exotic vegetables cannot grow well, but TLVs can grow (Schippers, 2000) allowing for the effective use of non-productive areas.

Traditional Leafy Vegetables in Africa

TLVs are plants which originate from the African continent and have a century-long history of cultivation and domestication subject to regional agro-climatic conditions (Ambrose-Oji, 2009). They are often more resilient to adverse weather conditions and withstand better unpredictable weather changes than varieties of staple crops and vegetables bred in industrialized countries, fitting very well to bioregional foods.

TLVs are naturally important to many people in Africa due to their inherent nutrient content and a vast range of phytochemicals (Yang & Keding, 2009). Generally, TLVs are more nutritious than commonly known exotic vegetables. Commonly consumed vegetables across Africa include Amaranth (*Amaranthus spp.*), Okra (*Abelmoschus esculentus*), Spider flower (*Cleome gynandra*), Pumpkin leaves (*Cucurbita spp.*), Cowpea leaves (*Vigna unguiculata*), Black jack (*Bidens pilosa*), Sweet potato leaves (*Ipomoea batatas*), Nightshade (*Solanum spp.*), Moringa (*Moringa oleifira*), African eggplant (*Solanum spp.*), Baobab leaves (*Adonsonia digitata*), Jute (*Corchorus olitorius*), and Cassava leaves (*Manihot esculentum*).

Table 1 shows laboratory results on the nutritional contents of TLVs found by the World Vegetable Center, TLVs have substantial amounts of micronutrients, vitamins, and minerals which are essential for a balanced and healthy diet. These vegetables often have higher amounts of nutrients than vegetables which are traditionally grown on other continents such as tomatoes and cabbage. For example, Moringa provides the highest quantity of micronutrients among the most used TLVs. TLVs are generally rich in vitamins A, C and E, zinc, calcium, iron, and antioxidants (Yang & Keding, 2009). In comparison, TLVs have higher nutrient amounts than other vegetables such as tomatoes and cabbages, Table 1.

Some TLVs can be produced or collected from forests or fallow lands. In Eastern and Western Africa, TLVs production is commercial in both urban and peri-urban areas (Weinberger & Pichop, 2009). Most farmers producing TLVs sell about 50% of the produce through different market channels, and the remainder is consumed within households. Table 2 shows that the total volume of TLVs produced by seven countries was 9,300 tons/year (Weinberger & Pichop, 2009). The quantity is relatively small given

Table 1. Micronutrient content of common TLVs

Vegetable type						
Micronutrient	Range	Tomato	Cabbage	Moringa	Amaranth	Sweet potato leaf
β- Carotene, mg	0.0 – 22	0.40	0.00	15.28	9.23	6.82
Vit C, mg	1.1 – 353	19	22	459	113	81
Vit E, mg	0.0 – 71	1.16	0.05	25.25	3.44	4.69
Iron, mg	0.2 – 26	0.54	0.30	10.09	5.54	1.88
Folates, mg	2.8 – 175	5	ND	93	78	39
Antioxidant activity, TE	0.6 – 82,000	323	496	2,858	394	870

Note: ND = Not Done

Source: Ojiewo, Tenkouano, Oluoch, & Yang, 2010

Table 2. Total TLVs production by African regions in 2016

Region (countries)	Total volume (tons)
Eastern Africa (Kenya, Tanzania, and Uganda)	6,900
Western Africa (Senegal, Benin, and Côte d'Ivoire)	2,400
Southern Africa (South Africa)	27
Total volume	9,327

Source: Authors based on IndigenoVeg survey data (Weinberger & Pichop, 2009)

that it is produced by seven countries. The total volume was largest in Eastern Africa, followed by Western Africa. The production of TLVs in these three parts of Africa is driven by generating extra income and home consumption, among others (Weinberger & Pichop, 2009). Given their nutritional potential, promotion campaigns such as road and cook shows, nutritional awareness and educational programs in hospitals, schools and markets have been used, for example in Tanzania to increase consumption of traditional vegetables by rural and urban consumers (Ochieng et al., 2016).

The quantities consumed per day vary across the African regions. For instance, in Southern Africa, 29–41 kg/year are consumed by South Africans/households depending on the location (rural or urban dwellers, with the latter consuming less) (Yang & Keding, 2009). A study of the TLVs consumption in seven countries of Africa showed that Tanzania consumed largest quantities of TLVs/capita/day, while Senegal consumed the least, as shown in Figure 3. On average, Eastern African countries consumed larger quantities, indicating the importance of nutrition awareness.

Over the last two decades in the context of the realization of the limitations of the Green Revolution, the national food programs and strategies, and the effects of climate change, researchers realized a need for a paradigm shift to look at underutilized plants such as TLVs. These neglected species have begun to attract considerable interest due to their multiple underexploited benefits in terms of nutritional value,

Figure 3. TLVs per capita per day (grams) by country
Source: Authors based on IndigenoVeg consortium data (Weinberger & Pichop, 2009)

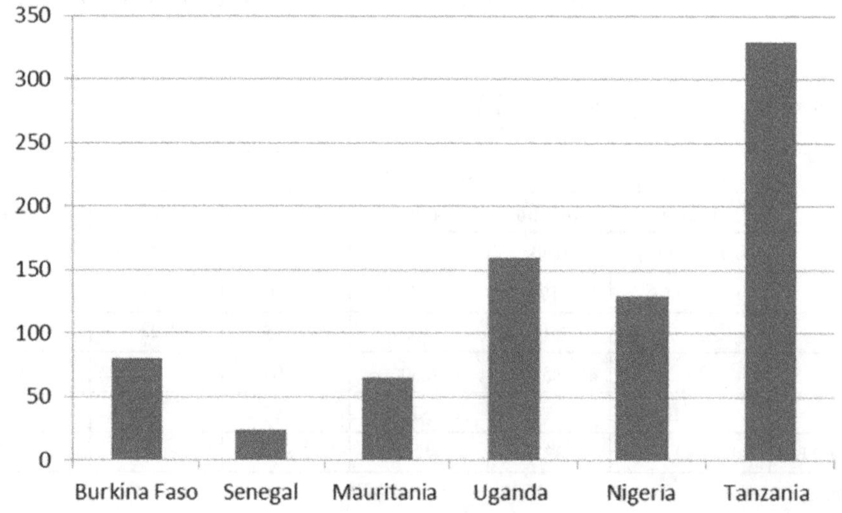

food security, income generation, medicinal value, and suitability for low input systems and for marginal environments (Shackleton, Pasquini, & Drescher, 2009). The interest in the underutilized species for research and development came during the late 1980s. For example, there was the establishment of the International Centre for Underutilised Crops in 1988, the Plant Resources of Tropical Africa in 2000, the Global Facilitation Unit of Underutilised Species in 2002, Consultative Group on International Agricultural Research in 2005 and the inclusion of indigenous vegetables for biodiversity, healthy diet and marketing opportunities as one of the World Vegetables Centre's strategic programme directions in 2002 (Shackleton et al., 2009). All these strategies aimed at increasing the use of the underutilized crop species such as TLVs to retard biodiversity losses, increase community-driven rural development, exploit the nutritional value, food security, income generation, and medicinal value and make marginal lands more productive.

Recent milestones in TLVs production and consumption include a significant growth of interest in re-activating traditional food and a reversal of an earlier decline in farming of indigenous crops that this food is based upon. Scientists in Africa and elsewhere have been engaged in research to develop TLVs in order to tap into their health benefits (Cernansky, 2015). These researches have created new opportunities for TLVs in many countries. A study on TLVs production has shown that Kenyan farmers have increased the TLVs production area by 25% between 2011 and 2013 (Cernansky, 2015). Also, the number of bunches sold in Kenya have been increasing rapidly, estimated at 164%, as depicted in Figure 4. The TLVs have gained a commercial status in large cities like Nairobi (Irungu, Mburu, Maundu, Grum, & Hoeschle-Zeledon, 2011). Some TLVs, e.g. African nightshade, have gained 32% in the total Nairobi vegetable markets share (Irungu et al., 2011) and 11% in Tanzania. The World Vegetable Center is currently budgeting annually around $20 million on TLVs research (Cernansky, 2015).

Figure 4. Supply trends of TLVs in Nairobi markets
Source: Authors based on Irungu et al., 2011

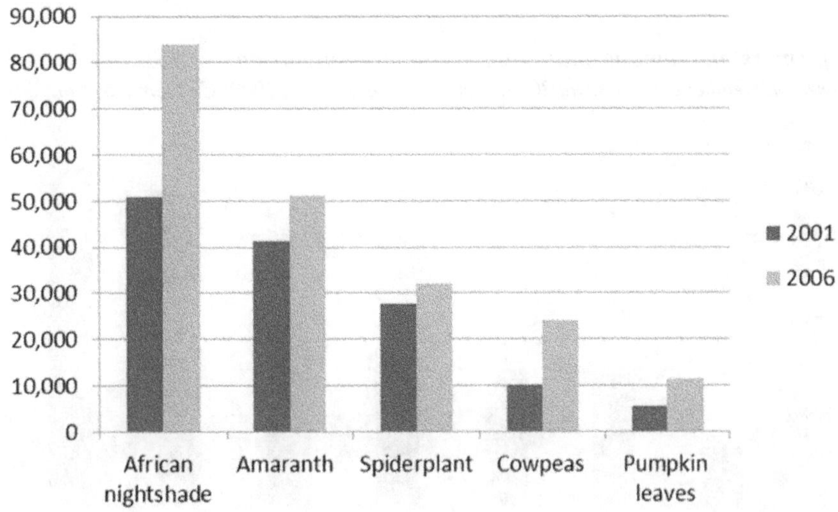

MAIN FOCUS OF THE CHAPTER

Current Status of Traditional Leafy Vegetables in Africa

TLVs play significant roles in the livelihoods of people along the supply chain in Africa. They contribute approximately 13%, 35%, and 30% of the smallholders' income in Tanzania, Malawi, and Mozambique, respectively (Chagomoka, Afari-Sefa, & Pitoro, 2014; Weinberger & Msuya, 2004), 29% to 76% in Uganda, Côte d'Ivoire, Kenya, Tanzania, Senegal, South Africa, and Benin (Weinberger & Pichop, 2009). Figure 5 depicts the share TLVs farmer receive of the consumer selling price. The data shows substantial differences between African countries. Higher shares are realized in South Africa and Benin. The differences in shares can be partially explained by prices. For example, a farmer in South Africa realizes better shares due to the price difference per kilogram of cowpea leaves selling at \$3.20 for South Africa compared to \$0.22 for Tanzania (Weinberger & Pichop, 2009).

A market study conducted in seven countries within Africa in 2009 shows the total volume and total turnover for the three most important TLVs traded at retail levels, as depicted in Figure 6. This study established that TLVs are relatively high-value crops generating a total turnover of \$5.5 million (Weinberger & Pichop, 2009). Assuming the same turnover being produced by 53 African countries, calculated, would yield approximately \$50 million generated by TLV supply chain markets.

Kenya shows the highest volume traded for the three most important TLVs, almost thrice the volume traded in Senegal yet with the total turnover lower than that of Senegal. South Africa exhibits the least total volume traded, but its total turnover is higher than that of Côte d'Ivoire. For the rest of the countries included in this data, only Senegal and South Africa have a total turnover higher than the total volume traded. These differences can be partially explained by the difference in selling prices (\$/kg) at retailers between these countries. For example, cowpea leaves selling per kilogram at \$3.64 and \$3.20 in Senegal and South Africa, respectively, compared to \$0.30, \$0.45, \$0.30, \$0.22 for Benin, Kenya, Uganda, and Tanzania, respectively (Weinberger & Pichop, 2009).

Figure 5. TLVs farmers' percentage share in consumer selling price
Source: Authors based on Weinberger & Msuya, 2004; Weinberger & Pichop, 2009; Chagomoka et al., 2014

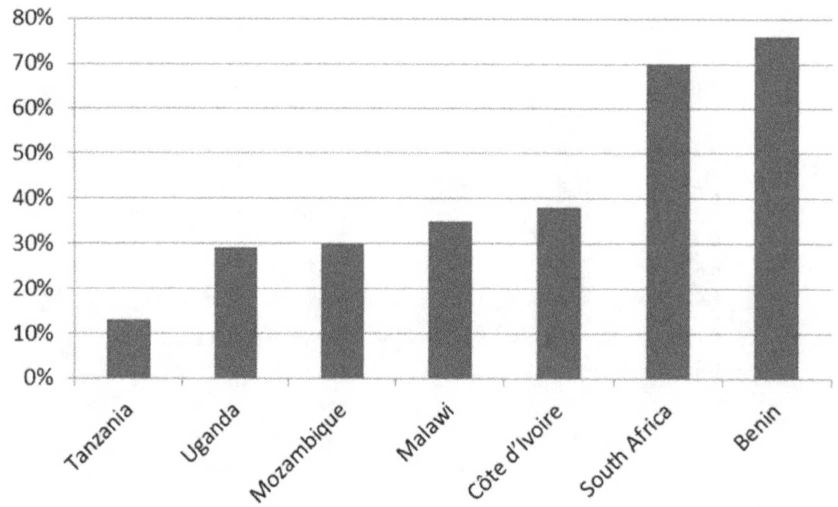

Figure 6. Total volume and total turnover for TLVs traded at retail/ country
Source: Authors based on IndigenoVeg survey data (Weinberger & Pichop, 2009)

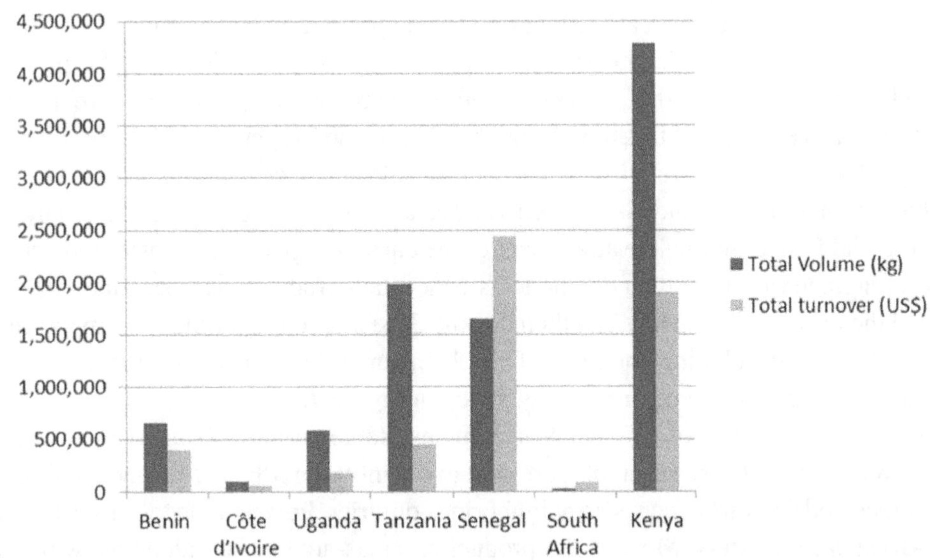

For TLVs, the trader and retailer share contribution at consumer selling point is the lowest in Benin at 24% and in South Africa at 30%. It is the highest in Uganda (71%), Côte d'Ivoire (62%), and Tanzania (around 55%) (Weinberger & Pichop, 2009). These differences can be partly explained by the share that is allocated to the farmers, e.g. the share received by farmers is relatively higher in Benin and South Africa, as shown in Figure 5.

Supply Chains of Traditional Leafy Vegetables

The authors assessed the role of TLVs in small-scale farmers' diets and incomes, the survey was conducted in five provinces of Zimbabwe (Mashonaland East, West and Central, Harare, Manicaland). The respondents were chosen using the snowball technique, starting with the seed production and supply segment upstream to traders, gathering data from five key informants within each segment. The authors developed question guides to help consistent data collection and interview technique at each segment. The information collected included crop management, marketing of the produce and information, among others. The content of this chapter is partly based on the results obtained from the survey.

The TLV supply chain is very short, due to high perishability. During the harvesting process, farmers normally place TLVs under the shade and at times sprinkle water on the vegetables. In Zimbabwe, farmers or intermediates pack TLVs in jute bags and then transport the product to the markets. Some bags are packed in big lorries or placed on top of the commuter buses and tied by a rope to avoid falling. Transportation is done during the evening when temperatures are low. Unfortunately, there is no appropriate storage during transportation and at the marketplaces, which can result in reduced shelf life of vegetables. To avoid losses through spoilages, the producers and traders are forced to bring or buy the TLVs in small enough quantities such that they can be sold within a day or two.

The producers minimize the transactional costs and losses due to high perishability by taking the TLVs directly to the markets. In such cases, it may be worthwhile for the producers to have market information before taking the TLVs to the markets. Alternatively, the producers sell their TLVs to intermediates moving around their farm gates and sourcing the vegetables. In that way, the producers would reduce transport costs and other costs that can be incurred, i.e. high market fees. For example, in Harare, the farmer is charged $10 per day, which means that if the farmer has unsold TLVs, the farmer either has to pay another $10 for the following day, is forced to dump the remaining product or give it away for free.

Figure 7 depicts an example of the movement of TLV seed and vegetables from the areas of production to markets. The TLV seeds are primarily produced, sold and exchanged locally, with many people deriving their incomes and livelihoods through various activities carried along the supply chain, thus having the potential for facilitating the attainment of the current imperative to foster food and seed sovereignty and strengthen various of their components. Since their production is local, farmers have control over seeds and the resulting vegetables can then be quickly supplied to consumers, supporting domestic marketing structures in developing countries. This chain supports the flow of purchasing power from more affluent urban regions to rural areas, as depicted in Figure 7.

The highly perishable nature of TLVs makes them unsuitable for long-distance transport. This statement is true across Africa. These vegetables are sold to neighboring urban areas and locals and in small quantities in processed form are exported to neighboring countries. For example for Zimbabwe, Botswana and South Africa are importers. Most of the production areas are near urban areas, while there exist others that are located more than 100 km away from the urban areas. In Zimbabwe, locations, where

Figure 7. TLVs seed production regions and markets in Zimbabwe
Source: Authors

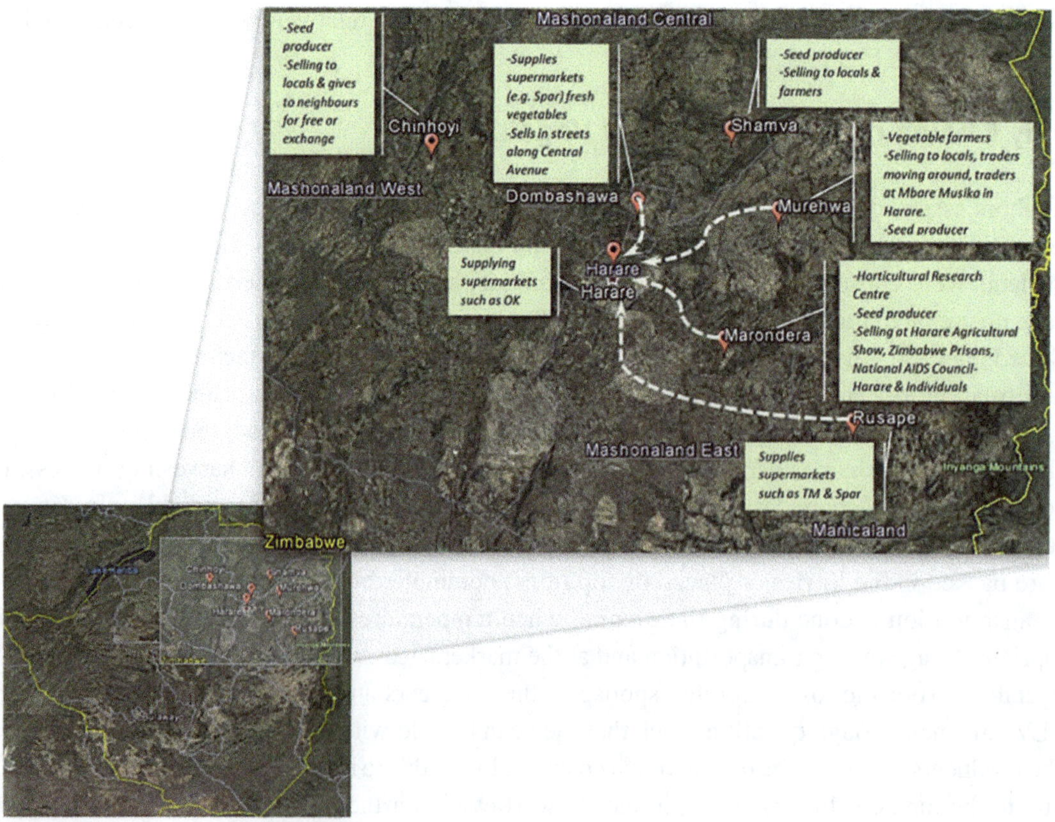

the authors collected data, are around 100 km away from the major markets, translating to an average of three to four hours to reach the market using public transport. In Kenya, the TLV production areas are near Nairobi, whereas others are more than 100 km away and other retailers had to walk 15 km daily for fresh TLVs (Nekesa & Meso, 1997). In Zambia, Lusaka supermarkets source their TLVs from local farmers in the peri-urban areas. In Uganda, the average distance traveled and time used to travel in order to sell the TLVs is not large. It rarely takes more than an hour and the traveling distance is around 14 km (Weinberger & Pichop, 2009). TLVs are very perishable and there is a need to produce them near or at the point of use.

In the community, the TLVs are sold at various market outlets, including at farm gates, on the side of the road and door-to-door. In this scenario, the traders are producers as well, as they sell their TLVs directly to consumers and to some intermediaries. Some of the intermediaries visit various farm gates, purchase the TLVs, transport them to the markets and eventually sell them to the consumers. It is mainly men who do the purchasing. In case the TLVs are not bought, in the case of Zimbabwe, the producers would process them and sell them later, some would be consumed in their households.

Women play significant roles in the TLV supply chain as producers, processors, and marketers (Weinberger & Pichop, 2009). In Uganda, men mainly act as intermediaries, as they move around to buy TLVs from producers (mostly women) and transport them to markets. In the markets, women buy them from the intermediaries and then sell these vegetables to the public. Women are, therefore, economically empowered as they are able to pay fees for their children and provide food their households.

Vegetable Varieties, Seed Production, and Farming Characteristics

Many TLV types are produced within homesteads, sold in local markets, exchanged as seeds or sold in the local markets, fostering localized developments. Despite their high potential in offering better opportunities for gender oriented developments, transforming the rural livelihoods through localized markets and addressing food insecurity problems, there are many challenges persisting along the supply chain. The availability of seed and undeveloped seed systems, depriving farmers' access to high-quality seed of preferred varieties. More so, with the supplies of vegetables now getting into big cities (increased demand for vegetables), this translates to increased demand for seed as well.

In Eastern Africa, mostly, TLV seed is produced by farmers. Seed stockists buy this seed and then sell it to many emerging TLV farmers. These seed producers are registered with renowned companies and are able to produce certified seed, giving the seed a commercial market value. They are capable of producing high-quality seed with the supervision of big seed companies. More so, the TLV seed markets are organized well, enabling seed producers to realize profits and incentivise. The farmers are assured of the ready markets and they know well before planting the types of TLVs preferred by their targeted markets. Unlike in Eastern Africa, Southern Africa, Western Africa and Central Africa, the production of TLV seed has not been formalized. There, no commercial seed company has shown interest in venturing into the seed production and marketing sector. One example where this has happened is Cameroon. Various international organizations funded the training and production of TLV seed, aiming at improving seed availability to farmers through the involvement of local seed companies in Eastern Africa and a few in Southern Africa. The top five selected for increased production, consumption and marketing include cowpea, amaranth, nightshade, spider flower, and African eggplant.

In Zimbabwe, TLV seed is produced by a government institute under the Department of Research & Specialist Services (Horticultural Research Centre) and by farmers dotted around the country. The most common TLV seed produced include spider flower, cowpea, amaranth, black jack, okra, tsunga, and pumpkin, with the most traded being spider flower and pumpkin seed. Similarly, findings by Chagomoka et al. (2014) revealed that TLV seed produced in Malawi included Ethiopian mustard, okra, pumpkin, and African eggplant. In Mozambique, the same authors found that most TLV farmers use returned seed, with some of the seed imported from Tanzania. The seed is usually packed in a plastic bag or plastic coated paper. In Zimbabwe, three main sections of TLV seed production were found, notably, the low section (subsistence farmer without knowledge of seed production), the middle section (commercialized farmer with relatively basic knowledge of seed production) or high section (commercialized seed producing company i.e. Horticultural Research Centre). The seed from the high section was more expensive than other seed sections, costing $50 per kilogram. The seed is marketed to different channels: to neighbors, at agricultural shows, and to various organizations promoting the use of TLVs for nutrition and health benefits.

The farmers generally measure out the seed in tablespoonfuls or pet coke cola leads. A tablespoonful is estimated to weigh approximately 15 grams and a pet coke cola lead is approximately 2 grams, marketed at $2.00 and $0.50 respectively or $50 per kilogram in Zimbabwe markets and $1.80 per 25g of amaranth seed in Kenya markets (Mumbi et al., 2006). When these seed weights and prices were calculated and estimated per hectare of spider flower and amaranth for one season, the gross income presented in Table 3 are derived. In comparison, spider flower seed's gross margin is higher than that of amaranth seed, valued at $6,500/hectare. The gross margin for farmers in the production of TLVs vegetables varies due to the TLV type. The least gross margin is obtained from cowpeas with $2,426. Nevertheless, in Zimbabwe, the TLV seed is rarely packaged, labeled and traded directly to the public without hiring the chain of people or companies to market the seed. Seed produced is sold locally to neighbors, friends, and relatives or nearby urban areas. The transactions are mostly based on the cash market.

Production of Traditional Leafy Vegetables

Similar to countries in Eastern and Western Africa, the production of TLVs can be commercialized in Zimbabwe, in particular areas close to major cities. For example, a Zimbabwean farmer who produces and sells the TLVs throughout the year as opposed to seasonally gets better prices when the market supplies are low, while also taking advantage of the fast moving of the TLVs when incoming supplies dwindle. The spider flower fetches the highest price and sells fastest. TLVs winter production is mainly carried

Table 3. Seasonal and annual gross margin analysis for TLV seed and TLVs ($)

Category	TLV grown	Country	Production cost	Gross income	Gross margins
Seed (per ha and season)	Spider flower	Zimbabwe	1,000	7,500	6,500
Seed (per ha and season)	Amaranth	Kenya	185	3,145	2,960
Vegetables (0.1 per ha/year)	Amaranth	Kenya	1,142	5,714	4,571
Vegetables (0.1 per ha/year)	Cowpea	Kenya	1,306	3,733	2,426
Vegetables (0.1 per ha/year)	Nightshade	Kenya	1,633	4,666	3,033

Source: Authors and Mumbi et al., 2006

along major rivers in some parts of Zimbabwe (Mpala, Dlamini, & Sibanda, 2013), concentrated in irrigation schemes during the dry seasons and in rural areas during the summer season. Most of the TLVs produced are consumed domestically, either by households or through local markets in nearby urban areas. Similarly, the production and marketing of TLVs has been increasing and becoming increasingly commercialized in Kenya, particularly within Nairobi and the peri-urban areas (Irungu et al., 2011). Women participation in the TLVs supply chain is constantly high in Kenya, Benin and South Africa (Table 4). Female participation in TLV production tends to be lowest to other levels of the supply chain. Table 4 shows that it is especially low in Côte d'Ivoire and Senegal.

However, production of TLVs is hindered by unavailability of certified or high-quality seed coupled with the underdeveloped seed systems. The underdeveloped seed systems limit the seed movement, subsequently, limits the farmers' variety of choices.

Processing of Traditional Leafy Vegetables

Processing and packaging is a way of minimizing spoilage of TLVs due to their high perishability. During the periods when the TLVs are abundant in the markets and in gardens, TLVs are partly processed for future use in the dry season or for distant marketing (Mpala et al., 2013). In Zimbabwe, 97% of the TLV farmers process parts of the vegetables harvested (Mpala et al., 2013), in Uganda 1.2%, in Côte d'Ivoire 9.3%, in Kenya 2.1%, in Senegal 0.8%, and in South Africa 10.3% (Weinberger & Pichop, 2009). Similarly, Chagomoka et al. (2014) note that TLVs are processed prior to selling as a way of minimizing quantity and quality losses. Only 12% of retailers in Malawi and 6% of retailers in Mozambique process the TLVs prior to selling. About 50% of the households in South Africa preserve vegetables for both selling and own consumption.

There are various methods of processing, including; sun-drying without any pre-treatment, blanching and sun-drying and blanching then freezing (Bioversity International, n.d.). Sun-drying without any pre-treatment method is the most preferred among the commonly preserving methods (Mpala et al., 2013). In Zimbabwe, cowpea tender leaves are harvested, boiled, and dried in direct sunlight. They are twisted by rolling them between the pumps. The processing varies across the nations, regions, and cultures. For example, in Malawi cowpea leaves are processed by sun-drying and/or blanching (Chagomoka et al., 2014), but not twisting. The pumpkin leaves are sold fresh across Africa. Except in some countries in Southern African, including Zimbabwe, they are traded in their dried form. They are dried without cooking and normally their flowers are added.

In Zimbabwe, drying is generally done in a semi-closed cabinet that protects the processed vegetables from fine dust particles that may settle on it in order to avoid making them gritty. Quality attributes such

Table 4. Share of women along the TLVs supply chain across countries (%)

Category	Country						
	Benin	Côte d'Ivoire	Senegal	Kenya	Tanzania	Uganda	South Africa
Farmers	40	16	14	59	34	37	65
Intermediaries	78	100	26	95	75	69	unknown
Retailers	100	100	58	86	58	68	86

Source: Weinberger & Pichop, 2009

as twisted and dried for cowpea, the absence of pathogens, i.e. dried without visible moulds, and neatly packaged with the labels showing shelf life, net weight and processors' details are mandatory. The concerns arise especially if the quality attributes are not met. This can result in negative perceptions among some urban dwellers and educated elite in response to their fear for their health. The worries related to the health risks were highlighted in the World Vegetable Center (2006) report on determination knowledge and use of TLVs by the elite urban dwellers. The average to middle-income earners consumed little in fear of risking their health.

Marketing Channels of Traditional Leafy Vegetables

Trading TLVs in Kenya dates back to the 1960s (Irungu et al., 2011). The trade of TLVs is highly specialized, as is the consumption. Nairobi has ten large markets, which sell TLVs in larger amounts. These markets are located in residential places and serve the surrounding consumers. Supermarkets and green grocery stores serve the affluent minority who have limited time to visit some of the city council markets, such as Harare and Nairobi City Council markets. In addition, there are several evening traders along the streets who target the people coming from work.

Similarities in trading of TLVs can be found across African regions. In Zimbabwe, the authors noted that three-quarters of traders interviewed purchased the TLVs from farmers moving around the market looking to sell to traders they know. While supermarkets in both Kenya and Zimbabwe buy directly from individuals, groups or processors. In comparison, Zimbabwean supermarkets mainly sell dried TLVs with the exception of pumpkin, spider plant, and amaranth leaves, while Kenyan supermarkets sell fresh TLVs.

The traders purchase a bundle of TLVs or two cups of dried TLVs for $0.50. A standard cup consists of around 60 g. The traders split the bundle into two and each is sold to consumers at $0.50. In Zimbabwe, a sachet weighing approximately 80 g sells for $2.00 in the green grocery, while a range of $2.00 to $2.35, depending on the TLVs type, for 60 g is sold in supermarkets. Similarly, in Kenya, the wholesaler buys a bundle of TLVs for KES2 ($0.04), splits the bundle into two and sells each for KES2 to the retailer. In turn, the retailer further splits the bundle before selling it to the consumer (Nekesa & Meso, 1997). In Zambia, a bundle of TLVs weighing 500 g is sold between 500 and 800 Zambian Kwacha ($0.25 – $0.30). The profit made on TLV sales is over 100% in Zimbabwe (Mpala et al., 2013), 33% in Tanzania, and over 75% in Kenya (Nekesa & Meso, 1997).

The major TLV dominate marketing channels across Africa, including city council markets, supermarkets, greengroceries, open markets and roadside, as reported by Irungu et al. (2011) and Schippers (2000). Table 5 shows some of the most popular TLVs consumed and found most tradeable in African regions. The types of TLVs popularly consumed and found most tradeable vary substantially by region, except for amaranth. Amaranth is sold fresh in all cities (Maundu, Achigan-Dako, & Morimoto, 2009). Contrary, spider flower is sold in the fresh or dried form in the Southern Africa region. These variations can be explained by different attributes, including cultural backgrounds, geographical location, and diversity. In comparison, in Western Africa, baobab leaves are some of the most popular TLVs consumed, whereas black jack is the most commonly consumed TLVs in the Southern African region. Northern Africa, due to its climate conditions, a few types of TLVs exist in the region.

Table 5. Some of the most popular TLVs consumed and traded in Africa

Region	Popular TLVs	Tradeable TLVs
Southern Africa	Amaranth, Okra, Spider flower, Pumpkin leaves, Cowpea leaves, Black jack, Sweet potato leaves	Spider flower, amaranth, black jack, cowpeas and pumpkin leaves, okra, African eggplant, amaranth, jute mallow, Ethiopian mustard, and wild cucumber
Eastern Africa	Amaranth, Cowpea vegetables, Pumpkin vegetables, Corchorus, Nightshade, Spider flower, Sweet potato leaves, Moringa	African nightshade, amaranth, Spider flower cowpea leaves, crotalaria, okra, Ethiopian kales, Ethiopian mustard, sweet potato leaves, cassava leaves
Western Africa	Amaranth, African eggplant, Corchorus, Okra, Sweet potato leaves, Cowpea leaves, Baobab leaves, Spider flower, Moringa,	Sweet potato leaves, African nightshade, cassava leaves, pumpkin leaves, jute, Amaranth
Central Africa	Amaranth, Pumpkin leaves, Sweet potato leaves, Okra, Jute, Spider flower, Cassava leaves	Moringa, jute, hibiscus, okra, amaranth, cowpea leaves

Source: Authors; Nekesa & Meso, 1997; Maundu et al., 2009; Irungu et al., 2011; Chagomoka et al., 2014

Challenges in Realising the Economic Potential of TLVs

TLVs have for decades not been recognized as a potential contributor to food and nutrition security by policy makers (Schippers, 2000). Only recently, TLVs have become important food crops and have attracted international spotlight because of their potential for human nutrition by providing essential micronutrients, phytochemicals, and antioxidants for achieving a balanced diet. The realization of their potential is subject to a number of challenges persisting along the TLV supply chain. Major challenges include the unavailability of quality-certified seed to enhance productivity, undeveloped seed systems depriving farmers' choices of varieties, limited preservation methods given their high perishability, resulting in poor quality, processed products, and negative perceptions among some of the young generation, educated elite, and urban dwellers.

Results from a study conducted by Weinberger & Pichop (2009) in three regions of Africa (Eastern, Southern, and Western), show that the median and inter-quartile range for urban areas dedicated to vegetable production is larger in Nairobi and Kisumu, for peri-urban areas of Cotonou, Durban, Lokossa, and Shoshangwe, reflecting increased demand of TLVs in the urban areas in Kenya (Figure 8 and Figure 9).

If the quality of TLV seed would be improved and become more homogenous, farmers might additionally benefit by having fewer expenditures for the seed and eventually get higher yields. However, this potential of increased earnings depends on the structure of the value chain, number of actors and their market power, and the value added at its various value chain nodes. The more diverse and immediate the marketing possibilities of the farmers are, the higher their benefits can be. The potential socio-economic and societal benefits of investing in the improvement of TLV seed quality by eventually incorporating TLV seed into the formal seed market crucially depends on the potential contributions which these indigenous food products can have towards the income of the stakeholders in the value chain on the one hand and the benefits for food security and the healthy nutrition of the consumers on the other. Most commercial seed companies consider the TLV seed production as uneconomic. Across Africa, TLV seed is marketed informally, with less than 10% of seed planted being purchased from the formal markets. Few seed companies have committed themselves to the TLV seed production, for example in Kenya and Tanzania farmers have benefited from the involvement of big seed companies with many of them changing from using own farm-saved seed to certified seed. Switching from own saved to certified seed increased the farmers market gross value by 213% between the period of 2001 and 2006 (Irungu et al., 2011).

Figure 8. Cultivated areas in urban and peri-urban for six cities
Notes: Median values and interquartile ranges are displayed
Source: Weinberger & Pichop, 2009

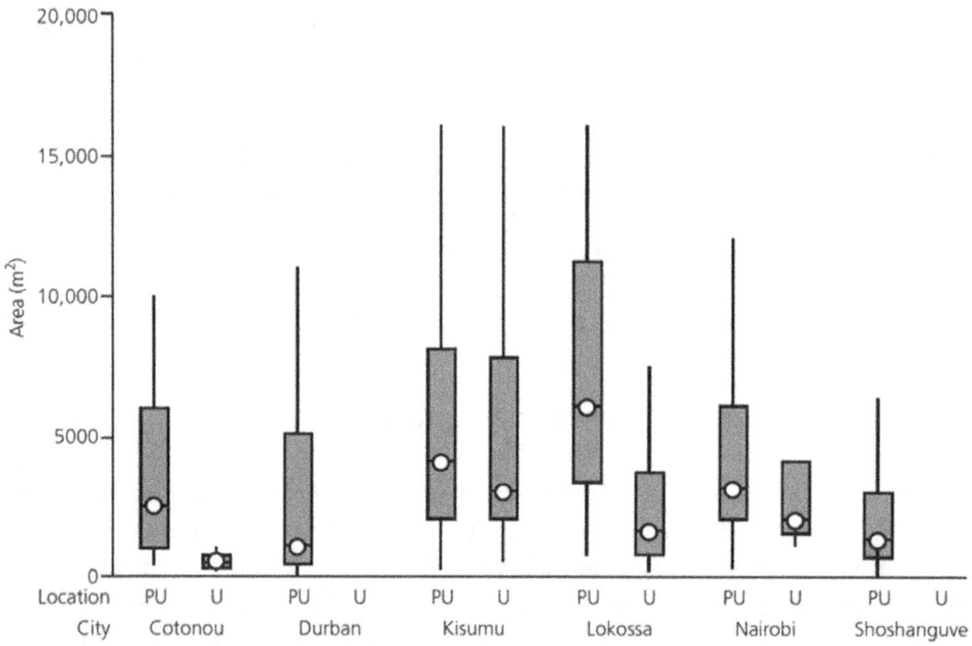

Figure 9. Vegetable production in urban and peri-urban areas of six major African cities
Notes: Median values and interquartile ranges are displayed
Source: Weinberger & Pichop, 2009

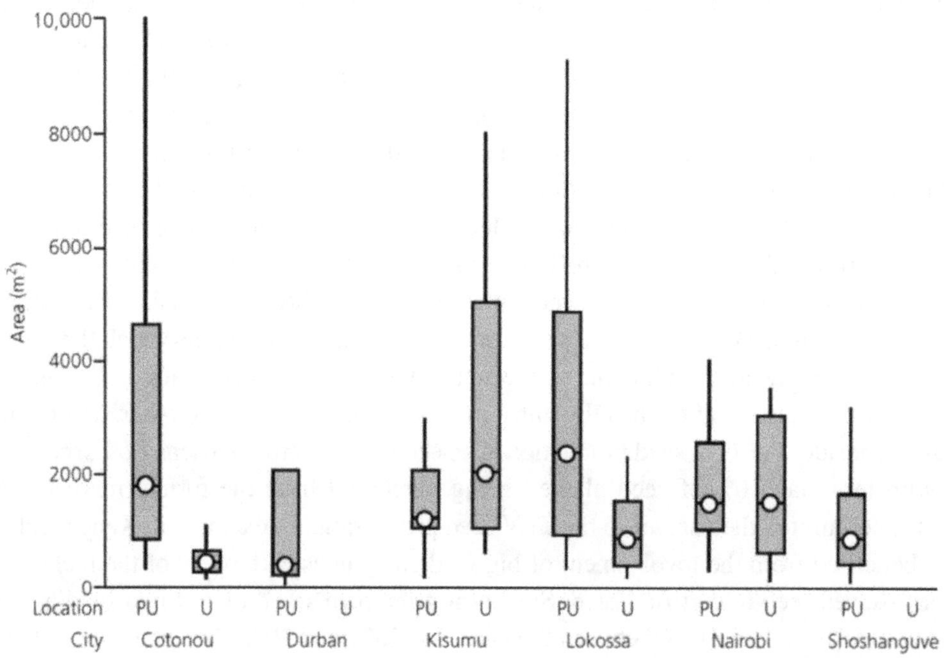

During the period of vegetable shortages, particularly during the dry seasons throughout the year, dried and processed TLVs become very important in household food security. Their abundance normally coincides with the peak of the rain season, making it difficult to preserve them well and for larger quantities to be marketed throughout the relish gap. TLVs, like some of the modernized horticultural crops, are highly perishable. They are produced by resource-poor rural sectors, who cannot afford the modernized preservation methods like refrigerating. The production of TLVs is relatively low especially during the dry seasons, meaning that the TLV market niche during this period can only be filled by the farmers using an irrigation mechanism. However, the quality of the processed TLVs is poor in some cases, can result in negative perceptions by the young generation, elite, and urban dwellers.

If the quality of the processed TLVs improves through preserving and turning them into products like cakes or flour, such products could increase the TLVs consumption by the young, elite, and urban dwellers. Also, the creation of awareness for the preparation or cooking of TLVs through cooking recipes can effectively and positively increase their consumption, particularly among the young, elite, and urban dwellers. Although the TLVs have the potential to facilitate the current imperative to foster food sovereignty and strengthen various of its components, this potential is not realized because they are classified as minor crops.

SOLUTIONS AND RECOMMENDATIONS

The availability of certified quality seeds coupled with the developed seed systems will enhance the production of TLVs. For example, Asfaw, Shiferaw, Simtowe, and Lipper (2012) reported better incomes were realized by smallholder farmers who adopted improved or certified seed varieties. The authors recommend the international organization such as the World Vegetable Center, already working in Eastern Africa to extend working with other countries in Southern and Western Africa to ensure quality seed production and develop TLV seed systems.

Different recipes and preparation methods that are passed through generations have also been promoted in Tanzania and should be scaled out in other countries in the continent to create awareness among the elite, young, and urban dwellers. For example, the World Vegetable Center is promoting consumption of TLVs and processing them into other products such as cakes and flour (Ochieng et al., 2016). The authors recommend an increase in production and processing of traditional leafy vegetables to enhance their consumption by the elite groups and children.

The Potential of TLVs to Improve Food and Nutrition Security

TLVs contribute significantly to food security in the dry regions of Zimbabwe, as they are consumed by more than 60% of the households in these regions (Mpala et al., 2013); about 21% and 30% of the produced indigenous vegetables are consumed at home in Malawi and Mozambique (Chagomoka et al., 2014). TLVs account for about 80% of the total food use during winter in South Africa. Another key important issue regarding food security: the cash obtained from selling TLVs contributes significantly to food security at the household level, as it can be used to buy other food related items. In Zimbabwe, the rural poor people rely on TLVs for a period of nine months (five months of the rainy season and four of the dry months). They are prime sources of minerals and are naturally rich in nutrients, i.e. iron, protein, vitamins, and other micronutrients needed by the human body for healthy growth. They have the potential to help address micronutrient deficiency problems in many African countries.

Some TLVs have relatively high levels of nutrients, e.g. cleome and cowpea leaves contain more than 4 g (4.5 g and 4.4 g, respectively) of protein and 275 mg and 186 mg of calcium, respectively, while sweet potato and cleome have 3.9 mg and 2.2 mg of iron, respectively, and cowpea leaves and cleome have 75 mg and 113 mg of Vitamin C, respectively per 100 g edible portion (Yang & Keding, 2009). If incorporated in the daily diets of people in Africa, this new diet can be a key step in reducing malnutrition and food security problems faced by the continent. Most of the nutrients found in TLVs are very important, especially for the poor population who consumes mostly staples and tubers with little or no meat. Khoury et al. (2014) reported a global change in the dietary patterns, people are consuming calories, protein and fat, relying increasingly on wheat, maize, and soybeans, along with meat and dairy products, showing increasingly shift towards monotonous and homogeneous diets. TLVs can serve both purposes, as vegetables and as a supplement to starchy foods, as some TLVs, such as sweet potato and cassava roots, provide much-needed energy as well, break the increasingly relying on monotonous diets.

TLVs are produced locally; farmers have control over seed and they can be quickly supplied to consumers, which ultimately supports domestic marketing structures in developing countries. Their marketing supports the flow of purchasing power from more affluent urban regions to rural areas. The development of TLV markets within African cities and other major urban centers could be considered a milestone in the re-introduction of underutilized local food crops for food security, food sovereignty, on-farm biodiversity conservation as well as halting biodiversity losses alluded to in SDG 15. Precious genetic variability of these domesticated plants, traditional foods and farming customs in developing countries can be preserved. From an agronomic perspective, TLVs might yield general beneficial effects for the cropping pattern and for fighting crop diseases.

Potential Socio-Economic Benefits for National Food Supply Chains

TLVs do not have high needs in terms of farming practices, production inputs, and weather conditions as they produce seed under the tropical condition, unlike the exotic vegetables. This opens up the potential for farmers to easily expand TLVs production and obtain higher incomes through increased marketing of TLV seed and of TLVs. TLVs, therefore, possess a substantial potential for sustainably improving the livelihoods of the poor rural population and provision of food for the poor urban dwellers.

The contribution of TLVs to household incomes have been reported to be higher than 20% in Malawi and Mozambique (Chagomoka et al., 2014). Besides income benefits, the production and trading of TLVs have been associated with women. In Zimbabwe and Kenya, the majority of the TLV traders and producers were women (60% and 65% of the sample, respectively). This proportion confirms the trend shown in other reports. Mpala et al. (2013) reported that most TLV traders are women. Women are predominant in the production and marketing of TLVs, accounting for 57% and 95%, respectively. Meaning that the women are empowered and therefore financially independent, consequently reducing their vulnerability. The incomes earned is very important, as it can be used for buying food for their families, provide housing and secure health care for the families among others (Schippers, 2000).

Potential Socio-Economic Benefits for Rural Areas

Only in the late 1990s health institutions increased awareness designed to reduce the diseases of the affluent (i.e. diabetes, high blood pressure, and ulcers, among others) (Irungu et al., 2011). These ailments were predominantly associated with people dwelling in urban areas because of their unhealthy food consumption pattern. Whereas for their counterparts in rural areas, many diseases caused by food insecurity (such as malnutrition) have been associated with an unbalanced diet. But note that other patients who suffer from HIV/AIDS could benefit from TLVs, as they are said to mitigate this type of ailment (Irungu et al., 2011).

For years, the rural people valued the TLVs for food security as well as medicinal, social, cultural, and income-generating item (Mpala et al., 2013). TLVs are important sources of essential macro- and micro-nutrients, hence their potential to reduce diseases of the affluent and diseases resulting from food insecurity. They are also important dietary components which are used to prepare sauces and relishes that accompany carbohydrate staples. Beyond their fundamental subsistence use, TLVs have the potential to address some of the Sustainable Development Goals (SDGs) that pertain to poverty, health, food and nutrition security in Africa.

Moreover, TLVs represent an inexpensive food of high and beneficial nutrient content (Schippers, 2000). With TLVs, a one-sided nutrition, especially for the poorer parts of the population, can be avoided by boosting this kind of crop production. In urban markets, TLVs command relatively better prices than the exotic vegetables and can be effectively produced by rural resource-limited farmers, hence the potential to foster rural development.

The Potential of Traditional Leafy Vegetables to Contribute to the Sustainable Development Goals

Recent food price volatility has resulted in poverty and food insecurity in many developing countries, in particular, Africa. The leaders from various countries created the MDGs to address some of the challenges, including poverty reduction and food insecurity problems. Several countries managed to reduce the number of the population deemed poor by half. Following the successful implementation of the MDGs, the leaders of various countries created a new set of goals "Sustainable Development Goals", aimed to end poverty and hunger by 2030.

TLVs have high nutritional values that can overcome malnutrition problems faced by Africa that still persist due to the lack of diversified diets. The consumption of TLVs has potential health benefits because they are rich in naturally inherent nutrients. TLVs can, therefore, contribute to the SDG 3 aimed at maintaining good health. These vegetables are resilient to adverse conditions prevailing in Africa. They can, therefore, provide food for many people across the continent, a trait making them excellent contributors to SDGs 1 and 2.

TLVs help smallholder farmers to diversify their limited income sources and provide an important source of employment in rural areas. TLVs command relatively high prices on the markets, giving them the potential to relieve rural poverty, in particular, for the vulnerable group, females. In most parts of Africa, TLVs are regarded as women crops, therefore, supporting their production and use can go a long way in the economic empowering of women, which in turn reduces the vulnerability of women and addresses gender inequality, as enshrined in SDG 5. Traditionally, African women are responsible for household food availability, preparation, and preservation. Women's involvement in the TLVs marketing is significantly high. Indigenoveg survey in Weinberger & Pichop (2009) reported that women play a central role in the production, peddling, and marketing of TLVs. They, therefore, offer a convenient pillar for lifting the women higher. Women dominate 100% in the retail markets and middle actors in Cote d'Ivoire, 68% in Uganda, 90% in Kenya, 65% in Tanzania, and 80% in South Africa (Indigenoveg in Weinberger and Pichop (2009)). These vegetables, therefore, form a crucial part of the female existence. TLVs have the potential to transform Africa's economy due to the high turnover and its effect on the income of women, as they play significant roles in the production and marketing of TLVs across Africa.

Currently, some countries are supplementing the vegetable demand shortfalls through importing vegetables from neighboring countries. For example, it is worth mentioning that Botswana and Zimbabwe have an import/export relationship. The local broadcasting company in Zimbabwe reported that Zimbabwe has spent more than $15 million in importing vegetables and citrus commodities in the first two months of 2016. Similarly, vegetables in Botswana are mostly imported from the neighboring South Africa. Thus, the limited foreign exchange is used to purchase products which could be supplied locally. The promotion of production, marketing, and consumption of TLVs enhances local markets, curbing the externalization of much needed foreign currency. It would also reduce enrichment of the richer nations enshrined in SDG number 10.

Nevertheless, TLVs can contribute to the environment through the reduction of biodiversity loss, enshrined in SDG 15. Some of the TLVs types, like the cowpea, can easily adapt to poor soils, making them useful staple crops for marginalized areas. They can be incorporated in the production of crops by farmers living in areas marked by water scarcity and prone to hot temperatures caused by climatic change. In addition, some TLVs are deep rooted, hence they help to stabilize the soils. They also grow faster and provide canopy cover to the soil, reducing soil erosions and loss of soil moisture.

TLVs, therefore, have the potential to contribute to the Sustainable Development Goals 1, 2, 3, 5, 10, 12,13, and 15 (UNDP, 2016). Biodiversity loss reduction, good health maintenance, and the reduction of poverty and hunger can be strongly achieved through activities and policies promoting the production and use of TLVs. They are traded locally, supporting purchasing power from more affluent urban regions to rural areas and involve women command in the supply chain. Many people derive their income from various activities along the supply chain. They are sources of micronutrients required by the human body, food sources during crop failures. African national food and health policies should spearhead the production as well as promote their consumption to help address food insecurity and micronutrient deficiency problems in Africa. The national programs are key in the implementation of such policies. A good example is the Malawi agricultural input program, from which the country succeeded in feeding its people in a short period of time. The Green Revolution in India also transformed the agriculture; the country produced a surplus and ended up exporting food.

FUTURE RESEARCH DIRECTIONS

The outcome of our chapter noted that relatively few countries in Africa have soundly developed TLV seed systems and supply chains, that incentivize the actors along the TLV chains, resulting in the limited investment towards research. If the quality of TLV seed and distribution would be improved, then actors might benefit, realizing that more profits can also attract investors and researchers. Such successes can only be realized if Africa regions work collectively to improve the seed systems and supply chains. In particular, countries which already have seed policies in place should collaborate with the ones lagging behind, in order to improve the quality and accessibility of good quality seed by TLVs farmers. More research on determining the quality of TLV seed would contribute to the availability of high-quality seed on the markets. Equally important, researches aimed at understanding how the TLV supply chains are organized, improved seed distribution channels considering TLVs are increasingly marketed in cities across Africa. Such researches could provide the basis of improved TLVs supplies across Africa.

There is still a lack of knowledge about the methods of preserving and drying TLVs. The presence of contaminants, like moulds or gritty-due fine sands, make TLVs at times unpalatable, due to abnormal smells. Much more attention needs to be given to improving the drying, preserving, and processing of the TLVs into products. Some TLV products such as cakes or flour may attract the elite and urban dwellers to consume them. Besides, more effort needs to be channeled towards how to prepare or cook these vegetables. Most of the elite and urban dwellers lack the knowledge on how to prepare TLVs. Preparation or cooking of TLVs awareness creation through cooking recipes may effectively and positively increase their consumption, especially by the elite and urban dwellers. Research on post-harvest practices is needed.

CONCLUSION

This chapter demonstrates the potential importance of TLVs in transforming the consumers' health and increasing farmers' incomes, particularly women as well as contributing to food security and the economies of African countries. TLVs have increasingly become crops of economic importance and their production cannot only feed an increasing and undernourished population but also make people living in the marginalized areas more productive, given that TLVs grow well even in such stressful environments. Besides offering a greater opportunity to utilize the marginal areas, these vegetables help in diversification of the food basket and breaks the monotonous diets, which is commonly missing in the daily diets of many households in Africa.

There is minimal attention channeled towards these vegetables, yet they could quickly contribute a reduction of the environmental biodiversity losses as well as contributing to a better food and nutrition security. In addition, relative to other field crops like maize, a conventional bred exotic vegetable, TLVs have received little attention in terms of research and funding from governments and international organizations. Just a few countries have put in place policies on TLV seed production, marketing, and distribution.

Nevertheless, TLVs are now getting into the cities, showing an increase in demand for both vegetables and seed. The TLV seed systems in Africa are undeveloped, depriving the farmers of a variety of vegetables, as seed cannot be moved from one region to another. Commercial seed companies also shun the multiplication of TLV seed because they are considered to be less economically viable.

The seasonality and high perishability of these vegetables reduce their potential to be an important tool for addressing food and nutrition insecurity in Africa. In some countries, the urban dwellers and elite consumers still have negative perceptions of TLVs, considering them as food for poor people. Researchers need to prepare tasty and appetizing TLV dishes that can entice the young generation to consume them as well as promote consumer awareness campaigns concerning the nutritional benefits of eating TLVs.

The supply chains of TLVs is not organized in many countries with exception of Kenya and Tanzania where farmers are benefiting from organized markets. In these countries, TLVs production and marketing has been transformed to be commercial and small-scale business entities and changing the livelihoods of many households. For example, in Kenya seed farmers under contract jointly produce seed of various TLV type with a total gross income of half a million US$ with farmers earning on average almost $5,000.

The women involvement in all activities along the TLV supply chain reduces women vulnerability and increases the women empowerment potential. TLVs have the potential to contribute to a number of SDGs, as they are traded locally, their consumption supports purchasing power from more affluent urban regions to rural areas, their market involves women command in the supply chain, their production, marketing and selling involve many people deriving their income through various activities along the supply chain, they are resources of micronutrients required by human body, they can be considered as an option for food during crop failures and their increase in production and consumption will reduce biodiversity.

Smallholder farmers particularly, women who engage in production and marketing TLV, would be more financially empowered. However, the upscaling of these national successes to the entire Africa will be challenging because of the different national perceptions among policy makers.

REFERENCES

Ambrose-Oji, B. (2009). Urban Food Systems and African Indigenous Vegetables: Defining the Spaces and Places for African Indigenous Vegetables in Urban and Peri-Urban Agriculture. In C. M. Shackleton, M. W. Pasquini, & A. W. Drescher (Eds.), *African Indigenous Vegetables in Urban Agriculture* (pp. 1–33). London: Earthscan.

Asfaw, S., Shiferaw, B., Simtowe, F., & Lipper, L. (2012). Impact of Modern Agricultural Technologies on Smallholder Welfare: Evidence from Tanzania and Ethiopia. *Food Policy*, *37*(3), 283–295. doi:10.1016/j.foodpol.2012.02.013

Bioversity International. (n.d.). *Nutritious Underutilized Species.* Retrieved February 15, 2017, from http://www.bioversityinternational.org//Nutritious_underutilized_species_-_Amaranth_1682.pdf

Cernansky, R. (2015). Super Vegetables Long Overlooked in Parts of Africa, Indigenous Greens Are Now Capturing Attention for their Nutritional and Environmental Benefits. *Nature*, *522*, 146–148. doi:10.1038/522146a PMID:26062494

Chagomoka, T., Afari-Sefa, V., & Pitoro, R. (2014). Value Chain Analysis of Traditional Vegetables from Malawi and Mozambique. *International Food and Agribusiness Management Review*, *17*(4), 59–86.

Food and Agriculture Organization of the United Nations. (2015). *The State of Food Insecurity in the World 2015. Meeting the 2015 International Hunger Targets: Taking Stock of Uneven Progress*. Rome: Food and Agriculture Organization of the United Nations.

Ihle, R., von Cramon-Taubadel, S., & Zorya, S. (2009). Markov-switching Estimation of Spatial Maize Price Transmission Processes between Tanzania and Kenya. *American Journal of Agricultural Economics, 91*(5), 1432–1439. doi:10.1111/j.1467-8276.2009.01360.x

Irungu, C., Mburu, J., Maundu, P., Grum, M., & Hoeschle-Zeledon, I. (2011). The Effect of Market Development On-farm Conservation of Diversity of African Leafy Vegetables around Nairobi. *International Journal of Humanities and Social Science*, *8*(1), 198–207.

Khoury, C. K., Bjorkmanc, A. D., Dempewolfd, H., Ramirez-Villegasa, J., Guarinof, L., Jarvisa, A., & Struik, P. C. et al. (2014). Increasing Homogeneity in Global Food Supplies and the Implications for Food Security. *Sustainability Science Agricultural Sciences*, *111*(11), 4001–4006. PMID:24591623

Luckmann, J., Ihle, R., Kleinwechter, U., & Grethe, H. (2014). World Market Integration of Vietnamese Rice Markets during the 2008 Food Price Crisis. *Food Security. The Science. Sociology and Economics of Food Production and Access to Food*, *6*(6), 1–17.

Maundu, P., Achigan-Dako, E., & Morimoto, Y. (2009). Biodiversity of African Vegetables. In C. M. Shackleton, M. W. Pasquini, & A. W. Drescher (Eds.), *African Indigenous Vegetables in Urban Agriculture* (pp. 65–104). London: Earthscan.

Mpala, C., Dlamini, M., & Sibanda, P. (2013). The Accessibility, Utilisation and Role of Indigenous Traditional Vegetables in Household Food Security in Rural Hwange District. *International Open & Distance Learning Journal International Research Conference Special Edition*, *3*(1), 40–48.

Mumbi, K., Karanja, N., Njenga, M., Kamore, M., Achieng, C., & Ngeli, P. (2006). *Investigative Market Research: Viable Market Opportunities and Threats for Urban and Peri-Urban Farmers, Farm Concern International*. Nairobi: Urban Harvest and International Potato Centre.

Nekesa, P., & Meso, B. (1997). Traditional African Vegetables in Kenya: Production, Marketing and Utilization. In *Proceedings of the IPGRI International Workshop on Genetic Resources of Traditional Vegetables in Africa: Conservation and Use*. Rome: International Plant Genetic Resources Institute.

Ochieng, J., Afari-Sefa, V., Karanja, D., Kessy, R., Rajendran, S., & Samali, S. (2016). How Promoting Consumption of Traditional African Vegetables Affects Household Nutrition Security in Tanzania. *Renewable Agriculture and Food Systems*, 1–11. doi:10.1017/S1742170516000508

Ojiewo, C., Tenkouano, A., Oluoch, M., & Yang, R. (2010). The Role of AVRDC – the World Vegetable Centre in Vegetable Value Chains. *African Journal of Horticultural Science*, *3*, 1–23.

Otsuka, K., Larson, D. F., & Hazell, P. B. R. (2013). An Overview. In K. Otsuka & D. F. Larson (Eds.), *An African Green Revolution. Finding Ways to Boost Productivity of Small Farms* (pp. 1–14). Amsterdam: Springer. doi:10.1007/978-94-007-5760-8_1

Sasson, A. (2012). Food Security for Africa: An Urgent Global Challenge. *Agriculture & Food Security*, *2*(1), 1–16.

Schippers, R. R. (2000). *African Indigenous Vegetables. An Overview of the Cultivated Species*. Chatham: Natural Resources Institute/ACP-EU Technical Centre of Agricultural and Rural Cooperation Book.

Shackleton, C. M., Pasquini, M. W., & Drescher, A. W. (Eds.). (2009). *African Indigenous Vegetables in Urban Agriculture*. London: Earthscan.

United Nations Development Programme. (2016). *Support to the Implementation of the 2030 Agenda for Sustainable Development.* Retrieved July 27, 2016, from http://www.undp.org/content/undp/en/home/librarypage/sustainable-development-goals/undp-support-to-the-implementation-of-the-2030-agenda/

Veninga, W., & Ihle, R. (in press). Import Vulnerability in the Middle East: Effects of the Arab Spring on Egyptian Wheat Trade. *Food Security: The Science, Sociology and Economics of Food Production and Access to Food.*

Weinberger, K., & Msuya, J. (2004). *Indigenous Vegetables in Tanzania- Significance and Prospects.* Shanhua: AVRDC – The World Vegetable Center.

Weinberger, K., & Pichop, G. N. (2009). Marketing of African Indigenous Vegetables along Urban and Peri-Urban Supply Chains in Sub-Saharan Africa. In C. M. Shackleton, M. W. Pasquini, & A. W. Drescher (Eds.), *African Indigenous Vegetables in Urban Agriculture* (pp. 225–244). London: Earthscan.

World Vegetable Center. (2006). *Empowering Small Scale and Women Farmers through Sustainable Production, Seed Supply and Marketing of African Indigenous Vegetables in Eastern Africa*. Bangkok: AVRDC – The World Vegetable Center and Family Concern.

Yang, R., & Keding, G. B. (2009). Nutritional Contributions of Important African Indigenous Vegetables. In C. M. Shackleton, M. W. Pasquini, & A. W. Drescher (Eds.), *African Indigenous Vegetables in Urban Agriculture* (pp. 105–143). London: Earthscan.

ADDITIONAL READING

Afari-Sefa, V., Rajendran, S., Kessy, R. F., Karanja, D. K., Musebe, R., Samali, S., & Makaranga, M. (2016). Impact of Nutritional Perceptions of Traditional African Vegetables on Farm Household Production Decisions: A Case Study of Smallholders in Tanzania. *Experimental Agriculture*, *52*(2), 300–313. doi:10.1017/S0014479715000101

Afari-Sefa, V., Tenkouano, A., Ojiewo, C.O., Keatinge, J.D.H., & Hughes, J.d'A. (2012). Vegetable Breeding in Africa: Constraints, Complexity and Contributions toward Achieving Food and Nutritional Security. Food Security: *The Science. Sociology and Economics of Food Production and Access to Food*, *4*(1), 115–127.

Durst, P., & Bayasgalanbat, N. (Eds.). (2014). *Promotion of Underutilized Indigenous Food Resources for Food Security and Nutrition in Asia and the Pacific*. Bangkok: Food and Agriculture Organization of the United Nations, Regional Office for Asia and the Pacific.

KEY TERMS AND DEFINITIONS

Gross Margin Analysis: An analysis conducted between the revenue and the cost of producing TLVs, in our case per hectare per season.

Marketing Channels: A set of activities that necessitate the movement of the produced TLVs to the end user, i.e. the consumer.

Processors: Dehydrating the fresh vegetables from a perishable form into one that encourages a long shelf-life and thereby allows later usage.

Supply Chain: A sequential process involved in the TLVs production to the distribution of the produce. The supply chain for TLVs is relatively short because of the high perishability of the vegetables.

Sustainable Development Goals: Goals that aim at transforming our world through ending poverty and hunger, thereby achieving food security and improving nutrition and promoting sustainable agriculture.

Traders: Persons who engage in the buying and selling TLVs.

Traditional Leafy Vegetables: Plants which originate on the African continent and have a century-long history of cultivation and domestication subject to regional agro-climatic conditions.

Traditional Leafy Vegetables Producers: Farmers that grow the TLVs.

Chapter 12
Effect of Climate Change on Crop Productivity and Prices in Benue State, Nigeria:
Implications for Food Security

Goodness C. Aye
University of Agriculture Makurdi, Nigeria

Ruth F. Haruna
University of Agriculture Makurdi, Nigeria

ABSTRACT

The chapter is aimed at assessing the effect of climate change on crop productivity and prices in Benue State, Nigeria. Time series data on selected output of crops (maize, rice, sorghum, yam, millet, groundnut, beans, and cassava), area planted, price, and climate variables such as rainfall, temperature, and sea level were used. Due to differing periods in data availability, this study used the time period 1995-2009 for analysis, in order to maintain a common period for all the series. First, the trend of productivity, prices, and climate change was analyzed using visual plots and results indicate some level of variability in these series over time. Second, a three stage least square regression was used to simultaneously analyze the effect of climate change on productivity and prices. Results show that climate change had significant impact on the productivity of millet, sorghum, cassava, and groundnut while it had significant impact only on the price of maize. These findings have important implications for food security situation in Benue State, Nigeria.

INTRODUCTION

Despite technological advances, such as improved varieties, genetically modified organisms, and irrigation systems, the weather is still a key factor in agricultural productivity, as well as soil properties and natural communities. Rough estimates suggest that over the next 50 years or so, climate change may likely have a serious threat to meeting global food needs than other constraints on agricultural systems

DOI: 10.4018/978-1-5225-2733-6.ch012

(Intergovernmental Panel on Climate Change [IPCC], 2007; Building Nigeria's Response to Climate Change [BNRCC], 2008). Specifically, population, income and economic growth could all affect these verities of climate change impacts in terms of food security, hunger, and nutritional adequacy. If climate change adversely affects agriculture, effects on human are likely to be more severe in a poorer world. Wolfe et al. (2005), Stige, Stave, and Chan (2006), and Orindi et al. (2006) were concerned that rising demand for food over the next century, due to population and real income growth, will lead to increasing global food scarcity, and a worsening of hunger and malnutrition problems particularly in developing countries. The Earth's average surface temperature has increased by 1.5°F (0.83°C) since 1880 (Wikipedia, 2013). This chapter presents an empirical analysis of the relationship between climate change, crop productivity, and prices in Benue State Nigeria with the intention of drawing implications for food security and hence make policy recommendations.

The specific objectives of this chapter are to: first, describe the productivity and price trend of major crops in the study area; second, describe the trend of climate change in the study area; third, analyze the effect of climate change on crop productivity in the study area and fourth, analyze the effect of climate change on prices in the study area.

BACKGROUND

There is no doubt that the earth is warming thus resulting in climate change over the past century and future change is envisaged. This has been attributed to human activities such as the use of fossil fuels to produce energy, coupled with unsustainable agricultural and land use practices. Climate change is a major environmental challenge to the world today, with significant threats to ecosystems, food security, water resources and overall economic stability (CAB International [CABI], 2010). IPCC (2007) indicates that rising temperatures, drought, floods, desertification and weather extremes will severely affect agriculture, especially in the developing world. Thus, changes in climatic factors like temperature, precipitation, snowfall, wind, windstorm, flooding etc. have the potential to irreversibly damage the natural resource base on which agriculture depends, and in general adversely affect agricultural productivities, which may consequently threaten the food security of nations. The last couple of years has been a period of increased food prices, thus increasing the number of people going to bed hungry at night to over one billion (CABI, 2010). The underlying causes of increases in food prices are complex. However, this has been partly attributed to poor harvests due to an increasingly variable climate and the use of food crops for biofuels. In agriculture, both temperature and precipitation are dominant climatic factors that affect crop yield and prices, which vary widely throughout the year and over time (Alexandrov & Hoogenboom, 2000). In a study by Brockett, Wang, and Yang (2005), it was indicated that for all industries including agriculture, manufacturing, and services; the role of change in temperature alone contributes more than 90% amongst all climatic factors followed by rainfall as a second dominant factor.

Reported extreme weather events such as floods, droughts, tornadoes, and thunderstorms are increasing over time and increase in climatic variability has been observed as well. This climatic variability is responsible for more severe periods of drought coupled with water periods which could be inevitable consequences of climate change (Christensen et al., 2007). The climatic factors are more vulnerable to agriculture in developing countries where they do not have irrigation facility, and also lack the more precise prediction facilities (Downing, 1996). Together with the advancement of agricultural technologies, developed nations are being able to minimize losses suffered from a change in temperature and precipitation (Almaraz, Mabood, Zhou, Gregorich, & Smith, 2008; Schlenker & Roberts, 2009). Greenhouse cultivation, resistance variety development, and cultivation are some practices to lessen the weather effect in agricultural production.

One of the important sources of variability in agricultural commodity prices is weather shock (Gilbert & Morgan, 2010). Although there is no evidence that weather shocks alone have played a major role in the food price spikes in 2008 (Headey & Fan, 2008), it seems, however, that they have contributed to the price spikes in interaction with other factors (von Braun & Tadesse, 2012). Climate change is likely to increase weather variability and incidences of extreme events (IPCC, 2012), potentially leading to growing impacts of weather shocks on agricultural productivity and food price (Torero & von Braun, 2010; von Braun et al., 2008). This may have considerable influence on poverty levels and economic performances of developing countries dependent on agricultural and agro-food sectors, especially if they are net food importers (Mirzabaev & Tsegai, 2012). The instrumental temperature record shows the average global surface temperature increase during the 20th century to have been 0.74°C or 1.33°F.

The melting of aquifers as a result of increased temperature could lead to increase in sea level and negative consequence on lowland human habitation. Research from the United States shows that if care is not taken one day the western part of Nigeria (e.g Lagos) surrounded by water may completely sink into the sea. There can be no drought, the earth climate is changing and this is just the beginning. The consequences would be felt by everyone. The action is crucial if the future of agriculture is to be secured (United Nations Framework Convention on Climate Change [UNFCCC], 2007). Most agronomists believe that agricultural production will be mostly affected by severity and pace of climate change, not so much by the gradual trend in climate. There may be enough change for adjustment if the change is gradual. However, the rapid global change could harm agriculture in many countries, especially those that are already suffering from rather poor soil and extreme climate conditions, because there is less time for optimum natural selection and adaptation.

African agriculture is already hampered by its dependence on weather elements such as rainfall, the sunshine etc. As drought and flooding worsen, temperature and growing season change and this may likely hit hard on farmers as they are forced off their land. There is a change in climate, from rainfall intensity, melting of the glaciers, increase and spread of agricultural diseases. At a time of financial and food crises, a new debate is emerging across the world about the future of agriculture.

A substantial amount of research has been conducted on the potential impacts of climate change on agriculture (Pearce et al., 1996; McCarthy, Canziani, Leary, Dokken, & White, 2001; Parry, Rosenzweig, Iglesias, Livermore, & Fisher, 2004; Stern 2007; Lobell & Burke, 2008; Deressa, Hassan, Alemu, Yesuf, & Ringler, 2008; BNRCC, 2008; Apata, Samuel, & Adeola., 2009; Parry et al., 2009; Deressa & Hassan, 2010). Attempts are made in these studies to link the state-of-the-art models developed by researchers in separate disciplines, including climatology, agronomy, and economics, in order to project future impact of climate change on agriculture. Some of the studies (Kane, Reilly, & Tobey, 1992; Rosenzweig, Parry, Fischer, & Frohberg, 1993; Rosenzweig & Parry, 1994; Reilly, 1999; Ayinde, Ajewole, Ogunlade, & Adewumi, 2010), used climate-induced changes in crop yields to estimate potential global economic impacts. Others have examined the indirect impact on economic variables such as farm revenue and income (e.g. Mendelsohn, Nordhaus, & Shaw (1994, 1996); Adams, Hurd, Lenhart, & Leary, 1998). The review of these studies helped to have an understanding of the physical and economic responses, and adjustments on climate change and agricultural production.

Studies by Liu, Guenther, and Laixiang (2004), De Wit and Stankiewicz (2006), Kurukulasuriya and Mendelsohn (2006), Deressa (2007), Mendelsohn and Dinar (2009), Yesuf, Di Falco, Deressa, Ringer, and Kohlin (2008), and Apata et al. (2009) are some of the economic studies that attempt to measure the impact of climate change on farm productivity. These studies imputed the cost of climate change as a proxy for capitalized land value and which are captured from farm net revenue. Although there are a number of empirical studies as it relates to agriculture and climate change as seen from the foregoing, there is, however, a dearth of studies that consider the effect of climate change on productivity and prices simultaneously.

MAIN FOCUS OF THE CHAPTER

Issues, Controversies, Problems

The global climate change crisis has become the most devastating human problem in recent times. Predictions on the scale of impacts on lives and property by climate change are very threatening and include widespread crop failures, reduced agricultural productivity, and increased hunger, malnutrition, and disease (World Development Report, 2010). This report emphasizes that developing countries will bear the brunt of the effects of climate change, even as they strive to overcome poverty and advance economic growth. In Sub-Saharan Africa, it is predicted that climate change will cause an estimated 145 million deaths due to famine, flooding, drought, erosion, biodiversity loss, and conflicts arising from competition for scarce resources in the next 25 years. Climate change further threatens to deepen vulnerabilities, erode hard-won gains, and seriously undermine prospects for development unless humans act now and differently (World Development Report, 2010).

The effect of climate on agriculture is related to variability in local climates rather than in global climate patterns. Consequently, agronomists consider any assessment to be individually considered for each local area. Although there the number of research on climate change impacts worldwide is growing, little or no research has been conducted for Benue State. Moreover, studies that consider the effect of climate change on productivity and prices simultaneously is lacking. This is the focus of the current study.

Prior to presenting the findings, background information on the study area, the population of study, sampling procedure and sample size, a method of data collection and method of data analysis are provided.

The Study Area

The study was conducted in Benue state, Nigeria. Benue State is within the middle belt region of Nigeria. It is located between Longitude 6°35'E and 10°E and between Latitude 6°30 and 8°10'N. The state covers 33,955 square kilometers of land out of which only 23,000 square kilometers are available for arable and tree crop cultivation. Only a little over 600 hectares of land are available under Fadama for dry season agriculture in seasonally flooded areas of the state (Benue State Agricultural and Rural Development Authority (BNARDA), cited in Daudu and Ajayi, 2009).

Two distinct seasons exist in the state, the wet and dry seasons, the wet, which spanks from April to October and the dry season which spanks from November to early March of each year respectively. The average rainfall is between 1500–1800 mm. The predominant climate experience is the tropical type. The state is made up of 23 Local Government Areas with a population of 4.2 million (National Population Commission of Nigeria [NPC], 2006). The Tiv tribe predominates followed by Idoma and Etulo. More than 80% of the populations are farmers.

Method of Data Collection

The study used secondary data. Time series data on selected output of crops (rice, sorghum, yam, and cassava) and area planted in Benue State were collected from BNARDA. Price data was collected from National Bureau of Statistics. Data on climate variables such as rainfall, temperature and sea level were collected from Nigeria Meteorological Station (Tactical Command), Airforce. Due to differing periods of data availability, this study used the time period 1995-2009 for analysis, in order to maintain a common period for all the series.

Method of Data Analysis

Descriptive statistics such as mean, standard deviation, kurtosis among others were used to achieve the first objective. Visual plots were used to realize the second objective. The third and fourth objectives were realized using a dynamic simultaneous two-equation model with a quantity (productivity) component (equation 1) and a price component (equation 2). The model considers the complex relationship between crop yield and price simultaneously. Assuming no inventories, i.e., a perfect market mechanism due to price adjustments, quantities supplied and demanded are equal and equations (1) and (2) can be interpreted as the crop supply and demand curves, respectively. Given the simultaneity between the price and quantity, equations (1) and (2) is estimated applying three-stage least squares (3SLS).The productivity equation is specified as:

$$\ln PROD_t = \alpha + \beta_1 \ln TEMP_t + \beta_2 \ln RAIN_t + \beta_3 \ln SEA_t + \ln \beta_4 PROD_{t-1} + \ln \beta_5 PRICE_t + \varepsilon \quad (1)$$

where $PROD_t$ = yield for a specific crop at time t in kg/ha;

$TEMP$ = annual mean value of maximum temperature in degree Celsius;

$RAIN$ = annual mean value of rainfall in milliliter;

SEA = annual mean value of vapour pressure (sea level) at 1500z (4:00pm) in Hecto-Pascal (HPA);

$PRICE_t$ = annual average market price per kg for a specific crop at time t;

$PRICE_{t-1}$ = previous year's annual market price per kg of a specific crop;

α = the intercept corresponding the productivity equation;

$\beta_1 - \beta_5$ = the coefficients which show the size of the effect of each variable on crop yield;

ε = the error or stochastic term corresponding the productivity equation;

ln is the natural logarithm. Log transforming the series helps to correct for potential heteroscedasticity and the dimensional difference between series.

The price equation is specified as:

$$\ln PRICE_t = \gamma + \delta_1 \ln TEMP_t + \delta_2 \ln RAIN_t + \delta_3 \ln SEA_t + \ln \delta_4 PROD_t + \ln \delta_5 PRICE_{t-1} + \xi \quad (2)$$

where γ = the intercept corresponding the price equation;

ξ = the error or stochastic term corresponding the price equation;

$\delta_1 - \delta_5$ = the coefficients which show the size of the effect of each variable on crop price;

The rest of the variables are as defined in equation (1).

Results and Discussion

Consistent with the specific objectives of this chapter, this section presents the findings.

Productivity and Price Trend of Major Crops in the Study Area

The productivity of major crops in the study area from 1995 to 2009 is shown in Figure 1, while the price trend of the crops is shown in Figure 2. The productivity and price trends of the crops are discussed individually.

Maize

The productivity of maize was stable at 7.26 kg/ha between 1995 and 1996 but declined rapidly to 7.09 kg/ha in 1997 probably due to rapid change in temperature. Productivity gradually increased between 1997 and 1999, pegging at 7.18 kg/ha in the year 2000. The productivity of maize was relatively stable between 2001 and 2009, fluctuating between 7.16 kg/ha and 7.2 kg/ha. With respect to price fluctuation, the price of maize increased from N2.0/kg to N2.81/kg between 1996 and 1998. The price between 1998 and 2009 has been stable between N2.81/kg and N3.0/kg.

Millet

The productivity of millet flourished between 1995 and 1996, rising from 7.45 kg/ha to 7.50 kg/ha. However, there was a decline in productivity between 1996 and 1997, bringing productivity to 7.325 kg/ha. The productivity of millet was relatively stable between 7.325 kg/ha and 7.330 kg/ha between 1997 and 2009. The price of millet rose from N2.0/kg to N2.8/kg between 1995 and 1998. Between 1998 and 2009, the price of maize fluctuated between N2.8kg and N3.0.

Figure 1. Productivity trend of major crops in Benue State
Source: Authors' development

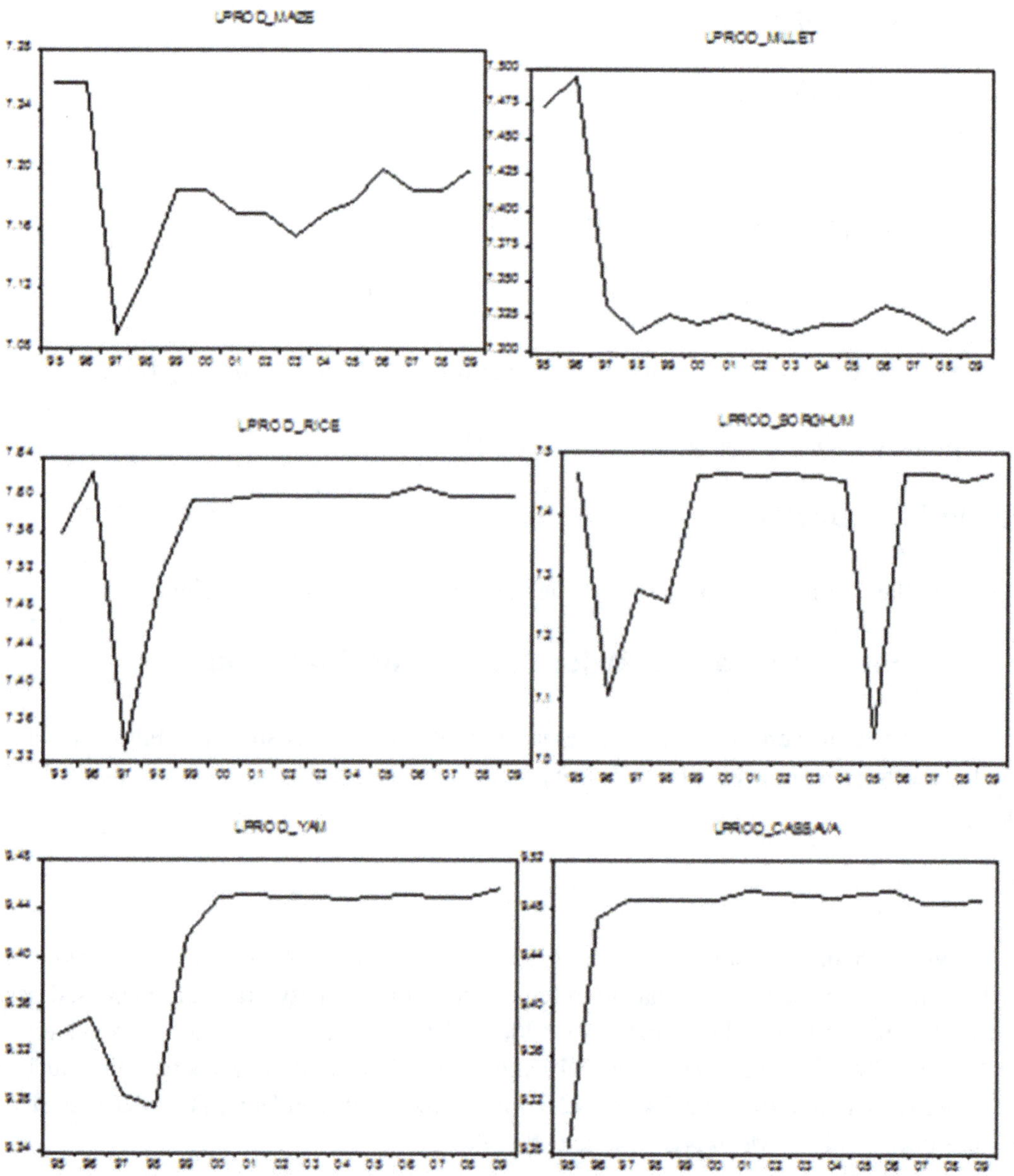

Rice

The productivity of rice rose steadily from 7.56 kg/ha to 7.63 kg/ha between 1995 and 1996. Thereafter, productivity declined to 7.33 kg/ha in 1997. Rice production then experienced a rapid rise to 7.6 kg/ha between 1997 and 1999 and remained at that level of production between 1999 and 2009. The price of rice rose from N2.2/kg to N3.0/kg between 1995 and 1998 and was relatively stable at N3.0/kg between 1998 and 2009.

Figure 2. Price trend of major crops in Benue State
Source: Authors' development

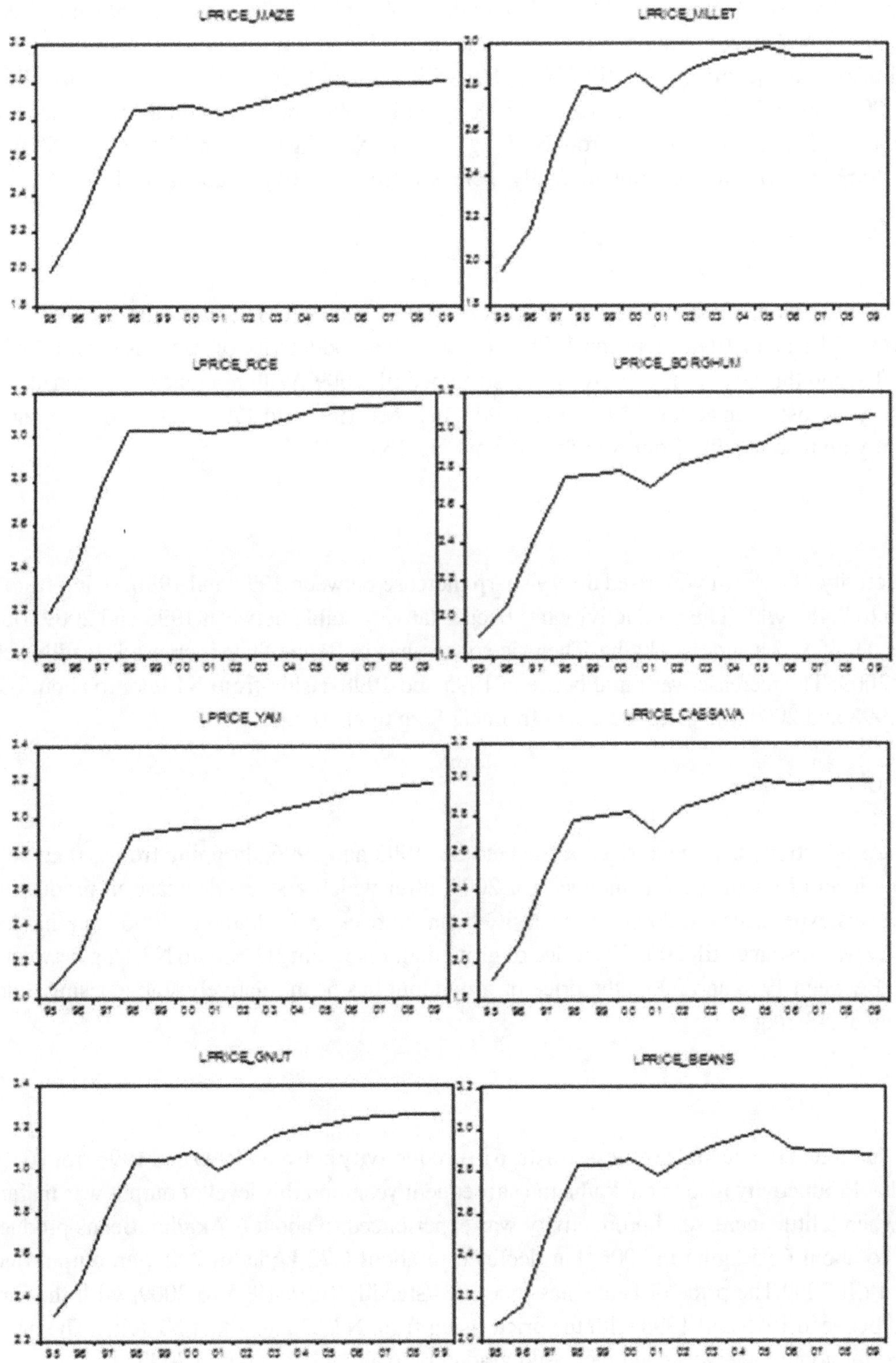

Sorghum

The productivity of sorghum declined from 7.48 kg/ha to about 7.1 kg/ha between 1995 and 1996. Productivity then experienced a little rise to 7.25 kg/ha in 1997, and a further rise to 7.48 kg/ha in 2000. This level of output was maintained till 2004, after which productivity declined sharply to 7.05 kg/ha between 2004 and 2005. Productivity rose to 7.48 kg/ha in 2006 and was maintained at that level till 2009.The price of sorghum increased from N1.9/kg to about N2.8/kg between 1995 and 1998. Between 1998 and 2009, the price of sorghum gradually increased from N2.8/kg to about N3.1/kg.

Yam

The productivity of yam was stable at 9.35 kg/ha between 1995 and 1996. Thereafter, productivity dropped steadily to 9.27 kg/ha in 1998. Between 1998 and 2001, the productivity of yam rose from 9.27 kg/ha to 9.45 kg/ha, and this level of productivity was sustained till 2009. With reference to price fluctuation, the price of yam rose from about N2.1/kg to N2.9/kg between 1995 and 1998. Between 1998 and 2009, the price of yam rose steadily from N2.9/kg to about N3.2/kg.

Cassava

The productivity of cassava witnessed a very sharp increase between 1995 and 1996, rising from about 9.28 kg/ha to 9.48 kg/ha. The productivity has been relatively stable between 1996 and 2009, fluctuating between 9.48 kg/ha and 9.49 kg/ha. The price of cassava in Benue State increased steadily between 1995 and 2009. The increase was rapid between 1995 and 1998, rising from N1.9/kg to about N2.8/kg between 1998 and 2009, the price increased from N2.8/kg to N3.0/kg.

Groundnut

Groundnut productivity experienced a decline between 1995 and 1996, dropping from 8.0 kg/ha to 7.4 kg/ha. This level of output was maintained till 2004, after which a sharp decrease in productivity to 7.05 kg/ha was experienced in 2005. Productivity bounced back to 7.4 kg/ha in 2006, and that level of productivity was sustained till 2009.The price of groundnut rose from N2.3/kg to N3.1/kg between 1995 and 1998. Between 1998 and 2009, the price of groundnut has been relatively stable, rising gradually and peaking at N3.2/kg.

Beans

Beans production in Benue State witnessed a drop in productivity between 1995 and 1996 from 6.8 kg/ha to 6.4 kg/ha. Productivity rose to 6.6 kg/ha the subsequent year, and this level of output was maintained till 1999 when a little increase of productivity was experienced to about 6.7 kg/ha. Beans productivity increased to about 6.95 kg/ha in 2006, but declined to about 6.72 kg/ha in 2007, an output that was maintained till 2009.The price of beans has increased steadily from 1995 to 2009, with the increase very rapid between 1995 and 1998 with the price rising from N2.1/kg to about N2.8/kg. The price was maintained for about three years and increased gradually to about N2.85/kg in 2009.

Trend of Climate Change in Benue State

The trend of climate change in the study area was studied using parameters such as the temperature fluctuations, fluctuation in rainfall and fluctuation in the level of the sea during the 1995-2009 period, and the results are presented in Figure 3.

Figure 3. Climate change trend in Benue State
Source: Authors' development

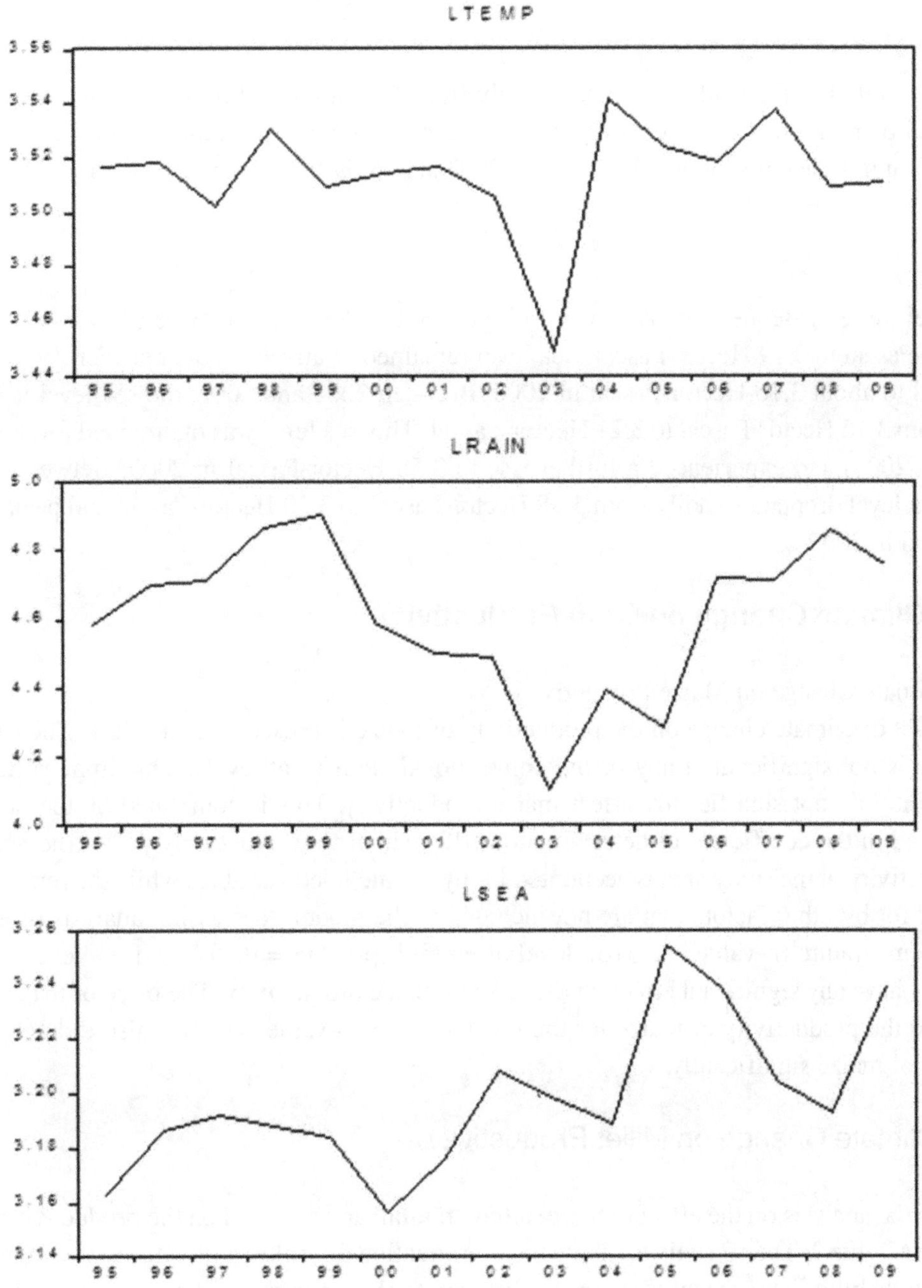

Temperature

The temperature of the study area was stable at 3.52°C between 1995 and 1996. It then dropped to 3.50°C in 1997 but rose to about 3.54°C in 1998. It dropped to 3.51°C in 1999, and this temperature was maintained till 2001. Between 2001 and 2003, there was a sharp fall in temperature from 3.51°C to 3.45°C but rose rapidly to 3.55°C the subsequent year. Temperature declined steadily between 2004 and 2006 but experienced another rise to 3.55°C in 2007. Between 2007 and 2009, temperature levels dropped and stabilized at 3.51°C.

Rainfall

The level of rainfall in the study area rose steadily from 4.6 mm to 4.9 mm between 1995 and 1999. Rainfall level then dropped significantly year after year, attaining an all-time low of 4.1 mm in 2003. The level of rainfall then rose gradually between 2003 and 2009 from 4.1 mm to 4.8 mm.

Sea Level

The sea level trend in Benue State is shown in Figure 3. Sea level rose between 1995 and 1996 from 3.16 Hector-Pascal to 3.19 Hector-Pascal. Sea level remained relatively stable between 1996 and 1999 and declined to about 3.16 Hector-Pascal in 2000. Between 2001 and 2002, the sea level in the study area rose from 3.16 Hector-Pascal to 3.21 Hector-Pascal. This sea level was maintained for the next two years (2002–2004) and experienced a further rise to 3.58 Hector-Pascal in 2005. Between 2005 and 2008, the sea level dropped steadily from 3.58 Hector-Pascal to 3.20 Hector-Pascal and peaked at 3.24 Hector-Pascal in 2009.

Effect of Climate Change on Crop Productivity

Effect of Climate Change on Maize Productivity

The impact of climate change on the productivity of maize is presented in Table 1. The Chi-square ($\chi 2$) statistic is not significant at any of the conventional significant levels. This implies that all the variables jointly do not significantly affect maize productivity. This is confirmed by the poor model fit as reflected in the coefficient of determination (R2) which shows that only 4% of the variation in maize productivity in the study area is accounted for by the included variables while the remaining 96% is accounted for by other factors that are not included in the model. Regression analysis revealed that changes in temperature (p-value = 0.516), level of rainfall (p-value = 0.797) and sea level (p-value = 0.808) do not have any significant effect on the level of maize productivity. The price of maize (p-value = 0.627) and the productivity of maize for the previous year (p-value = 0.344) also did not affect the productivity of maize significantly.

Effect of Climate Change on Millet Productivity

The result of the analysis on the effect of temperature, rainfall and sea level on the productivity of millet is presented in Table 2. The $\chi 2$ statistics show that the equation for millet productivity was significant at 1% level of probability. This means that jointly, the climate change variables have a significant effect on

Table 1. Impact of climate change on maize productivity

Variable	Coefficient	Std. error	T-statistic	P-value
Intercept	3.481	2.889	1.200	0.228
lrain	0.012	0.048	0.260	0.797
lsea	0.101	0.416	0.240	0.808
lprice_maize	0.034	0.071	0.490	0.627
lprod_maize (t-1)	0.290	0.306	0.950	0.344
R2	0.04			
$\chi 2$	1.760			0.882

Source: Authors' development

Table 2. Impact of climate change on millet productivity

Variable	Coefficient	Std. error	T-statistic	P-value
Intercept	7.206	1.813	3.980	0.000
ltemp	0.287	0.232	1.240	0.216
lrain	-0.021	0.023	-0.910	0.362
lsea	0.382	0.204	1.870	0.062
lprice_millet	-0.232	0.065	-3.550	0.000
lprod_millet (t-1)	-0.184	0.217	-0.850	0.397
R2	0.87			
$\chi 2$	59.990			0.000

Source: Authors' development

the productivity of maize. The coefficient of determination for the equation (R2) is 0.87, indicating that 87% of the variation in the productivity of millet in the study area is accounted for by climate change variables. Results indicated that only the level of sea rise (p-value = 0.062) significantly affected the productivity of millet at 10% level of probability among the climate variables. Changes in temperature (p-value = 0.216) and level of rainfall (p-value = 0.362) had no significant effect on the productivity of millet. The price of millet significantly affected productivity at 1% level of probability (p-value = 0.000), while the previous productivity of maize had no significant effect on the present productivity (p-value = 0.397). Millet is a crop that has acclimatized to the warm weather of the tropics, and that could be the possible reason why it is not affected by changes in temperature and rainfall.

Effect of Climate Change on Rice Productivity

Research findings, as presented in Table 3 indicate that climate change has no significant effect on rice productivity in Benue State. Temperature (p-value = 0.627), rainfall (p-value = 0.455) and sea rise (p-value = 0.803) had no significant effect on the productivity of rice. However, the price of rice significantly affected productivity at 10% level of probability (p-value = 0.075), while the productivity of the previous year did not affect the productivity of the present year (p-value = 0.998).The variables in the

Table 3. Impact of climate change on rice productivity

Variable	Coefficient	Std. error	T-statistic	P-value
Intercept	6.259	4.055	1.540	0.123
Ltemp	0.481	0.990	0.490	0.627
Lrain	-0.070	0.094	-0.750	0.455
Lsea	-0.195	0.783	-0.250	0.803
lprice_rice	0.191	0.108	1.780	0.075
lprod_rice (t-1)	0.000	0.042	0.000	0.998
R2	0.039			
χ2	197.570			0.000

Source: Authors' development

equation contributed to changes in rice productivity by 3.9 percent (R2 = 0.039). Most of the farmers cultivate rice in swamps and do not depend on so much rainfall for cultivation since the swamps last for the period of farming. The temperature of those swampy areas is also almost constant, and this could account for the insignificant effect of climate change on rice production.

Effect of Climate Change on Sorghum Productivity

The 3SLS regression was used to test the effect of climate change on the productivity of sorghum in the study area, and the result obtained is presented in Table 4. Research findings reveal that only the level of the sea (p-value = 0.001) significantly affected the productivity of sorghum at 1% level of probability. Changes in temperature (p-value = 0.144) and level of rainfall (p-value = 0.615) did not affect the productivity of sorghum. The price of sorghum affected productivity significantly (p-value = 0.000), but the previous productivity had no significant effect on the present productivity. The coefficient of determination (R2) for the regression equation is 0.657, indicating that the variables entered into the equation accounted for 65.7 percent of the variations in productivity of sorghum. While the Chi square (χ2) value obtained for the regression (29.170) was statistically significant at 1% level of probability. The overall equation was therefore statistically significant.

Table 4. Impact of climate change on sorghum productivity

Variable	Coefficient	Std. error	T-statistic	P-value
Intercept	24.203	5.097	4.750	0.000
ltemp	-1.752	1.199	-1.460	0.144
lrain	0.064	0.127	0.500	0.615
lsea	-3.588	1.038	-3.460	0.001
lprice_sorghum	0.511	0.104	4.940	0.000
lprod_sorghum (t-1)	-0.124	0.185	-0.670	0.505
R2	0.657			
χ2	29.170			0.000

Source: Authors' development

Effect of Climate Change on Yam Productivity

Analysis of the impact of climate change on yam productivity in the study area showed that temperature (p-value = 0.866), rainfall (p-value = 0.657) and sea level (p-value = 0.396) had no significant effect on yam productivity. The price of yam significantly affected its productivity (p-value = 0.008) at 1% level, while the productivity of the previous year had no effect on the present productivity. Yam productivity is therefore affected only by price and not by changes in the climate and can be considered as a hardy crop. The result of the analysis is presented in Table 5.

Effect of Climate Change on Cassava Productivity

The productivity of cassava is affected by one of the element of climate (rainfall) in the study area (Table 6). Research findings reveal that the level of rainfall (p-value = 0.017) affected the productivity of cassava at 5% level of probability. Temperature (p-value = 0.851) and sea level (p-value = 0.219) had no significant effect on the productivity of cassava. The price of cassava had no significant effect on its productivity (p-value = 0.630), but research findings reveal that the previous productivity of cassava

Table 5. Impact of climate change on yam productivity

Variable	Coefficient	Std. error	T-statistic	P-value
Intercept	6.777	2.900	2.340	0.019
ltemp	-0.088	0.521	-0.170	0.866
lrain	-0.027	0.061	-0.440	0.657
lsea	-0.359	0.423	-0.850	0.396
lprice_yam	0.150	0.057	2.650	0.008
lprod_yam (t-1)	0.402	0.252	1.590	0.111
R2	0.634			
χ2	30.500			0.000

Source: Authors' development

Table 6. Impact of climate change on cassava productivity

Variable	Coefficient	Std. error	T-statistic	P-value
Intercept	8.503	0.357	23.810	0.000
ltemp	0.006	0.034	0.190	0.851
lrain	-0.008	0.003	-2.390	0.017
lsea	0.038	0.031	1.230	0.219
lprice_cassava	-0.003	0.007	-0.480	0.630
lprod_cassava (t-1)	0.094	0.029	3.210	0.001
R2	0.822			
χ2	63.210			0.000

Source: Authors' development

affected the present productivity significantly (p-value = 0.001) at 1% level. The implication of this finding is that a decline in rainfall will significantly affect the productivity of cassava. Also, some of the farmers may be discouraged from producing going by the productivity of the previous year.

Effect of Climate Change on Groundnut Productivity

Groundnut productivity is greatly affected by changes in the climate of the study area. Research results as presented in Table 7 shows that rainfall (p-value = 0.077) affected the productivity of groundnut at 10% level of productivity, while sea level (p-value = 0.018) affected groundnut productivity at 5% level of probability. Temperature (p-value = 0.186) had no significant effect on the productivity of groundnut. While the price of groundnut (p-value = 0.238) had no significant effect on productivity, the previous productivity affected the present at 10% level of probability (p-value = 0.051). The overall equation was statistically significant ($\chi 2$= 568.820) at 1% level of probability, and the variables in the equation accounted for 42.5% of the variation in the productivity of groundnut (R2= 0.425). The implication of this finding is that if the effort is not put in place to halt climate change and its effects, the cultivation and productivity of groundnut will continue to drop. This may be the reason why the present productivity is affected by the productivity of the previous year.

Effect of Climate Change on Beans Productivity

Research findings on the effect of climate change on beans production show that beans is a hardy crop, and is not affected by changes in temperature, rainfall, sea level and the previous productivity. Productivity was however affected by the price of beans (p-value = 0.000) at 1% level of significance. Beans productivity is therefore not affected by changes in climate and productivity, but only by changes in price. This may be because beans do not require much water, and its production cycle is short and not affected by changes in temperature. The result of the analysis and the p-value of the variables are presented in Table 8.

Table 7. Impact of climate change on groundnut productivity

Variable	Coefficient	Std. error	T-statistic	P-value
Intercept	18.928	4.681	4.040	0.000
ltemp	-1.510	1.141	-1.320	0.186
lrain	0.194	0.110	1.770	0.077
lsea	-2.242	0.946	-2.370	0.018
lprice_groundnut	0.163	0.138	1.180	0.238
lprod_groundnut (t-1)	-0.054	0.028	-1.950	0.051
R2	0.425			
$\chi 2$	568.820			0.000

Source: Authors' development

Table 8. Impact of climate change on beans productivity

Variable	Coefficient	Std. error	T-statistic	P-value
Intercept	3.668	3.398	1.080	0.280
ltemp	-0.541	0.895	-0.600	0.545
lrain	0.102	0.085	1.200	0.230
lsea	1.051	0.677	1.550	0.121
lprice_beans	0.448	0.093	4.820	0.000
lprod_beans (t-1)	-0.026	0.157	-0.170	0.867
R2	0.683			
$\chi 2$	33.520			0.000

Source: Authors' development

Effect of Climate Change on Crop Prices

Effect of Climate Change on the Price of Maize

Research findings, as presented in Table 9 show that the price of maize is not significantly affected by changes in temperature (p-value = 0.334) and rainfall (p-value = 0.317). The price of maize is however affected by the sea level (p-value = 0.073) and the productivity of maize (p-value = 0.069) at 10% level of probability, and the previous price of maize at 1% level of probability (p-value = 0.000).The price of maize is therefore affected by the price of the previous year and by the productivity of the present year, not by changes in climate. The level of productivity could affect price, as an increase in output could lead to a decline in price, going by the laws of demand and supply.

Effect of Climate Change on the Price of Millet

The result of the regression analysis shows that the price of millet is not affected by any of the variables entered into the equation (Table 10). It, therefore, implies that changes in climate, as well as changes in price and productivity, do not significantly affect the current price of millet.

Table 9. Impact of climate change on the price of maize

Variable	Coefficient	Std. error	T-statistic	P-value
Intercept	12.909	8.115	1.590	0.112
ltemp	0.749	0.775	0.970	0.334
lrain	0.065	0.065	1.000	0.317
lsea	1.054	0.587	1.800	0.073
lprod_maize	-2.528	1.390	-1.820	0.069
lprice_maize (t-1)	0.640	0.046	13.860	0.000
R2	0.95			
$\chi 2$	238.190			0.000

Source: Authors' development

Table 10. Impact of climate change on the price of millet

Variable	Coefficient	Std. error	T-statistic	P-value
Intercept	-1.375	35.002	-0.040	0.969
ltemp	-0.132	2.158	-0.060	0.951
lrain	-0.007	0.148	-0.050	0.962
lsea	0.295	1.682	0.180	0.861
lprod_millet	0.254	5.921	0.040	0.966
lprice_millet (t-1)	0.684	0.703	0.970	0.331
R2	0.87			
χ2	93.03			0.000

Source: Authors' development

Effect of Climate Change on the Price of Rice

The finding on the effect of climate change on the price of rice is similar to that of the finding on millet. The result of the analysis, as presented in Table 11 shows that the price of rice is not affected by any of the variables (Temperature, rainfall, sea level, the productivity of rice and previous price of rice) entered into the regression equation. The price of rice is therefore relatively constant, in spite of the threat of climate change and change in the level of productivity.

Effect of Climate Change on the Price of Sorghum

The effect of climate change on the price of sorghum is presented in Table 12. The result shows that the price of sorghum is not significantly affected by variables in the equation. The price of sorghum is therefore expected to be stable in the face of changes in climate and productivity.

Table 11. Impact of climate change on the price of rice

Variable	Coefficient	Std. error	T-statistic	P-value
Intercept	-67.475	705.82	-0.100	0.925
ltemp	0.234	12.819	0.020	0.985
lrain	0.052	1.098	0.050	0.962
lsea	0.416	7.421	0.060	0.955
lprod_rice	-0.182	6.601	-0.030	0.978
lprice_rice (t-1)	0.679	1.008	0.670	0.501
R2	0.933			
χ2	248.980			0.000

Source: Authors' development

Table 12. Impact of climate change on the price of sorghum

Variable	Coefficient	Std. error	T-statistic	P-value
Intercept	-23.942	20.779	-1.150	0.249
ltemp	1.686	2.414	0.700	0.485
lrain	-0.045	0.201	-0.220	0.824
lsea	3.957	2.712	1.460	0.145
lprod_sorghum	0.999	0.914	1.090	0.275
lprice_sorghum (t-1)	0.364	0.329	1.110	0.269
R2	0.846			
χ2	85.160			0.000

Source: Authors' development

Effect of Climate Change on the Price of Yam

The price of yam is significantly affected by the price of the previous year (p-value = 0.000) at 1% level of probability (Table 13). All the other variables in the equation had no significant effect on the price of yam. The price of yam is therefore not affected by changes in climate.

Effect of Climate Change on the Price of Cassava

The effect of climate change on the price of cassava is presented in Table 14. Research findings reveal that the price of cassava is not significantly affected by elements of climate, but by the level of productivity (p-value = 0.011) and the previous price of cassava (p-value = 0.000) at 1% level of probability. This goes to confirm that cassava is a hardy crop that is not affected by changes in climate.

Table 13. Impact of climate change on the price of yam

Variable	Coefficient	Std. error	T-statistic	P-value
Intercept	5.048	9.860	0.510	0.609
ltemp	-0.298	0.704	-0.420	0.672
lrain	0.040	0.093	0.430	0.664
lsea	0.602	0.573	1.050	0.294
lprod_yam	-0.579	0.956	-0.610	0.545
lprice_yam (t-1)	0.798	0.148	5.410	0.000
R2	0.959			
χ2	321.530			0.000

Source: Authors' development

Table 14. Impact of climate change on the price of cassava

Variable	Coefficient	Std. error	T-statistic	P-value
Intercept	-192.522	75.456	-2.550	0.011
ltemp	0.220	1.258	0.180	0.861
lrain	0.178	0.132	1.350	0.178
lsea	0.604	0.993	0.610	0.543
lprod_cassava	20.076	7.919	2.540	0.011
lprice_cassava (t-1)	0.475	0.111	4.290	0.000
R2	0.862			
χ2	97.960			0.000

Source: Authors' development

Effect of Climate Change on the Price of Groundnut

Research findings, as presented in Table 15 reveal that the price of groundnut is significantly affected by the price of the previous year (p-value = 0.075) at 10% level of probability. All the other variables in the equation had no significant effect on the price of groundnut. The price of groundnut is therefore not affected by changes in climate.

Effect of Climate Change on the Price of Beans

The result of the analysis of the effect of climate change on the price of beans is presented in Table 16. Going by the results as presented, none of the variables entered into the regression equation (Temperature, rainfall, sea level, the level of productivity and the price of beans from the previous year) had any significant effect on the price of beans. The beans farmers could, therefore, ignore the threats of climate change, as it would not affect the price of their product.

Table 15. Impact of climate change on the price of groundnut

Variable	Coefficient	Std. error	T-statistic	P-value
Intercept	-85.618	389.490	-0.220	0.829
Ltemp	-0.646	4.012	-0.160	0.872
Lrain	0.040	0.521	0.080	0.939
Lsea	0.835	3.473	0.240	0.810
lprod_groundnut	0.057	1.148	0.050	0.961
lprice_groundnut (t-1)	0.688	0.387	1.780	0.075
R2	0.912			
χ2	927.470			0.000

Source: Authors' development

Table 16. Impact of climate change on the price of beans

Variable	Coefficient	Std. error	T-statistic	P-value
Intercept	-7.224	8.938	-0.810	0.419
ltemp	1.143	1.970	0.580	0.562
lrain	-0.210	0.210	-1.000	0.315
lsea	-2.070	1.992	-1.040	0.299
lprod_beans	2.007	1.298	1.550	0.122
lprice_beans (t-1)	0.073	0.404	0.180	0.856
R2	0.623			
$\chi 2$	33.450			0.000

Source: Authors' development

SOLUTIONS AND RECOMMENDATIONS

This study examined the effect of climate change on the productivity and prices of selected major crops in Benue State. Results show that sea level has a significant effect on millet, sorghum, cassava and groundnut productivity. Rainfall had a significant effect on groundnut productivity. The impact of climate change on prices is very minimal as only the price of maize was significantly affected by sea level. The previous year's productivity and the current price had a significant impact on current productivity for some crops while the previous year's price and current productivity had a significant impact on the current prices for some crops. In line with the findings of this research work, it is recommended that climate research institutes such as NIMET should take up the mandate to increase awareness on climate change, its causes, and possible solutions. Research institutes and extension agents should make climate change education as part of their program and make information dissemination easy for assimilation by illiterates and local farmers. Television and radio should be used to disseminate information to farmers using local dialects where necessary. Telecommunication providers should be encouraged by the government to send short messages on climate to their customers at regular intervals. Farmers should be educated on agricultural practices that minimize climate change.

FUTURE RESEARCH DIRECTIONS

As the earth gets warmer, there is a reduction in rainfall pattern and this may affect productivity and prices of crops. Since the current study did not cover issues about causality and forecasting of future impacts of climate change on productivity and prices, future research may consider these areas. It is also possible that there may be some nonlinearities in the data arising from structural breaks and extreme events. Future studies can confront the data with nonlinear models that are capable of accounting for structural breaks and also models that look at how the effect of climate change varies over time, unlike the current study that assumed the effect is constant over time. Exploring these avenues may add value to future research in this area.

CONCLUSION

Climate change may affect food availability, quality, and access. For instance, projected increases in temperatures, changes in extreme weather events, reductions in water availability and changes in precipitation patterns, may lead to reduced agricultural productivity. This study empirically investigated the effect of climate change variables, specifically temperature, rainfall and sea level on the productivity and prices of selected major crops in Benue State. This study used time series data from 1995-2009 for analysis. First, a trend analysis was conducted for productivity, prices and climate change variables using visual plot and results show fluctuations in the series over time. Next, the three-stage least square (3SLS) regression was used to estimate a simultaneous equation involving both productivity and price of each selected crop. Overall, climate change was found to play a significant role in the productivity and prices of crops in Benue State. It is, therefore, important to take necessary policy actions, mitigation and adaptation measures to curtail climate change if the world's future in general and that of Benue State, in particular, is to be secured. This is to avoid a situation whereby the indigenes of the State are exposed to food security challenges arising from reduced productivity and prices increases. Such situations may range from inability to purchase food at affordable prices to issues of instability in the food market. Therefore, the proactive involvement of all stakeholders (policy makers, non-governmental organizations, farmers, extension agents, communities and community leaders among others) in curbing the effect of climate change cannot be overstressed.

REFERENCES

Adams, R. M., Hurd, B. H., Lenhart, S., & Leary, N. (1998). Effects of Global Climate Change on Agriculture: An Interpretative Review. *Climate Research*, *11*(1), 19–30. doi:10.3354/cr011019

Alexandrov, V. A., & Hoogenboom, G. (2000). Vulnerability and Adaptation: Assessments of Agricultural Crops under Climate Change in the Southeastern USA. *Theoretical and Applied Climatology*, *67*(1–2), 45–63. doi:10.1007/s007040070015

Almaraz, J., Mabood, F., Zhou, X., Gregorich, E., & Smith, D. (2008). Climate Change, Weather Variability and Corn Yield at a Higher Latitude Locale: Southwestern Quebec. *Climatic Change*, *88*(2), 187–197. doi:10.1007/s10584-008-9408-y

Apata, T. G., Samuel, K. D., & Adeola, A. O. (2009). *Analysis of Climate Change Perception and Adaptation among Arable Food Crop Farmers in South Western Nigeria.* Paper presented at the International Association of Agricultural Economists' 2009 Conference, Beijing, China.

Ayinde, O. E., Ajewole, O. O., Ogunlade, I., & Adewumi, M. O. (2010). Empirical Analysis of Agricultural Production and Climate Change: A Case Study of Nigeria. *Journal of Sustainable Development in Africa*, *12*(6), 345–253.

Brockett, P. L., Wang, M., & Yang, C. (2005). Weather Derivatives and Weather Risk Management. *Risk Management & Insurance Review*, *8*(1), 127–140. doi:10.1111/j.1540-6296.2005.00052.x

Building Nigeria's Response to Climate Change. (2008). *The Recent Global and Local Action on Climate Change*. Paper presented at the 2008 Annual Workshop of Nigerian Environmental Study Team (NEST), Abuja.

Christensen, J. H., Hewitson, B., Busuioc, A., Chen, A., Gao, X., Held, I., … Whetton, P. (2007). Regional Climate Projections. In S. Solomon, D. Qin, M. Manning, Z. Chen, M. Marquis, K.B. Averyt, … H.L. Miller (Eds.), Climate Change 2007: The Physical Science Basis (pp. 847–940). Cambridge, UK: Cambridge University Press.

Daudu, S., & Ajayi, A. R. (2009). Assessment of the Performance of the National Special Food Security Programme in Benue State, Nigeria. *Journal of Social Sciences*, *21*(1), 25–32.

De Wit, M., & Stankiewicz, J. (2006). Changes in Surface Water Supply across Africa with Predicted Climate Change. *Science*, *311*(5769), 1917–1921. doi:10.1126/science.1119929 PMID:16513946

Deressa, T. (2007). *Measuring the Economic Impact of Climate Change on Ethiopian Agriculture: Ricardian Approach.* World Bank Policy Research Paper No. 4342. Washington, DC: World Bank.

Deressa, T., Hassan, R., Alemu, T., Yesuf, M., & Ringler, C. (2008). *Analyzing the Determinants of Farmers' Choice of Adaptation Measures and Perceptions of Climate Change in the Nile Basin of Ethiopia.* Discussion Paper No. 00798. Washington, DC: International Food Policy Research Institute (IFPRI).

Deressa, T., & Hassan, T. (2010). Economic Impact of Climate Change on Crop Production in Ethiopia: Evidence from Cross-Section Measures. *Journal of African Economies*, *18*(4), 529–554. doi:10.1093/jae/ejp002

Downing, T. E. (Ed.). (1996). *Climate Change and World Food Security*. Berlin: Springer-Verlag. doi:10.1007/978-3-642-61086-8

Gilbert, C. L., & Morgan, C. W. (2010). Food Price Volatility. *Philosophical Transactions of the Royal Society of London. Series B, Biological Sciences*, *365*(1554), 3023–3034. doi:10.1098/rstb.2010.0139 PMID:20713400

Headey, D., & Fan, S. (2008). Anatomy of a Crisis: The Causes and Consequences of Surging Food Prices. *Agricultural Economics*, *39*, 375–391. doi:10.1111/j.1574-0862.2008.00345.x

Intergovernmental Panel on Climate Change. (2007). *Climate Change: Impacts, Adaptation and Vulnerability. Contribution of Working Group II to the Fourth Assessment Report of the Intergovernmental Panel on Climate Change*. Cambridge, UK: Cambridge University Press.

Intergovernmental Panel on Climate Change. (2012). *Managing the Risks of Extreme Events and Disasters to Advance Climate Change Adaptation. Special Report of the Intergovernmental Panel on Climate Change*. Cambridge, UK: Cambridge University Press.

International, C. A. B. (2010). *Climate Change and Crop Production*. Oxfordshire, UK: CAB International.

Kane, S., Reilly, J., & Tobey, J. (1992). An Empirical Study of the Economic Effects of Climate Change on World Agriculture. *Climatic Change*, *21*(1), 17–35. doi:10.1007/BF00143251

Kurukulasuriya, P., & Mendelsohn, R. (2006). *A Ricardian Analysis of the Impact of Climate Change on African Cropland.* CEEPA Discussion Paper No. 8. Pretoria: Centre for Environmental Economics and Policy in Africa, University of Pretoria.

Liu, H., Guenther, F., & Laixiang, S. (2004). Study on the Impacts of Climate Change on Chinas Agriculture. *Climatic Change*, *65*(1/2), 125–148. doi:10.1023/B:CLIM.0000037490.17099.97

Lobell, D. B., & Burke, M. B. (2008). Why are Agricultural Impacts of Climate Change so Uncertain? The Importance of Temperature Relative to Precipitation. *Environmental Research Letters*, *3*(3), 034007. doi:10.1088/1748-9326/3/3/034007

McCarthy, J., Canziani, O. F., Leary, N. A., Dokken, D. J., & White, C. (Eds.). (2001). *Climate Change 2001: Impacts, Adaptation, and Vulnerability. Contribution of Working Group II to the Third Assessment Report of the Intergovernmental Panel on Climate Change*. Cambridge, UK: Cambridge University Press.

Mendelsohn, R., & Dinar, A. (Eds.). (2009). *Climate Change and Agriculture – An Economic Analysis of Global Impacts, Adaptation and Distributional Effect, New Horizons in Environmental Economics*. Cheltenham, UK: Edward Elgar.

Mendelsohn, R., Nordhaus, W. D., & Shaw, D. (1994). The Impact of Global Warming on Agriculture: A Ricardian Analysis. *The American Economic Review*, *84*(4), 753–771.

Mendelsohn, R., Nordhaus, W. D., & Shaw, D. (1996). Climate Impacts on Aggregate Farm Values: Accounting for Adaptation. *Agricultural and Forest Meteorology*, *80*(1), 55–67. doi:10.1016/0168-1923(95)02316-X

Mirzabaev, A., & Tsegai, D. (2012). *Effects of Weather Shocks on Agricultural Commodity Prices in Central Asia. ZEF Discussion Papers on Development Policy No. 171*. Bonn: Center for Development Research.

National Population Commission of Nigeria. (2006). *Population and Housing Census of the Federal Republic of Nigeria*. Retrieved November 7, 2016, from http://www.population.gov.ng/images/Priority%20Tables%20Volume%20I-update.pdf

Orindi, V., Ochieng, A., Otiende, B., Bhadwal, S., Anantram, K., Nair, S., … Kelkar, U. (2006). Mapping Climate Vulnerability and Poverty in Africa. In P.K. Thornton, P.G. Jones, T. Owiyo, R.L. Kruska, M. Herrero, P. Kristjanson, … A. Omolo (Eds.), Report to the Department for International Development. Nairobi: ILRI.

Parry, M., Arnell, N., Berry, P., Dodman, D., Fankhauser, S., & Hope, C. … Wheeler, T. (2009). Assessing the Costs of Adaptation to Climate Change – A Review of UNFCCC and other Recent Studies. London: International Institute for Environment and Development and Grantham Institute for Climate Change, Imperial College London.

Parry, M., Rosenzweig, C., Iglesias, A., Livermore, M., & Fisher, G. (2004). Effects of Climate Change on Global Food Production under SRES Emissions and Socio-Economic Scenarios. *Global Environmental Change*, *14*(1), 53–67. doi:10.1016/j.gloenvcha.2003.10.008

Pearce, D., Cline, W., Achanta, A., Fankhauser, S., Pachauri, R., Tol, R., & Vellinga, P. (1996). The Social Costs of Climate Change: Greenhouse Damage and Benefits of Control. In J. Bruce, H. Lee, & E. Haites (Eds.), *Climate Change 1995: Economic and Social Dimensions of Climate Change* (pp. 179–224). Cambridge, UK: Cambridge University Press.

Reilly, J. (1999). Climate Change – Can Agriculture Adapt? Choices. In C. Rosenzweig & A. Iglesias (Eds.), *Implications of Climate Change for International Agriculture: Crop Modeling Study* (pp. 4–8). Washington, DC: US EPA Office of Policy, Planning and Evaluation, Climate Change Division, Adaptation Branch.

Rosenzweig, C., & Parry, M. L. (1994). Potential Impacts of Climate Change on World Food Supply. *Nature, 13*(6459), 133–138. doi:10.1038/367133a0

Rosenzweig, C., Parry, M.L., Fischer, G., & Frohberg, K. (1993). *Climate Change and World Food Supply*. Oxford, UK: Environmental Change Unit, University of Oxford.

Schlenker, W., & Roberts, M. J. (2009). Nonlinear Temperature Effects Indicate Severe Damages to us Crop Yields under Climate Change. *Proceedings of the National Academy of Sciences of the United States of America, 106*(37), 15594–15598. doi:10.1073/pnas.0906865106 PMID:19717432

Stern, N. (2007). *The Economics of Climate Change: The Stern Review*. Cambridge, UK: Cambridge University Press. doi:10.1017/CBO9780511817434

Stige, L. C., Stave, J., Chan, K., Ciannelli, L., Pettorelli, N., Glantz, M., & Stenseth, N. C. et al. (2006). The Effect of Climate Variation on Agro-pastoral Production in Africa. *Proceedings of the National Academy of Sciences of the United States of America, 103*(9), 3049–3053. doi:10.1073/pnas.0600057103 PMID:16492727

Torero, M., & von Braun, J. (2010). *Alternative Mechanisms to Reduce Food Price Volatility and Price Spikes*. Paper presented at the International Food and Agricultural Trade Policy Council, Barcelona, Spain.

United Nations Framework Convention on Climate Change. (2007). *Report on the Workshop on Climate Related Risks and Extreme Events*. Note by the Secretariat. FCCC/SBSTA/2007/7. Bonn: UNFCCC.

Von Braun, J., Ahmed, A., Asenso-Okyere, K., Fan, S., Gulati, A., Hoddinot, J., & Von Grebmer, K. (2008). *High Food Prices: The What, Who, and How of Proposed Policy Actions*. Washington, DC: International Food Policy Research Institute.

Von Braun, J., & Tadesse, G. (2012). *Global Food Price Volatility and Spikes: An Overview of Costs, Causes, and Solutions*. ZEF Discussion Papers, 161.

Wikipedia. (2013). *Climate Change and Agriculture*. Retrieved November 6, 2016, from http://en.wikipedia.org/wiki/Climate_change_and_agriculture

Wolfe, D. W., Schwartz, M. D., Lakso, A. N., Otsuki, Y., Pool, R. M., & Shaulis, N. J. (2005). Climate Change and Shifts in Spring Phenology of Three Horticultural Woody Perennials in Northeastern USA. *International Journal of Biometeorology, 49*(5), 303–309. doi:10.1007/s00484-004-0248-9 PMID:15592880

World Development Report. (2010). *Development and Climate Change*. Washington, DC: The World Bank. Retrieved November 7, 2016, from https://openknowledge.worldbank.org/bitstream/handle/10986/4387/9780821379875_fm.pdf?sequence=104&isAllowed=y

Yesuf, M., Di Falco, S., Deressa, T., Ringer, C., & Kohlin, G. (2008). *The Impact of Climate Change and Adaptation on Food Production in Low-Income Countries: Evidence from Nile Basin, Ethiopia*. International Food Policy Research Institute (IFPRI) Discussion Paper, 00828.

ADDITIONAL READING

Brooks, S., & Loevinsohn, M. (2011). Shaping Agricultural Innovation Systems Responsive to Food Insecurity and Climate Change. *Natural Resources Forum*, *35*(3), 185–200. doi:10.1111/j.1477-8947.2011.01396.x

Faraco, R., Silva, C., Amorim, W., Silveira, A., Neiva, S., & Baltazar, J. (2016). Food Security, Agriculture and Climate Change Mitigation Strategies: A Scientific Production Panorama. *Scholedge International Journal of Multidisciplinary and Allied Studies*, *3*(2), 34–62.

Food and Agriculture Organization of the United Nations. (2016). *Climate Change and Food Security: Risks and Responses*. Rome: Food and Agriculture Organization of the United Nations.

Gregory, P. J., Ingram, J. S. I., & Brklacich, M. (2005). Climate Change and Food Security. *The Royal Society*, *360*, 2139–2148. PMID:16433099

Intergovernmental Panel on Climate Change. (2014). *Climate Change 2014: Impacts, Adaptation, and Vulnerability. Part B: Regional Aspects. Contribution of Working Group II to the Fifth Assessment Report of the Intergovernmental Panel on Climate Change*. Cambridge, New York, NY: Cambridge University Press.

Misra, A. K. (2014). Climate Change and Challenges of Water and Food Security. *International Journal of Sustainable Built Environment*, *3*(1), 153–165. doi:10.1016/j.ijsbe.2014.04.006

US Environmental Protection Agency. (2016). *Climate Impacts on Agriculture and Food Supply*. Retrieved November 6, 2016, from https://www.epa.gov/climate-impacts/climate-impacts-agriculture-and-food-supply

KEY TERMS AND DEFINITIONS

Climate Change: A statistically significant variation either in the mean state of the climate or in its variability persisting for an extended period (typically decades or longer).

Climate Mitigation: Any action taken to permanently eliminate or reduce the long-term risk and hazards of climate change to human life and property.

Crop Productivity: Is the crop production per hectare, otherwise known as crop yield.

Food Prices: The average annual market value of food commodities.

Food Security: Access by all people at all times to sufficient food for a healthy and productive life.

Impact: The powerful effect that one variable, e.g. climate change has on another variable e.g. crop productivity.

Poverty: It is a state of being extremely poor. It is a general scarcity of dearth or the state of one who lacks a certain amount of material possession or money.

Precipitation: Distribution of water, i.e. rainfall.

Temperature: A measure of hotness or coldness of the weather.

Weather: The day-to-day changes in the conditions of the atmosphere.

Chapter 13
Irrigation and Food Security:
Case of Soybean Production in Serbia

Marko Jeločnik
Institute of Agricultural Economics, Serbia

Jovan Zubović
Institute of Economic Sciences, Serbia

ABSTRACT

Agriculture is among the sectors strongly affected by climate changes (the largest damage in agriculture is usually caused by drought), which simultaneously influence the volume of production and safety of products. Besides, soybean is one of the most important crops used for human and animal nutrition, and certain branches of light industry. In climate conditions of Serbia, soybean is predominantly produced in the system of dry land farming. The chapter has several closely related goals. Primarily, it presents the detailed analysis of the economic importance of soybean production for the Republic of Serbia (with a special focus on sustainability of national food security). One of the goals is the assessment of cumulative losses incurred in the soybean production in the climatic conditions of the Republic of Serbia (dominantly caused by drought) in the last several years. Also, it was estimated the possible positive impact of irrigation on the current state of food security on the macro level, as well as the higher profitability of soybean production on the micro level.

INTRODUCTION

Soybean is a legume, which belongs to the group of thermophilic plants, with wide area of spreading (natural habitat). Its vegetation varies, and it can last from 100 to 145 days. Depending on the variety used, it requires temperature sum from 1,600°C to 3,200°C (Miladinović, Hrustić, & Vidić, 2011). This is one of the oldest plants on the Earth. It originates from Asia (specifically, from China – Manchuria region), from where it was spread to the East and Southeast Asia (Korea, Japan, Vietnam, Indonesia, and India), and then to other continents. Its presence in America was firstly recorded in 1804. As an arable crop, it began to be cultivated in Europe in the XIX century, while in Serbia it was introduced at the beginning of the XX century. Globally, its intensive production intensified just before the World War

DOI: 10.4018/978-1-5225-2733-6.ch013

II. Optimal areas for its growing are between 56° north and 45° south latitude (Stamenković, Bošnjak, & Stamenković, 2007).

The specific chemical composition of soybean (it is the only plant which contains all essential amino acids necessary for the functioning of the human body) ranks it among the most important crops (oilseeds) intended for human and animal nutrition. According to its usability, soybean proteins are equivalent to the proteins of animal origin. For that reason, it can be used for proteins synthesis in the human body. It also contains vitamins, especially thiamine, riboflavin, vitamin B6, niacin, and to some extent vitamins C, D, E, and K, as well as folic and pantothenic acid. The grain also contains substantial amount of microelements such as calcium, phosphorus, potassium, iron, sulphur, manganese, sodium, molybdenum, copper, boron, iodine, cobalt, and zinc, as well as around 18% of oil that contains almost 85% of unsaturated fatty acids (Roljević, Subić, & Jeločnik, 2009).

Diversification of it potential use (food and feed industries, pharmaceuticals and cosmetics, up to the production of biodiesel) has resulted in the growing demand and expansion of the harvested areas under the soybean. In human nutrition (adequate substitute for meat), it is usually used in the form of seed, bleached germ, or soybean oil. It increases the overall nutritional and biological value of some food products, and it is directly used in alimentation as soybean steaks, milk, cheese, pate, or flour. Contemporary medicine recommends soybeans as a foodstuff that prevents the occurrence of cardiovascular diseases and cancer, as well as for regulation of digestion and maintaining of hormonal balance.

Soybean is an essential component for animal feeding. It can be a part of concentrated mixtures (usually as soybean meal or some other waste obtained during the industrial processing of soybean, rich in proteins and minerals), or used as voluminous (bulky) feed (use of vegetative organs – stem and leaves). Besides all, soybean has a large agro-technical importance, as it favorably affects the maintaining and improvement of soil structure with its well-developed and deep root system. Also, in symbiosis with specific bacteria (nitrogen fixers), after meeting its own need in nitrogen, it enriches the topsoil layer with above-mentioned microelement. That is why soybean is highly recommended preceding crop in the crop rotation for many plant species, especially for cereals (Đorđević & Nenadić, 1980).

Areas with optimal conditions for soybean production are the regions with average summer temperatures of around 19°C, and night temperatures above 13ºC. Warm periods are necessary for the phase of germination and emergence (during this period, it is tolerant to low temperatures) up to the phase of flowering and bean (grains) establishment. Longer intervals with temperatures of above 32°C and high deficiency of moisture can lead to the poor blooming, deficient beans and grains establishment, directly influencing the decline in yields. The optimal temperatures for the phase of flowering are up to 25°C, while during the establishment and grain filling, they are up to 23°C. During the maturation, the ideal temperature is around 20°C. The plant does not require much fresh water, and it is tolerant to drought until the phase of flowering. Yields are significantly more dependent on soil moisture, and relative air humidity, while the greatest requirements for moisture are in the phases of flowering and grain filling (Jančić, 2015). In other words, soybean is less water demanding in the initial stages of plant growth and development. According to water availability, critical months are July and August. In Serbia's environment, overall water requirements vary from 390-550 mm. Similar to other field crops, the impact of irrigation on soybean yield depends on the amount and distribution of rainfalls during the growing season (Dragović, 2008). Contemporary advances in genetics and plant selection have contributed to the creation of new, more resistant varieties. Existing varieties are improved to be resistant to plant diseases, pests, and climatic conditions.

BACKGROUND

Soybean is being grown in more than 35 countries. The major producers are the USA, Brazil, Argentina, China, and India (Kovačević, 2014). At the global level, there is an increasing trend in total harvested areas of soybean (by over 20% during the previous decade), while oscillations in gained yields and total production are mostly caused by a change in climatic conditions (shortage in rainfalls and occurring of droughts) (Table 1).

During the previous period, the biggest exporters of soybean were Argentina (about 39% of world export), the USA, and Brazil, while the biggest importers were China (about 41% of world import), the EU, Japan, and Mexico (Song, Marchant, & Xu, 2006).

Commercial introduction of genetically modified (GM) crops made soybean the leader in GM sector, with an estimation that it occupies around 51% of the areas under the GM crops worldwide, mostly in North and South America (Papić Brankov & Lovre, 2011).

In Serbia (Appendix 1), soybean yields are approximately at the level of the European average (Kovačević, 2014). Sowing or precise harvesting areas under the soybean are rather stable and mostly concentrated in the AP Vojvodina (Northern Province of Serbia).

In Serbian climate conditions, variations in yields and changes in total soybean production are usually caused by insufficient rainfall, as well as by relatively small area under irrigation. Production significantly declined in dry years of 2007 and 2012.

Compared to other countries in the region (former Yugoslavia and other neighboring countries) by all production factors (especially gained average yields), Serbia is a leader in the region. The leadership is especially notable in cumulative output volume during 2006-2014, where the ratio varies from 1:2.58 as compared to Romania up to 1:1,000 as compared to Slovenia. Due to the long tradition in crop production and the size of the country, it is especially important to compare Serbian data with Romania, Hungary, and Croatia.

At the global level, trade in soybean is usually contracted at commodity exchanges in several locations. The most usual commodities traded are soybean grain and soybean meal. In Serbia, trade is organized at the commodity exchange in Novi Sad, or in the form of pre-arranged contracts with processing companies (oil refineries). When trading at commodity exchange, it is highly desirable to have a predefined quality standard for grain which satisfies the needs of a buyer (generally, a buyer makes a blind buying transactions of a product, where product characteristics are usually provided by storage industry or independent certification agencies). Soybean oil takes the biggest share in Serbia's soybean exports. It is exported to Austria, Slovenia, Hungary, Bulgaria, and countries of the Central European Free Trade Agreement (CEFTA) (Kovačević, 2014).

Table 1. Soybean production in the world in 2004-2013

Element	Year									
	2004	2005	2006	2007	2008	2009	2010	2011	2012	2013
Area*, million ha	91.6	92.6	95.3	90.2	96.5	99.3	102.8	103.8	104.9	111.3
Yield**, tons/ha	2.2	2.3	2.3	2.4	2.4	2.2	2.6	2.5	2.3	2.5
Production*, million tons	205.5	214.6	222.0	219.7	232.3	223.4	265.0	261.9	241.1	276.4

Note: * Calculated data; ** Aggregated data.
Source: FAOSTAT, 2016

It is important to note that according to the Law on Genetically Modified Organisms (Government of the Republic of Serbia, 2009), production, import, and marketing of products containing over 1% of GM organisms have been forbidden in Serbia since 2009. That gives Serbia a competitive advantage for export of soybean to the countries where demand for this crop is high and which legislation does not allow GM organisms.

Ensurance of both direct and indirect food safety for people in Serbia (throughout the animal husbandry) requires analysis of commodity balance in relation to soybean and its main products (soybean meal and soybean oil) at the national level (see Table 2 for soybean, Table 4 for soybean meal, and Table 4 for soybean oil).

Data presented in Table 2 show variation of harvested areas and achieved yields in soybean production. The decrease in harvested areas is primarily a result of lowering soybean prices in the previous period (soybean is mostly purchased by processing industries which dictate prices). Instability of yields is usually a direct consequence of climate variations and unfavorable weather conditions (most often, droughts), having in mind that very small number of plots under soybean crops are equipped with irrigation systems. Although small in volumes, import of soybean products to Serbia was recorded in the years when conditions for domestic production were unfavorable. It was usually consumed to satisfy the needs of the processing industry.

Soybean is mainly used in the processing industry. Very small quantity of fresh product is used for animal feeding. It has to be noted that significant loss in grains produced (in average, about 5% of the total production) happens due to poor storage conditions. Only small portion of soybean grain is exported. Soybean grain and soybean meal are mainly traded at the national commodity exchange.

Table 2. Commodity balance of soybean production/consumption in 2010-2014

Category	Unit of measure	2009/2010	2010/2011	2011/2012	2012/2013	2013/2014
Production						
Harvested areas	thousand hectares	144	170	165	163	160
Yield	tons/ha	2.4	3.2	2.7	1.7	2.4
Supply						
Starting reserves	thousand tons	64	19	81	31	14
National production	thousand tons	349	541	441	281	385
Import	thousand tons	51	14	10	78	75
Total available quantities	thousand tons	464	574	532	390	474
Consumption						
Industrial processing	thousand tons	409	359	431	324	395
Feed (unprocessed)	thousand tons	14	24	30	32	32
Loss	thousand tons	17	27	22	14	19
Export	thousand tons	5	83	18	6	23
Ending reserves	thousand tons	19	81	31	14	5
Total consumption	thousand tons	464	574	532	390	474

Note: Economic year is a period from October to September of the following year.
Source: Ministry of Agriculture and Environmental Protection of the Republic of Serbia [MAEPRS], 2016

Table 3. Commodity balance of soybean meal in 2010-2014

Category	Unit of measure	2009/2010	2010/2011	2011/2012	2012/2013	2013/2014
Supply						
Starting reserves	thousand tons	15	12	14	13	25
National production	thousand tons	319	280	307	252	308
Import	thousand tons	30	14	12	78	76
Total available quantities	thousand tons	364	306	333	343	409
Consumption						
Feed	thousand tons	325	244	238	270	270
Export	thousand tons	27	48	82	48	47
Ending reserves	thousand tons	12	14	13	25	92
Total consumption	thousand tons	364	306	333	343	409

Note: Economic year is a period from October to September of the following year.
Source: MAEPRS, 2016

Table 4. Commodity balance of soybean oil in 2010-2014

Category	Unit of measure	2009/2010	2010/2011	2011/2012	2012/2013	2013/2014
Supply						
Starting reserves	thousand tons	7	15	10	7	10
National production	thousand tons	74	65	71	58	71
Import	thousand tons	0	0	0	1	0
Total available quantities	thousand tons	81	80	81	66	81
Consumption						
Food processing industry	thousand tons	22	24	24	24	24
Industrial consumption	thousand tons	1	1	1	1	1
Export	thousand tons	43	45	49	31	31
Ending reserves	thousand tons	15	10	7	10	25
Total consumption	thousand tons	81	80	81	66	81

Note: Economic year is a period from October to September of the following year.
Source: MAEPRS, 2016

According to the commodity balances of industrial processing of soybean (Table 3 and Table 4), most of the soybean oil is used by the food industry or exported. On the other hand, production of soybean meal does not match the overall needs of livestock production, what causes import purchases.

METHODOLOGY AND DATA SOURCES

This chapter has several closely related goals. According to those goals, all indicators related to global soybean production will be presented. Further, detailed analysis of the economic importance of soybean production for the Republic of Serbia will be conducted, with a special accent on the sustainability of national food security.

One of the goals of the chapter is to estimate the potential cumulative loss incurred in soybean production in Serbia during the period of 2006-2015 due to the variation of climatic conditions, primarily lack of rainfalls and extensive drought. Methodology of assessment of the recorded losses in soybean production caused by drought is based on the following approach: harvested areas and achieved yields in observed period (2006-2015) are presented, and according to data of the Hydro-meteorological Service of the Republic of Serbia (RHMSS) on agrometeorological conditions, all dry years (appearance of drought) that initiate decrease in yields are considered.

The authors have tested two scenarios. Scenario I (optimistic) – all negative deviations of achieved yields caused by drought are compared to the average yield. Based on the average selling price of soybean in 2015, the difference between the achieved and potentially possible yield in each year is calculated. By multiplying the gained value with the harvested area (additional elimination of eventual indirect impacts of floods and storms), the total loss for each year is determined, while the sum of all losses represents cumulatively expressed losses in the period under observation. Scenario II (pessimistic variant) methodologically follows the same pattern but assumes the deviation of achieved yield from the highest (maximal) yield gained within the period under observation (the best year for soybean production in terms of climatic conditions). As there are no previously completed regional researches which results are suitable for comparison. Conducted research can be used as the one by which other researches could be compared with. Besides, gained results can be a good recommendation to policy makers and producers to introduce the wider implementation of irrigation.

In accordance with the above, one of the chapter objectives is the analysis of irrigation effects in soybean production according to achieved contribution margins (analytical calculation based on variable costs) in the production systems with or without irrigation. All necessary data are obtained from the in-depth interviews with members of selected family agricultural holdings in the city of Pančevo (South Banat District, Republic of Serbia), which is dominantly oriented on crop production. Although the selected holdings differ in their economic power, the level of mechanization, available equipment, and land area under agricultural production, they are basically very similar in used agrotechnical and technological approach to farming, as well as inconsistency in good agricultural practice related to soybean production. Relatively insignificant deviations are observed in necessary inputs for purchase and sale of final products. Data obtained refer to production season of 2014/2015. All elements in the analytical calculations are presented per unit of production area (hectare). In order to facilitate the comparison, all values are expressed in USD (according to the current exchange rate of the National Bank of Serbia).

Contribution margin (gross financial result) is calculated by subtracting total variable costs from total incomes gained in the observed line of crop production per unit of production area. Contribution margin can be expressed by the following equations (Subić, Jeločnik, & Zubović, 2015):

$$MP = VP - VT, \tag{1}$$

whereby $VP = (q * c) + p,$ (2)

where: MP – contribution margin;

VP – total production value;

VT – total variable costs;

q – quantity of products per unit of production area;

c – price of product per ha;

p – subsidy or premium per ha.

Simplicity and practicality of this analytical approach give an excellent analytical base to decision makers at agricultural holdings for efficient management of production costs. Therefore, using this analytical tool, holdings can easily assess the economic performance of existing production lines, as well as the sustainability of technology utilized and yields achieved (Vasiljević & Subić, 2010). The value of contribution margin is a good indicator for producers to understand what financial assets are disposed for covering fixed costs and gaining profits in observed lines of crop production in the defined production cycle. For evaluation of irrigation effects in soybean production, the authors have conducted a comparative analysis of change in the structure of variable costs after the implementation of agro-technical tools.

In certain lines of crop production, with or without irrigation, evaluation of results in uncertain environment can be done by different methods and techniques. Use of the following methods has been proven as a good methodological approach to assessment of the results of production organised in open field by the critical price, critical yield, and critical variable costs, where indicators represent critical values at which the contribution margin equates to zero. This can be presented as follows (Nastić, Jeločnik, & Subić, 2014):

$$KC = (VT - S) / OP \quad (3)$$

$$KP = (VT - S) / OC \quad (4)$$

$$KVT = (OP * OC) + S \quad (5)$$

where: KC – critical price;

KP – critical yield;

KVT – critical variable costs;

OP – expected yield;

OC – expected price;

S – subsidies;

VT – variable costs.

MAIN FOCUS OF THE CHAPTER

During recent decades, the impact of climate on agriculture has been investigated by several authors (Arnell et al., 2016; Reidsma et al., 2015; Nelson et al., 2009; Fuhrer et al., 2006; Smit & Skinner, 2002). Research results show that its strongest negative impact is in the sectors of agriculture, energetics, and tourism, as well as in activities carried out to support the sustainability of the existing ecosystems (e.g. in activities of forest fires prevention) (Giannakopoulos et al., 2009). Climate change and climate variations simultaneously influence the volume of primary production (food security) and safety of agricultural and food products (food safety) (Food and Agriculture Organization of the United Nations [FAO], 2003; Balambal, 2015).

A deeper analysis of the impact and decision-making related to the application of appropriate measures and activities, as a response to a trend of climatic changes, need to be conducted and harmonized with the value assessment. Simulation of global climatic changes caused by the growth of annual temperature by 1°C, as well as the growth of world sea level by 0.2 m in the first half of the XXI century, predicts annual expected aggregate damages in equivalent of approximately $522 billion, or 2.7% of total world income (Tol, 2002).

In agriculture, precisely, in crop production, the biggest damages according to frequency, intensity, and territorial scope are usually caused by droughts (lack of precipitation in conjunction with high temperatures). Such climatic phenomenon has a direct or indirect impact on all segments of societies and environment (National Drought Mitigation Centre, 2016).

Some researchers have emphasized the impact of climatic changes on the decrease of yields in crop production, whereby precipitation and changes in temperature have been recognized as key factors. For example, Lobell and Field (2007) indicate the growth of average annual temperatures of around 0.4°C during the past three decades (in some regions, growth is several times higher). They assess that global production of some common crops (wheat, rice, maize, soybean, barley, and sorghum) is featured by fluctuations in the value of average yields up to 30%, which can be explained by variations of total precipitations and temperature. When comparing the impact of climate change and technological progress in agriculture on achieved yields, technology is more influential. However, the impact of climate variations should not be ignored. Besides the fact that recorded losses in global crop production due to climate changes during the last two decades of the XX century were just several percent points, they were very big in absolute terms (Lobell & Field, 2007).

As opposed to the fact that current worldwide level of food production can cover the nutrition needs of the global population, almost 850 million people are unable to have adequate nutrition both by energy and volume of meals (Papić Brankov & Milovanović, 2015). Some of the recent researches show that the global demand for food will increase by 50% in the next few decades (Nelson et al., 2014). Therefore, in order to maintain food security, with expected growth of major agricultural crop yields by 40%, the above requires the use of additional 10% of land areas for agricultural production (excluding biofuels production). Such estimations are usually based on the evaluation of the impact of different groups of factors which also include climate change and weather extremes (Spiertz, 2012). Therefore, one of the global scenarios of climate change which predicts an impact on current level of agricultural production by 2080, states that it can be expected to have additional 80 million of people exposed to the risk of hunger, where majority of them will originate from the arid and sub-humid tropical regions (primarily from the African continent) (Parry, Rosenzweig, Iglesias, Fischer, & Livermore, 1999).

On the other hand, estimates for some territories go in the direction of positive effects caused by an increase in global temperature (primarily according to the extension of growing seasons, crops spreading over the certain territory and yields growth). Generally, all crops that predominate in southern Europe (corn, sunflower, and soybeans) will have a rapid transfer to the northern territories or territories at the south of Europe located at higher altitudes (Audsley et al., 2006). Therefore, it is expected that winter (oilseed rape or field beans) and spring (soybean or sunflower) crops will spread toward colder regions. Specifically for soybean, it will come to increase in yields in Western Europe (Wolf, 2000).

The largest damage for agriculture is usually caused by drought. In Serbia, like in many other countries, the relatively high impact of climate changes on national agricultural production is recorded Due to a lack of water and spatially and temporally unfavorable distribution of rainfalls, as well as due to extremely warm summers and more frequent and longer heat waves, crop production is declining because of drought.

Some authors estimated that during the period 1994-2014, drought caused the loss of around $1.7 billion in the production of basic agricultural products (wheat, corn, soybean, sunflower, and sugar beet). Over 80% of those losses were maize and wheat (Lalić, Stričević, Janković, Jaćimović, & Firanj, 2015).

Within the conducted research, in order to estimate cumulative losses in soybean production during the period 2006-2015 caused by drought, each year considered as critical in terms of agro-meteorological conditions has to be considered. Therefore, a brief review of all factors in each soybean production season was made:

1. Season 2014/2015 – season was moderately warmer with the achievement of average rainfalls on the territory of Serbia (normal level – the multiannual average of 1981/2010). The winter period was slightly warmer, with high precipitation (50% higher compared to normal), what created significant reserves of moisture in the soil. During spring, temperatures were average, while rainfalls were notably above the average. The summer (June-August) was moderately warmer, with a large deficit in rainfalls, which created unfavorable conditions of soil moisture (drought), particularly during the period from mid-July to mid-August. Autumn brought the slightly higher amount of precipitation. Spring crops, primarily corn and soybeans, were the most vulnerable (yields lower by 30-40% compared to the multiannual average). Even such yields may be considered as good ones due to the fact that there were 56 days with the temperature above 30°C, out of which 17 days with the temperature over 35°C, as well as throughout very high negative deviations of rainfall from multiannual average (79%), (Republican Hydrometeorological Service of Serbia [RHMSS], 2015).
2. Season 2013/2014 – season was moderately warmer with abundant rainfalls. Winter was much warmer and dryer than average, while the spring was characterized by warm weather with heavy rainfalls. In general, during the growing period, significant deviations were recorded only for rainfalls. It was the rainiest year over the past 45 years, with rainfalls more by 190% as compared to normal level. It caused extreme problems, since excess moisture made the soil unsuitable for normal growth and development of most of the crops, what resulted in lower yields of soybean compared to the multiannual average (RHMSS, 2014).
3. Season 2012/2013 – during the season, there were some deviations of weather indicators from the multiannual averages. In relation to crop production, adverse effects were identified because of spring frosts, snow cover, the cold soil in the pre-sowing period, above average temperatures, and low rainfalls from mid-July till mid-August. Summer drought in the second half of the growing season especially threatened spring crops, primarily, maize and soybean (RHMSS, 2013).

4. Season 2011/2012 – in general, the season was moderately warmer, with significantly fewer rainfalls in comparison to the average value for the period of 1971-2000. In addition, weather extremes resulted in significantly lower crop yields, especially in spring crops (yields failed by almost 40% compared to the multi-year averages). There were recorded extreme droughts during autumn 2011, cold waves and high snow cover during winter, strong spring frosts in April, and very intense drought during summer (June-August) and at the beginning of autumn 2012. Three heat waves occurred (in total, there were 68 days with temperature over 30°C, 19 of which with temperature over 35°C) with distinct deviation from the average maximum (up to 13°C) and significant shortage of precipitation throughout the growing season (only 77% of average rainfalls). Quantity and quality (for example, maize was contaminated by aflatoxin) of achieved yields would have been much better if the agro-technical measure of irrigation was applied on larger areas during the sensitive phases of vegetation (RHMSS, 2012).
5. Season 2010/2011 – season was moderately warmer, with precipitation below the multiannual average (83%). Winter was characterized by average temperatures, with moderately more precipitation at the end of winter, which led to the formation of moisture reserves in the soil. Spring was slightly warmer with a high deficit in rainfalls. Similar conditions remained during summer. In the period of vegetation, there were 47 days with temperature over 30°C, with precipitation lower than average (65%). Confrontation of heat and moisture waves in the mid-July led to drought in Serbia. It resulted in significantly lower soybean yields than expected (RHMSS, 2011).
6. Season 2009/2010 – season was moderately warmer with higher rainfalls (40% above average). Winter and spring had temperatures above average and significant precipitation surplus. Similar conditions remained during summer, and intensive precipitations ended in mid-August. During vegetation period, there were 32 days with temperature over 30°C and amount of rainfalls above average (128%). Despite the precipitation surplus, soybean yields were slightly above the average (RHMSS, 2010).
7. Season 2008/2009 – in general, weather conditions were within the normal range (with greater deviations during spring and summer). Winter was moderately warmer with slight precipitation surplus, while spring was characterized by droughts in April and May. In summer, there were unstable conditions with the expressed unevenness in rainfalls, which resulted in moderate to severe drought during late July – late August. Despite severe constraints to farming in the form of adverse weather conditions during the vegetation period, production of soybean achieved relatively good yields (RHMSS, 2009).
8. Season 2007/2008 – season was characterized by intensive changes in weather conditions. The period from October till March was moderately warmer, with precipitation above average (135%). The period from April till September was moderately warmer with uneven rainfalls (89% of the multiannual average). Great lack of precipitation and high temperatures during the first half of July and entire August led to drought in Serbia. In a certain degree, it affected the achieved yields of many agricultural crops (RHMSS, 2008b).
9. Season 2006/2007 – the season was characterized by several climate extremes. The winter was much warmer, accompanied by a moderate surplus in precipitation, while in spring there was recorded significant influx of rainfalls. During the period of vegetation, the weather was significantly warmer along with a high deficit in rainfalls (drought during April, July, and August). Weather extremes (high temperatures and long heat waves of considerable intensity) were particularly intense in July and August (strong to extreme drought). September brought colder weather and heavy precipitation.

Temperature changes and drought in the critical stages of vegetation contributed to the significant decrease in achieved yields, with the maximum declines recorded for maize and soybean (RHMSS, 2008a).

10. Season 2005/2006 – season with extreme temperature instability, as well as shifts of rainy and dry periods. There were extremely low temperatures in winter, while spring brought heavy rainfalls (floods and sodden soil) that resulted in a delay in sowing of spring crops. During summer, from mid-June and throughout entire July, there was dry and very warm weather (heat waves and severe drought), while August was moderately warm, but accompanied with heavy rains. Lower achieved yields in spring crops were caused by cold winter, spring floods, and July drought (RHMSS, 2006).

Based on the given facts related to weather conditions during 2006-2015, and according to the trends in indicators related to soybean production in Serbia (Table 5), there has been conducted the assessment of cumulative loss in this line of crop production caused by drought (deficiency and/or uneven temporal and spatial distribution of precipitation).

During the observed period, soybean achieved the average yield of 2.6 tons/ha, where the maximal yield was in 2014 (3.5 tons/ha), and minimal in 2012 (1.7 tons/ha).

Estimation of losses requires determination of average selling price of soybean. According to the SORS (2016b), the average purchase price for soybean for 2015 is presented in Table 6.

The total estimated amount of losses caused by the drought in soybean production in a period of observation, according to Scenario 1, were around $115 million. The estimated cumulative loss is expressed in the current price of soybean for 2015 in Table 7.

On the other hand, according to Scenario 2, the total loss in soybean production caused by rainfalls deficit was almost $520 million (Table 8). The ratio between the levels of loss calculated by presented scenarios was 1:4.5.

Table 5. Trends in soybean production indicators in Serbia in 2006-2015

Year	Harvested area, ha	Yield, tons/ha	Total production, million tons
2006	156,680	2.7	0.43
2007	146,988	2.1	0.30
2008	143,684	2.4	0.35
2009	144,386	2.4	0.35
2010	170,255	3.2	0.54
2011	165,253	2.7	0.44
2012	162,714	1.7	0.28
2013	159,724	2.4	0.38
2014	154,250	3.5	0.55
2015	184,841	2.5	0.45
Average	158,878	2.6	0.41

Source: Statistical Office of the Republic of Serbia [SORS], 2008, 2011, 2014, 2015, 2016a

Table 6. Average selling price of soybean in 2015, $/kg

Selling price of soybean (monthly)												Average price
I	II	III	IV	V	VI	VII	VIII	IX	X	XI	XII	
0.35	0.37	0.36	0.36	0.36	0.35	0.35	0.34	0.33	0.35	0.35	0.35	0.35

Note: All calculations were based on exchange rate of 1 USD = 110.4 RSD (according to the middle exchange rate of the National Bank of Serbia as on September 1, 2016).

Source: SORS, 2016b.

Table 7. Estimated losses in soybean production caused by drought in Serbia in 2006-2015 (Scenario 1)

Year	Harvested area, ha	Difference compared to the average yield, tons/ha	Additional production, tons	Price, $/ton	Potential loss, $ million
2006	156,680	*	*	*	*
2007	146,988	0.5	73,494	350	25,723
2008	143,684	0.2	28,737	350	10,058
2009	144,386	0.2	28,877	350	10,107
2010	170,255	*	*	*	*
2011	165,253	*	*	*	*
2012	162,714	0.9	146,443	350	51,255
2013	159,724	0.2	31,945	350	11,181
2014	154,250	*	*	*	*
2015	184,841	0.1	18,484	350	6,649
Total					114,973

Note: All calculations were based on exchange rate of 1 USD = 110.4 RSD (according to the middle exchange rate of the National Bank of Serbia as on September 1, 2016).

Source: Authors' own calculations based on Table 6 and Table 7

Table 8. Estimated losses in soybean production caused by drought in Serbia in 2006-2015 (Scenario 2)

Year	Harvested area, ha	Difference compared to the average yield, tons/ha	Additional production, tons	Price, $/ton	Potential loss, $ million
2006	156,680	0.8	125,344	350	43,870
2007	146,988	1.4	205,783	350	72,024
2008	143,684	1.1	158,052	350	55,318
2009	144,386	1.1	158,825	350	55,589
2010	170,255	0.3	51,077	350	17,877
2011	165,253	0.8	132,202	350	46,270
2012	162,714	1.8	292,885	350	102,510
2013	159,724	1.1	175,696	350	61,494
2014	154,250	*	*	*	0
2015	184,841	1.0	184,841	350	64,694
Total					519,646

Note: All calculations were based on exchange rate of 1 USD = 110.4 RSD (according to the middle exchange rate of the National Bank of Serbia as on September 1, 2016).

Source: Authors' own calculations based on Table 6 and Table 7

Over the past centuries, agricultural production has shown a high level of ability to adapt to changed environment. The majority of adaptations occurred spontaneously, without the impact on agricultural producers (Brooks, Adger, & Kelly, 2005). According to the European Commission [EC] (2007), changing conditions represent a set of activities carried out in order to mitigate the adverse impact of current or upcoming climate changes, as well as to utilize the potential benefits. The term includes strategic documents at all levels, as well as actions (measures) taken on the level of local community, household or individuals, in order to provide sustainability of undertaken investments during acclimatization to expected new climate change (EC, 2007). Focusing on agriculture, most important are the measures of autonomous adaptation, as they assume change of used inputs, implementation of new technical and technological solutions for optimal management of all elements of the production process, modifications in desirable agro-technical time limits within the growing season, implementation of new economic activities, etc. They also include both the mechanisms and activities that improve the management under the production risks identified in elements of climate change (Forsyth & Evans, 2013). It should be noted that every adaptation follows the certain level of system limitation in order to adapt to new or expected climatic conditions, which achievement initiates loss of required level of protection against expected negative impact. In other words, predefined instruments of protection will start to lose its primary function (Klein et al., 2014).

In the light of food security, irrigation as an agro-technical instrument has been recognized as one of the best measures for adaptation of crop production to existing climate variability and expected climate change, unlike insurance as a financial instrument that has the main effect on the financial viability of production (Barenklau, 2001).

Unfortunately, despite the given climate conditions in the Republic of Serbia, soybean is predominantly produced in the system of dry land farming. For this reason, it is necessary to assess the effects of irrigation (by calculating variable costs) on achieved yields in soybean production in Serbian climatic environment (Table 9 and Appendix 2), in other words to estimate if the implementation of irrigation initiates positive impact on current state of food security on macro level, as well as higher profitability of soybean production on micro level.

In both systems of soybean production (with and without irrigation), positive contribution margins were gained. Since variable costs in crop production are assigned to costs of inputs, variable part of the costs of mechanization and implemented irrigation system, costs of engaged labor, and variable part of general costs, it can be noted that in both cases costs of mechanization are dominant, while costs of fertilizers and pesticides have significant share. In the following tables, the authors presented the structure of certain segments of variable costs (Appendix 3, for the costs of mechanization and labor; Appendix 4, for fertilizers and pesticides; Table 10 for soybean irrigation).

Out of total variable costs, around 10.4% belong to the costs of irrigation, while costs of energy prevail in the structure of total irrigation costs with over 68% (Table 10).

Table 9. Baselines of the analysis

Line and production season: soybean, 2014/2015.	Territory: South Banat region / Pančevo city
Production unit: 1 ha	Production system: without/with irrigation

Source: Institute of Agricultural Economics [IAE], 2015

Table 10. Costs of soybean irrigation

Description	Quantity	Unit of measurement	Price per unit of measurement	Total, $/ha
Energy and other variable costs related to irrigation system	-	-	-	83.8
Tax for water used in irrigation*	1,575.0	m^3	0.001	1.6
Tax for use of public water facilities related to irrigation*				18.5
- for installed capacity	1	ha	11.2	11.2
- for engaged capacity	1,575.0	1,000 m^3	4.7	7.4
Total				122.5

Note: All calculations were based on the exchange rate of 1 USD = 110.4 RSD (according to the average exchange rate of the National Bank of Serbia as of September 1, 2016);

* Government of the Republic of Serbia, 2015b

Source: IAE, 2015

Based on the obtained production results, implementation of irrigation (Table 11 and Table 12) contributed to increase in yields by around 71%, increase in production value by over 62%, increase in variable costs by about 49%, and achieving almost 2.7 times higher contribution margin.

Use of irrigation in soybean production can greatly affect the increased sustainability of agricultural holdings, considering that essential financial assets to cover fixed costs and achieve positive financial result remain after covering variable costs. Besides, wider implementation of the observed agro-technical instrument can significantly contribute to ensuring food security at the national level.

Critical values of all elements in soybean production (with and without irrigation) at which the contribution margin equates to zero are presented in Table 16.

Table 11. Effects of irrigation use in soybean production

Description	Total	
	$/ha	%
Production without irrigation		
VP_0 (Production value)	896.1	*
VT_0 (Variable costs)	791.6	*
MP_0 [Contribution margin ($VP_0 - VT_0$)]	104.5	*
Production with irrigation		
VP_1 (Production value)	1,456.1	*
VT_1 (Variable costs)	1,176.7	*
MP_1 [Contribution margin ($VP_1 - VT_1$)]	279.4	*
Calculation of effects of irrigation		
$VP_1 - VP_0 = VP_p$ (Increase of production value)	560.0	62.5
$VT_1 - VT_0 = VT_p$ (Increase of variable costs)	385.1	48.6
$MP_1 - MP_0 = MP_p$ (Increase of contribution margin)	174.9	167.4

Note: All calculations were based on the exchange rate of 1 USD = 110.4 RSD (according to the average exchange rate of the National Bank of Serbia as of September 1, 2016).

Source: Calculations based on the data from the Table 14

Table 12. Critical values in soybean production (without or with irrigation)

Description	USD (kg)/ha	
	With irrigation	Without irrigation
Expected yield (OP)	3,850.0	2,250.0
Expected price (OC)	0.35	0.35
Subsidy (S)	108.6	108.6
Variable costs (VT)	1,176.7	791.6
Critical price: KC = (VT – p) / OP	0.28	0.30
Critical yield: KP = (VT – p) / OC	3,051.7	1,951.4
Critical variable costs: KVT = (OP * OC) + p	1,456.1	896.1

Note: All calculations were based on the exchange rate of 1 USD = 110.4 RSD (according to the average exchange rate of the National Bank of Serbia as of September 1, 2016).

Source: Calculations based on the data from the Table 14

It can be noticed that irrigation reduces the risks of business uncertainty that are potentially related to agricultural holdings (or to entire soybean production sector), especially risks imposed by the variability of weather conditions.

SOLUTIONS AND RECOMMENDATIONS

Throughout the chapter, it has been noted that systematic support in Serbia to introduce irrigation is limited. The authors have shown that contribution margin, income, and yields confirm necessity to deeper systematic approach to this issue. Having in mind growing world population and limited land resource, the only solution is to improve existing production system. Due to a fact that the average farm size is small in Serbia, landowners would find it difficult to finance irrigation on their own. Hence, the authors recommend the following measures to be undertaken:

1. Policymakers:
   a. Conduct in-depth research on irrigation needs;
   b. Provide larger incentives for irrigation;
   c. Facilitate gaining loans in direct cooperation with banks.
2. Ministry of Agriculture:
   a. Provide education to farm owners in the field of irrigation and financing;
   b. Continuously seek for the EU support through Instrument for Pre-Accession Assistance (IPA) and structural funds;
   c. Look for additional budgetary funds for support.
3. Landowners:
   a. Get educated on soy production, irrigation, and financial support;
   b. Introduce new varieties of soy into production;
   c. Combine irrigation system for production of different crops.

FUTURE RESEARCH DIRECTIONS

In this chapter, a basic model for assessment of losses caused by climate change (dominantly, drought) in crop production (soybean) has been analyzed. The effects of the introduction of irrigation (by use of calculation of contribution margin) on the potentially achieved production results at micro level were also calculated. There is a potential for future research in finding out the size of total losses caused by drought in entire crop production (all production lines) at the national level, as well as to estimate to what extent they can be reduced after implementation of irrigation systems at suitable areas, in order to better assess the national potentials (production limits) for maintaining the food security in crop products.

CONCLUSION

There is no doubt that plant production in the Republic of Serbia is increasingly hampered by negative impacts of climate change. In crop production, changes in temperature regime, as well as changes in level and spatial and temporal distribution of rainfalls are particularly important. They are primarily reflected in a number of crop losses caused by the occurrence of more frequent and more intense droughts, and indirectly by the decrease of the national level of food security.

The example of negative impact is given by the results gained in soybean production. Estimations show that cumulative losses in production of this particular crop during the previous decade range from \$115 million to \$520 million (between 0.3–1.5 million tons). Based on the analysis of the obtained results in the soybean production at the micro (farm) level, it can be concluded that the implementation of irrigation would greatly contribute to the growth of profitability of certain agricultural holdings which dominantly produce this crop.

The claim is based on the fact that implementation of the irrigation system initiates a significant increase in yields (by about 62%), income (by almost 49%), and contribution margin (more than 2.5 times). It assures that after covering total variable costs there will be enough money for coverage of fixed costs and gaining positive financial results. It should also be mentioned that variable costs of irrigation in the structure of total variable costs have a relatively low share (about 10%).

On the other hand, the wider use of irrigation in the production of soybean and other field crops (wherever it is in line with terrain configuration, suitable access to water resources, state of physical infrastructure, size and compactness of the holding, etc.) will certainly contribute to reduction of potential losses caused by drought in the future, as well as give an impulse to improvement of national food security, strengthening and better utilization of processing capacities, livestock production, and agri-food export.

ACKNOWLEDGMENT

This chapter is a part of project III 46006 and 179015, funded by the Ministry of Education, Science and Technological Development of the Republic of Serbia, for the project period 2011-2017.

REFERENCES

Arnell, N. W., Brown, S., Gosling, S. N., Hinkel, J., Huntingford, C., Lloyd-Hughes, B., & Zelazowski, P. et al. (2016). Global-Scale Climate Impact Functions: The Relationship between Climate Forcing and Impact. *Climatic Change, 134*(3), 475–487. doi:10.1007/s10584-013-1034-7

Audsley, E., Pearn, K. R., Simota, C., Cojocaru, G., Koutsidou, E., Rousevell, M. D. A., & Alexandrov, V. et al. (2006). What Can Scenario Modelling Tell Us about Future European Scale Agricultural Land Use, and What Not? *Environmental Science & Policy, 9*(2), 148–162. doi:10.1016/j.envsci.2005.11.008

Balambal, R. V. (2015). Ethical Values of Food Safety. In S. Hongladarom (Ed.), *Food Security and Food Safety for the XXI Century* (pp. 33–44). Heidelberg, Germany: Springer.

Barenklau, K. E. (2001). *Agricultural Safety.* Boca Raton, FL: Lewis Publishers, CRC Press LLC. doi:10.1201/9781420032772

Brooks, N., Adger, W. N., & Kelly, P. M. (2005). The Determinants of Vulnerability and Adaptive Capacity at the National Level and Implications for Adaptation. *Global Environmental Change, 15*(2), 151–163. doi:10.1016/j.gloenvcha.2004.12.006

Cooperation Union of Vojvodina. (2013). *Pricelist of Services with Mechanisation in 2013.* Novi Sad: Cooperation Union of Vojvodina.

Đorđević, V., & Nenadić, N. (1980). *Soja.* Belgrade: Nolit.

Dragović, S. (2008). Effect of Irrigation on Field Crops Yield under the Variable Agro-Climatic Conditions of Serbia. *Agriculture & Forestry, 54*(1), 25–40.

European Commission. (2007). *Adapting to Climate Change in Europe – Options for EU Action.* Brussels: European Commission.

FAOSTAT. (2016): *Soybean Production at the World Level.* Retrieved November 21, 2016, from http://faostat.fao.org/site/567/DesktopDefault.aspx?PageID=567#ancor

Food and Agriculture Organization of the United Nations. (2003). *Trade Reforms and Food Security: Conceptualising the Linkages.* Rome: Food and Agriculture Organization of the United Nations.

Forsyth, T., & Evans, N. (2013). What is Autonomous Adaption? Resource Scarcity and Smallholder Agency in Thailand. *World Development, 43*, 56–66. doi:10.1016/j.worlddev.2012.11.010

Fuhrer, J., Beniston, M., Fischlin, A., Frei, C., Goyette, S., Jasper, K., & Pfister, C. (2006). Climate Risks and Their Impact on Agriculture and Forests in Switzerland. *Climatic Change, 79*(1–2), 79–102. doi:10.1007/s10584-006-9106-6

Giannakopoulos, C., Le Sager, P., Bindi, M., Moriondo, M., Kostopoulou, E., & Goodess, C. M. (2009). Climatic Changes and Associated Impacts in the Mediterranean Resulting from a 2C Global Warming. *Global and Planetary Change, 68*(3), 209–224. doi:10.1016/j.gloplacha.2009.06.001

Government of the Republic of Serbia. (2009). *Law on Genetically Modified Organisms (Publication no. 41/09).* Belgrade: Official Gazette of the Republic of Serbia.

Government of the Republic of Serbia. (2015a). *Regulation on Subsidies Allocation in Agriculture and Rural Development in 2015 (Publication no. 19/15).* Belgrade: Official Gazette of the Republic of Serbia.

Government of the Republic of Serbia. (2015b). *Regulation on Water Fees for 2015 (Publication no. 15/15).* Belgrade: Official Gazette of the Republic of Serbia.

Institute of Agricultural Economics. (2015). *Results of In-depth Interview with Members of Selected Agricultural Holdings. Internal documentation.* Belgrade: Institute of Agricultural Economics.

Jančić, M. (2015). *Uticaj klimatskih promena na biljnu proizvodnju* (Doctoral dissertation). University of Novi Sad, Novi Sad.

Klein, R. J., Midgley, G. F., Preston, B. L., Alam, M., Berkhout, F. G. H., Dow, K., & Shaw, M. R. (2014). Adaptation Opportunities, Constraints, and Limits. In C. B. Field et al. (Eds.), *Climate Change 2014: Impacts, Adaptation, and Vulnerability, Part A: Global and Sectoral Aspects. Contribution of Working group II to V Assessment Report of IPCC* (pp. 899–943). Cambridge, UK: Cambridge University Press.

Kovačević, V. (2014). *Razvoj robnih berzi u funkciji upravljanja rizikom poslovanja poljoprivrednih preduzeća u Srbiji* (Doctoral dissertation). University of Belgrade, Belgrade, Serbia.

Lalić, B., Stričević, R., Janković, D., Jaćimović, G., & Firanj, A. (2015). *Prilagođavanje na nove klimatske uslove: Zagrevanje useva.* Report presented at Dialogue on Climate Change in Serbia: Towards the Conference in Paris, Belgrade. Retrieved March 28, 2016, from http://www.klimatskepromene.rs/uploads/useruploads/Documents/Zagrevanje-useva_B.Lalic.pdf

Lobell, D., & Field, C. (2007). Global Scale Climate–Crop Yield Relationships and the Impacts of Recent Warming. *Environmental Research Letters, 2*(1), 1–7. doi:10.1088/1748-9326/2/1/014002

Miladinović, J., Hrustić, M., & Vidić, M. (2011). *Soybean.* Novi Sad: Institute of Field and Vegetable Crops.

Ministry of Agriculture and Environmental Protection of the Republic of Serbia. (2016). *Balance of Soybean and Basic Soybean Products Production/Consumption at the Level of the Republic of Serbia.* Retrieved November 21, 2016, from www.mpzzs.gov.rs/dokumenti/

Nastić, L., Jeločnik, M., & Subić, J. (2014). Analysis of Calla Lily and Cucumber Production in Greenhouse. *Ekonomika, 60*(4), 209–217.

National Drought Mitigation Centre. (2016). *Types of Drought Impacts.* Lincoln, NE: National Drought Mitigation Centre, University of Nebraska.

Nelson, G. C., Rosegrant, M. W., Koo, J., Robertson, R., Sulser, T., Zhu, T., … Lee, D. (2009). Climate Change: Impact on Agriculture and Costs of Adaptation. Washington, DC: International Food Policy Research Institute.

Nelson, G. C., Van der Mensbrugghe, D., Ahammad, H., Blanc, E., Calvin, K., Hasegawa, T., & Willenbockel, D. et al. (2014). Agriculture and Climate Change in Global Scenarios: Why dont the Models Agree. *Agricultural Economics, 45*(1), 85–101. doi:10.1111/agec.12091

Papić Brankov, T., & Lovre, K. (2011). Investiranje istraživanja i razvoja u oblasti poljoprivredne biotehnologije. *Ekonomika*, *57*(1), 30–36.

Papić Brankov, T., & Milovanović, M. (2015). Measuring Food Security in the Republic of Serbia. *Economics of Agriculture*, *62*(3), 801–812.

Parry, M. L., Rosenzweig, C., Iglesias, A., Fischer, G., & Livermore, M. (1999). Climate Change and World Food Security: A New Assessment. *Global Environmental Change: Human and Policy Dimensions*, *9*(supplement 1), 51–67. doi:10.1016/S0959-3780(99)00018-7

Reidsma, P., Wolf, J., Kanellopoulos, A., Schaap, B. F., Mandryk, M., Verhagen, J., & Van Ittersum, M. K. (2015). Climate Change Impact and Adaptation Research Requires Integrated Assessment and Farming Systems Analysis: A Case Study in the Netherlands. *Environmental Research Letters*, *10*(4). doi:10.1088/1748-9326/10/4/045004

Republican Hydrometeorological Service of Serbia. (2006). *Agrometeorological Conditions in Production Cycle 2005/2006 at the Territory of the Republic of Serbia*. Belgrade: Republican Hydrometeorological Service of Serbia.

Republican Hydrometeorological Service of Serbia. (2008a). *Agrometeorological Conditions in Production Cycle 2006/2007 at the Territory of the Republic of Serbia*. Belgrade: Republican Hydrometeorological Service of Serbia.

Republican Hydrometeorological Service of Serbia. (2008b). *Agrometeorological Conditions in Production Cycle 2007/2008 at the Territory of the Republic of Serbia*. Belgrade: Republican Hydrometeorological Service of Serbia.

Republican Hydrometeorological Service of Serbia. (2009). *Agrometeorological Conditions in Production Cycle 2008/2009 at the Territory of the Republic of Serbia*. Belgrade: Republican Hydrometeorological Service of Serbia.

Republican Hydrometeorological Service of Serbia. (2010). *Agrometeorological Conditions in Production Cycle 2009/2010 at the Territory of the Republic of Serbia*. Belgrade: Republican Hydrometeorological Service of Serbia.

Republican Hydrometeorological Service of Serbia. (2011). *Agrometeorological Conditions in Production Cycle 2010/2011 at the Territory of the Republic of Serbia*. Belgrade: Republican Hydrometeorological Service of Serbia.

Republican Hydrometeorological Service of Serbia. (2012). *Agrometeorological Conditions in Production Cycle 2011/2012 at the Territory of the Republic of Serbia*. Belgrade: Republican Hydrometeorological Service of Serbia.

Republican Hydrometeorological Service of Serbia. (2013). *Agrometeorological Conditions in Production Cycle 2012/2013 at the Territory of the Republic of Serbia*. Belgrade: Republican Hydrometeorological Service of Serbia.

Republican Hydrometeorological Service of Serbia. (2014). *Agrometeorological Conditions in Production Cycle 2013/2014 at the Territory of the Republic of Serbia*. Belgrade: Republican Hydrometeorological Service of Serbia.

Republican Hydrometeorological Service of Serbia. (2015). *Agrometeorological Conditions in Production Cycle 2014/2015 at the Territory of the Republic of Serbia*. Belgrade: Republican Hydrometeorological Service of Serbia.

Roljević, S., Subić, J., & Jeločnik, M. (2009). Importance of Soybean Production as Safe Food in Conditions of Economic Crisis. In *Proceedings of Competitivitatea economiei agrolimentare si rurale in conditile crizei mondiale*. Bucharest: Academiei de Studii Economice.

Smit, B., & Skinner, M. W. (2002). Adaptation Options in Agriculture to Climate Change: A Typology. *Mitigation and Adaptation Strategies for Global Change*, *7*(1), 85–114. doi:10.1023/A:1015862228270

Song, B., Marchant, M. A., & Xu, S. (2006). Competitive Analysis of Chinese Soybean Import Suppliers: U.S., Brazil, and Argentina. In *Proceedings of the American Agricultural Economics Association Annual Meetings*. Long Beach, CA: American Agricultural Economics Association.

Spiertz, H. (2012). Avenues to Meet Food Security. The Role of Agronomy on Solving Complexity in Food Production and Resource Use. *European Journal of Agronomy*, *43*, 1–8. doi:10.1016/j.eja.2012.04.004

Stamenković, I., Bošnjak, D., & Stamenković, A. (2007). Geographic Characteristics of Soya Production in Agricultural Areas of Vojvodina Region. *Geographical Journal – Geographica Pannonica*, *10*, 78–84.

Statistical Office of the Republic of Serbia. (2008). *Statistical Yearbook of the Republic of Serbia – 2008*. Belgrade: Statistical Office of the Republic of Serbia.

Statistical Office of the Republic of Serbia. (2011). *Statistical Yearbook of the Republic of Serbia – 2011*. Belgrade: Statistical Office of the Republic of Serbia.

Statistical Office of the Republic of Serbia. (2014). *Statistical Yearbook of the Republic of Serbia – 2014*. Belgrade: Statistical Office of the Republic of Serbia.

Statistical Office of the Republic of Serbia. (2015). *Statistical Yearbook of the Republic of Serbia – 2015*. Belgrade: Statistical Office of the Republic of Serbia.

Statistical Office of the Republic of Serbia. (2016a). *Basic Data of Soybean Production in 2015*. Retrieved November 21, 2016, from http://webrzs.stat.gov.rs/WebSite/public/ReportView.aspx

Statistical Office of the Republic of Serbia. (2016b). Monthly Statistical Bulletin, vol. LXV, no. 12/2015. Belgrade: Statistical Office of the Republic of Serbia.

Subić, J., Jeločnik, M., & Zubović, J. (2015). Primena navodnjavanja kao agrotehničke mere – analiza marže pokrića u proizvodnji kukuruza. *Ecologica*, *78*(22), 245–251.

Tol, R. S. J. (2002). Estimates of the Damage Costs of Climate Change. Part 1: Benchmark Estimates. *Environmental and Resource Economics*, *21*(1), 47–73. doi:10.1023/A:1014500930521

Vasiljević, Z., & Subić, J. (2010). Importance of the Costs Calculation at the Family Farms in Serbia. In D. Tomic & M. Sevarlic (Eds.), *Agriculture in Late Transition – Experience of Serbia* (pp. 123–138). Belgrade: Serbian Association of Agricultural Economists.

Wolf, J. (2000). Modelling Climate Change Impacts at the Site-Scale on Soybean. In T. E. Downing, P. A. Harrison, R. E. Butterfield, & K. G. Lonsdale (Eds.), *Climate Change, Climatic Variability and Agriculture in Europe* (pp. 103–116). Oxfordshire, UK: Environmental Change Unit, Oxford University.

ADDITIONAL READING

Adams, R. M., Hurd, B. H., Lenhart, S., & Leary, N. (1998). Effects of Global Climate Change on Agriculture: An Interpretative Review. *Climate Research, 11*, 19–30. doi:10.3354/cr011019

Balaghi, R., Badjeck, M. C., Bakari, D., De Pauw, E., De Wit, A., Defourny, P., & Tychon, B. et al. (2010). Managing Climatic Risks for Enhanced Food Security: Key Information Capabilities. *Procedia Environmental Sciences, 1*, 313–323. doi:10.1016/j.proenv.2010.09.020

Cassman, K. G., & Harwood, R. R. (1995). The Nature of Agricultural Systems: Food Security and Environmental Balance. *Food Policy, 20*(5), 439–454. doi:10.1016/0306-9192(95)00037-F

Hamelin, A. M., Habicht, J. P., & Beaudry, M. (1999). Food Insecurity: Consequences for the Household and Broader Social Implications. *The Journal of Nutrition, 129*(2), 525–528. PMID:10064323

Hanjra, M. A., & Qureshi, M. E. (2010). Global Water Crisis and Future Food Security in an Era of Climate Change. *Food Policy, 35*(5), 365–377. doi:10.1016/j.foodpol.2010.05.006

Ingram, J. (2011). A Food Systems Approach to Researching Food Security and its Interactions with Global Environmental Change. *Food Security, 4*(3), 417–431. doi:10.1007/s12571-011-0149-9

Jančić, M., Lalić, B., Mihailović, D., & Jaćimović, G. (2015). Impact of Climate Change and Carbon Dioxide Fertilization Effect on Irrigation Water Demand and Yield of Soybean in Serbia. *The Journal of Agricultural Science, 153*(08), 1365–1379. doi:10.1017/S0021859615000179

Kang, Y., Khan, S., & Ma, X. (2009). Climate Change Impacts on Crop Yield, Crop Water Productivity and Food Security – A Review. *Progress in Natural Science, 19*(12), 1665–1674. doi:10.1016/j.pnsc.2009.08.001

Müller, C., Elliott, J., Chryssanthacopoulos, J., Deryng, D., Folberth, C., Pugh, T., & Schmid, E. (2015). Implications of Climate Mitigation for Future Agricultural Production. *Environmental Research Letters, 10*(12), 125004. http://iopscience.iop.org/article/10.1088/1748-9326/10/12/125004 RetrievedNovember212016 doi:10.1088/1748-9326/10/12/125004

Sage, C. (2013). The Interconnected Challenges for Food Security from a Food Regimes Perspective: Energy, Climate, and Malconsumption. *Journal of Rural Studies, 29*(1), 71–80. doi:10.1016/j.jrurstud.2012.02.005

Vermeulen, S. J., Aggarwal, P. K., Ainslie, A., Angelone, C., Campbell, B. M., Challinor, A. J., & Wollenberg, E. et al. (2012). Options for Support to Agriculture and Food Security under Climate Change. *Environmental Science & Policy*, *15*(1), 136–144. doi:10.1016/j.envsci.2011.09.003

Wiebe, K., Lotze Campen, H., Sands, R., Tabeau, A., van der Mensbrugghe, D., & Biewald, A. … Willenbockel, D. (2015). Climate Change Impacts on Agriculture in 2050 under a Range of Plausible Socioeconomic and Emissions Scenarios. *Environmental Research Letters, 10*(8). Retrieved November 21, 2016, from http://iopscience.iop.org/article/10.1088/1748-9326/10/8/085010/meta

Wilcox, J. R. (Ed.). (1987). *Soybeans Improvement, Production and Uses* (2nd ed.). Madison, WI: American Society of Agronomy, University of Wisconsin.

KEY TERMS AND DEFINITIONS

Adaptation Measures to Climate Change in Agriculture: Agriculture is inherently sensitive to climate conditions, representing one of the most vulnerable sectors to the risks and impacts of global climate change. In previous decades, number of researches determined possible adaptation options (measures) in agriculture that have to increase its adaptive capacity by modifying farming practices, improving crops and livestock through breeding and investing in new technologies, management approaches, and used infrastructure. Nowadays, it is considered as an important element of climate change impact and vulnerability assessment, as well as one of the policy options in response to climate change impacts.

Climate Change: It refers to a change in the state of the climate that can be identified (after using statistical tests) by changes in the mean and/or the variability of its properties, and that persists for an extended period, typically decades or longer. It refers to any change in climate over time, whether due to natural variability or because of human activity. Recently, the definition implies a change of climate that is attributed directly or indirectly to human activity that alters the composition of the global atmosphere and that is in addition to natural climate variability observed over comparable time periods.

Drought: As a phenomenon, the drought could be described from an aspect of meteorology as a prolonged period of abnormally low rainfall in some region, leading to a shortage of available water. Agricultural drought relates to the availability of required amount of water during the different growing stages of plants. Hydrological drought is usually linked with meteorological drought and refers to persistently low water volumes in surface and underground water sources. Socioeconomic drought occurs when the demand for an economic good exceeds supply because of a weather-related shortfall in water supply.

Food Safety: It refers to the conditions and practices that preserve the quality of food. These practices prevent contamination and foodborne illnesses. Food safety is about producing, handling, storing, and preparing food to prevent infection and help to make sure that food keeps enough nutrients for people in terms of healthy diet. Unsafe food and water mean that they have been exposed to mechanical and chemical dirt and germs, or may even be rotten, which can cause many infections or diseases in humans and animals.

Food Security: A condition in which all people, at all times, have physical, social, and economic access to sufficient safe and nutritious food that meets their dietary needs and food preferences for an active and healthy life. Over the coming decades, changing the climate, growing global population, rising food prices, and environmental stressors will have significant yet highly uncertain impacts on food security.

Irrigation: An application of water to land in order to supply crops and other plants with necessary water (sometimes including nutrients). This is an artificial application of water to land to assist in the growing of crops and pastures.

Soybean: Glycine max L., is an annual herbaceous plant in the Fabaceae (legume or bean family) grown in diverse climates worldwide that originated from South-Eastern Asia. It was domesticated more than 3,000 years ago for its edible seeds and young pods, and now is the world's most important legume crop (the most widely produced oilseed and protein crop) used in human and animal nutrition.

APPENDIX 1

Table 13. Soybean production in the Republic of Serbia and other countries in 2006-2014

Country/ Element	Year									Average	Cumulative
	2006	2007	2008	2009	2010	2011	2012	2013	2014		
Albania											
Harvested area (ha)*	200	200	300	300	300	324	300	100	207	248	*
Yield (tons/ha)**	2.0	2.0	2.0	1.7	1.7	1.8	1.3	1.0	1.7	1.7	*
Production (tons)*	400	400	600	500	500	600	400	100	350	428	4,278
Macedonia											
Harvested area (ha)*	600	300	100	118	306	528	332	97	140***	280	*
Yield (tons/ha)**	1.5	1.3	2.0	1.6	2.3	1.6	2.2	2.4	2.4	1.9	*
Production (tons)*	900	400	200	192	699	840	717	237	340***	503	5,028
Montenegro											
Harvested area (ha)*	*	*	*	*	*	*	*	*	*	*	*
Yield (tons/ha)**	*	*	*	*	*	*	*	*	*	*	*
Production (tons)*	*	*	*	*	*	*	*	*	*	*	*
Bosnia and Herzegovina											
Harvested area (ha)*	6,355	5,416	4,187	3,802	4,037	3,884	5,325	4,474	4,186	4,630	*
Yield (tons/ha)**	2.0	1.5	2.0	2.1	2.0	1.7	1.3	1.8	2.1	1.8	*
Production (tons)*	12,842	8,084	8,412	8,193	8,007	6,748	6,708	7,964	9,020	8,442	84,420
Croatia											
Harvested area (ha)*	62,810	46,506	35,789	44,292	56,456	58,896	54,109	47,156	47,104	50,346	*
Yield (tons/ha)**	2.8	1.9	3.0	2.6	2.7	2.5	1.8	2.4	2.8	2.5	*
Production (tons)*	174,214	90,637	107,558	115,159	153,580	147,271	96,718	111,316	131,424	125,320	1,253,197
Slovenia											
Harvested area (ha)*	226	126	49	66	109	107	140	278	404	167	*
Yield (tons/ha)**	2.3	2.8	2.8	2.9	2.7	2.7	2.4	1.7	2.6	2.5	*
Production (tons)*	527	348	137	193	290	292	343	463	1,046	404	4,043
Hungary											
Harvested area (ha)*	35,911	32,900	28,259	31,508	37,688	41,009	40,910	42,350	42,980	37,057	*
Yield (tons/ha)**	2.4	1.6	2.6	2.4	2.3	2.3	1.7	1.9	2.7	2.2	*
Production (tons)*	85,035	53,982	74,000º	75,587	85,440	94,955	67,730	78,760	115,600	82,136	813,225
Romania											
Harvested area (ha)*	177,481	109,314	46,135	48,249	63,424	71,861	77,927	67,409	79,275	82,342	*
Yield (tons/ha)**	1.9	1.2	2.0	1.7	2.4	2.0	1.3	2.2	2.6	1.9	*
Production (tons)*	344,909	136,094	90,579	84,268	149,940	142,636	104,330	149,931	202,892	156,175	1,561,754
Bulgaria											
Harvested area (ha)*	88	113	134	414	725	588	248	336	305	328	*
Yield (tons/ha)**	1.5	0.6	1.3	1.0	2.3	1.2	0.8	1.8	2.4	1.4	*
Production (tons)*	137	68	179	400	1,647	680	196	604	736	516	5,163

continued on following page

Table 13. Continued

Country/ Element	Year									Average	Cumulative
	2006	2007	2008	2009	2010	2011	2012	2013	2014		
Serbia											
Harvested area (ha)*	156,680	146,988	143,684	144,386	170,359	165,253	162,714	159,724	154,249	156,004	*
Yield (tons/ha)**	2.7	2.1	2.4	2.4	3.2	2.7	1.7	2.4	3.5	2.6	*
Production (tons)*	429,639	303,950	350,946	349,193	540,859	440,847	280,638	385,214	545,898	403,020	4,030,204

Note: * Official data; ** Calculated data; *** estimation of the FAO; ° Unofficial data.
Source: FAOSTAT, 2016

APPENDIX 2

Table 14. Contribution margin in soybean production with or without irrigation in 2014-2015, $

Description	Quantity	Unit of measurement	Price per unit of measurement, $	Total, $/ha
Production in the system of dry land farming				
I – Income (VP_0)				
Soybean grain[1]	2,250	kg	0.35[2]	787.5
Subsidy (total)[4]	*	*	108.6	108.6
Total				896.1
II – Variable costs (VT_0)				
Seed (with Nitragin)[3]	100.0	kg	0.55	55.0
Fertilisers				213.0
Pesticides				137.8
Costs of mechanization				369.7
Costs of engaged labor				16.1
Total				791.6
Contribution margin (MP_0) (I-II)				104.5
Production with irrigation				
I – Income (VP_1)				
Soybean grain[1]	3,850	kg	0.35[2]	1,347.5
Subsidy (total)[4]	*	*	108.6	108.6
Total				1,456.1
II – Variable costs (VT_1)				
Seed (with Nitragin)[3]	100.0	kg	0.55	55.0
Fertilisers				283.5
Pesticides				244.4
Costs of mechanization				452.9
Costs of engaged labor				18.4
Costs of irrigation				122.5
Total costs				1,176.7
Contribution margin (MP_1) (I-II)				279.4

Note: All calculations were based on the exchange rate of 1 USD = 110.4 RSD (according to the average exchange rate of the National Bank of Serbia as of September 1, 2016);

[1]Weather conditions which generally do not suit soybean production;

[2]Expected average purchase price of soybean in the year under consideration;

[3]Seeds are delivered by distributor;

[4]Basic subsidies in plant production (54.3 $/ha) increased with incentives for fuel and/or fertilizers (54.3 $/ha) (Government of the Republic of Serbia, 2015a).

Source: IAE, 2015

APPENDIX 3

Table 15. Costs of mechanization and labor (without and with irrigation)

Description	Quantity	Unit of measurement	Price per unit of measurement	Total, $/ha
Production within the system of dry land farming				
Costs of mechanization*				
Ploughing (30 cm)	1	ha	82.4	82.4
Transportation of mineral fertilizers	2	tour	4.5	9.0
Dispersal of mineral fertilizers	2	ha	13.1	26.2
Pre-sowing soil treatment	1	ha	22.2	22.2
Sowing	1	ha	19.9	19.9
Pesticide treatment	3	ha	22.6	67.8
Inter-row tilling	2	ha	17.7	35.4
Harvesting	1	ha	98.7	98.7
Transportation of grain (tractor trailer of 3t)	1	tour	8.1	8.1
Total				369.7
Costs of engaged labor				
Loading and unloading of mineral fertilizers	2	h	2.3	4.6
Dispersal of mineral fertilizers	2	h	2.3	4.6
Pesticide treatment (spraying)	3	h	2.3	6.9
Total				16.1
Production with irrigation				
Costs of mechanization*				
Manure loading with front tractor loader**	2	h	14.1	28.2
Transportation and dispersal of manure**	1	ha	28.5	28.5
Ploughing (30 cm)	1	ha	82.4	82.4
Transportation of mineral fertilizers	2	tour	4.5	9.0
Dispersal of mineral fertilizers	2	ha	13.1	26.2
Pre-sowing soil treatment	1	ha	22.2	22.2
Sowing	1	ha	19.9	19.9
Pesticide treatment	4	ha	22.6	90.4
Inter-row tilling	2	ha	17.7	35.4
Harvesting	1	ha	98.7	98.7
Transportation of grain (tractor trailer of 5t)	1	tour	12.0	12.0
Total				452.9
Costs of engaged labor				
Loading and unloading of mineral fertilizers	2	h	2.3	4.6
Dispersal of mineral fertilizers	2	h	2.3	4.6
Pesticide treatment (spraying)	4	h	2.3	9.2
Total				18.4

Note: All calculations were based on the exchange rate of 1 USD = 110.4 RSD (according to the average exchange rate of the National Bank of Serbia as of September 1, 2016);

* Decrease of costs differences (caused by non-unified mechanization and equipment) was done by calculation of mechanization costs based on the values expressed in Cooperation Union of Vojvodina (2013);

** As fertilization with manure is done every third year, each production cycle is burdened with one-third of costs of all activities related to manure.

Source: IAE, 2015

APPENDIX 4

Table 16. Costs of fertilizers and pesticides (without and with irrigation)

Description	Quantity	Unit of measurement	Price per unit of measurement	Total, $/ha
Production within the system of dry land farming				
Mineral fertilizers*				
NPK 15:15:15	350.0	kg	0.42	147.0
KAN 27%	200.0	kg	0.33	66.0
Total				213.0
Pesticides**				
Velton WG	0.4	kg	37.3	14.9
Dual Gold 960EC	1.4	1	21.7	30.4
Savazon 480	1.5	1	11.5	17.2
Kvazar	1.2	1	29.9	35.9
Okvir	0.008	kg	600.1	4.8
ES Plus	0.2	1	10.0	2.0
Gallant Super	1	1	32.6	32.6
Total				137.8
Production with irrigation				
Mineral fertilizers*				
NPK 15:15:15	300.0	kg	0.42	126.0
UREA	100.0	kg	0.43	43.0
KAN 27%	150.0	kg	0.33	49.5
Manure***	6,500.0	kg	0.01	65.0
Total				283.5
Pesticides**				
Sencor Plus	0.4	1	34.4	13.8
Dual Gold 960EC	1.4	1	21.7	30.4
Pulsar 40	1.2	1	57.1	68.5
Harmony 75 WG	0.008	kg	628.4	5.0
Trend	0.2	1	10.0	2.0
Fusilade Forte	1.2	1	41.7	50.0
Basta 15	2.5	1	29.9	74.7
Total				244.4

Note: All calculations were based on the exchange rate of 1 USD = 110.4 RSD (according to the average exchange rate of the National Bank of Serbia at September 1, 2016);

* Mineral fertilizers are usually purchased from distributor of agricultural inputs from a territory of Pančevo city;

** Pesticides are usually purchased in agricultural pharmacy from a territory of Pančevo city;

*** Fertilization with bovine manure is done each third year in the volume of 20 ton/ha. Each production cycle is burdened with one-third of purchased manure value.

Source: IAE, 2015

Chapter 14

Local Production-Based Dietary Supplement Distribution in Emerging Countries: Bienestarina Distribution in Colombia

Jesus Gonzalez-Feliu
Mines Saint-Etienne, France

Laura Palacios-Arguello
Mines Saint-Etienne, France

Carlos Osorio-Ramírez
National University of Colombia, Colombia

Carlos Alberto Talamantes
Autonomous University of Ciudad Juarez, Mexico

ABSTRACT

The production and distribution of Bienestarina to the vulnerable population of Colombia is one of the strategies of the Colombian Institute of Familiar Wellness (ICBF) to fight malnutrition, especially among children. This case is a good example of establishing food security and social improvement logistics that merits particular attention. The chapter presents an analysis of the Bienestarina supply chain based on the four elements: steering, organization, development, and financial issues. First, an overview of social improvement logistics and the Bienestarina context is provided. Second, theoretical frameworks related to the case are presented. Third, the case is described on the basis of the proposed analysis framework. Finally, generalization issues and conclusions allow the authors proposing the first characterization of social improvement logistics.

INTRODUCTION

Food safety is one of the commitments which the Colombian government wants to ensure as a right of people (Dirección Nacional de Planeación, 2007). Indeed, the government must help the part of the Colombian population who is under malnutrition (i.e. due to nutrient deficiencies in their food), where children are considered as the most affected. Malnutrition of children generates several problems such as lack of concentration, and energy, decreased learning ability, and general delay in physical and mental

DOI: 10.4018/978-1-5225-2733-6.ch014

development, among others (Cuevas García, 2005). To fight malnutrition, different strategies can be implemented by national and regional entities (Food and Agriculture Organization of the United Nations [FAO], 2010). Since several initiatives focus on bringing unused food to the most sensible families (Maldonado & Moya, 2013) or on increasing families revenues, others (mainly in Latin America) deploy dietary complements or enriched food production-distribution systems addressed to those families.

In this context, the main products used to fight malnutrition are based on flours mixtures obtained from cereals that can be locally produced. Indeed, the rates of local (at least national or regional) production for those products are high (Rozo, 2000). However, several reports state that the distribution systems related to those products present deficiencies. It is important to consider those systems in a supply chain management perspective, in order to identify the main processes but also observe the evolutions of the integrated supply-production-distribution chain.

Over 30 years, Colombia has been implementing the strategy to improve food consumption of high nutritional value through the production and distribution of food complement based on mixtures of plant origin with high nutritional content, which is called Bienestarina. The purpose of this complement is to arrive in time to children, young, elderly, poor families, ethnic groups, and other population who require sufficient nutrients which their basic food do not provide sufficiently. Production and distribution of Bienestarina to the vulnerable population of Colombia is one of the strategies of the Colombian Institute of Familiar Wellness (ICBF) to combat malnutrition, especially among children.

The aim of this chapter is to investigate the current state of the Bienestarina logistics process from a supply chain viewpoint and focus on social aspects of logistics which are not related to commercialization of a product but to make it available to sensible populations. The authors start from the theoretical framework of the four pillars of the viability of a logistics project (Gonzalez-Feliu, Malhéné, Morganti, & Morana, 2014) and examine them in the case of the Bienestarina distribution network in Colombia.

This chapter is organized as follows. First, the background and context of the research are presented. Then, the methodological issues are provided. After that, the main results of the research are summarized and discussed. In the conclusion section, practical implications of those results and further developments are proposed.

BACKGROUND

Food security is a major issue since decades and takes a special interest in developing countries, where malnutrition and hunger are one of the first causes of mortality (Valdes, 1981; Reutlinger, 1986). However, food security is in general in competition to the industrialization and performance-making of productive systems and agro-industrial developments of such countries, mainly related either to feed developed countries or to produce non-food agricultural and agro-industrial products, such as biofuels (Ewing & Msangi, 2009) and textile fibers (Fortucci, 2002), among others. In opposition to the development of such industries, the needs of local inhabitants to nourish and develop themselves, in seeking higher welfare status, feeds the public debate on the place of social improvement in the economic and sustainable development of a country (Braun & Kennedy, 1994).

The issue of social improvement is a major strategic point in emerging countries. Indeed, with the high levels of poverty and malnutrition that characterize those countries, one of the main objectives of governments, along with economic development, is an improvement of the quality of life of people. Moreover, those countries receive particular attentions of non-governmental organizations (NGOs) and other humanitarian organizations. Although humanitarian logistics is a popular subject of research (Kovács & Spens, 2007, 2009; Holguin-Veras, Jaller, Van Wassenhove, Pérez, & Wachtendorf, 2012; Holguin-Veras, Wachtendorf, Jaller, & Jefferson, 2013), it focuses on emergency response logistics and post-disasters organization, with a low scientific interest for social improvement logistics. Indeed, the first statement on the importance of logistics in supporting social improvement is found in Orbell and Dawes (1993). After that, only a few works deal with logistics contributing to the social improvement. To the best of the authors' knowledge, Stock (1990) was the first who emphasized the poor contribution of logistics research in relation to social welfare. Adivar, Atan, Sevil Oflaç, and Örten (2010) focus on the notion of the value chain for social improvement and welfare. Finally, Maldonado and Moya (2013) show the importance of combining social improvement policies with reverse logistics in developing and improving food banks actions and logistics schemes. Those authors identify different ways to reduce hunger and malnutrition, as, for example, development of food banks, different ways of increasing incomes or reducing food acquisition costs, or free distribution of dietary complements of different nature and actions.

According to the FAO (2010, 2011), deployment of the actions on improvement dietary composition of daily meals in families belonging to sensible population is crucial to their development since they decrease hunger and malnutrition. To this fact, an initiative called Latin America and the Caribbean without Hunger was launched by the FAO (2011). In this initiative, several types of stakeholders and actions were deployed. From this initiative, it was concluded that several measures could be implemented to combat malnutrition (Maldonado & Moya, 2013). The initiatives may be aggregated into five major categories:

1. Primary economic improvement: in this category, the authors included initiatives that allow the families' income increase to purchase basic food and nutritional elements. The main initiatives in this category are subsidies and income transfer actions. Those initiatives are related to financial transfers, without direct relation to logistics.
2. Cost reduction to increase accessibility to food: initiatives in this category aim to reducing the cost of basic foods for sensible groups of populations by giving income not to families but to vendors in order to make basic foods less expensive for beneficiaries. The most known examples in this category are subsidies to retailers, tax exemptions, or programs of free (or cheaper) basic baskets. Anyway, those initiatives do not have (or little) impact on logistics and freight transportation. They can indirectly increase logistic flows of retailers but do not need to deploy specific distribution channels and supply chains.
3. Food access initiatives: they aim to increase access of sensible groups of population to basic food by providing them a set of basic products. In this category, the initiatives are more heterogeneous, and the main examples are the programs on delivering healthy food to hospitals, nurseries, and schools, deployment of social retailers (which have a specific logistics based on reverse flow collection), food banks, or distribution of dietary complements to families. In all those cases, specific supply chains are needed (or improvements of existing channels to take into account reverse flows or new products). They have a direct impact on supply, production, and distribution flows.
4. Education and monitoring initiatives: this category includes all initiatives that support and promote good habits, but also those that aim to steer and follow the risk of families and sensible groups of the population regarding malnutrition. They are, like categories 1 and 2, not directly related to logistics.

5. Promotion and development of self-production: the last, but not the least, the category is related to those actions that help families in sensible groups of population to satisfy some part or entire of their nutritional needs by themselves by providing them the agricultural resources (or access to them) to develop self-production. Although both creation and distribution of those resources need a logistical organization (sometimes being specific to those initiatives), the focus of this category is not on the distribution of products but on promotion their production for self-consumption.

As seen above, from those five categories of initiatives, two of them are related to logistic issues, but only one directly deploys specific logistic channels with an aim to distribute products in order to help sensible groups of populations, not to make profits. This chapter focuses on one of the initiatives in this category, i.e. distribution of dietary complements. The authors focus on this strategy for two reasons. The first is that it is one of the main practical solutions for emerging countries to reduce malnutrition. The second is that it has received small attention in the scientific literature.

Moreover, the case of Bienestarina in Colombia (a public-supported dietary complement addressed to children, mothers, and other inhabitants of sensible groups of the population) has received a particular attention recently (Palacios, Morana, Gonzalez-Feliu, & Devia, 2016; Peñaloza, Palacios, & Gonzalez-Feliu, 2016). The present chapter synthesizes those works and proposes a systemic vision of the distribution system of Bienestarina in relation to local production and development of local economies.

In Colombia, the government, via the Colombian Institute of Family Welfare (Instituto Colombiano del Bienestar Familiar, or ICBF), developed a series of programs to combat malnutrition by providing the families with the highest risk in terms of their dietary complement, mainly addressed to childhood. Indeed, since 1976, the ICBF produces the Bienestarina product as a dietary complement of high nutritional value as a strategy to strengthen the fundamental right to food security in the country (Dirección Nacional de Planeación, 2006). In early 2013, the ICBF launched the Bienestarina Más program, which proposed an improved product reformulation (Presidencia de la República, 2013), aimed at vulnerable Colombian population, especially among children. This product benefits children, teenagers, pregnant women, mothers, different ethnic groups, families, and seniors belonging to levels 1 and 2 of the System for the Selection of Beneficiaries for Social Programs (SISBEN), the Colombian national system for identification of beneficiaries of social subsidies (Peñaloza et al., 2016). It classifies people according to their socio-economic level into six levels, where level 1 is for homeless people and extreme poverty, while level 6 is the highest level of affluence.

The importance of this research is based on the fact that in recent years there have been irregularities in the management of Bienestarina involving distribution processes, product traceability, lead time distribution to the beneficiary (end consumer), among others. One example is the case of product expiration, according to El Nuevo Siglo (2013). 190 expired product packages were reported in the period of January-June 2012. Moreover, opposing European and North complementary food, which is based on extraction and biochemical products, Bienestarina and other Latin-American dietary complements are based on flours. The particularity of Colombian Bienestarina is that almost all flour is of domestic origin, which promotes national production and ensures a basic level of food safety (for example, Colombian corn cannot be of transgenic nature by law).

The program is entirely steered by the ICBF, a public body which directly depends on the Colombian Ministry of Social Protection. Moreover, it has the particularity that the product is entirely produced in Colombia and then distributed by a combined network of national and regional facilities, showing an important contribution to local food production and distribution. For those reasons, the authors propose to describe the case of Bienestarina following a deductive case study methodology and focusing on the description of the distribution logistic network.

MAIN FOCUS OF THE CHAPTER

Theoretical and Methodological Framework

There are many frameworks to identify and characterize supply chains. Among the most known are the SCOR model (Supply Chain Council, 2008) and the Supply Chain characterization framework by Lambert, Cooper, and Pagh (1998). The first model represents various echelons of supply chain and main categories of the processes, the processes themselves, related needs, and main tasks derived from those needs and resources of companies. The second is a global conceptual representation which includes a structure of supply chain, its main stakeholders, and its technical and managerial components (Lambert, 2001).

Those models are mainly related to business and intra-organization logistics but remain the valid tools to examine supply chain management in organizations of any type. However, they focus on logistic processes and other associated activities only, completed at private and contractual levels only. An alternative to those frameworks can be found in industrial and system engineering, where a framework to examine the viability of industrial system can be related to four main elements: steering, organization, development, and financing (Gonzalez-Feliu, Taniguchi, & d'Arcier, 2014). The authors identified those four elements on a structure-based and economic viability goal for urban logistics initiatives, proposing the first analysis framework of urban freight distribution that addressed those four elements. The authors aim to extend this framework to food distribution, by re-defining the four elements as follows:

1. Steering structure defines various stakeholders involved in steering and monitoring the deployment of food distribution systems, as well as their relations and communication processes. Different actions aimed at consultation and main tools for steering are also included in this category.
2. Organization of the process is the set of elements that allow the distribution system to be conceived and designed, as well as to ensure its daily operations. In other words, this category defines the organization and management of the whole distribution chain and its related processes.
3. Development of the distribution system is the way in which the system has spread and/or evolved during the period from its starting point to the situation when it is operational. Therefore, it is important to focus on the actions performed and constraints to which the distribution system is a subject to better understand its evolution.
4. Financial strategies need to be addressed as well. Although in private systems those strategies are well set in advance and once the system is operated, it is clear how the investments will be refunded, but not so clear in public-based systems. Moreover, it is important to know who finances and pays for the system (mainly in social improvement logistics systems) to better understand the other three components.

From those four elements, the authors orient the analysis of the Bienestarina distribution to define its organizational model, as summarized in Figure 1.

Figure 1. Proposed framework for defining social improvement logistics systems
Source: Adapted from Gonzalez-Feliu, Malhéné, Morganti & Morana (2014)

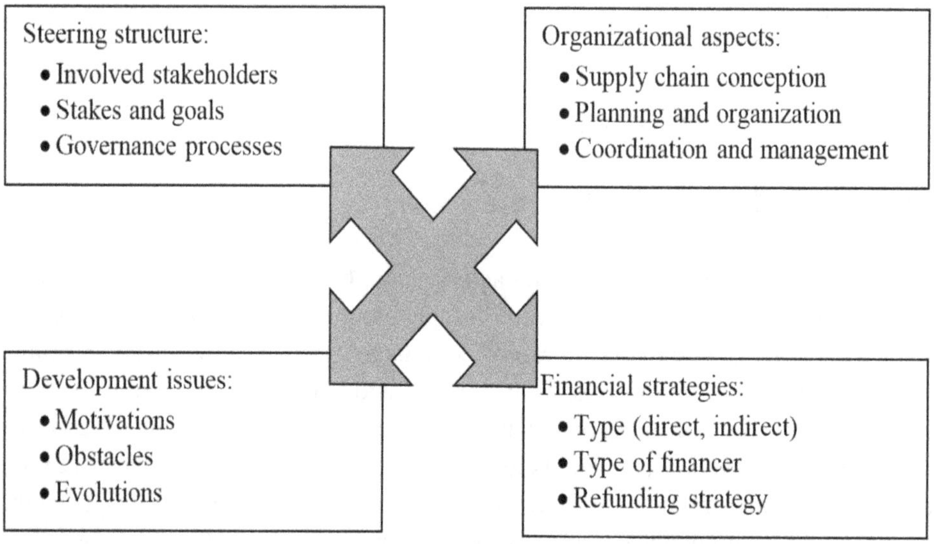

The case study has been developed via the collection of secondary and primary data to first characterize the distribution system (more details on the data collection and analysis for the characterization of the system is found in Peñaloza et al. (2016)), completed by the second data collection operation, mainly on the basis of interviews to the ICBF. Those data have been aggregated and synthesized following the theoretical framework above to build the deductive case study presented below.

The Proposed Case Study

In order to relate the theoretical framework proposed above to practice, the authors aim to develop the case study of Bienestarina distribution. Over 30 years, Colombia has been implementing the strategy to improve food consumption of high nutritional value through the production and distribution of food complements based on mixtures of plant origin with high nutritional content. This product, called Bienestarina, is addressed to populations under danger of malnutrition. In other words, the purpose of the production and distribution of this complement is to arrive in time to children, young, elderly, poor families, ethnic groups, and other population who require sufficient nutrients which their basic food do not provide sufficiently. Production and distribution of Bienestarina to the vulnerable population of Colombia is one of the strategies of the ICBF to combat malnutrition, especially among children.

This food complement, which is constituted of a mixture of flours of vegetable origin, such as corn starch, wheat, rice, soy, and milk powder, is a highly enriched source of calories, carbohydrates, proteins, natural fats, calcium, vitamins A and C, and other nutrients. It is a reinforcement to attack the problems of child malnutrition as a dietary complement and should not replace the basic food or breast milk. Bienestarina is also used as an input in the preparation of different foods such as cakes, cookies or bread, among others, which are also aimed at the target population in the different programs of the ICBF. The nutritional information about the current form of Bienestarina is presented in Table 1.

Table 1. Nutritional information for Bienestarina (per 100 g portion)

Component	Content
Energy (kcal)	360
Carbohydrates (g)	64.79
Protein (g)	20.145
Fat (g)	3.082
Total dietary fiber (g)	1.806
Calcium (mg)	700
Phosphorus (mg)	550
Iron (mg)	14.1
Vitamin A (UI)	2,000
Vitamin C (mg)	45
Niacin (mg)	7.3
Thiamine (mg)	0.5
Riboflavin (mg)	0,6
Vitamin B6 (mg)	0.9
Folic acid (mg)	160
Vitamin B12 (mg)	1.4
Zinc (mg)	8.3

Source: Colombian Institute of Familiar Wellness, 2009

The Bienestarina program was launched in 1976 by the Colombian Ministry of Social Protection (Dirección Nacional de Planeación, 2006). After that, a research on different vegetal mixed formulations, the first Bienestarina formulation, was produced and distributed in 1976. In 1989, the process of flour pre-cooking was deployed and operated. Other changes, including the ones in the formulation and the raw materials used, were introduced in 2000, 2002, 2004, and 2013. All those evolutions implied important changes in the production chain but not in the distribution one. Indeed, the distribution system is strongly dependent on the nature and activity of the subcontractor.

The entity in charge of programming, selling (to collective bodies), and managing the production and distribution of Bienestarina is the ICBF, and this for all programs. However, there are different programs and functions, and the ICBF is not able to manage every process by its own. It is why the logistic operations (for the entire supply chain of Bienestarina) are subcontracted to a private company. This subcontract is made by a public concession. Since December 2007, the ICBF granted the concession to the Industrias Del Maiz S.A. Corn Product Andina (IDM) Company, which deployed a branch, named Ingredion, which manages production and distribution of Bienestarina and other foods of high nutritional value.

The most recent program launched in 2013 is Bienestarina Más, which proposed an improved product, mainly enriched with vitamins and other essential elements. The ICBF defines the number and locations of beneficiaries, together with public bodies. Indeed, the ICBF defines a yearly production which is derived from a number of “credits”. Those credits have to be acquired by public entities (mainly municipalities, but also nurseries, hospitals, or other public institutions) who define a number of beneficiaries according to the number of credits acquired. Then, the ICBF plans a demand (making the Bienestarina supply chain being on pulled flows steering mode) and its characteristics (a type of products, quantities, and frequencies).

Knowing the demand, Ingredion is in charge of the production, which implies supply of raw materials and entire distribution process. Final destinations of the distribution process steered by Ingredion are delivery points where executing units (in general, operators in charge of bringing Bienestarina to beneficiaries) pick up the required demand.

Currently, the subcontractor (Ingredion S.A.) has two production plants, one located in Sabanagrande (Atlantico region, in the north of Colombia, near Barranquilla on the Atlantic side of the country) and Cartago (Valle region, in the west of the country, near Pereira and between Bogotá and the Pacific Ocean). The main raw products needed to produce Bienestarina are corn starch, wheat, rice, soy, and milk converted into milk powder before being mixed with other ingredients. All raw products are derived from national producers, which implies local processing and manufacturing for flours and milk powder, as well as for additives. Although it would be possible to use raw products issued from non-Colombian producers to make flours and additives, the specifications for the subcontracting impose criteria of quality, proximity, and reactivity which promote local production. However, taken into account the geography of Colombian agriculture, "local" production needs to be considered at a macro-regional level. This is also the case of milk powder, where production needs to be local to ensure the freshness of milk taken into account the need of not adding conservatives. From each production macro-region, local manufacturers produce flours and milk powder and supply them to the Bienestarina mixing plants to produce the final product.

Due to the location of the two plants, and the current number of beneficiaries (about 9 million inhabitants (Peñaloza, Palacios & Gonzalez-Feliu, 2016) spread on the entire territory of the country), the ICBF uses a division in eight macro-regions, according to the issues of regional distribution programs and operating costs of logistic, as shown in Figure 2.

To each macro-region, a number of first-tier warehouses (called primary distribution points by the ICBF) is associated. Those warehouses have the function of reception and managing the inventories to distribute the assigned beneficiaries for the entire macro-region. The demand management is made by the ICBF in function of the number of credits bought by public entities at each final destination, and then the monthly production and distribution plans. However, the inventorying operations take place at first-tier warehouses to ensure a higher inventorying capacity and a better reactivity, taken into account the geography of Colombia.

After preparing the commands in regional warehouses, the products are transported to second-tier facilities, which are called delivery points and are the interface between the distribution system and the executing units that deliver the product to the beneficiaries. There is approximatively a set of 5,000 delivery points, spread throughout the country. The Bienestarina supply chain uses different mechanisms and means of transportation (land, air, and sea) to bring a product to the remote parts of the country and achieve full coverage of the territory, as shown in Table 2.

From the delivery points, products are either delivered to the executing units or those units go to the delivery points to pick up the commands. Those units are in charge of distribution of the products to the beneficiaries. Each unit has a set of beneficiaries that are assigned to it and a weekly distribution plan. Depending on a type of the beneficiaries (hospitals, nurseries, households, etc.), demand is either shipped or collected by a beneficiary. This final echelon of the chain is difficult to be steered since no traceability exists after the executing units pick up the goods. A recent study of the Universidad Nacional de Colombia shows that the production and distribution chain until the executing units result on negligible

Figure 2. Division of logistic distribution macro-regions (MR)
Source: Peñaloza, Palacios & Gonzalez-Feliu, 2016

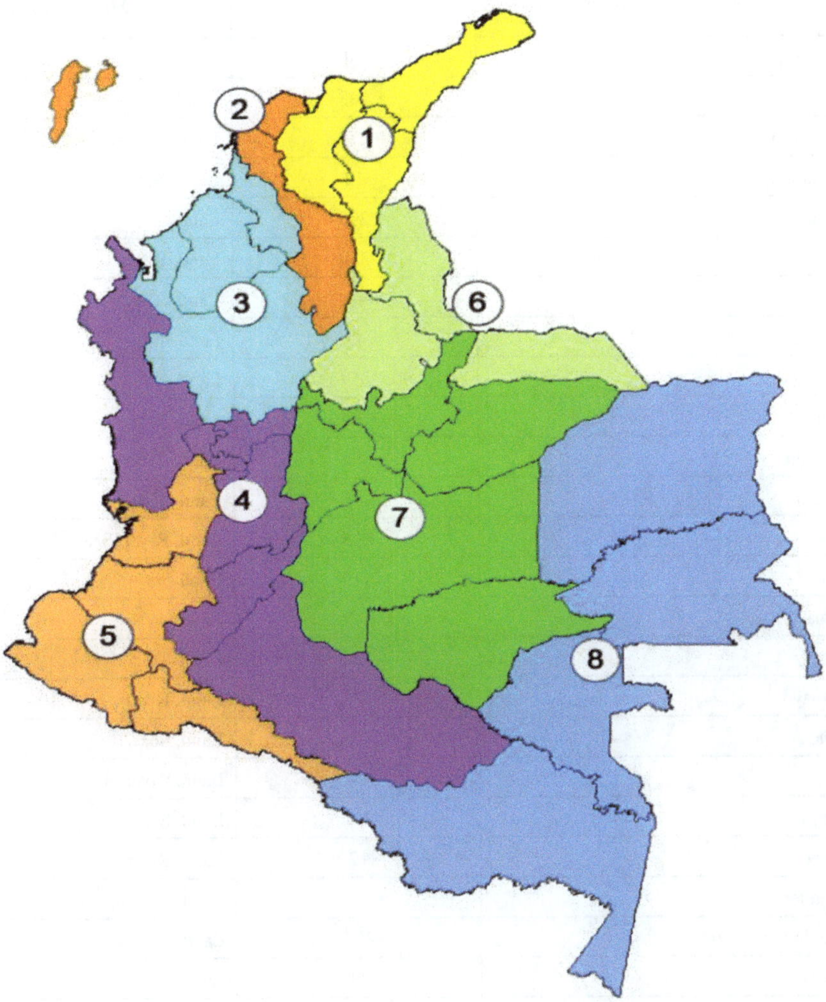

product loses. However, the interviews of a selected number of representatives of the executing units made the authors to identify some dysfunctions in this last echelon. The first is that several beneficiaries do not collect the products systematically, and a reassignment is not expected by the plans. The second is a non-negligible rate of self-consumption of the products, mainly those not claimed by the beneficiaries. The third is a difficulty to have an in-deep centralized knowledge of the entire final distribution system (an information system had not been implemented when the case study was constructed), which do not allow to have a complete product traceability. Moreover, local press reports some unfair practices, like using the product to feed farming animals instead of the population under malnutrition, but those practices have not been observed and confirmed (since they are made by a set of executing units and, being illegal, are hidden to any surveying action). However, in order to improve the efficiency of this last echelon and the entire supply chain, the ICBF has recently deployed an information system (which comes in full operation in 2017) and contributes to several research projects with Colombian leading universities.

Table 2. Logistic information about the Bienestarina distribution process in Colombia

Macro-region	Regional warehouse	Number of delivery points	Main transport modes
1	La Guajira	24	Land
	Cesar	238	Land, River
	Magdalena	279	Land, River
2	San Andrés	12	Land, Sea
	Atlántico	295	Land
	Bolívar	350	Land, River
3	Sucre	13	Land, River
	Córdoba	47	Land
	Antioquia	359	Land
4	Caquetá	31	Land, River
	Risaralda	44	Land, River
	Quindío	44	Land
	Chocó	115	Land, River, Sea
	Caldas	119	Land
	Tolima	177	Land
	Huila	188	Land
5	Putumayo	9	Land, River
	Nariño	72	Land, Sea
	Cauca	97	Land, River, Sea
	Valle	341	Land, Sea
6	Arauca	17	Land, River
	Santander	150	Land
	Norte de Santander	397	Land
7	Casanare	44	Land
	Guaviare	47	Land, River, Air
	Meta	124	Land, River, Air
	Boyacá	156	Land
	Cundinamarca	185	Land
	Bogotá	626	Land
8	Guainía	2	Land, River, Air
	Vaupés	7	River, Air
	Amazonas	9	Land, River, Air
	Vichada	14	Land. River, Air

Source: Authors' elaboration from Colombian Institute of Familiar Wellness (2009)

Although the Bienestarina powder is the main product under distribution, there are other products provided to the beneficiaries. There is liquid Bienestarina (with three flavors) based on milk also produced and distributed. Moreover, some artisanal products are also produced locally. Indeed, some primary distribution points and executing units are engaged in the manufacture of food products using Bienestarina as an ingredient (cakes, biscuits, juices, etc.). Those products which need Bienestarina powder are produced locally and mainly use local agricultural products (fruits but also some flours and sugar cane products) resulting in the development of local industries. Beneficiaries of the product have a possibility to go to the executing units or corresponding primary distribution points to receive the product, just as the executing units has to go to the primary distribution point to receive the product, as shown by the purple arrows in Figure 2 with opposite way to the regular supply chain.

To summarize, the Bienestarina supply chain can be characterized as follows. The authors define the longitudinal structure of Bienestarina supply chain as shown in Figure 3. The Bienestarina supply chain counts six echelons when considering various steps from raw material (RM) suppliers to final consumers (beneficiaries):

- Only one echelon is related to the suppliers since the factories do the complete production process, so the suppliers provide the factories with raw materials only, without any industrial transformation before arriving at the factories. The only exception to this statement is that of vitamins and oligo-elements, which can be acquired from specific industrial producers.
- One echelon is related to production, although it could be decoupled in other three echelons: one for flour production, one for Bienestarina powder production (resulting in the mix of different flours, with the addition of vitamins and oligo-elements), and the third, when applicable, related to the production of elaborated products with the addition of Bienestarina. Currently, two main types of elaborated products are proposed: enriched liquid milk and cookies.
- Three echelons are related to the distribution system, which shows the importance and main place of the distribution system in the overall Bienestarina supply chain.
- The last echelon is related to the beneficiaries.

In order to illustrate the proposed characterization framework, the authors synthesized the four planning elements (steering, organization, development, and financing) (Table 3).

This exploratory overview of the Bienestarina system shows that the Bienestarina logistics follows a supply chain management structure, but there are substantial differences with respect to commercial logistics chains. The steering of the supply chain is ensured by the customer (the ICBF and the end-consumer (beneficiaries) are not customers). The rates of beneficiaries are established by the ICBF with local authorities (customer of the customer), adding a complex commercial/use chain with little impact on supply chain organization and steering. However, the logistics operations need to be ensured, and although there is not an evident relation between the service quality and the satisfaction of the consumer, the steering organization (ICBF) evaluates the global satisfaction of the system.

But is this system a specificity of Bienestarina or Colombian context? To give a preliminary set of elements to the discussion, the authors propose to explore other three systems: two in Mexico and one in Peru.

Figure 3. Scheme of Bienestarina distribution chain
Source: Authors' own elaboration based on Colombian Institute of Familiar Wellness (2009)

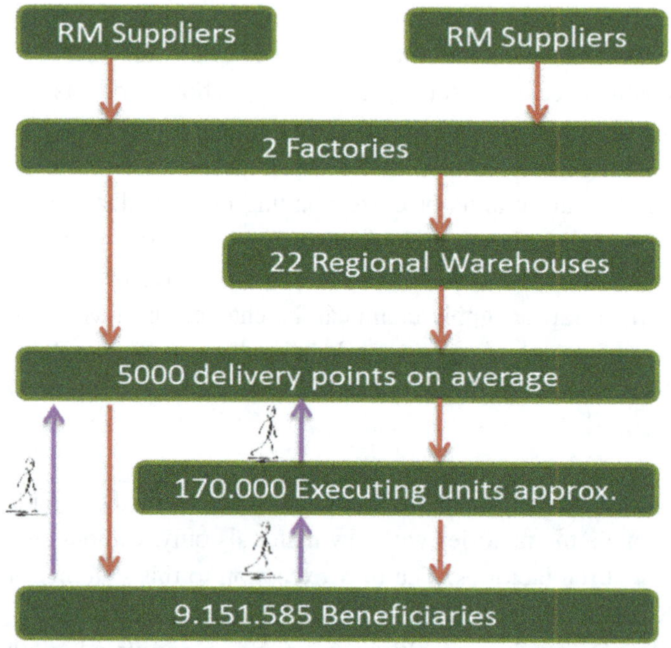

SOLUTIONS AND RECOMMENDATIONS

This product, exclusive of Colombia and developed and distributed under welfare programs of the Colombian Ministry of Familiar Protection, cannot be found in any other country. However, the logistics system presents similarities to other products and programs in the world. For example, in Mexico, two similar programs exist. The first is LICONSA which is aimed at production and distribution of a nutritionally-improved milk. It is not free, but its price is much lower than those of classical milk. The product is mainly distributed to sensible population locations. The second program is DICONSA which is a public company that produces and distributes flour (mainly of corn, but other variants are also produced) to public restaurants (mainly nurseries, schools, and hospitals).

The LICONSA program was established in 1944 with the objective of increasing supply of milk by ensuring of its good quality and affordable prices for the population in a vulnerable situation.

Taking into account the results of this work and comparing them with the analysis of the logistics processes of the LICONSA supply chain (Mexican nutritional enriched-milk), some differences and similarities have been found.

Differences:

- Percentage of population served is 5.97% in Mexico and 19.14% in Colombia. In Columbia, 1 out of 5 citizens is served, while in Mexico, only 1 out of 20 citizens, respectively.
- In Mexico, the LICONSA distribution process is supported by the use of the geographic location system (SISGE). This system allows an optimization of a sale point location, in order to be more efficient in terms of delivery time, distance, routes, among others.

Table 3. Synthesis of planning elements in the Bienestarina supply chain

Methods and structures	Steering structure	Organizational aspects	Development issues	Financial elements
Planning and control methods	The system is entirely steered and planned by the ICBF. Ingredion adjusts and executes those plans.	The entire supply chain is planned and controlled by Ingredion following global planning methods.		
Workflow and activity structure		SCM is made by Ingredion. Operations are managed by Ingredion, and only follow-up reports for yearly updating of the contract are made by the ICBF. Local agriculture at the national level is promoted.	Supply chain activities were firstly managed by the ICBF, then subcontracted. From 2007, the current system is implemented.	The service is paid by public entities through the beneficiary tickets to the ICBF, which then uses the funds to finance the supply chain costs
Organization structure		ICBF is a very hierarchical structure with many departments. Ingredion is more specialized but also has a hierarchical structure.	The organizational structure has not been strongly evolved over the years	
Information and communication structure	There is no traceability system. The steering of the supply chain is decentralized. The information exchange between the ICBF and Ingredion concerns monthly and yearly indicators of quantities delivered.	Exchanges on a number of beneficiaries between the ICBF and Ingredion, linear structure following the supply chain structure for internal transactions.		
Product flow structure	-	6-echelon supply chain: raw material suppliers, production plants, regional warehouses, delivery points, executing units, and beneficiaries	Number and variety of products and diversity of flows grow. The structure is similar for all of them.	
Management methods	Management is facilitated by Ingredion. Vertical management seems to be predominant.	Executing units are mono-personal or very small structures (no need of strong management methods); plants and warehouses use classic SCM methods.	Management has evolved over the last years towards the adoption of international SCM standards.	
Power and leadership structure	ICBF is a “customer” configuration since it purchases services from Indredion.	The supply chain has a very decentralized leadership structure which embarrasses the identification of dysfunctions and improvement opportunities.		
Risks and rewards structure	Risks are forecasted by the ICBF, but operational issues are steered and managed by Ingredion.	No rewards structure has been identified. Main risks are related to transport/traffic conditions, ruptures in raw materials, production problems and losses/bad use of products in the different echelons of the supply chain.		Risks are assumed by Ingredion but can be shared with the ICBF.
Culture and attitude	The decentralized steering and the lack of traceability have an impact on losses at the final levels.		With the increase of information about the system, collective conscience and solidarity have increased.	Not all public authorities have the same sensibility to promote the system and buy services.

Source: Authors’ own elaboration

- Similarities:
- Production and distribution processes are managed by the public entities (the ICBF in Colombia and the SEDESOL in Mexico) which are in charge of ensuring the well-being of the population.
- Beneficiaries are the people who are in the conditions of food vulnerability.
- The last mile distribution is similar. There are established final points (sale points, executing units) which are the ones that have contact with the target population.

The authors could also establish a parallel with DICONSA (a flour-based product similar to Bienestarina, also for Mexican families) or with the Peruvian Glass of Milk program (Programa Vaso de Leche) (Ministerio de Desarrollo e Inclusión Social, 2013), which both aim to distribution of nutritionally-enriched menus, mainly for breakfasts or pre-dinner meals, to schools. In both cases, the main difference is that the beneficiaries are not families, but public entities, making the product traceability higher and the process easier to monitor and plan. However, in all four cases, the authors observe similar patterns, related to the four elements defined above. More precisely:

- On the financing viewpoint, all those systems start from public concern and need public support. However, the percentages of public financing capacity, as well as the mechanisms of refunding or distributing funds, are not standardized. However, it is clear that an efficient organization is needed to ensure the good use of those funds.
- On the development viewpoint, their evolution depends on that of the beneficiaries. In other words, those systems have a vocation to support the nutrition development of families. More they contribute to this nutritional development, fewer products are needed. The disappearance of those supply chains as useless would be a consequence of the good use and deployment of the food distribution systems for social improvement. However, this final stage is far to arrive, but the evolution in the number and nature of beneficiaries is a good indicator of the global performance of the social improvement logistic schemes.
- On the steering viewpoint, those systems are promoted and guided by public entities, mainly in the form of yearly or multi-yearly plans. The way those plans are actuated and deployed depends on the national (or regional) context, but in general, the logistics management is made by the experts who act as subcontractors of the public entity. The role of public authorities is to monitor the subcontractors, finance the distribution system, and verify correct actuation of the yearly plans.
- Finally, the organization of the supply chain seems to be a consequence of the other three elements. Different choices can be made, but Bienestarina is following an integrated supply chain with the consequent management choices, steered by a subcontractor. This seems to be the case of LICONSA, DICONSA, and Programa Vaso de Leche, so the first statement that can be induced is the need of deploying efficient and integrated supply chains even if there is no product to sell and no economic value. However, the social value of social improvement of supply chains is clearly identifiable, and in a context of limited resources and high needs in terms of malnutrition reduction, this efficiency seems crucial.

FUTURE RESEARCH DIRECTIONS

All four initiatives seem to incite the use of local producers and are based on the extraction and transformation of raw products which are cultivated or produced in a country. However, the details on the contribution of such systems to local production economics need to be studied further.

The work completed by the authors remains preliminary. A wider comparison with other initiatives existing in Latin America and a qualitative interview-based analysis on key stakeholders of the Bienestarina supply chain would give more evidence on the characteristics of social improvement supply chains and define the main analysis framework of those systems. Moreover, further researches should be focused on traceability and operations management in order to identify the major critical points of the Bienestarina supply chain and propose management guidelines to improve their efficiency and sustainability.

CONCLUSION

The ICBF program of Bienestarina is considered one of the main actions to combat the problems of malnutrition in Colombia, which demonstrates reductions in the rates of chronic malnutrition and underweight (Colombian Institute of Familiar Wellness, Instituto Nacional de Salud, Profamilia, Ministerio de Protección Social, & DANE, 2010). This is why it is important to characterize the supply chain deployed to bring Bienestarina just in time and making compliance with the fundamental rights of the vulnerable population. For the crucial contribution of logistics to reach the main objectives of the Bienestarina plans, this case is a good example of social improvement logistics.

Via adaptation of the analysis framework of Gonzalez-Feliu et al. (2014) to Bienestarina, and a synthetic comparison of this system to similar social improvement logistics initiatives in Latin America, the authors have defined the main patterns of agroindustrial-based social improvement logistics and highlighted the importance of logistics efficiency in non-economically deployed logistics networks.

The authors observe that the logistics chain of Bienestarina presents particularities with respect to commercial logistics (by the fact that the end consumer is not the customer who pays for the product), resulting in the central role of the public entity managing and contracting the logistics system. The operator is making a service (close to that of commercial logistics, with lower costs), but the indicators for evaluating the performance of this supply chain seem different (since the customer's satisfaction is not measured, and instead the capacity of reaching families is preferred as the main measure of satisfaction). On the other hand, and with respect to humanitarian logistics in disasters, the urgent nature of distribution and suffering is perceived differently. The aim of social improvement logistics is to improve the quality of life in a medium-long term, whereas that of humanitarian logistics in disasters is to reduce suffering as quickly as possible.

Without entering to compare in-depth those three types of logistics, the authors observe that social improvement logistics seems to have particularities that make it different from other humanitarian logistics and commercial logistics. This chapter gives a first analysis by applying a combined analysis framework but remains exploratory. More in-depth analyses will be required to characterize social improvement logistics. Moreover, the question of evaluation, then of deploying suitable dashboards, on the line of current works for urban logistics (Gonzalez-Feliu & Morana, 2014; Morana & Gonzalez-Feliu, 2015) specific to dietary complement supply chains for social improvement would be an important future development of this research.

REFERENCES

Adivar, B., Atan, T., Sevil Oflaç, B., & Örten, T. (2010). Improving Social Welfare Chain Using Optimal Planning Model. *Supply Chain Management: An International Journal, 15*(4), 290–305. doi:10.1108/13598541011054661

Braun, J., & Kennedy, E. T. (1994). *Agricultural Commercialization, Economic Development, and Nutrition*. Baltimore, MD: Johns Hopkins University Press.

Colombian Institute of Familiar Wellness. (2009). *Bienestarina: Distribución, cuidado y uso de un recurso sagrado*. Bogotá: ICBF.

Colombian Institute of Familiar Wellness. Instituto Nacional de Salud, Profamilia, Ministerio de Protección Social, & DANE. (2010). Encuesta Nacional de La Situación Nutricional En Colombia 2010 – ENSIN. Bogotá: Ministerio de Protección Social.

Cuevas García, R. (2005). *El diseño de los programas de alimentación escolar y la función de la industria alimentaria*. Retrieved November 26, 2016, from http://www.fao.org/docrep/008/y5906m/Y5906M05.htm

Dirección Nacional de Planeación. (2006). *Conpes 3443. Contratación del operador para la producción y distribución del componente nutricional (Bienestarina) en el Instituto Colombiano de Bienestar Familiar*. Bogotá: ICBF, Dirección Nacional de Planeación.

Dirección Nacional de Planeación. (2007). *Conpes 113. Política Nacional de Seguridad Alimentaria Y Nutricional (PSAN)*. Bogotá: Dirección Nacional de Planeación.

El Nuevo Siglo. (2013). *Lupa oficial a la Bienestarina.* Retrieved March 2, 2017, from http://www.elnuevosiglo.com.co/articulos/5-2013-lupa-oficial-bienestarina.html

Ewing, M., & Msangi, S. (2009). Biofuels Production in Developing Countries: Assessing Tradeoffs in Welfare and Food Security. *Environmental Science & Policy, 12*(4), 520–528. doi:10.1016/j.envsci.2008.10.002

Food and Agriculture Organization of the United Nations. (2010). *La Iniciativa América Latina y Caribe sin Hambre*. Rome: Food and Agriculture Organization of the United Nations.

Food and Agriculture Organization of the United Nations. (2011). *Global Food Losses and Food Waste.* Study Conducted for the International Congress SAVE FOOD! Rome: Food and Agriculture Organization of the United Nations.

Fortucci, P. (2002). *The Contribution of Cotton to Economy and Food Security in Developing Countries.* Retrieved March 2, 2017, from https://www.icac.org/meetings/cgtn_conf/documents/11_fortucci.pdf

Gonzalez-Feliu, J., Malhéné, N., Morganti, E., & Morana, J. (2014). The Deployment of City and Area Distribution Centers in France and Italy: Comparison of Six Representative Models. *Supply Chain Forum: An International Journal, 15*(4), 84–99.

Gonzalez-Feliu, J., & Morana, J. (2014). Assessing Urban Logistics Pooling Sustainability via a Hierarchic Dashboard from a Group Decision Perspective. In C. Macharis, S. Melo, J. Woxenius, & T. Van Lier (Eds.), *Sustainable Logistics* (pp. 113–135). Bingley, UK: Emerald Group Publishing. doi:10.1108/S2044-994120140000006004

Gonzalez-Feliu, J., Taniguchi, E., & d'Arcier, B. F. (2014). Financing Urban Logistics Projects. In J. Gonzalez-Feliu, F. Semet, & J.-L. Frédéric (Eds.), *Sustainable Urban Logistics: Concepts, Methods and Information Systems* (pp. 245–265). Berlin: Springer. doi:10.1007/978-3-642-31788-0_13

Holguin-Veras, J., Jaller, M., Van Wassenhove, L. N., Pérez, N., & Wachtendorf, T. (2012). On the Unique Features of Post-Disaster Humanitarian Logistics. *Journal of Operations Management*, *30*(7), 494–506. doi:10.1016/j.jom.2012.08.003

Holguin-Veras, J., Wachtendorf, T., Jaller, M., & Jefferson, T. (2013). Logistics and the Management of Critical Supplies Following Catastrophes. In R. Bissell (Ed.), *Preparedness and Response for Catastrophic Disasters* (pp. 131–150). Boca Raton, FL: CRC Press.

Kovács, G., & Spens, K. M. (2007). Humanitarian Logistics in Disaster Relief Operations. *International Journal of Physical Distribution & Logistics Management*, *37*(2), 99–114. doi:10.1108/09600030710734820

Kovács, G., & Spens, K. M. (2009). Identifying Challenges in Humanitarian Logistics. *International Journal of Physical Distribution & Logistics Management*, *39*(6), 506–528. doi:10.1108/09600030910985848

Lambert, D. M. (2001). The Supply Chain Management and Logistics Controversy. In A. M. Brewer, K. M. Button, & D. A. Hensher (Eds.), *Handbook of Logistics and Supply Chain Management* (pp. 99–126). Oxford, UK: Pergamon Press.

Lambert, D. M., Cooper, M. C., & Pagh, J. D. (1998). Supply Chain Management: Implementation Issues and Research Opportunities. *The International Journal of Logistics Management*, *9*(2), 1–20. doi:10.1108/09574099810805807

Maldonado, M., & Moya, S. (2013). The Challenge of Implementing Reverse Logistics in Social Improvement: The Possibility of Expanding Sovereignty Food in Developing Communities. In J. Cheung & H. Song (Eds.), *Logistics: Perspectives, Approaches and Challenges* (pp. 1–34). New York, NY: Nova Science Publishers.

Ministerio de Desarrollo e Inclusión Social. (2013). *Programa Vaso de Leche*. Lima: Gobierno de la República del Perú.

Morana, J., & Gonzalez-Feliu, J. (2015). A Sustainable Urban Logistics Dashboard from the Perspective of a Group of Operational Managers. *Management Research Review*, *38*(10), 1068–1085. doi:10.1108/MRR-11-2014-0260

Orbell, J. M., & Dawes, R. M. (1993). Social Welfare, Cooperators Advantage, and the Option of not Playing the Game. *American Sociological Review*, *58*(6), 787–800. doi:10.2307/2095951

Palacios, L., Morana, J., Gonzalez-Feliu, J., & Devia, J. (2016). Traceability for a More Sustainable Supply Chain Management: The Case of Bienestarina in Colombia. In *Proceedings of the Third International Conference on Green Supply Chain Management*. London: Loughborough University.

Peñaloza, C., Palacios, L., & Gonzalez-Feliu, J. (2016). Characterization of the Collection and Distribution Processes of Bienestarina in Bogotá. In *Proceedings of the International Workshop on Franchising & Distribution Networks in Emerging Countries*. Saint-Etienne: Jean Monet University.

Presidencia de la República. (2013). *Bienestarina Más' llegará gratuitamente este año a 6,5 millones de beneficiarios, especialmente niños y niñas de la primera infancia*. Retrieved November 23, 2016, from http://wsp.presidencia.gov.co/Prensa/2013/Febrero/Paginas/20130228_07.aspx

Reutlinger, S. (1986). *Poverty and Hunger: Issues and Options for Food Security in Developing Countries. A World Bank Policy Study*. Washington, DC: World Bank.

Rozo, C. (2000). Complementary Foods in Colombia. *Food and Nutrition Bulletin*, *21*(1), 55–61. doi:10.1177/156482650002100109

Stock, J. R. (1990). Logistics Thought and Practice: A Perspective. *International Journal of Physical Distribution & Logistics Management*, *20*(1), 3–6. doi:10.1108/09600039010140845

Supply Chain Council. (2008). *Supply-Chain Operations Reference-Model. Overview of SCOR version, 5(0)*. Chicago, IL: Supply Chain Council.

Valdes, A. (1981). *Food Security for Developing Countries*. Boulder, CO: Westview Press.

ADDITIONAL READING

Colombian Institute of Familiar Wellness. (2013). *Logística de distribución en el territorio nacional*. Bogotá: ICBF.

Diaz-Bonilla, E., Thomas, M., Robinson, S., & Cattaneo, A. (2000). *Food Security and Trade Negotiations in the World Trade Organization*. Washington, DC: International Food Policy Research Institute.

Eisenhardt, K. M. (1989). Building Theories from Case Study Research. *Academy of Management Review*, *14*(4), 532–550.

Food and Agriculture Organization of the United Nations. (2016a). *Global Food Composition Database for Pulses. Version 1.0 -uPulses1.0 User guide*. Rome: FAO.

Food and Agriculture Organization of the United Nations. (2016b). *International Network of Food Data Systems: International Food Composition Tables/Database Directory*. Rome: FAO.

Godfray, H. C. J., Beddington, J. R., Crute, I. R., Haddad, L., Lawrence, D., Muir, J. F., & Toulmin, C. et al. (2010). Food Security: The Challenge of Feeding 9 Billion People. *Science*, *327*(5967), 812–818. doi:10.1126/science.1185383 PMID:20110467

IA Alimentos. (2013). *El ICBF producirá 25 mil toneladas de bienestarina en el 2013*. Retrieved January 27, 2017, from http://www.revistaialimentos.com.co/news/1223/443/El-ICBF-producira-25-mil-toneladas-de-bienestarina-en-el-2013.htm

Lutter, C. K. (2003). Macrolevel Approaches to Improve the Availability of Complementary Foods. *Food and Nutrition Bulletin*, *24*(1), 83–103. doi:10.1177/156482650302400105 PMID:12664528

Pinstrup-Andersen, P. (2009). Food Security: Definition and Measurement. *Food Security*, *1*(1), 5–7. doi:10.1007/s12571-008-0002-y

KEY TERMS AND DEFINITIONS

Bienestarina Program: A production-distribution program of a dietary complement, called Bienestarina, steered by the Colombian government (more precisely, the Colombian Institute of Familiar Welfare (ICBF)) to fight malnutrition on sensible populations.

Commercial Logistics: A set of techniques and methods dealing with planning, design, and support of business operations that deal with procurement, purchase, inventory, warehousing, distribution, and transportation, among others, in order to bring a product or service to its final user.

Dietary Complement: A product intended for ingestion that contains a "dietary ingredient" intended to add further nutritional value to the diet. A dietary complement is ingested alone (or diluted with water) and aims to supply to deficiencies in diets by substitution to traditional foods on only the concerning elements, like some vitamins or minerals). Dietary complements may be found in many forms such as tablets, capsules, soft gels, gelcaps, liquids, or powders.

Dietary Ingredient: One or any combination of the following substances: vitamins, minerals, herbs (or botanical products), amino acids, concentrates, metabolite, constituents, or extracts. It can be used to complement diets and meals.

Humanitarian Logistics: A set of logistics operations deployed to react to emergencies (natural or human-based), and which follow a completely different logic than commercial logistics.

Social Improvement Logistics: A set of logistics operations developed to improve the social status of populations, which are of lower urgency than those of humanitarian logistics but follow aims and issues which remain different than those of commercial logistics.

Supply Chain Management: An integrated management strategy that includes the planning and management of all activities involved in sourcing and procurement, conversion, and all logistics management activities, from the extraction of raw materials to the delivery at the retailers' or end- consumers' locations, and including all the different stages.

Chapter 15
Food Safety From Consumer Perspective:
Consumer Confidence in Food Safety

Ivana Domazet
Institute of Economic Sciences, Serbia

Nenad Djokić
University of Novi Sad, Serbia

ABSTRACT

Consumer perception is one of the aspects from which food safety can be considered. This view of food safety, named in the literature as subjective safety, does not necessarily correspond to objective food safety and is also considered as not being easily influenced by education of consumers. Having that in mind, the first goal of this chapter is to examine different approaches to food safety and the results of previous research dealing with consumer perspective of food safety. Besides giving broad literature review of the topic, the goal of this chapter is to implement Consumer confidence in food safety scale in Serbian consumer research. The validity, reliability, and one-dimensionality of this scale, which consists of optimism and pessimism (two dimensions of confidence), are tested and determinants of different rating on it are examined. The research from 2016 includes 300 participants in the main survey. Finally, starting from profiling consumers with different levels of general confidence in food safety, managerial recommendations are provided.

INTRODUCTION

The importance given to food safety and quality can be understood as a part of competition among actors in agribusiness based upon adding value rather than relying only on efficiency and quality control, what is caused by a number of food scares in the focus of public attention, segments of the general public paying greater attention to manners of producing food at the farm and processing level as well as by consumers becoming more fragmented in their food choices, especially in developed countries, according to Grunert (2005, pp. 369–391).

DOI: 10.4018/978-1-5225-2733-6.ch015

Ritson and Mai (1998) give a broad interpretation of food safety, which includes nutritional quality of the diet and concerns about novel foods, as well as chemical and microbiological safety, then what binds these characteristics together from an economics perspective is that they all reflect a potentially adverse impact on an individual as a consequence of the consumption of food. Hereby, such impact is structured of two components: hazard, which represent the severity of the impact, and risk, which considers the probability of the hazard occurring. Hazards can be classified into sources of risk: microbiological, chemical, and technological (Yeung & Morris, 2001).

Ritson and Mai (1998) identify four manners for potential market economy "failures" in providing the optimum degree of food safety: asymmetry in the knowledge of risks, aspects of food safety (public goods), social costs of food safety, and the divergence between objective scientific evidence and consumer perception of food safety risks. When it comes to consumer behavior in regard to food safety risk, Verbeke, Frewer, Scholderer, & De Brabander (2007) point out that they tend to systematically overestimate risks in connection to the technical probability of harm occurring, while to underestimate in a large extent other risks. Despite distinguishing broad categories of risks, they do not find enough differences between various types of risk within a particular food group. Even in conditions of existence of information focused on reducing food safety uncertainty, consumers do not process this information. It is in accordance with differentiation of objective food safety (assessment of the risk of consuming a certain food by scientists and food experts) and subjective food safety (in the mind of the consumer), their mutual deviation and failures of attempts to educate consumers in regard to these questions (Grunert, 2005). It is also in accordance with Yeung and Morris (2001) pointing out consumer risk perception of food safety not being largely determined by the hazard per se, but more with its social and psychological characteristics. Finally, this is similar to suggestions of Frewer, De Jonge, & Van Kleef (2010) that evaluation performed by consumers regarding the severity of different risks appear actually differ from valuations given by persons with knowledge about a specific hazard domain.

Having all this in mind, there are several goals of this chapter:

1. Examine different approaches to food safety and the results of previous research dealing with consumer perspective of food safety,
2. Implement consumer confidence in food safety scale, developed by de jonge et al. (2008a), in serbian consumer research,
3. Provide managerial implications (when considering the results of the research conducted in serbia, as well as considering similar researches abroad).

The first goal is implemented in the Background section of the chapter, the second in the Main Focus section, and the third in Solutions and Recommendations and Conclusions sections.

BACKGROUND

As for the consumer relation to food safety, De Jonge et al. (2004) argue that in the conditions characterized by the absence of any incidents in regard to food products, food safety will be "taken for granted" by consumers. Ritson and Mai (1998) stress that even if food safety was easily identifiable when existing the inverse relationship between it and some positive attribute (e.g. taste, appearance, price), consumers are not seeking for 100% safety. Grunert (2005) points out that a way in which food safety affects consumer food choice is different from those of the other dimensions of quality (taste, health, convenience, and process). When considering the consumer purchase decision-making process, it is dominant in two conditions: when major safety problems are perceived and in regard to certain production technologies. Thereby, food safety perceptions are described as a "sleeping giant" – they do not influence quality perceptions in regular conditions, but this influence becomes substantial in the conditions of crisis.

The relation between food quality and safety is also studied by Van Rijswijk and Frewer (2008). The research was conducted in four European countries: Germany, France, Italy, and Spain and included 163 consumers interviewed in qualitative semi-structured interviews. These interviews yielded in consumers' own definitions of food quality and safety, as well as their relations. The greatest number of consumers sees food quality and food safety as interlinked concepts and there are overlapping of definitions of the two concepts. They pay more attention to food quality during the purchase. Hereby, from the consumer perspective, traceability was not only in connection with food safety but with food quality as well.

By the confidence in food safety, De Jonge et al. (2004) understand the general expectation of consumers that food products will not be the cause of any harm for their health or for the environment. The authors identify five potential determinants of consumer confidence in food safety: consumer trust in regulatory institutions and actors in the food chain, the occurrence of food safety incidents, media coverage of food risks, consumer recall of food safety incidents, and consumer confidence in the safety of particular product groups. As for the general confidence in food safety, the authors argue that it is likely to have both a cognitive and an affective dimension. They also identify possible behavioral consequences of changes in general consumer confidence in food safety: information search, brand choice, retail choice, product substitution, and reduced consumption.

Yeung and Morris (2001) propose a conceptual model of consumer food purchase relating to food safety. The combination of perceptions of the degree of exposure to a hazard and the consequences of such exposure influence the perception of food safety risk. Bearing in mind the nature of food safety risk, risk perception (strengthened by uncertain characteristics of food safety risks) shows a tendency to weight the perceived potential severity of unhealthy food more in comparison to the probability of exposure (Zubovic & Domazet, 2009).

Verbeke et al. (2007) explore the reasons for consumer behavior in relation to food safety. They first identify the gap between lay and expert opinions and find its causes in a so-called perception filter related to this bias. Secondly, they explain pathway from risk to crisis in relation to the theory of social amplification of risk. Three factors are identified, providing insight into the problem of why some relatively minor risks frequently become crucial for evolution of risk into a crisis: fright factors (individual's perception of the seriousness of a risk), panic elements (related to the nature of the risk itself, whether it is universal, new, believable, uncertain, etc.), and societal triggers (e.g. media information). Thirdly, the authors make an opposition between lifestyle and technological hazards, whereas optimistic biases are more expressed for lifestyle. Finally, they explain the reluctance to information processing in the context of potential irrelevance or usefulness of information to consumers since failing in targeting particular needs or interests.

Frewer et al. (2010) research similar topics. They point out the importance of adding additional criteria to judging risks to classical factors (level of probability and degree of possible harm): the degree to which a hazard is unknown and the degree of dread experienced by the individual evaluating a particular hazard. In consideration to individuals' perception of risk, there should be mentioned a psychological phenomenon of optimistic bias or unreal optimism. This phenomenon is related to judgment that negative events are less likely to happen to the person making the judgment than to an average member of society. When explaining often-mentioned opposition to emerging food technologies, the authors stress that this opposition does not have to come only from evaluations of the potential personal benefits and health effects, but also from moral concerns, beliefs, and values.

These authors stress the need for deeper understanding individual responses to food issues, especially in the situations in which habits, emotions, and information processing heuristics have an effect on consumer decision-making. Hereby, they notice the connection of individual differences and consumer perceptions of food safety, not only in regard to socio-demographic characteristics (gender, age, education level, number of children in the household, etc.) with lack of consistent results, but also in regard to other variables – personal experience of food poisoning (which reduces optimistic bias), general openness to worry and stress, being neophobic, as well as the locus of control ("a measure of attribution of control of external events which potentially affect people's health status").

Verbeke et al. (2007) also suggest strategies for restoring consumer confidence in food safety: establishment of traceability and labeling, segmented communication approaches, and public involvement in risk management decision-making.

De Jonge et al. (2004) report results regarding a nationally representative sample of Dutch consumers from 2002, whereas data collection was performed in three "waves" (January, May, and December) with around 500 respondents (aged 16 or above) participating in each. Hereby, out of all respondents, 85% considered that, when it comes to food safety, there was more than sufficient attention being paid, as well as actions being taken. As for their perception of the responsibility for the safety of food, 43% considered Government as primarily responsible, 27% perceived manufacturers, and 26% farmers in this role, while only 3% stated consideration of supermarkets as primarily responsible for food safety.

In the same research, consumer confidence in food safety was measured by four items: "I think the quality of food keeps improving", "In recent years my confidence in food products has decreased" (recorded in the analysis), "I am worried about the safety of food products" (recorded in the analysis), and "Food products have never been as safe as nowadays". When analyzing the mean rating of statements, several conclusions were reached. The mean for all statements by all respondents was 3.1, on a scale from 1 (low confidence) to 5 (high confidence), but there were significant differences across gender and age segments. Hereby, men were more confident in the safety of food than women. However, when it comes to age, there were differences in several directions regarding consideration of food safety by different groups. When evaluating the statement "I think the quality of food keeps improving", the youngest age group (16-24) accepted it the most, while the age group 35-49 was the most pessimistic concerning it. Considering the second and the third statement, there was a negative relation between age and their evaluation. However, the oldest respondents accepted the most that "Food products have never been as safe as nowadays", while this was strongly rejected by the middle age groups (25-34 and 35-49).

When asking the same respondents about their primary concerns with regard to food, 16% of them had no concern, whereas more men gave this answer in comparison to women, while there were no differences across age groups when answering. The five most common concerns were: additives, unnatural ingredients, genetic manipulation, lack of information on the product label, and meat. During the evaluation of their confidence in 18 different product categories, 42% of respondents reported no concerns. Confidence was especially high for agricultural products (cheese, dairy, and fresh vegetables) while especially low for the products that have been processed extensively (ready-to-eat products and meals), as well as for the products that have a chemical character (vitamin supplements, sport, and energy drinks). This implied that consumers were more concerned regarding appliance of new technologies in food production than regarding products produced without extensive processing.

De Jonge et al. (2008b) conducted another research in the Netherlands – this time among Dutch consumers responsible for the daily grocery shopping. The authors used data from December 2003 that were collected among 3,669 consumers out of 4,437 consumers belonging to a panel of Dutch consumers. The aim of the research was to evaluate the level in which general consumer confidence in food safety (represented by optimism and pessimism) is caused by their trust in general, trust in specific actors (food manufacturers, the Government, farmers, and retailers), specific trust dimensions (the extent to which consumers perceived actors to take care of public well-being, perceived competence, and perceived openness), as well as specific combinations between actors and trust dimensions. Hereby, separate regression analyses were performed for two dimensions of consumer confidence in food safety – optimism and pessimism, while having in mind that there were identified four actors and three trust dimensions, twelve variables were used within the model as independent variables.

The results have shown that although consumer general trust strengthened their optimism and diminished their pessimism when it came to food safety in general, the strength of such relation varied in the context of the actor being trusted and the trust dimension taken into consideration. Generally, consumer confidence in the safety of food is most strongly influenced by their trust in food manufacturers. The extent of that influence is much larger in comparison to the influence of the Government, farmers, and retailers.

When considering trust dimensions in regression analysis where optimism was the dependent variable, it was concluded that the extent to which consumers perceived actors to take care of public well-being is the most important dimension. Besides, there were significant interaction effects for perceived openness and perceived competence dimensions in the relation of trusting in food manufacturers and the Government. However, while for food manufacturers there was positive interaction effect with perceived competence and negative interaction effect with perceived openness, the results for the Government are opposite since perceived competence was less important and perceived openness more important. When considering trust dimensions in a regression where pessimism was the dependent variable, the greatest difference from the previous analysis, was that the extent to which consumers perceived actors to take care of public well-being is not the most important dimension. All other results could be understood as "the mirror image of the results from the model with optimism as the dependent variable".

Table 1 presents shortly some of the main findings from the previously listed literature.

When it comes to Serbian researches, according to the authors' knowledge, there was no research regarding consumer view on food safety with the implementation of instruments developed abroad. However, having in mind the possible connection of food quality and safety from a consumer perspective, and the results of some previous research of organic and functional food, it can be expected that food safety is an important issue for Serbian consumers. Among such researches can be listed the one of Grubor and Djokic (2016) on a segment of Serbian consumers who prefer organic food. These respondents among other characteristics strongly value the importance of a diet for their health and believe that a disease can be prevented by proper diet. There is also the research of Grubor, Djokic, Milicevic, & Kovac-Znidersic (2014), which identifies segments of consumers preferring food with "health claims", including functional food.

Researching similar topics in domestic conditions as well as in other developing countries can be significantly from two perspectives. Firstly, it can be useful for food producers to try to differentiate and target premium segments. Secondly, if considering exporting food products, researching foreign markets concerning food safety from a consumer perspective is very important.

Table 1. Food safety from consumer perspective

Author(s)	Findings
Grunert (2005)	There is a difference between objective food safety (assessment of the risk of consuming a certain food by scientists and food experts) and subjective food safety (in the mind of the consumer).
Frewer et al. (2010)	A valuation performed by consumers regarding the severity of different risks appear actually differ from valuations given by persons with knowledge about a specific hazard domain.
Yeung and Morris (2001)	- Consumer risk perception of food safety is not being largely determined by the hazard per se, but more with its social and psychological characteristics. - Combination of perceptions of the degree of exposure to a hazard and the consequences of such exposure influence the perception of food safety risk.
Verbeke et al. (2007)	- Consumers tend to systematically overestimate risks in connection to the technical probability of harm occurring, while to underestimate in a large extent other risks. - Three factors are identified, providing insight into the problem of why some relatively minor risks frequently become crucial for evolution of risk into a crisis: fright factors (individual's perception of the seriousness of a risk), panic elements (related to the nature of the risk itself, whether it is universal, new, believable, uncertain, etc.), and societal triggers (e.g. media information).
Frewer et al. (2010)	The authors point out the importance of adding additional criteria to judging risks to classical factors (level of probability and degree of possible harm): the degree to which a hazard is unknown and the degree of dread experienced by the individual evaluating a particular hazard.
De Jonge et al. (2004)	The authors identify five potential determinants of consumer confidence in food safety: consumer trust in regulatory institutions and actors in the food chain, the occurrence of food safety incidents, media coverage of food risks, consumer recall of food safety incidents, and consumer confidence in the safety of particular product groups.

Source: Authors' compilation

In the following sections of the chapter, the results of primary research are presented considering the implementation of consumer confidence in food safety scale in Serbia.

MAIN FOCUS OF THE CHAPTER

Research Methodology

Data collection was carried out in May 2016. The respondents took up to 15 minutes on the average to complete the questionnaires. The questionnaire consisted of two parts. The first part was Consumer confidence in food safety scale, developed by De Jonge et al. (2008a). Hereby, all items were rated on seven-point Likert scales from "strongly disagree" to "strongly agree". During the translation of questionnaire into Serbian, back-translation procedure was applied.

The second part of the questionnaire was related to respondents' socio-demographic characteristics. These characteristics were taken into account having in mind that being implicitly present in the behavior theories in relation to consumers' food choices: the Theory of Reasoned Action and its later modification – the Theory of Planned Behaviour Theory (Aertsens, Verbeke, Mondelaers, & Van Huylenbroeck, 2009). Included socio-demographic characteristics were: gender, age, education, marital status, the presence of children in the family, household size, self-assessed household income (from 1 – the lowest mark, to 5 – the highest mark), and place of residence. The levels of these variables that are nominal and ordinal are presented in Table 2.

Table 2. Description of the sample

Characteristics of the sample		Measures	Data
Gender	Male	Percentage	38.0
	Female		62.0
Marital status	Married		59.0
	Single		41.0
Presence of children in family	Families with no children		47.3
	Families with children		52.7
Education	Finished secondary school		54.3
	Student		17.0
	Finished college or faculty		28.7
Place of residence	Urban		59.7
	Rural		40.3
Age		M (SD)	38.44 (13.906)
Household size			2.84 (1.396)
Self-assessed household income			2.87 (0.816)

Note: M – mean; SD – standard deviation.
Source: Authors' development

Within the research method, structured interviews by means of personal communication with the respondents with the use of questionnaire were conducted. The respondents were chosen from the population aged between 18 and 65 years. In an attempt to reach average Serbian consumer, they were approached near green markets, large shopping centers, and universities. The interviews were performed simultaneously in several big Serbian cities: Belgrade, Novi Sad, Subotica, Kragujevac, Kraljevo, and Niš. The interviewed persons were those who were in charge for performing at least 50% of buying food for their households. Convenience sampling method was implemented.

After formal and logical analysis of questionnaires and elimination of questionnaires containing answers of respondents scoring low or high on all items, 300 questionnaires were included in final analysis. The socio-demographic characteristics of those respondents are summarized in Table 1.

On the first stage, reliability, validity, and unidimensionality of the questionnaires were tested according to the recommendations by Churchill (1979), Anderson and Gerbing (1982), Danes and Mann (1984), and Steenkamp and van Trijp (1991). In operationalisation, this considered that Cronbach's alpha, Cronbach's alpha if item deleted, and Corrected item-total correlation were calculated – separately for two dimensions of confidence – optimism and pessimism. On the next stage, exploratory factor analysis by maximum likelihood factor analysis method and Promax rotation method were conducted for testing validity and included all the question of the scale. Prior to Exploratory factor analysis, adequacy of conducting was tested by Kaiser-Meyer-Olkin (KMO) test and Bartlett's Test of Sphericity. Besides, confirmatory factor analysis with maximum likelihood method was used within each of two dimensions of confidence.

Next, after omission of certain items in accordance with the results of the previous tests, two dimensions of Consumer confidence in food safety scale were analyzed in the context of respondents' socio-demographic characteristics. Hereby, both optimism and pessimism were presented by means of ratings of all respondents in regard to the scales. When testing whether there were statistically significant differences between men and women, married and single respondents, families with and without children, people living in urban and rural areas, Independent samples t-test was implemented in relation to the ratings. On the other hand, when testing whether there were statistical significant relations between consumer education and respondents' ratings of these scales, Spearman rank correlation coefficient was used, while when establishing whether there were statistical significant relations between consumer age, household size, and self-assessed household income with respondents' ratings of scales, Pearson correlation coefficient was implemented.

On the third stage, the influence of all researched socio-demographic characteristics of the respondents (gender, age, education, marital status, presence of children in family, household size, self-assessed household income, and place of residence) on both dimensions of consumer confidence in food safety (optimism and pessimism) was analysed by using multiple regression with General Linear Model. The dependent variables (consumer optimism and pessimism) were presented by means of ratings of all respondents in regard to the items remained in the scales after testing their reliability, validity, and unidimensionality. The results were observed at the level of average age, average household size, and average self-assessed household income. Coding was performed, so that the value of 0 was assigned to the following levels of respondents' socio-demographic characteristics: female, single, living in family with no children, finished secondary school, and living in rural area, while the value of 1 was assigned to the following levels of the characteristics: male, married, living in family with children, student or finished faculty or college, and living in urban area.

Data were processed in the SPSS and AMOS statistical packages.

Reliability, Validity, and Unidimensionality of Scales

The results of the analysis showed that the values of Cronbach's alpha, both for the Optimism dimension of the Consumer confidence in food safety scale (0.932) and the Pessimism dimension of the Consumer confidence in food safety scale (0.866) were above 0.7 suggested by Nunnally (1978). Hereby, although removal of an item "As a result of the occurrence of the food safety incidents, I am suspicious about certain food products" could increase the value of Cronbach's alpha for Pessimism dimension of the Consumer confidence in food safety scale (from 0.866 to 0.901), the option was kept in the questionnaire. Besides, all the values of Corrected item-total correlation were above the threshold of 0.3 (Nunnally & Bernstein, 1994).

The result of Kaiser-Meyer-Olkin test of 0.842 was above 0.6 (Kaiser, 1970, 1974), and the level of significance of the Barthlett's sphericity test of 0.000 suggested adequacy of conducting exploratory factor analysis. Two factors with Eigenvalue above 1 (Kaiser, 1960) were identified. Identified factors explained 77.496% of the total variance.

When analyzing the Patern matrix, it was noticed that the first four items belonging to the Optimism dimension of the Consumer confidence in food safety scale loaded on the first factor, while the other three items belonging to the Pessimism dimension of the Consumer confidence in food safety scale loaded on the second factor. Hereby, although all the loadings on appropriate factor were higher than 0.35, which is the minimum recommended for a sample of 250 respondents, and 0.30, which is the minimum recommended for a sample of 350 respondents (Hair, Tatham, Anderson, & Black, 1998). The item with the weakest loading "Generally, food products are safe" was excluded from further analysis.

When comparing two factor loadings of each item, it was identified that all the differences were above 0.2 (Matsunaga, 2010), what suggested no cross-loadings. Previous results of testing reliability and validity of the scale are presented in Table 3.

Following Hair et al. (1998), all items were retained when considering that their factor loadings in confirmatory factor analysis were higher than the minimum recommended for that sample size, but the last item was omitted from further analysis because its R^2 was lower than 0.5 (Table 4).

After removal of the named item and correlating residuals' covariances, according to modification indices, the results of confirmatory factor analysis showed acceptable model fit (compared to thresholds given by Hu and Bentler (1999)) regarding most of the indicators (Table 5).

Even when considering indicators that suggest that model fit is not acceptable, one can have in mind that probability of p-value being significant increases as the sample size increases, no matter what the model-specific features are (Hooper, Coughlan, & Mullen, 2008).

Table 3. Reliability and validity analysis of the scale

Scale	Cronbach's alpha	Items	Cronbach's alpha if item deleted	Corrected item-total correlation	F1	F2
Optimism	0.932	I am optimistic about the safety of food products.	0.919	0.816	**0.899**	0.064
		I am confident that food products are safe.	0.899	0.876	**1.052**	0.179
		I am satisfied with the safety of food products.	0.896	0.884	**0.834**	-0.128
		Generally, food products are safe.	0.928	0.786	**0.569**	-0.451
Pessimism	0.866	I worry about the safety of food.	0.718	0.839	0.025	**0.980**
		I feel uncomfortable regarding the safety of food.	0.788	0.770	0.020	**0.860**
		As a result of the occurrence of the food safety incidents, I am suspicious about certain food products.	0.901	0.638	0.080	**0.714**

Source: Authors' development

Table 4. Unidimensionality analysis of the scale

Scale	Items	Factor loadings	R^2
Optimism	I am optimistic about the safety of food products	0.961	0.923
	I am confident that food products are safe	0.890	0.792
	I am satisfied with the safety of food products	0.955	0.913
Pessimism	I worry about the safety of food	0.978	0.956
	I feel uncomfortable regarding the safety of food	0.840	0.705
	As a result of the occurrence of the food safety incidents, I am suspicious about certain food products	0.668	0.446

Source: Authors' development

Table 5. Results of confirmatory factor analysis – model fit

	Chi-square / df	p – value for the model	CFI	GFI	AGFI	SRMR	RMSEA	PCLOSE
Consumer confidence in food safety scale	4.498	0.000	0.977	0.959	0.905	0.0499	0.108	0.02
Thresholds (Hu and Bentler, 1999)	< 3 good; < 5 sometimes permissible	> 0.05	> 0.95 great; > 0.9 traditional; > 0.8 sometimes permissible	> 0.95	> 0.8	< 0.09	< 0.05 good; 0.05-0.10 moderate > 0.10 bad	> 0.05

Source: Authors' development

Consumer Confidence in Food Safety and Their Socio-Demographic Characteristics

When considering dimensions of the Consumer confidence in food safety scale at the total level, it can be seen that average optimism is 4.08 (standard deviation is 1.45), while average pessimism is 3.52 (standard deviation is 1.76). Hereby, there is moderate, negative, and statistically significant correlation among them: $r = (-0.467)$; $n = 300$; $p = 0.000 < 0.05$.

In Table 6, both optimism and pessimism regarding the consumer confidence in food safety are observed in the context of socio-demographic characteristics of respondents. Hereby, besides the results of appropriate tests, when considering nominal and ordinal variables, for each of their levels the averages for both dimensions are given.

When considering gender, there are no statistically significant differences for both optimism and pessimism in regard to the consumer confidence in food safety. As for the presence of children in the family, there are also no statistically significant differences for food safety optimism and pessimism between the families with and without children. The same conclusion can be reached regarding the place of residence since consumers living in urban areas are not statistically different from consumers living in rural areas when it comes to their optimism and pessimism related to food safety. However, during the analysis of marital status, it can be concluded that there is no statistically significant difference between married and single respondents in relation to optimism. However, there are statistically significant differences between these two segments in relation to pessimism. Married people statistically significantly differ from the single ones by being more pessimistic about food safety.

There is no statistically significant relation between the level of education of consumers and their rating optimism scale considering food safety, as well as between level of education and rating pessimism scale. Also, there is no statistically significant relation between rating optimism scale in food safety and respondents' age, the size of their households, and their self-assessed household income. However, when considering rating pessimism scale in food safety, there are statistically significant relations with the size of households their self-assessed household income of the respondents. These significant relations confirm that pessimism in food safety increases along with the increase of consumers' household size, while decreases along with the increase of self-assessed household income.

Table 6. Consumer confidence in food safety scale and socio-demographic characteristics

Characteristics of the sample		Measure	Optimism	Results of the tests	Pessimism	Results of the tests
Gender	Male	Percent	4.06	t(df = 298) = -0.217, p = 0.828 > 0.05	3.27	t(df = 298) = -1.840, p = 0.067 > 0.05
	Female		4.09		3.66	
Marital status	Married		4.08	t(df = 298) = -0.051, p = 0.960 > 0.05	3.81	t(df = 298) = 3.346, p = 0.001 < 0.05
	Single		4.08		3.10	
Presence of children in family	Families with no children		4.20	t(df = 298) = 1.557, p = 0.121 > 0.05	3.62	t(df = 298) = 1.032, p = 0.303 > 0.05
	Families with children		3.94		3.40	
Education	Finished secondary school		4.13	p = -0.047, n = 300, p = 0.412 > 0.05	3.56	p = -0.035, n = 300, p = 0.546 > 0.05
	Student		4.09		3.65	
	Finished college / faculty		3.98		3.37	
Place of residence	Urban		4.15	t(df = 298) = 1.096, p = 0.274 > 0.05	3.41	t(df = 298) = -1.307, p = 0.192 > 0.05
	Rural		3.97		3.68	
Age			r = 0.087, n = 300, p = 0.134 > 0.05		r = -0.005, n = 300, p = 0.934 > 0.05	
Household size			r = -0.022, n = 300, p = 0.700 > 0.05		r = 0.295, n = 300, p = 0.000 < 0.05	
Self-assessed household income			r = 0.016, n = 300, p = 0.783 > 0.05		r = -0.187, n = 300, p = 0.001 < 0.05	

Source: Authors' development

Results of the General Linear Model Application

By implementing multiple regression with the General Linear Model procedure, the results for the dependent variable of the Consumer optimism regarding confidence in food safety (OPT) can be expressed by a regression equation as follows:

OPT = 3.893 – 0.085 * gender + 0.006 * age – 0.084 * marital status + 0.324 * presence of children in family – 0.077 * household size – 0.029 * education + 0.022 * self-assessed household income + 0.186 * place of residence

Table 7 shows the values of parameters for the dependent variable of the Consumer optimism regarding confidence in food safety at the level of average age, average household size, and average self-assessed household income.

It may have sense to interpret the above equation only in the context of coding stated in the procedures section: 0 is assigned to female, single, living in family with no children, finished secondary school, living in rural area, while 1 is assigned to male, married, living in family with children, student or finished faculty or college, living in urban area. Hereby, there is no any socio-demographic characteristic of respondents that statistically significantly influences Consumer optimism regarding confidence in

Table 7. Values of parameters for the dependent variable of the Consumer optimism regarding confidence in food safety

Parameter	B	Std. Error	t	Sig.	95% Confidence Interval	
					Lower Bound	Upper Bound
Intercept	3.893	0.266	14.641	0.000	3.369	4.416
gender	-0.085	0.186	-0.457	0.648	-0.451	0.281
age	0.006	0.007	0.793	0.428	-0.009	0.021
marital status	-0.084	0.220	-0.383	0.702	-0.517	0.349
presence of children in family	0.324	0.230	1.409	0.160	-0.129	0.776
household size	-0.077	0.085	-0.909	0.364	-0.245	0.090
education	-0.029	0.189	-0.152	0.879	-0.401	0.343
self-assessed household income	0.022	0.116	0.188	0.851	-0.206	0.250
place of residence	0.186	0.176	1.056	0.292	-0.160	0.531

Source: Authors' development

food safety. However, it can be stated that at the level of average age, average household size, average self-assessed household income, and under other initial conditions (all other levels except the discussed coded with 0), in the units of seven-point Likert scale, men are less optimistic regarding food safety in relation to women (by 0.085), married people are less optimistic regarding food safety in relation to single ones by (0.084), people living in families with children are more optimistic regarding food safety in comparison to people living in families with no children by (0.324), students or respondents that graduated faculty or college are less optimistic regarding food safety compared to the respondents who graduated secondary school by (0.029), and people living in urban areas are more optimistic regarding food safety in relation to people living in rural areas by (0.186).

Besides, when it comes to age, it can be seen that increase in age by one year in relation to average age (at other initial conditions) increases Consumer optimism regarding confidence in food safety by 0.006 on average on a seven-point Likert scale. As regards household size, it can be stated that the increase by one family member in relation to average household size (at other initial conditions) decreases Consumer optimism regarding confidence in food safety by 0.077 on average on a seven-point Likert scale. When it comes to self-assessed household income, it can be found that the growth by 1 on a five-point Likert scale in comparison with an average grade, at other initial conditions, increases the Consumer optimism regarding confidence in food safety by 0.022 on a seven-point Likert scale.

The level of the Consumer optimism regarding confidence in food safety can be considered as a combination of levels of respondents' socio-demographic characteristics (gender, education, marital status, presence of children in family, and place of residence) at the level of average age, average household size, and average self-assessed household income (Table 8).

At the level of average age, average household size, and average self-assessed household income, Consumer optimism regarding confidence in food safety is mostly manifested by females, single respondents, people living in families with children, respondents who finished secondary school and living in urban places (4.40).

Table 8. Consumer optimism regarding confidence in food safety (decreasing values) and socio-demographic characteristic

Gender	Marital status	Presence of children in family	Education	Place of residence	Consumer optimism regarding confidence in food safety
female	single	with children	graduated secondary school	urban	4.40
female	single	with children	student or graduate faculty or college	urban	4.37
female	married	with children	graduated secondary school	urban	4.32
male	single	with children	graduated secondary school	urban	4.32
female	married	with children	student or graduate faculty or college	urban	4.29
male	single	with children	student or graduate faculty or college	urban	4.29
male	married	with children	graduated secondary school	urban	4.23
female	single	with children	graduated secondary school	rural	4.22
male	married	with children	student or graduate faculty or college	urban	4.20
female	single	with children	student or graduate faculty or college	rural	4.19
female	married	with children	graduated secondary school	rural	4.13
male	single	with children	graduated secondary school	rural	4.13
female	married	with children	student or graduate faculty or college	rural	4.10
male	single	with children	student or graduate faculty or college	rural	4.10
female	single	no children	graduated secondary school	urban	4.08
female	single	no children	student or graduate faculty or college	urban	4.05
male	married	with children	graduated secondary school	rural	4.05
male	married	with children	student or graduate faculty or college	rural	4.02
female	married	no children	graduated secondary school	urban	3.99
male	single	no children	graduated secondary school	urban	3.99
female	married	no children	student or graduate faculty or college	urban	3.97
male	single	no children	student or graduate faculty or college	urban	3.96
male	married	no children	graduated secondary school	urban	3.91
female	single	no children	graduated secondary school	rural	3.89
male	married	no children	student or graduate faculty or college	urban	3.88
female	single	no children	student or graduate faculty or college	rural	3.86
female	married	no children	graduated secondary school	rural	3.81
male	single	no children	graduated secondary school	rural	3.81
female	married	no children	student or graduate faculty or college	rural	3.78
male	single	no children	student or graduate faculty or college	rural	3.78
male	married	no children	graduated secondary school	rural	3.72
male	married	no children	student or graduate faculty or college	rural	3.69

Source: Authors' development

Accordingly to the previous results, by implementing multiple regression with the General Linear Model procedure, the results for the dependent variable of the Consumer pessimism regarding confidence in food safety (PES) can be expressed by a regression equation as follows:

PES = 3.496 – 0.024 * gender – 0.006 * age + 0.290 * marital status – 0.353 * presence of children in family + 0.370 * household size + 0.202 * education – 0.171 * self-assessed household income – 0.079 * place of residence.

The values of parameters for the dependent variable of the Consumer pessimism regarding confidence in food safety at the level of average age, average household size, and average self-assessed household income are presented in Table 9.

It is reasonable to interpret the above equation, as well as the previous one, only in the context of coding stated in the procedures section. Hereby, the only socio-demographic characteristic of respondents that statistically significantly influences the Consumer pessimism regarding confidence in food safety is household size in a manner that its increase by one member in relation to average household size (at other initial conditions) increases the Consumer pessimism regarding confidence in food safety by 0.370 on average on a seven-point Likert scale.

However, when analysing other characteristics, it can be stated that at the level of average age, average household size, average self-assessed household income, and under other initial conditions, in the units of seven-point Likert scale, men are less pessimistic regarding food safety in relation to women (by 0.024), married people are more pessimistic regarding food safety in relation to single ones (by 0.290), people living in families with children are less pessimistic regarding food safety in comparison to people living in families with no children (by 0.353), students or respondents that graduated faculty or college are more pessimistic regarding food safety than respondents who graduated secondary school (by 0.202), and people living in urban areas are less pessimistic regarding food safety in relation to people living in rural areas (by 0.079).

Table 9. Values of parameters for the dependent variable of the Consumer pessimism regarding confidence in food safety

Parameter	B	Std. Error	t	Sig.	95% Confidence Interval	
					Lower Bound	Upper Bound
Intercept	3.496	0.306	11.433	0.000	2.894	4.098
gender	-0.024	0.214	-0.112	0.911	-0.445	0.397
age	-0.006	0.009	-0.715	0.475	-0.023	0.011
marital status	0.290	0.253	1.146	0.253	-0.208	0.788
presence of children in family	-0.353	0.264	-1.337	0.182	-0.873	0.167
household size	0.370	0.098	3.790	0.000	0.178	0.563
education	0.202	0.217	0.929	0.354	-0.226	0.630
self-assessed household income	-0.171	0.133	-1.283	0.200	-0.434	0.091
place of residence	-0.079	0.202	-0.392	0.696	-0.477	0.318

Source: Authors' development

Besides, when it comes to age, it can be concluded that increase in age by one year in relation to average age (at other initial conditions) decreases the Consumer pessimism regarding confidence in food safety by 0.006 on average on a seven-point Likert scale. When analyzing self-assessed household income, it can be seen that its growth by 1 on a five-point Likert scale in comparison with an average grade, at other initial conditions, decrease the Consumer pessimism regarding confidence in food safety by 0.079 on a seven-point Likert scale.

The level of Consumer pessimism regarding confidence in food safety can also be considered as a combination of levels of respondents' socio-demographic characteristics (gender, education, marital status, presence of children in family, and place of residence) at the level of average age, average household size, and average self-assessed household income (Table 10).

At the level of average age, average household size, and average self-assessed household income, the Consumer pessimism regarding confidence in food safety is mostly manifested by females, married respondents, people living in families with no children, students or people graduated faculty or college and living in rural areas (3.99).

SOLUTIONS AND RECOMMENDATIONS

When analysing the results of research showing average optimism being 4.08, while average pessimism 3.52, in the context of seven-point Likert scale used within this research, it can be seen that in both cases respondents are in average not sure whether they have reasons to be optimistic or pessimistic (in the second case this mark is between "Disagree" and "Neither agree nor disagree"). These results resemble the results of the research by De Jonge et al. (2004), by which the mean for all statements considering consumer confidence in food safety at the level of all respondents was 3.1 on a scale from 1 (low confidence) to 5 (high confidence).

When considering the results in the context of socio-demographic characteristics of the respondents, it can be concluded that married people statistically significantly differ from single ones by being more pessimistic in regard to food safety. Pessimism in food safety significantly increases along with the increase in the size of households, while significantly decreases with the increase of self-assessed household income. These results are also in accordance with De Jonge et al. (2004), where there are significant differences across socio-demographic segments in regard to consumer confidence in food safety. Besides, the results are in accordance with Frewer et al. (2010) who notice the connection of individual differences and consumer perceptions of food safety in regard to socio-demographic characteristics (gender, age, education level, and a number of children in the household). Finally, differences existing when it comes to one of the dimensions of the Consumer confidence in food safety are in accordance with De Jonge et al. (2008b) when stating that such differences suggest that optimism and pessimism are distinct concepts that should not be considered as two end poles of a unidimensional concept.

When it comes to the application of the General Linear Model and consideration of influences of the researched socio-demographic characteristics of respondents on two dimensions of the Consumer confidence in food safety (optimism and pessimism), it is concluded that the only statistically significant influence is the one of household size on the Consumer pessimism in a manner that increases in household size influences increase of pessimism. Such results furthermore support discussion stated above, although some differences occur as a consequence of different statistical approaches.

Table 10. Consumer pessimism regarding confidence in food safety (decreasing values) and socio-demographic characteristics

Gender	Marital status	Presence of children in family	Education	Place of residence	Consumer optimism regarding confidence in food safety
female	married	no children	student or graduate faculty or college	rural	3.99
male	married	no children	student or graduate faculty or college	rural	3.96
female	married	no children	student or graduate faculty or college	urban	3.91
male	married	no children	student or graduate faculty or college	urban	3.89
female	married	no children	graduated secondary school	rural	3.79
male	married	no children	graduated secondary school	rural	3.76
female	married	no children	graduated secondary school	urban	3.71
female	single	no children	student or graduate faculty or college	rural	3.70
male	married	no children	graduated secondary school	urban	3.68
male	single	no children	student or graduate faculty or college	rural	3.67
female	married	with children	student or graduate faculty or college	rural	3.63
female	single	no children	student or graduate faculty or college	urban	3.62
male	married	with children	student or graduate faculty or college	rural	3.61
male	single	no children	student or graduate faculty or college	urban	3.59
female	married	with children	student or graduate faculty or college	urban	3.56
male	married	with children	student or graduate faculty or college	urban	3.53
female	single	no children	graduated secondary school	rural	3.50
male	single	no children	graduated secondary school	rural	3.47
female	married	with children	graduated secondary school	rural	3.43
female	single	no children	graduated secondary school	urban	3.42
male	married	with children	graduated secondary school	rural	3.41
male	single	no children	graduated secondary school	urban	3.39
female	married	with children	graduated secondary school	urban	3.35
female	single	with children	student or graduate faculty or college	rural	3.34
male	married	with children	graduated secondary school	urban	3.33
male	single	with children	student or graduate faculty or college	rural	3.32
female	single	with children	student or graduate faculty or college	urban	3.27
male	single	with children	student or graduate faculty or college	urban	3.24
female	single	with children	graduated secondary school	rural	3.14
male	single	with children	graduated secondary school	rural	3.12
female	single	with children	graduated secondary school	urban	3.06
male	single	with children	graduated secondary school	urban	3.04

Source: Authors' development

Besides the managerial implications of the results of the research, whereas optimism in regard to food safety should be increased and pessimism decreased, in the latter case, there can be identified the managerial implication coming from the researches listed in the literature review, i.e. establishment of traceability and labelling, segmented communication approaches, and public involvement in risk management decision-making.

FUTURE RESEARCH DIRECTIONS

Future researches could include a more representative sample, connection to certain product categories as well as including additional variables in relation to food safety. These researches could also explore in detail the relation of food safety and quality in the perception of domestic consumers and their relation to organic and functional food.

CONCLUSION

The topic of this chapter is related to the divergence between objective scientific evidence of food safety and consumer perception of food safety risks, i.e. differentiation of objective food safety (assessment of a risk of consuming a certain food) and subjective food safety (existing in consumer's mind). The chapter listed the authors who emphasized consumers' behavior regarding food safety stating that they systematically overestimated risks in relation to the technical probability of harm occurring, while underestimated other risks. Despite distinguishing broad categories of risks, the authors have not found enough differences between various types of risk within a particular food group. Even in the conditions of existence of information focused on reducing food safety uncertainty, consumers do not process this information. In this regard, there was given a review that consumer risk perception of food safety was not largely determined by the hazard per se, but more by its social and psychological characteristics, i.e. that valuation performed by consumers regarding the severity of different risks actually differed from valuations given by the persons with knowledge about a specific hazard domain.

The chapter also seeks in the literature for the explanation of such consumer behavior. Hereby, there were cited the authors stating that the causes of the gap between expert opinions were found in a so-called perception filter. They explained pathway from risk to crisis in relation to the theory of social amplification of risk by identifying three factors, providing insight into the problem of why some relatively minor risks frequently become crucial for the evolution of risk into a crisis: fright factors, panic elements, and societal triggers.

Besides giving a broad description of the researches regarding consumer relation to food safety conducted abroad, the chapter presents the results of the first application of the Consumer confidence in food safety scale in Serbia. The results show that average optimism is 4.08, while average pessimism is 3.52, in the context of the seven-point Likert scale used within this research. The results are considered in the context of socio-demographic characteristics of the respondents firstly by using tests of differences and relations and secondly by using multiple regression by applying General Linear Model. In the first case, it is shown that married people statistically significantly differ from single ones by being more pessimistic in regard to food safety. Pessimism in food safety significantly increases along with the increase of household size, while significantly decreases along with the increase of self-assessed household income of consumers. Secondly, in the application of General Linear Model, it is concluded that the only statistically significant influence is the one of household size on pessimism in a manner that increases in household size cause an increase of pessimism.

The chapter included managerial implications from both primary and secondary research.

ACKNOWLEDGMENT

This chapter is written as a part of the research projects III47009 (European Integrations and Social and Economic Changes in Serbian Economy on the Way to the EU) and OI179015 (Challenges and Prospects of Structural Changes in Serbia: Strategic Directions for Economic Development and Harmonisation with EU Requirements), financed by the Ministry of Education, Science, and Technological Development of the Republic of Serbia.

REFERENCES

Aertsens, J., Verbeke, W., Mondelaers, K., & Van Huylenbroeck, G. (2009). Personal Determinants of Organic Food Consumption: A Review. *British Food Journal, 111*(10), 1140–1167. doi:10.1108/00070700910992961

Anderson, J. C., & Gerbing, D. W. (1982). Some Methods for Respecifying Measurement Models to Obtain Unidimensional Construct Measurement. *JMR, Journal of Marketing Research, 19*(4), 453–460. doi:10.2307/3151719

Churchill, G. A. Jr. (1979). A Paradigm for Developing Better Measures of Marketing Constructs. *JMR, Journal of Marketing Research, 16*(1), 64–73. doi:10.2307/3150876

Danes, J. E., & Mann, O. K. (1984). Unidimensional Measurement and Structural Equation Models with Latent Variables. *Journal of Business Research, 12*(3), 337–351. doi:10.1016/0148-2963(84)90016-X

De Jonge, J., Frewer, L., Van Trijp, H., Renes, R. J., De Wit, W., & Timmers, J. (2004). Monitoring Consumer Confidence in Food Safety: An Exploratory Study. *British Food Journal, 106*(10/11), 837–849. doi:10.1108/00070700410561423

De Jonge, J., Van Trijp, J. C. M., Van der Lans, I. A., Renes, R. J., & Frewer, L. J. (2008a). *A Scale for Consumer Confidence in the Safety of Food.* Discussion Paper no. 44. Wageningen: Mansholt Graduate School of Social Sciences.

De Jonge, J., Van Trijp, J. C. M., Van der Lans, I. A., Renes, R. J., & Frewer, L. J. (2008b). How Trust in Institutions and Organizations Builds General Consumer Confidence in the Safety of Food: A Decomposition of Effects. *Appetite, 51*(2), 311–317. doi:10.1016/j.appet.2008.03.008 PMID:18450326

Frewer, L. J., De Jonge, J., & Van Kleef, E. (2010). Consumer Perceptions of Food Safety. In B. P. Mansourian, A. Wojtczak, & B. McA. Sayers (Eds.), *Medical Sciences* (pp. 244–264). Oxford, UK: EOLSS Publishers.

Grubor, A., & Djokic, N. (2016). Organic Food Consumer Profile in the Republic of Serbia. *British Food Journal*, *118*(1), 164–182. doi:10.1108/BFJ-06-2015-0225

Grubor, A., Djokic, N., Milicevic, N., & Kovac-Znidersic, R. (2014). Acceptance of Food with "Health Claims" in Serbia: Consumers' Profile Research and Recommendations for Promotion. *Custos e Agronegócio Online*, *10*(4), 199–220.

Grunert, K. G. (2005). Food Quality and Safety: Consumer Perception and Demand. *European Review of Agriculture Economics*, *32*(3), 369–391. doi:10.1093/eurrag/jbi011

Hair, J. H., Tatham, R. L., Anderson, R. E., & Black, W. (1998). *Multivariate Data Analysis.* Upper Saddle River, NJ: Prentice Hall.

Hooper, D., Coughlan, J., & Mullen, M. (2008). Structural Equation Modelling: Guidelines for Determining Model Fit. *Electronic Journal of Business Research Methods*, *6*(1), 53–60.

Hu, L., & Bentler, P.M. (1999). Cutoff Criteria for Fit Indexes in Covariance Structure Analysis: Conventional Criteria versus New Alternatives. *Structural Equation Modelling: A Multidisciplinary Journal*, *6*(1), 1–55.

Kaiser, H. F. (1960). The Application of Electronic Computers to Factors Analysis. *Educational and Psychological Measurement*, *20*(1), 141–151. doi:10.1177/001316446002000116

Kaiser, H. F. (1970). A Second Generation Little Jiffy. *Psychometrika*, *35*(4), 401–415. doi:10.1007/BF02291817

Kaiser, H. F. (1974). An Index of Factorial Simplicity. *Psychometrika*, *39*(1), 31–36. doi:10.1007/BF02291575

Matsunaga, M. (2010). How to Factor-Analyse your Data Right: Do's Dont's, and How-To's. *International Journal of Psychological Research*, *3*(1), 97–110.

Nunnally, J. C. (1978). *Psychometric Theory*. New York, NY: McGraw-Hill.

Nunnally, J. C., & Bernstein, I. H. (1994). *Psychometric Theory*. New York, NY: McGraw-Hill.

Ritson, C., & Mai, L. W. (1998). The Economics of Food Safety. *Nutrition & Food Science*, *98*(5), 253–259. doi:10.1108/00346659810224163

Steenkamp, J. E. M., & van Trijp, H. C. M. (1991). The Use of LISREL in Validating Marketing Constructs. *International Journal of Research in Marketing*, *8*(4), 283–299. doi:10.1016/0167-8116(91)90027-5

Van Rijswijk, W., & Frewer, L. J. (2008). Consumer Perceptions of Food Quality and Safety and Their Relation to Traceability. *British Food Journal*, *110*(10), 1034–1046. doi:10.1108/00070700810906642

Verbeke, W., Frewer, L. J., Scholderer, J., & De Brabander, H. F. (2007). Why Consumers Behave as They Do with Respect to Food Safety and Risk Information. *Analytica Chimica Acta*, *586*(1/2), 2–7. doi:10.1016/j.aca.2006.07.065 PMID:17386689

Yeung, R. M. W., & Morris, J. (2001). Food Safety Risk – Consumer Perception and Purchase Behavior. *British Food Journal*, *103*(3), 170–186. doi:10.1108/00070700110386728

Zubovic, J., & Domazet, I. (2009). Organic Food Market Development: The Case of Serbia. In J.-P. Guichard (Ed.), *Challenges of Economic Sciences in the 21st Century* (pp. 179–189). Belgrade: Institute of Economic Sciences.

ADDITIONAL READING

Alhakami, A. S., & Slovic, P. (1994). A Psychological Study of the Inverse Relationship between Perceived Risk and Perceived Benefit. *Risk Analysis*, *14*(6), 1085–1096. doi:10.1111/j.1539-6924.1994.tb00080.x PMID:7846317

Anderson, E. S., Winett, R. A., & Wojcik, J. R. (2000). Social-Cognitive Determinants of Nutrition Behavior among Supermarket Food Shoppers: A Structural Equation Analysis. *Health Psychology*, *19*(5), 479–486. doi:10.1037/0278-6133.19.5.479 PMID:11007156

Barnett, J., & Breakwell, G. M. (2001). Risk Perception and Experience: Hazard Personality Profiles and Individual Differences. *Risk Analysis*, *21*(1), 171–177. doi:10.1111/0272-4332.211099 PMID:11332545

Baron, J., Hershey, J. C., & Kunreuther, H. (2000). Determinants of Priority for Risk Reduction: The Role of Worry. *Risk Analysis*, *20*(4), 413–427. doi:10.1111/0272-4332.204041 PMID:11051067

Berg, L. (2004). Trust in Food in the Age of Mad Cow Disease: A Comparative Study of Consumers Evaluation of Food Safety in Belgium, Britain and Norway. *Appetite*, *42*(1), 21–32. doi:10.1016/S0195-6663(03)00112-0 PMID:15036780

Dosman, D. M., Adamowicz, W. L., & Hrudey, S. E. (2001). Socioeconomic Determinants of Health- and Food Safety-Related Risk Perceptions. *Risk Analysis*, *21*(2), 307–317. doi:10.1111/0272-4332.212113 PMID:11414539

Eiser, R. J., Miles, S., & Frewer, L. J. (2002). Trust, Perceived Risk, and Attitudes toward Food Technologies. *Journal of Applied Social Psychology*, *32*(11), 2423–2433. doi:10.1111/j.1559-1816.2002.tb01871.x

Fischhoff, B., Slovic, P., Lichtenstein, S., Read, S., & Combs, B. (1978). How Safe is Safe Enough? A Psychometric Study of Attitudes towards Technological Risks and Benefits. *Policy Sciences*, *9*(2), 127–152. doi:10.1007/BF00143739

Frewer, L. J., Howard, C., Hedderley, D., & Shepherd, R. (1996). What Determines Trust in Information about Food-Related Risks? Underlying Psychological Constructs. *Risk Analysis*, *16*(4), 473–485. doi:10.1111/j.1539-6924.1996.tb01094.x PMID:8819340

Frewer, L. J., Miles, S., & Marsh, R. (2002). The Media and Genetically Modified Foods: Evidence in Support of Social Amplification of Risk. *Risk Analysis*, *22*(4), 701–711. doi:10.1111/0272-4332.00062 PMID:12224744

Frewer, L. J., & Salter, B. (2002). Public Attitudes, Scientific Advice and the Politics of Regulatory Policy: The Case of BSE. *Science & Public Policy*, *29*(2), 137–145. doi:10.3152/147154302781781092

Frewer, L. J., & Salter, B. (2003). The Changing Governance of Biotechnology: The Politics of Public Trust in the Agri-Food Sector. *Applied Biotechnology, Food Science and Policy*, *1*(4), 1–8.

Green, J. M., Draper, A. K., & Dowler, E. A. (2003). Short Cuts to Safety: Risk and Rules of Thumb in Accounts of Food Choice. *Health Risk & Society*, *5*(1), 33–52. doi:10.1080/1369857031000065998

Miles, S., Brennan, M., Kuznesof, S., Ness, M., Ritson, C., & Frewer, L. J. (2004). Public Worry about Specific Food Safety Issues. *British Food Journal*, *106*(1), 9–22. doi:10.1108/00070700410515172

Slovic, P. (1987). Perception of Risk. *Science*, *236*(4799), 280–285. doi:10.1126/science.3563507 PMID:3563507

KEY TERMS AND DEFINITIONS

Consumer Confidence: Confidence which consumers have in terms of certain phenomenon or product (food they consume), based on their own or other consumers' experience.

Different Levels of General Confidence: Differences in the degree of confidence which consumers have in certain products depending on their own experience, gender, age, education, and other social, economic, demographic, psychological, and other characteristics.

Dimensions of Confidence: Dimensions of confidence which range from fully positive (optimism) to utterly negative (pessimism).

Food Safety Scale: Measuring scale which shows the degree of confidence of consumers in safety of certain food.

Functional Food: Food which biological characteristics have a positive impact on human health and certain physiological functions in order to provide necessary energy and nutrients, thus preventing certain diseases and improving physical condition.

Organic Food: Food produced in a natural way, without genetically modified or chemical additives, pesticides, and chemical fertilizers.

Profiling Consumers: Defining narrow segments of producers based on similarities in food consumption, while the difference between the defined segments is significant and refers to attitudes, values, and perception of food consumption.

Chapter 16
Improvement of Food Security Through Reforming of Domestic Veterinary Service:
Case of Russia

Anna Ivolga
Stavropol State Agrarian University, Russia

Natalia Bannikova
Stavropol State Agrarian University, Russia

Vladimir Trukhachev
Stavropol State Agrarian University, Russia

Anzhelika Baicherova
Stavropol State Agrarian University, Russia

ABSTRACT

The chapter gives an overview of current issues of achievement of food security through reforming of veterinary service in the light of the social, ecological, and economic development of the society. The authors analyze certain challenges of food security existing in emerging countries, including expansion of market relations within veterinary service. The results of organizational and economic reforms of veterinary service are assessed on the case of Russia in a form of the survey of the heads of regional veterinary centers. Such issues as governmental and public regulations of quarantine operations and the most dangerous animal diseases are also discussed. In order to improve food security, the authors justify the measures for competition and demand stimulation, smoothing consequences of market imperfections, encouraging veterinarians' efforts on reducing the negative impact of livestock breeding on the environment.

INTRODUCTION

Livestock breeding is an important sector of the global agriculture. In many countries, this sector is more advanced compared to crop production. One-third of cropland is occupied by forage crops for the purposes of animal husbandry.

In the modern conditions, livestock breeding largely determines food security of the countries. Livestock products provide a full range of nutritional value. Namely, they are the sources of energy, bio-digestible

DOI: 10.4018/978-1-5225-2733-6.ch016

minerals and vitamins, and high-quality animal protein containing all essential amino acids. Each person should consume above 20 grams of animal protein per day for a healthy diet. This can be achieved by annual consumption of 33 kg of meat, or 45 kg of fish, or 60 kg of eggs, or 230 kg of milk, respectively. Both production and consumption of animal products grow rapidly throughout the world. According to the forecast of the Food and Agriculture Organization of the United Nations [FAO] (2015), world meat production is expected to double by 2050, mainly owing to developing and emerging countries. The growth rate of meat production could exceed the growth rate of the world population. The possibility of advancing growth of meat sector in the developing and emerging countries was revealed. Such countries will be able to satisfy their domestic demand at the expense of their own production (FAO, 2015).

However, the growing scale of livestock is accompanied by a wide range of problems. One of the main problems is an environmental impact in the forms of degradation of pastures, irreversible changes of the local genetic resources of animals and plants, increasing water consumption, soil and water pollution with animal waste, as well as drugs used in animals breeding. Thus, ensuring animal health is one of the most important challenges of combating negative environmental impact. This is supported by ever-growing attention of scientists and experts, who consider the development of veterinary service in terms of its contribution to the sustainable development of the society and ensurance of food security (Davis, 2008; Gerber, Mottet, Opio, Falcucci, & Teillard, 2015; Ipema, Bleumer, & Lokhorst, 2011). Over the years, veterinary professionals have played significant roles not only in animal health, but also in human health and welfare, food quality, food safety, and food security (Caceres, 2012). Modern veterinary practice includes a series of activities that lead to environmental pollution, soil degradation, and contamination. The inevitable consequences of the veterinary interventions are a collection, recycling, and disposal of biological waste. The problem of biological waste disposal and destruction is relevant for a number of countries, especially emerging ones (McLean, Watson, & Muswema, 2007; Traverse & Aceto, 2015; Osipova, 2008; Popkova, Dubova, Yakovleva, Azarova, & Titova, 2014). In addition, negative impact on the environment may be caused by the veterinary treatment of cattle, disinfection, disinfestation, and deratization. This also contributes to soil degradation as a result of the chemical effect of disinfectants.

Many publications are devoted to the research of the possibilities and ways to reduce antibiotic treatments in veterinary (Trevisi et al., 2014; Mohring et al., 2009; Caruso et al., 2013). In the developed countries, much attention is paid to the issues of ensuring animal welfare as an aspect of decisions about whether animal-usage systems are sustainable (Velarde, Fàbrega, Blanco-Penedo, & Dalmau, 2015; Broom, 2010; Da Silva & Naas, 2012).

Based on the vision of "one medicine", scientists consider the role of veterinary medicine in protecting food security and safety (Pappaioanou, 2004; Sargeant, 2008; Steele, 2008). Dividing the concepts of food security and food safety, in this chapter, the authors address the food security in relation to the veterinary service understanding that the activities of veterinary services have a significant impact on food security which is relatively homogeneous over the four components: availability, accessibility, utilization, and sustainability. According to Zinsstag, Schelling, Waltner Toews, and Tanner (2011), the concept of "one medicine" should be transformed into the concept of "one health", and even "ecosystem health". Considering the intensifying processes of globalization, it should assume a global scale.

At the same time, the above-mentioned problems related to veterinary service remain relevant for the majority of the emerging countries because they have not yet found any optimal solutions of economic issues of veterinary care. In many countries, demand for veterinary services is quite low because of the low income of farmers and lack of qualified veterinarians, insufficient technical, financial, and administrative support of their work. This situation adversely affects the stability of livestock breeding, environment, and rural development.

The solution to this problem is largely connected to the reform of the national veterinary services (Ahuja, Umali-Deininger, & de Haan, 2003; Rutabanzibwa, 2011). Experts note that a flexible combination of state support to the veterinary services and active stimulation of the commercial sector of the veterinary medicine is required (Leonard, 1987; Ekboir, 1999; Turkson, Slenning, & Brownie, 1999; Rubyogo, Murithii, Agumbah, & Obhai, 2005; Bhandari & Wollen, 2008; Amankwah et al., 2014). In emerging countries, institutional reforms run at different speeds and with different efficiency. A success can be achieved through analysis of international experience, as well as regular monitoring of anticipated and unanticipated effects of privatization and decentralization to enable policy adjustment.

The main idea of the research presented in this chapter is to investigate the experience of reforming of the veterinary services in one of the emerging countries. The authors studied the case of Russia and surveyed the heads of the regional departments of the veterinary service in one of the Russian regions.

Research tasks included:

- Identification of the strengths and weaknesses of the reform, evaluation of its impact on different divisions;
- Expert survey to study the opinion of the veterinarians on the features and results of the further development of market relations in veterinary medicine, government impact on the improvement of the veterinary service, opportunities of stimulating demand for veterinary services, the role of veterinary medicine in ecosystem health improvement;
- Generalization and ranking of the survey results, theoretical justification of the level of government intervention;
- Development of solutions and recommendations for improving the work of the veterinary services departments based on the identified problems.

BACKGROUND

Russian animal breeding is one of the largest in the world. Russia takes the 5th place in the world in meat production, 6th place in milk production, 9th place in wool clipping (Khokhlov, 2014). However, during the 1990s, Russian livestock breeding was damaged by the market reforms and transition processes happened after the collapse of the Soviet Union in all post-soviet countries. The main damaging factor was the fall in purchasing power, which affected the most income-elastic meat and dairy products. In addition, the efficiency of livestock production and yield per unit of forage in Russia during the 1970-1980s were much lower compared to the developed countries. Therefore, import of meat and dairy products to the country rose sharply and created a competitive position to the Russian producers giving them no time for fundamental modernization of the industry.

Along with the sharp decline in output, significant structural changes occurred. Before the 1990s, the basic commodity production was concentrated in the large public agricultural enterprises. As of 2015, 54.7% of cattle and 53.4% of sheep and goats are produced by small-scale production units: farms and private subsidiary farms.

Such transformation required substantial changes to the system of veterinary care. Both before the market transformation and now, only a few veterinarians are responsible for the health condition of livestock at large agricultural enterprises. However, in small-scale farms, animal management required the establishment of a special service. In Russia, this service was established on the basis of public institutions – district centers for animal disease control that existed in each administrative area. Such centers became the major veterinary units engaged in rendering services to the owners of small animal farms within the territory of an administrative district, as well as in the coordination of the work of veterinarians at large enterprises and private veterinarians engaged in entrepreneurial activity.

The economic basis for the veterinary services has also undergone significant changes. In the planned economy, the cost of keeping and organizing the veterinary service was financed by the government. The liberalization of economic relations in the sphere of animal husbandry and veterinary medicine started in the 1990s along with the national market reforms. In 1991, public veterinary institutions financed by the state were allowed to render commercial veterinary services. However, for over 20 years, state veterinary authorities had no right to administer their earnings. Money obtained from the rendered commercial services was transferred to the regional budgets and then distributed centrally between all district centers for animal disease control located in the area. A number of budget allocations were calculated based on the performance achieved by the district centers for animal disease control centers during previous years. Thus, the degree of state economic regulation of the veterinary service remained high.

The situation changed in 2013 when the new model of market activities of district centers for animal disease control was introduced. The amount of budget money transferred to the accounts of the veterinary centers is determined based on the performance of the governmental plan on rendering veterinary services to small private farms, not the operational plan of the center itself. A governmental plan is developed based on livestock number on farms and smallholdings within the territory of an administrative district and is aimed at the prevention of quarantine and dangerous diseases. It includes scheduled preventive vaccinations and treatments of animals, scheduled diagnostic tests, a sampling of biological material for diagnostic studies on quarantine and especially dangerous animal diseases, as well as registration and issuing certificates of compliance of foodstuffs with the requirements of the veterinary standards of the food market.

The fact that such types of veterinary services are financed and controlled by the state is justified. Thus, both Lin et al. (2003) and Trevisi et al. (2014) emphasize the priority of preventive measures for the formation of the so-called "sustainable medicine" (SM) as one of the important components of a wide range of possible sustainable approaches to peaceful co-existence.

On the other hand, the involvement of a government in the facilitation of preventive measures for small private farms has been remaining top-priority in emerging countries for a long time (Leonard, 1987; Ekboir, 1999; Amankwah et al., 2014). The reasons for that have been low-income of farmers and lack of experience and industrial practices which both have led to underestimation of preventive measures and consequences of dangerous animal diseases to the society and environment.

The second feature of the new model of the market functioning of district centers for animal disease control is associated with the use of funds from commercial services. Previously, money earned by the veterinary centers was transferred to the regional budget. Currently, the money remains on the centers' accounts and is spent according to the centers' needs. Thus, the transition to the new finance system promoted the commercial component of veterinary centers' activities, strengthening their autonomy and financial independence.

To assess the results of the implementation of the new finance system of public veterinary institutions the authors conducted an expert survey on the case of Stavropol region of Russia. The region was selected as a case-study because of its well-developed animal husbandry industry. The share of animal production in total agricultural output in Stavropol region is 31%. Stavropol region ranks the second in Russia by a number of sheep and output of wool, the ninth by poultry stock, and the tenth by a number of cows and output of milk. At the same time, 69% of cattle and 82% of sheep are kept on small farms (Territorial Authority of the Federal State Statistics Service of the Russian Federation in Stavropol Region, 2014). Veterinary care of this livestock is carried out by the experts from the district centers for animal disease control. Each year, they complete over 640 million actions (vaccination, research, and animal decontamination), over 10 million laboratory and diagnostic studies, and 120 million of veterinary and sanitary examinations.

31 heads of district centers for animal disease control took part in the survey. 73.3% of them have more than 15 years of experience in veterinary medicine. The questionnaire forms were forwarded to them via e-mails and then completed forms were returned to the authors by post.

MAIN FOCUS OF THE CHAPTER

The transition to the new model of district centers of animal disease control pursued economic aims related to saving budget funds, strengthening financial independence of disease control centers, and stimulating demand for veterinary services. The most obvious consequence of this transition was a reduction of staff. Before the reform, the average number of staff per a disease control center was 70 people, now it is about 53 people.

One of the main hypothesis of the current study is that the results of transition are not similar for the veterinary centers in different districts. In order to prove it, the authors grouped the questionnaires by the size of animal population served by the district centers. Grouping the survey results allowed the authors to distinguish four groups of disease control centers. The first group includes large district veterinary networks that manage a large number of livestock (over 40 thousand livestock units). Over 70% of livestock is managed by small private farms. There is 21% of such administrative districts in the region. During the staff optimization, the number of employees per center was reduced by 14 persons in average, but the average total number of staff remained high – 61 persons. Veterinary centers are characterized by the high rate of government task performance (100% and more) and high level of income earned from commercial services.

The second group comprises veterinary centers where the livestock number amounts to 20-40 thousand units. Its considerable part (40-70%) is managed by small private farms. The share of such disease control centers is 24%. The staff was cut by 11 persons, and now the average number totals 52 persons. The majority of disease control centers in this group do not perform the government task in full volume.

The third group of the veterinary centers (42%) includes districts where livestock number is below 20 thousand units, but 70% of which is managed by small private farms. The average staff number of such centers is 47 persons. The reforms mainly affected this type of veterinary centers as 16 persons on average were fired. Underdevelopment of livestock in such administrative districts has an adverse effect on the volume of commercial services. The government task in not performed unevenly, the main targeted indicators are not met.

The fourth group of veterinary units includes city veterinary centers (about 13% of the total number). They do not treat livestock, but control processes of livestock products processing and sale. Their main function is to provide safety control.

The opinions of the heads of Stavropol district centers for animal disease control on the reform are very diverse. 40% of the respondents from first group mention did not find any disadvantages in the new system, whereas 25% of the respondents from the third group did not see any advantages. The analysis of the survey results shows that the main controversial issue is the most obvious result of the reform – staff optimization with an account of the livestock managed. Almost 47% of the respondents mentioned it as the main advantage, while 43% recognized it as the main drawback of the new system. Some experts explain that both the threat of dangerous disease spread and growth of livestock number result to the need of improving response activity, timely diagnostics, and veterinary care. This is the reason why the reduction of the staff number may endanger veterinary security.

The reduction of staff number also increased the load on the remained employees. On the other hand, it created conditions for the increase of salaries and incentives for specialists. The analysis shows that the largest number of staff optimization supporters is in the first group (60%) and the third group (54%). For the respondents from the third group, the main factor is saving limited means of such type of disease control centers. 87.5% of the respondents from the second group assess the staff layoff as the main drawback of the new system.

The lack of interest of disease control centers in environmental protection was considered as a disadvantage only by 7% of respondents. This highlights inadequate attention of experts to the issues related to sustainable development of rural areas. Strengthening of the financial independence of the centers as a positive result of the reform was mentioned by 43.3% of the respondents. It was particularly noted by the respondents in the second group (70%). Herewith, 51.7% of the respondents from the second group think that the new financing system enables a complete coverage of livestock with preventive measures and reveals new opportunities for the development of physical infrastructure.

Calculation of volume of state subsidies based on livestock number led to the change of the amount of state funding. In general, every fifth respondent mentioned the significant increase of budget funding after the transition to the new system. Herewith, there is a trend to decrease in state funding in 40 districts. 20% of the respondents from the first group mentioned the significant increase in state funding, 40% said it had slightly increased. That means that the new system improved financial situation in 60% of large district centers for animal disease control. Even bigger share of the respondents from the second group (71.4%) noted an increase in funding, while the situation in the third group was different. The reduced state financing was observed in 54% of veterinary centers. Every third respondent mentioned significant decrease of subsidy rates.

The volume of such preventive measures as vaccination, sample selection, and allergic response study is related to the current livestock number and the total level of livestock development in an administrative area. Herewith, over a half of the respondents mentioned the underperformance of the government task implementation. Only 13% of the centers succeeded to meet the targets. The authors believe that the main reason for that is an evasion of compulsory preventive veterinary and sanitary measures. This is connected to the underestimation of the importance of preventive measures for livestock health and environment as well as to reluctance of farmers to show the real livestock number. As a result, the governmental plan does not always correspond with the situation in the particular center. As a rule, the unaccounted livestock does not participate either in compulsory (free) or in commercial sanitary and veterinary treatment.

The threat of dangerous disease outbreaks remains relevant. It is difficult for the veterinarians to manage such diseases in the conditions of low awareness of their origin and course. Consequently, this situation can be regarded as a constraint for ensurance of food safety and food security in the area. The situation is complicated by organizational challenges. The current system of material and technical supply of the centers for animal disease control allows cases of untimely supply of vaccines and diagnostic materials. More than 53% of the respondents consider the problems of veterinary service supply with veterinary preparations as one of the most urgent ones.

The second important source of financing of the district centers for animal disease control is commercial services. Commercial services include all types of veterinary activities not included in the governmental plan. Commercial services are priced by each disease control center individually according to the guidelines developed by the regional state agency – Veterinary Authority. There is a unified mechanism of pricing, however, the guidelines do not provide any price limits or any other regulatory tools. As a result, the price of the same type of service varies from one district to another. This leads to the dissatisfaction between the owners of the farms located in neighboring areas, causes social conflicts, and reduces demand for veterinary services.

The problems related to pricing affect the amount of money received from rendering commercial services. Its volume reduced in 50% of the veterinary centers, while increased in 40% of those veterinary centers included to the first and the second groups. That means that the large veterinary centers have succeeded to strengthen their commercial component and increase their financial independence under new conditions.

The respondents note that the full list of commercial services is implemented only by a half of the veterinary centers. In most cases, this relates to the special features of veterinary network performance in a certain district. Thus, for example, not all the districts have veterinary laboratories. Another issue is a lack of legally formalized markets as it limits the opportunities for veterinary and sanitary examination of food products sold on the market. As the result, 42% of veterinary centers cannot cover their economic needs with money received from rendering commercial services. The main share of this money is spent for salaries and bonuses for employees. The other significant item is utility payments.

The same sort of situation is observed in the case of fixed assets purchase. Only 22.3% of the respondents considered equipment procurement as the main item for spending money received from rendering commercial services. According to the survey results, 13% of the veterinary centers do not spend their earnings on new equipment. As a result, in the districts with underdeveloped livestock breeding veterinary service is backward. There is a lack of own means for veterinarian vocational training.

Respondents were asked about the possible ways of solving existing problems. The majority of respondents believe that in order to improve veterinary service it is necessary to develop the system of financial incentives and bonuses in the case when a center systematically overperforms the governmental plan. Additional financial rewards of veterinarians are considered to be fair by 80% of the respondents from the first group, 71.4% of the respondents from the second group, and 38.5% of the respondents from the third group.

About 67% of the respondents mention the acute need for improving price policy of commercial services. This measure seems to be urgent for 80% of the respondents from the first and the second groups. These results are confirmed by the data obtained by Tregubov et al. (2012) carried out during the survey of Stavropol veterinarians on the procedural and institutional issues of veterinary service operation. In 2012, 74.1% of the respondents mentioned the problem of the inadequate price level.

Providing centers with modern equipment and vehicles is also considered as a significant issue. The importance of measures for equipment renovation was mentioned by 85.7% of the respondents from the second group and 61.5% of the respondents from the third group.

Many of the above-mentioned problems are typical for other Russian regions (Ilyinykh, 2013; Nikitin, Akmullin, & Trofimova, 2012), so the measures proposed based on the survey results are relevant for the improvement of the veterinary service.

Summarizing the survey results, the authors conclude that the reform of the veterinary service has undeniable positive organizational and economic results. The most significant of them are strengthening the commercial base of the veterinary service, staff optimization, financial incentives for the employees, more opportunities for earning financial resources, and use of new technologies. Undoubtedly, all those factors will contribute to the development of rural areas and environmental health. On the other hand, an extension of market relations caused further stratification of the network of district centers for animal disease control by financial and economic condition, opportunities for sustainable development, and ensurance of food security.

Generally, there is no opportunity to render commercial services in sufficient volume in the areas with a few livestock due to the low demand a lack of veterinary laboratories and laboratories of veterinary and sanitary examination. In such centers for animal disease control, possibilities of adequate salary rates are limited and technical backwardness and employees' dissatisfaction with the working conditions are observed. Those problems may be further complicated by a lack of transport in rural veterinary services, untimely provision of drugs, etc. As the result, preventive measures may be adversely affected, and the volume of medicines used in a case of animal disease outbreaks increases.

The lack of measures aimed at the adjustment of the working conditions of those centers can lead to the spiral development of adverse consequences, worsening conditions for veterinary security, and, as a result, further reductions in the number of livestock that taken together negatively affect food security in the area. Therefore, it is advisable to limit the differentiating features of the market, which can eventually become the signifying one. That is unacceptable in terms of the social, ecological, and economic importance of the veterinary service.

In the emerging economies, the need for state regulation of organizational and economic aspects of the veterinary service is caused by its affinity to infrastructure sectors. The issue of the market mechanism development and increasing its effectiveness is important and is still an unsolved problem in production infrastructure within the system of market economy in Russia. One of the main reasons is that many of the branches of production can be regarded as natural monopolies. Consequently, the rate of liberalization and its borders in various industries, as well as the resulting effect, are different.

According to Sulakshin et al. (2007), natural monopoly as an object of state regulation is the sphere of economic activity where competition is limited due to the specifics of the production process or contradicts the public interest. In order to identify the sectors that could be attributed to natural monopolies, the authors propose to allocate four types of market boundaries: economic, technological, social, and strategic.

Under such approach, it can be assumed that the veterinary infrastructure also has certain features of a natural monopoly. The leading element is the social boundaries of the market, according to which the social importance of the veterinary security relates to the issues of ensurance of food security. In this context, the economic boundaries of the market are also relevant, as competition in the field of quarantine and dangerous animal diseases prevention under conditions of market imperfection, which is usually observed in the emerging countries, could have negative economic consequences for agricultural production. Thus, competition in quarantine and dangerous animal diseases prevention is possible but requires public or government regulation in terms of creating conditions for ensurance of food security. It should be noted that in the developed countries the government always played an important role in the development and regulation of the veterinary service (Enticott, 2014; Kellar, 2012).

The degree of danger of animal diseases, which have to be eliminated by the veterinary service, can be regarded as social and technology boundaries of the market, beyond which the competition is rather effective. However, it should be noted that the concept of technological boundaries of the market is closely linked to the territorial limitation. In the field of veterinary medicine, this factor is quite important due to the territorial dispersal of a livestock. Therefore, in order to avoid local monopolies, a certain degree of state intervention and public control should be presented in the area of veterinary practice, which is not associated with dangerous animal diseases. This intervention should apply to regulation of tariffs on services, development of veterinary network entities, adjustment of differentiating and sanitizing functions of the market, i.e. those aspects that were mentioned by the respondents during the survey.

Based on the survey results, the authors formed a diagram of causes and effects in improving the efficiency of the veterinary services that reflects the importance of various factors (Figure 1).

The description of elements presented in Figure 1 and links between them is given in Table 1.

The system of factors presented in Figure 1 allows the authors to outline the priority goals and propose measures for further improvement of the new system of operation of the veterinary centers in terms of environmental and economic issues (Figure 2).

The first goal is connected with mitigation of the consequences of differentiative market function. Its solution should be aimed at the balance of commercial and public interests, primarily, in the area of pricing of veterinary services. According to the survey results, the issue of regulation, dynamics, and limits of prices for the veterinary services is of particular importance. There are the following drawbacks:

Figure 1. Causes and effects in improving efficiency of the veterinary services
Source: Authors' development

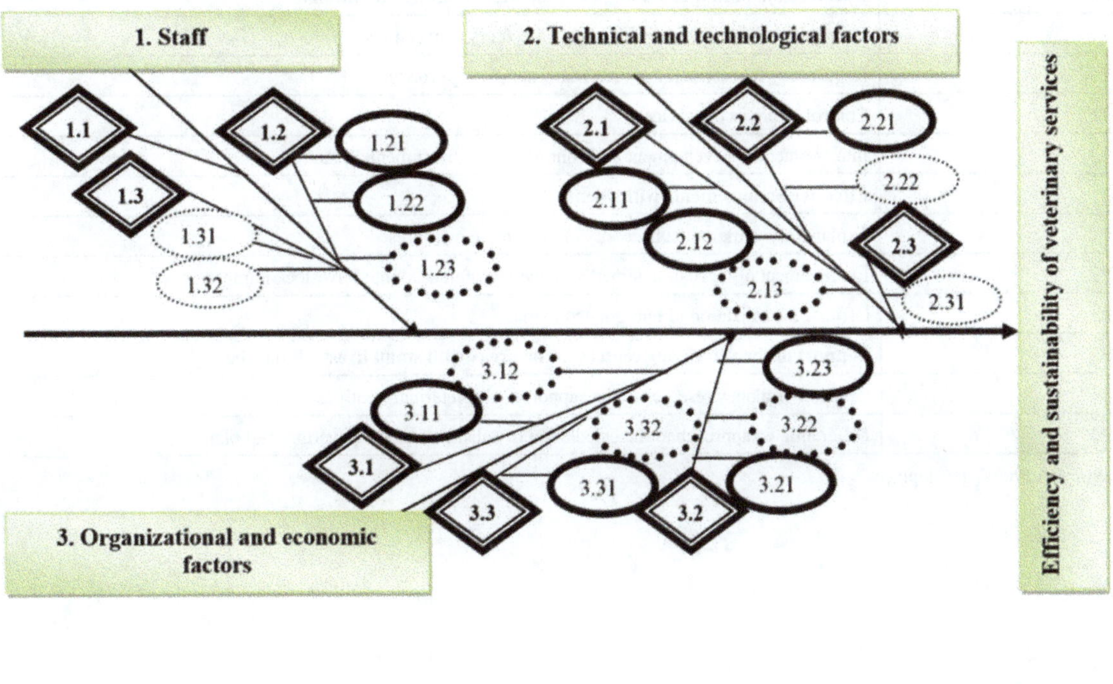

Table 1. Description of causes and effects in improving efficiency of veterinary services

Element	Description
	Resulting factors of the second level
	Resulting factors of the third level mentioned by more than 60% of the respondents
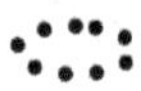	Resulting factors of the third level mentioned by 20-60% of the respondents
	Resulting factors of the third level mentioned by less than 20% of the respondents
1.1	**Targeted subsidizing of the veterinarians in rural areas**
1.2	**Staff management improvement**
1.21	Bonuses for the veterinarians for better performance
1.22	Optimization of staff structure of the veterinarian service
1.23	Improved availability of the qualified professionals
1.3	**Advanced training of the veterinary service employees**
1.31	Entrepreneurship training for the veterinarians
1.32	Exchange of experience with the veterinarians from other regions and countries, vocational training
2.1	**Modern equipment for the veterinary entities**
2.11	Availability of modern equipment
2.12	Availability of transport
2.13	Reconstruction of buildings belonging to the veterinary services
2.2	**Availability of veterinary preparations**
2.21	Well-timed provision of veterinary preparations to the district centers for animal disease control
2.22	Improvement of medication dosing techniques
2.3	**Introduction of new technologies to medical treatment of animals**
2.31	Cooperation with scientific and research centers, universities
3.1	**Regulatory framework for veterinary service operation**
3.11	Control of prices of services
3.12	Improvement of development and control of the governmental plan
3.2	**Effective arrangements with clients**
3.21	Explanatory work on the necessity of preventive measures
3.22	Improvement of legislation aimed at reduction of unaccounted livestock number
3.23	Official formalization of unregulated markets
3.3	**Support of the veterinary centers in the areas with small livestock number**
3.31	Creation of money reserves to the support of the veterinary entities
3.32	Differentiated approach to the calculation of subsidies for the governmental plan

Source: Authors' development

Figure 2. Goals and measures for operation of the district centers for animal disease control
Source: Authors' development

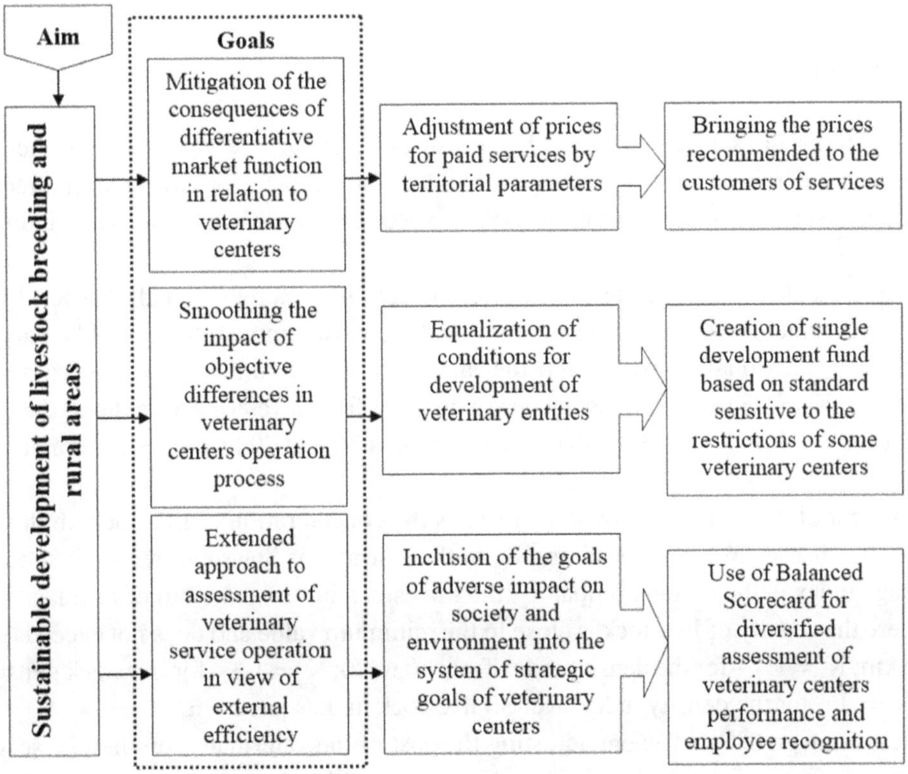

- Prices of the main veterinary and sanitary services significantly vary from one administrative district to another;
- Absence of the system of price regulation taking into account the objective differences in the performance of the veterinary centers;
- Price lists have not been updated for several years.

In order to eliminate the above-mentioned drawbacks, it is advisable to readjust recommended prices of the most common veterinary services once or twice a year. Moreover, the differences in operation conditions of the district centers for animal disease control should be noted. These should primarily include a spatial dispersion of animals. One of its major parameters is a distance from the main livestock breeding units to a district center for animal disease control calculated using a size and configuration of an area, as well as a density of livestock managed in numerous small farms.

The following integral coefficient comprehensively considers all these factors:

$$K_{in} = K_f * \bullet K_d \quad (1)$$

where

K_{in} – integral coefficient;

K_f – coefficient of distance;

K_d – coefficient of livestock density.

The cost of transport services incurred by a district veterinary service also significantly affects responsiveness to emerging livestock disease outbreaks. Thus, this factor having an indirect impact on the efficiency and stability of the veterinary services should be considered when calculating the prices of services.

In order to calculate the coefficient of distance, the authors differentiated all districts by a distance between a district center for animal disease control and the most remote settlement (the value of "transport leg"). Thus, it is reasonable to determine the value of K_f for the administrative districts of Stavropol region as follows: $K_f = 1.5$ for the districts where a distance from a veterinary center to the most remote settlement exceeds 70 km; $K_f = 1.3$ if a distance is between 50 and 70 km; $K_f = 1.1$ if a distance is less than 50 km.

The coefficient of the density of livestock reflects the concentration of livestock on the territory of an administrative district. When determining K_d in the districts of Stavropol region it should be noted that the average index in the region is equal to 15.64 livestock units/square km of farmland. $K_d = 1.5$ in the areas where the density of livestock is close to the minimum value and does not exceed ten livestock units/square km; $K_d = 1.4$ – for the density index from 10 to 20; $K_d = 1.3$ – for livestock density between 20 and 30; $K_d = 1.3$ for the density index over 30 livestock units/square km.

The values of integrated coefficient adjusting the cost of the veterinary commercial services in the administrative districts of Stavropol region are presented in Table 2.

As a case, the authors studied an example of mastitis treatment in Kochubeevskii and Grachevskii districts of Stavropol region. Currently, the prices of such service in Grachevskii and Kochubeevskii districts are 105 RUB and 205 RUB, respectively. However, the price recommended by the Veterinary Authority of Stavropol region is 72.4 RUB. The value of "transport leg" in Grachevskii district is 60 km, livestock density corresponds to the district average and does not exceed 15 livestock units/square km. In Kochubeevskii district, these parameters are 70 km and 42.4 livestock units/square km, respectively. Taking into account the inflation rate in 2014 (6.7%), the base price of mastitis treatment should be 77.25 RUB. Considering the value of K_{in}, the prices in Grachevskii and Kochubeevskii districts should be 140.6 RUB and 110.5 RUB, respectively.

It would be reasonable for the Veterinary Authority of Stavropol region informing the consumers about the price change. In order to facilitate this process, the recommended prices should be posted on the official website of the Veterinary Authority of Stavropol region, as well as in each district veterinary center so that to be available for the consumers of veterinary services.

The second goal of improving economic conditions in the district centers for animal disease control relates to the smoothing of conditions for their development. Sharp disparities in funding of the veterinary centers of Stavropol region for the purposes of the governmental plan performance on preventive measures and rendering commercial services should not affect the availability of necessary equipment and material resources. In order to overcome the imbalance, it is necessary to create conditions to support district centers for animal disease control, which belong to the third group, the most disadvantaged one according to the survey results. They are located in the administrative districts with small livestock

Table 2. Value of the integrated coefficient adjusting the cost of the veterinary commercial services in the districts of Stavropol region

Districts	K_r	K_d	K_{in}
Aleksandrovskii	1.1	1.5	1.65
Andropovskii	1.3	1.4	1.82
Apanasenkovskii	1.3	1.4	1.82
Arzgirskii	1.1	1.5	1.65
Blarodarnenskii	1.3	1.5	1.95
Budennovskii	1.3	1.5	1.95
Georgievskii	1.3	1.3	1.69
Grachevskii	1.3	1.4	1.82
Izobilnenskii	1.1	1.4	1.54
Ipatovskii	1.5	1.4	2.10
Kochubeevskii	1.3	1.1	1.43
Krasnogvardeyskii	1.1	1.3	1.43
Kirovskii	1.1	1.4	1.54
Kurskii	1.5	1.5	2.25
Levokumskii	1.5	1.3	1.95
Mineralovodskii	1.3	1.5	1.95
Neftekumskii	1.5	1.3	1.95
Novoaleksandrovskii	1.1	1.4	1.54
Novoselitskii	1.1	1.4	1.54
Petrovskii	1.1	1.5	1.65
Predgornii	1.1	1.4	1.54
Sovetskii	1.5	1.5	2.25
Stepnovskii	1.1	1.4	1.54
Trunovskii	1.1	1.4	1.54
Turkmenskii	1.5	1.5	2.25
Shpakovskii	1.5	1.1	1.65

Source: Authors' development

number, so the possibilities of increasing the volume of services and improving financial and economic conditions are limited. Thus, all these factors can result in worsening conditions of the veterinary security and further development of adverse trends in such centers, which can have a multiplier effect and cut the number of livestock in the area. Despite the current complicated situation, those centers for animal disease control are still responsible for the veterinary safety assurance and livestock breeding. Currently, the state support in providing the centers with the basic means is carried out mainly through the targeted funding from the regional budget, but this process is now complicated due to the budget deficit.

Under such circumstances, it would be advisable to create a regional fund for all veterinary centers. Its main purpose would be the formation of a reserve to meet the needs of the centers in modern equipment and vehicles. By paying a fixed percentage of income the centers receive from rendering commercial services, they could get the necessary amount of money in priority sequence and consider the urgency of their needs. Such system of co-financing of the veterinary network would also contribute to the strengthening of relations between district veterinary centers, which are located in the areas dependent on the veterinary security.

The authors suggest differentiation of the payments made by the veterinary centers to this fund based on their technical equipment, staffing, and availability of additional income sources, such as veterinary laboratories. Therefore, the authors propose the following procedure for determining the ratio of payment to the fund for the development of the veterinary service:

$$H_s = H_b + H_d + H_a + H_t + H_p \tag{2}$$

where

H_s – ratio of subscription of a district center for animal disease control to the development fund;

H_b – basic ratio;

H_d – ratio of livestock density;

H_a– ratio related to the absence of a laboratory;

H_t– ratio related to the lack of financial and logistical support;

H_p– ratio related to the insufficient availability of staff.

The structure of the ratios was determined based on the opinions of the experts surveyed. The ratio of livestock density reflects the efficiency level of preventive measures carried out by the veterinary centers.

The ratio related to the absence of a veterinary laboratory reflects the opportunities of a veterinary center for additional income that could be received from laboratory examinations. The veterinary laboratory is an important element of the veterinary infrastructure of an area. Its availability and appropriate equipment enable carrying out timely diagnostics and possessing the relevant information on the local epizootic situation. Under the new financing system, those district veterinary centers which cannot run their own laboratories are forced to close them down and devolve the power of examinations to the laboratories in neighboring areas. It reduced their incomes and increased the timing of the diagnostic procedures.

The ratio related to the lack of financial and logistical support reflects the level of veterinary center's equipment necessary for effective veterinary practice. Availability of vehicles for the timely veterinary care, adequate equipment of veterinary centers and clinics for dangerous disease elimination are necessary to ensure the veterinary security of an area and contribute to food security and sustainable development of animal husbandry.

The ratio related to the insufficient availability of staff determines the degree of availability of the veterinary specialists for a district center for animal disease control for carrying out therapeutic and preventive measures. In the result of the staff structure optimization under the new financing system, some veterinary centers became unable to keep staff and cut it down to a minimum, which negatively affected the results of their work.

In order to determine the values of the ratios under discussion, it is necessary to define the common standards for the material and technical base renewal, which primarily reflect the amount of annual depreciation. In 2014, the aggregate amount of depreciation charges paid by the district centers for animal disease control in Stavropol region amounted to 15.2 million RUB. Corrected on the rate of inflation, the total amount of resource required to ensure the replacement was 16.2 million RUB. As the total value of income received from rendering commercial services amounted to 225.4 million RUB in 2014, the capital required for a creation of the development fund was 7.2% of the total income amount. Resources of the fund devoted for the replacement of normal assets, extended replacement, or equipment renewal.

Currently, the main source of funds covering the costs of technical equipment for the veterinary service are federal and regional target programs. Thus, using both targeted funding and reserves of the development fund, it would be possible to meet not only the most urgent needs of the veterinary centers in fixed assets but also to carry out assets replacement process in an innovative way. In order to create a fund that could completely finance the needs of the district veterinary centers, it should also receive funds allocated from the regional budget for target needs of the veterinary centers and determine the basic ratio for deductions to a fund. This ratio should be similar for all veterinary centers and should be supported by the differentiated ratios of payments presented in Table 3.

In the previous three years, the amount of the targeted funds allocated from the regional budget was 8.7 million RUB. When supplemented by the funds collected involving the differentiated coefficients (5.8 million RUB), the fund totals 14.5 million RUB, that covers 89.5% of needs. The rest of the funds can be raised using the basic ratio, which would be equal to 0.75%.

SOLUTIONS AND RECOMMENDATIONS

The ultimate goal of improving economic conditions of a district center for animal disease control should be related to the objective evaluation of the veterinary service and respective employee recognition. When assessing the performance of the veterinary service based on "ecosystem health" concept both the results of livestock breeding and social and ecological effects should be considered. It is reasonable to distinguish internal and external efficiency of infrastructure industries, which include veterinary medicine as well. The internal efficiency of an infrastructure industry is expressed in achieving its own goals. For the veterinary infrastructure, the internal efficiency means the rational use of resources and application of appropriate methods and tools for ensuring the health of a livestock.

The external efficiency is considered as a creation of economic benefits for the society, enabling sustainable development of an area. In relation to the veterinary infrastructure, such benefits include the prevention of dangerous animal disease outbreaks, reduction of human morbidity with anthropozoonosis, and mitigation of adverse environmental impact.

The Balanced Scorecard management system is considered to be an appropriate tool for the extended assessment of the performance results of a veterinary center (Kaplan & Norton, 2007). Many organizations worldwide adopt Balanced Scorecard as a strategic planning and performance management methodology. One of its main advantages is a transfer of focus from activities and projects on impacts when determining long-term strategy.

Table 3. Grouping of the districts by differentiated components of ratio of deductions to the development fund

<table>
<tr><td colspan="3">1. Grouping by a livestock density</td></tr>
<tr><td>Up to 15 livestock units/km^2 – ratio = 0.5%</td><td>15–30 livestock units/km^2 – ratio = 1.0%</td><td>Over 30 livestock units/km^2 – ratio = 1.5%</td></tr>
<tr><td>Aleksandrovskii, Andropovskii, Apanasenkovskii, Arzgirskii, Blagodarnenskii, Budennovskii, Grachevskii, Ipatovskii, Kirovskii, Kurskii, Mineralovodskii, Novoaleksandrovskii, Novoselitskii, Petrovskii, Predgornii, Sovetskii, Stepnovskii, Trunovskii, Turkmenskii</td><td>Georgievskii, Izobilnenskii, Krasnogvardeyskii, Levokumskii, Neftekumskii</td><td>Kochubeevskii, Shpakovskii</td></tr>
<tr><td colspan="3">2. Grouping by an availability of a veterinary laboratory</td></tr>
<tr><td>Not available – ratio = 0%</td><td colspan="2">Available – ratio = 0.5%</td></tr>
<tr><td>Andropovskii, Arzgirskii, Grachevskii</td><td colspan="2">Aleksandrovskii, Apanasenkovskii, Blagodarnenskii, Budennovskii, Georgievskii, Izobilnenskii, Ipatovskii, Kochubeevskii, Krasnogvardeyskii, Kirovskii, Kurskii, Levokumskii, Mineralovodskii, Neftekumskii, Novoaleksandrovskii, Novoselitskii, Petrovskii, Predgornii, Sovetskii, Stepnovskii, Trunovskii, Turkmenskii, Shpakovskii</td></tr>
<tr><td colspan="3">3. Grouping by a level of financial and logistical support</td></tr>
<tr><td>Low – ratio = 0%</td><td>Middle – ratio = 0.5%</td><td>High – ratio = 1.5%</td></tr>
<tr><td>Andropovskii, Arzgirskii, Georgievskii, Kurskii, Sovetskii, Trunovskii</td><td>Aleksandrovskii, Apanasenkovskii, Budennovskii, Georgievskii, Izobilnenskii, Kirovskii, Novoaleksandrovskii, Novoselitskii, Predgornii, Turkmenskii, Shpakovskii</td><td>Blagodarnenskii, Ipatovskii, Kochubeevskii, Krasnogvardeyskii, Levokumskii, Mineralovodskii, Neftekumskii, Stepnovskii, Petrovskii</td></tr>
<tr><td colspan="3">4. Grouping by a level of staffing support</td></tr>
<tr><td>Insufficient – ratio = 0%</td><td colspan="2">Sufficient – ratio = 0.5%</td></tr>
<tr><td>Andropovskii, Arzgirskii, Blagodarnenskii, Grachevskii, Kirovskii, Novoselitskii, Sovetskii, Trunovskii, Turkmenskii</td><td colspan="2">Aleksandrovskii, Apanasenkovskii, Budennovskii, Georgievskii, Izobilnenskii, Ipatovskii, Kochubeevskii, Krasnogvardeyskii, Neftekumskii, Novoaleksandrovskii, Petrovskii, Predgornii, Stepnovskii, Shpakovskii</td></tr>
</table>

Source: Authors' development

The authors recommend including targets of each veterinary center into a system of strategic goals in order to reduce the adverse environmental impact (including social aspects of this problem). For example, it may be a target of reducing human morbidity with such a dangerous disease as brucellosis, which is common for animals and human beings. Prevention of this disease remains relevant for many countries (Marcotty et al., 2013). According to the Balanced Scorecard concept, each target has a description (specific, measurable, achievable, realistic, and time-bound). When determining a target one should consider an area under assessment, level of morbidity, and time for achieving this target.

The second advantage of the Balanced Scorecard is an opportunity to translate strategy into action. This is implemented by involving employees into solving strategic goals by developing an adequate system of indicators. According to the survey made by the authors, veterinarians do not pay enough attention to the environmental effects. This case requires additional stimulation and cascade strategic targets for each team member. For this purpose, the system of measures for each target and metrics is developed. Thus, in the case of reducing human morbidity with brucellosis, such measures are immunization of the animals and detection of unaccounted livestock. The number of livestock examined and treated and promptness of the quarantine operations can be considered as metrics.

As stated earlier, the majority of the respondents think that improvement of the veterinary service in Stavropol region under the new conditions requires the development of financial incentives. It is advisable to pay bonuses to employees when they achieve the targeted values of the metrics, which reflect the contribution of veterinarians to the ensurance of food security and sustainable development of an area.

FUTURE RESEARCH DIRECTIONS

Veterinary care of livestock is of crucial importance for the successful livestock breeding. It also influences sustainable rural development. Scientists from all over the world study issues of a decrease in environmental pressure caused by the veterinary service including reduction of antibiotic treatments, disposal of biological waste, expansion of a variety of preventive measures, etc. However, the solution of these problems in emerging countries is complicated by a low level of veterinary medicine caused by an inadequate demand for veterinary services and lack of qualified staff and equipment. That is the reason why the further research of such issues is important in the light of their influence on food security and development of market relations in the sphere of veterinary medicine.

Another future research direction which is perspective in terms of the relations between the development of veterinary service and ensurance of food security is a link between the veterinary sector and the food chains, which is rather weak in the emerging countries. As of Bonnet, Lancelot, Martinez, and Seegers (2011), any health problems that occur in animal production have complex repercussions all the way along the food chains. That is why, in the emerging countries, the role of the veterinary service as one of the guarantors of the stability of the food chains has to be investigated.

The authors founded their research on the case of one administrative region of Russia. The methodology implemented may be spread on the investigation of the roles of the veterinary services in establishing food security in other regions of Russia, as well as other emerging countries.

CONCLUSION

Emerging countries make their first steps towards liberalization of their veterinary services. The expert survey held on the case of one of the regions of Russia demonstrated that the new system brought some positive results. However, some district centers for animal disease control faced their drawbacks. According to the survey results, veterinary medicine should be managed using the tools of competition promotion, demand stimulation, and smoothing consequences of market imperfections.

In order to improve the efficiency of the organizational and economic reform of the district centers for animal disease control, it is necessary to solve such problems as development and adjustment of prices of the commercial services differentiated according to the spatial parameters of an area. In addition, in order to smooth the conditions for manageable asset replacement process, it is advisable to create a fund for the development of veterinary medicine. A number of deductions to this fund would be calculated based on the ratio, which considers constraints of some district centers for animal disease control in the sphere of food security.

The effectiveness of a veterinary service in the broad sense should be regarded as an indicator of veterinary and food security of an area both in terms of internal and external effects. The former displays themselves in the results of the medical and health-promoting measures of reducing the livestock morbidity, the latter are reflected in the spheres which do not directly relate to the veterinary activities (public health service and environment). Practical application of the extended approach to the veterinary service assessment, taking into account its contribution to food security is possible through the use of the Balanced Scorecard. Strategic goals in the area of environmental impact can be transferred to a system of balanced targeting metrics, which promote motivation of veterinarians.

REFERENCES

Ahuja, V., Umali-Deininger, D., & de Haan, C. (2003). Market Structure and the Demand for Veterinary Services in India. *Agricultural Economics*, *29*(1), 27–42. doi:10.1111/j.1574-0862.2003.tb00145.x

Amankwah, K., Klerkx, L., Sakyi Dawson, O., Karbo, N., Oosting, S. J., Leeuwis, C., & Zijpp, A. J. (2014). Institutional Dimensions of Veterinary Services Reforms: Responses to Structural Adjustment in Northern Ghana. *International Journal of Agricultural Sustainability*, *12*(3), 296–315. doi:10.1080/14735903.2014.909635

Bhandari, D. P., & Wollen, T. S. (2008). Community-Based Animal Health Care: A Sustainable Solution for Small-Scale Farmers around the World. *Annals of the New York Academy of Sciences*, *1149*(1), 9–11. doi:10.1196/annals.1428.054 PMID:19120163

Bonnet, P., Lancelot, R., Martinez, D., & Seegers, H. (2011). *When Veterinary Surgeons Start Talking about Food Security!* Retrieved February 21, 2017, from www.cirad.fr/en/news/all-news-items/press-releases/2011/food-security

Broom, D. M. (2010). Animal Welfare: An Aspect of Care, Sustainability, and Food Quality Required by the Public. *Journal of Veterinary Medical Education*, *37*(1), 83–88. doi:10.3138/jvme.37.1.83 PMID:20378884

Caceres, S. B. (2012). The Roles of Veterinarians in Meeting the Challenges of Health and Welfare of Livestock and Global Food Security. *Veterinary Research Forum, 3*(3), 155–157.

Caruso, D., Lusiastuti, A. M., Taukhid, , Slembrouck, J., Komarudin, O., & Legendre, M. (2013). Traditional Pharmacopeia in Small Scale Freshwater Fish Farms in West Java, Indonesia: An Ethnoveterinary Approach. *Aquaculture (Amsterdam, Netherlands)*, *416–417*, 334–345. doi:10.1016/j.aquaculture.2013.09.048

Da Silva, R. B. T. R., & Naas, I. D. A. (2012). *A Survey to Investigate Awareness of People about Animal Welfare in Brazil.* Paper presented at the 9th International Livestock Environment Symposium, Valencia.

Davis, D. K. (2008). Brutes, Beasts and Empire: Veterinary Medicine and Environmental Policy in French North Africa and British India. *Journal of Historical Geography*, *34*(2), 242–267. doi:10.1016/j.jhg.2007.09.001

Ekboir, J. M. (1999). The Role of the Public Sector in the Development and Implementation of Animal Health Policies. *Preventive Veterinary Medicine*, *40*(2), 101–115. doi:10.1016/S0167-5877(99)00015-X PMID:10384947

Enticott, G. (2014). Relational Distance, Neoliberalism and the Regulation of Animal Health. *Geoforum*, *52*, 42–50. doi:10.1016/j.geoforum.2013.12.004

Food and Agriculture Organization of the United Nations. (2015). *Animal Production and Health.* Retrieved August 8, 2015, from http://www.fao.org/ag/againfo/themes/ru/animal_production.html

Gerber, P. J., Mottet, A., Opio, C. I., Falcucci, A., & Teillard, F. (2015). Environmental Impacts of Beef Production: Review of Challenges and Perspectives for Durability. *Meat Science*, *109*, 2–12. doi:10.1016/j.meatsci.2015.05.013 PMID:26117397

Ilyinykh, P. A. (2013). Economy under Conditions of Veterinary Service Optimization in Sverdlovsk Region. *Agricultural Bulletin of the Urals*, *115*(9), 91–94.

Ipema, A. H., Bleumer, E. J. B., & Lokhorst, C. (2011). *Precision Livestock Farming Claims for Sustainable Dairy Production.* Paper presented at the 5th European Conference on Precision Livestock Farming, Prague.

Kaplan, R. S., & Norton, D. P. (2007). Using the Balanced Scorecard as a Strategic Management System. *Harvard Business Review*, *85*(7–8), 150–161.

Kellar, J. A. (2012). Animal Health Surveillance: Navigation amidst the Flotsam of Human Frailty and Fiscal Inertia. *Preventive Veterinary Medicine*, *105*(3), 169–175. doi:10.1016/j.prevetmed.2011.12.009 PMID:22310236

Khokhlov, A. V. (2014). *Reference Materials on Geography of World's Economy.* Retrieved May 30, 2015, from http://vlant-consult.ru/projects/materials/

Leonard, D. K. (1987). The Supply of Veterinary Services: Kenyan Lessons. *Agricultural Administration and Extension*, *26*(4), 219–236. doi:10.1016/0269-7475(87)90061-4

Lin, J. H., Kaphle, K., Wu, L. S., Yang, N. Y. J., Lu, G., Yu, C., & Rogers, P. A. M. et al. (2003). Sustainable Veterinary Medicine for the New Era. *OIE Revue Scientifique et Technique*, *22*(3), 949–964. doi:10.20506/rst.22.3.1451 PMID:15005552

Marcotty, T., Thys, E., Conrad, P., Godfroid, J., Craig, P., Zinsstag, J., & Boelaert, M. et al. (2013). Intersectoral Collaboration between the Medical and Veterinary Professions in Low-Resource Societies: The Role of Research and Training Institutions. *Comparative Immunology, Microbiology and Infectious Diseases*, *36*(3), 233–239. doi:10.1016/j.cimid.2012.10.009 PMID:23260374

McLean, M., Watson, H. K., & Muswema, A. (2007). Veterinary Waste Disposal: Practice and Policy in Durban, South Africa (2001–2003). *Waste Management (New York, N.Y.)*, *27*(7), 902–911. doi:10.1016/j.wasman.2006.05.004 PMID:16876398

Mohring, S. A. I., Strzysch, I., Fernandes, M. R., Kiffmeyer, T. K., Tuerk, J., & Hamscher, G. (2009). Degradation and Elimination of Various Sulfonamides during Anaerobic Fermentation: A Promising Step on the Way to Sustainable Pharmacy? *Environmental Science & Technology*, *43*(7), 2569–2574. doi:10.1021/es802042d PMID:19452918

Nikitin, I. N., Akmullin, A. I., & Trofimova, E. N. (2012). Pricing of Paid Veterinary Services in Volgograd Region. *Scientific Notes of Kazan State Academy of Veterinary Medicine*, *211*, 423–429.

Osipova, N. I. (2008). Sanitary and Bacteriological Examination of Organic Wastes of Livestock Enterprises. *Veterinary Medicine*, *1*, 4–5.

Pappaioanou, M. (2004). Veterinary Medicine Protecting and Promoting the Publics Health and Well-Being. *Preventive Veterinary Medicine*, *62*(3), 153–163. doi:10.1016/j.prevetmed.2003.11.001 PMID:15041202

Popkova, E. G., Dubova, Y. I., Yakovleva, E. A., Azarova, N. A., & Titova, E. V. (2014). Role of Ecological Marketing in Formation and Development of Ecological Cluster. *Asian Social Science*, *23*(10), 1–8.

Rubyogo, J. C., Murithii, P. M., Agumbah, G. J. O., & Obhai, G. (2005). Sustainability of a Privatized Community-Based Animal Health Worker System in Mwingi District, Kenya. *Tropical Animal Health and Production*, *37*(4), 253–266. doi:10.1007/s11250-005-4164-8 PMID:15934633

Rutabanzibwa, A. P. (2011). *Veterinary Legal Reform in Tanzania.* Retrieved February 12, 2017, from http://sites.tufts.edu/capeipst/files/2011/03/Rutabanzibwa-Mombasa.pdf

Sargeant, J. M. (2008). The Influence of Veterinary Epidemiology on Public Health: Past, Present and Future. *Preventive Veterinary Medicine*, *86*(3–4), 250–259. doi:10.1016/j.prevetmed.2008.02.011 PMID:18375003

Steele, J. H. (2008). Veterinary Public Health: Past Success, New Opportunities. *Preventive Veterinary Medicine*, *86*(3–4), 224–243. doi:10.1016/j.prevetmed.2008.02.014 PMID:18417229

Sulakshin, S. S., Bardasaryan, V. E., Vilisov, M. V., Zachesova, Y. A., Pak, N. K., Seredkina, O. A., & Chirva, A. N. (2007). *State Competition Policy and Competition Stimulation in the Russian Federation.* Retrieved February 12, 2017, from http://rucont.ru/file.ashx?guid=af71340c-4bee-41a2-8723-7514a217fb50

Territorial Authority of the Federal State Statistics Service of the Russian Federation in Stavropol Region. (2014). *Livestock and Poultry Population – Livestock Production*. Stavropol: Territorial Authority of the Federal State Statistics Service of the Russian Federation in Stavropol Region.

Traverse, M., & Aceto, H. (2015). Environmental Cleaning and Disinfection. *The Veterinary Clinics of North America. Small Animal Practice*, *45*(2), 299–330. doi:10.1016/j.cvsm.2014.11.011 PMID:25555560

Tregubov, V. I., Efimov, Y. G., Kononov, A. N., Zaerko, V. I., Svetlakova, E. V., & Ozheredova, N. A. (2012). *Procedural and Institutional Issues of Veterinary Service.* Retrieved February 12, 2017, from http://www.science-education.ru/102#61

Trevisi, E., Zecconi, A., Cogrossi, S., Razzuoli, E., Grossi, P., & Amadori, M. (2014). Strategies for Reduced Antibiotic Usage in Dairy Cattle Farms. *Research in Veterinary Science*, *96*(2), 229–233. doi:10.1016/j.rvsc.2014.01.001 PMID:24508188

Turkson, P. K., Slenning, B. D., & Brownie, C. F. (1999). Perceptions of Veterinarians Regarding Privatization of Veterinary Services Delivery in Ghana and Jamaica. *Preventive Veterinary Medicine*, *40*(3–4), 221–232. doi:10.1016/S0167-5877(99)00023-9 PMID:10423776

Velarde, A., Fàbrega, E., Blanco-Penedo, I., & Dalmau, A. (2015). Animal Welfare towards Sustainability in Pork Meat Production. *Meat Science*, *109*, 13–17. doi:10.1016/j.meatsci.2015.05.010 PMID:26013042

Zinsstag, J., Schelling, E., Waltner Toews, D., & Tanner, M. (2011). From One Medicine to One Health and Systemic Approaches to Health and Well-Being. *Preventive Veterinary Medicine*, *101*(3–4), 148–156. doi:10.1016/j.prevetmed.2010.07.003 PMID:20832879

ADDITIONAL READING

Buntain, B. J. (2004). Emerging Challenges in Public Health Protection, Food Safety and Security: Veterinary Needs in the USDAs Food Safety and Inspection Service. *Journal of Veterinary Medical Education*, *31*(4), 334–340. doi:10.3138/jvme.31.4.334 PMID:15551227

DeHaven, W. R., & Goldberg, R. (2006). Animal Health: Foundation of a Safe, Secure, and Abundant Food Supply. *Journal of Veterinary Medical Education*, *33*(4), 496–501. doi:10.3138/jvme.33.4.496 PMID:17220485

Kelly, A. M., & Marshak, R. R. (2007). Veterinary Medicine, Global Health. *Journal of the American Veterinary Medical Association*, *231*(12), 1806–1808. doi:10.2460/javma.231.12.1806 PMID:18081517

Minor, R. (2011). Food Security and Veterinary Services in Developing Countries. *The Veterinary Record*, *168*(25), 674. doi:10.1136/vr.d3918 PMID:21705512

Murai, K., Lehenbauer, T. W., Champagne, J. D., Glenn, K., & Aly, S. S. (2014). Cost-Effectiveness of Diagnostic Strategies Using Quantitative Real-Time PCR and Bacterial Culture to Identify Contagious Mastitis Cases in Large Dairy Herds. *Preventive Veterinary Medicine*, *113*(4), 522–535. doi:10.1016/j.prevetmed.2014.01.001 PMID:24485275

Oudshoorn, F. W., & De Boer, I. J. M. (2005). Survey of the Sustainability Aspects of Automatic Milking Systems (AMS) in Organic Dairy Production. In S. Cox (Ed.), *Precision Livestock Farming* (pp. 193–199). Wageningen: Wageningen Academic Publishers.

Reist, M., Jemmi, T., & Stärk, K. D. C. (2012). Policy-Driven Development of Cost-Effective, Risk-Based Surveillance Strategies. *Preventive Veterinary Medicine*, *105*(3), 176–184. doi:10.1016/j.prevetmed.2011.12.014 PMID:22265642

Rich, K. M., & Perry, B. D. (2011). The Economic and Poverty Impacts of Animal Diseases in Developing Countries: New Roles, New Demands for Economics and Epidemiology. *Preventive Veterinary Medicine*, *101*(3–4), 133–147. doi:10.1016/j.prevetmed.2010.08.002 PMID:20828844

KEY TERMS AND DEFINITIONS

Balanced Scorecard: A system for strategic planning and management used to align business activities to the vision and strategy of an organization, improve communications, and monitor organization performance.

Biological Waste: A biological material obtained because of clinical work and elimination of hotbeds of animal diseases.

Environmental Pressure: An activity that causes environmental change. It includes land use, extraction of resources, greenhouse gas emission, water use, and energy use.

Food Security: Availability of adequate food supplies of basic foodstuffs to sustain a steady expansion of food consumption and to offset fluctuations in food production and food prices.

Sustainability: The quality of not being harmful to the environment or depleting natural resources, and thereby supporting long-term ecological balance.

Veterinary Reform: A reform which objectives are to improve the quality, performance, and status of veterinary service.

Veterinary Service: A network of organizations which provide vital veterinary aid and ensure the quality of animal husbandry products.

Compilation of References

Abbasov, I.D. (2013). *Azərbaycanın və Dünya Ölkələrinin Kənd Təsərrüfatı.* Baku: Şərq-Qərb Publishing House.

Abiad, A., Bluedorn, J., Guajardo, J., & Topalova, P. (2012). *The Rising Resilience of Emerging Market and Developing Economies.* IMF Working Paper, WP/12/300. Washington, DC: International Monetary Fund.

Aceleanu, M. I., Serban, A. C., & Burghelea, C. (2015). Greening the Youth Employment – A Chance for Sustainable Development. *Sustainability, 7*(3), 2623–2643. doi:10.3390/su7032623

Adams, R. M., Hurd, B. H., Lenhart, S., & Leary, N. (1998). Effects of Global Climate Change on Agriculture: An Interpretative Review. *Climate Research, 11*(1), 19–30. doi:10.3354/cr011019

Adesina, A. (2012). *Agricultural Transformation Agenda: Repositioning Agriculture to Drive Nigeria's Economy.* Abuja: Ministry of Agriculture and Rural Development of the Federal Republic of Nigeria.

Adivar, B., Atan, T., Sevil Oflaç, B., & Örten, T. (2010). Improving Social Welfare Chain Using Optimal Planning Model. *Supply Chain Management: An International Journal, 15*(4), 290–305. doi:10.1108/13598541011054661

Aertsens, J., Verbeke, W., Mondelaers, K., & Van Huylenbroeck, G. (2009). Personal Determinants of Organic Food Consumption: A Review. *British Food Journal, 111*(10), 1140–1167. doi:10.1108/00070700910992961

African Countries Trade Agreements. (n.d.). *Focus Africa.* Retrieved November 18, 2016, from http://focusafrica.gov.in/Trade_Agreements_Africa.html

African Economic Outlook. (n.d.). *African Economic Outlook – Nigeria 2013.* Retrieved November 17, 2016, from http://www.africaneconomicoutlook.org/fileadmin/uploads/aeo/2013/PDF/Nigeria%20-%20African%20Economic%20Outlook.pdf

Agricultural Market Information System. (n.d.). *AMIS Market Database.* Retrieved February 8, 2017, from http://statistics.amis-outlook.org/data/index.html#COMPARE

Ahadov, D. (2009). *Land Reform in Azerbaijan.* Paper presented at the FIG Working Week 2009, Surveyors Key Role in Accelerated Development, Eilat.

Ahuja, V., Umali-Deininger, D., & de Haan, C. (2003). Market Structure and the Demand for Veterinary Services in India. *Agricultural Economics, 29*(1), 27–42. doi:10.1111/j.1574-0862.2003.tb00145.x

Aiking, H. (2011). Future Protein Supply. *Trends in Food Science & Technology, 22*(2), 112–120. doi:10.1016/j.tifs.2010.04.005

Aiking, H., & De Boer, J. (2004). Food Sustainability: Diverging Interpretations. *British Food Journal, 106*(5), 359–365. doi:10.1108/00070700410531589

Aker, J. C., & Lemtouni, A. (1999). A Framework for Assessing Food Security in the Face of Globalization: The Case of Morocco. *Agroalimentaria*, *8*, 13–26.

Akyüz, Y. (2005). *Trade, Growth and Industrialization: Issues, Experiences and Policy Challenges*. Penang, Malaysia: Third World Network.

Alexandrov, V. A., & Hoogenboom, G. (2000). Vulnerability and Adaptation: Assessments of Agricultural Crops under Climate Change in the Southeastern USA. *Theoretical and Applied Climatology*, *67*(1–2), 45–63. doi:10.1007/s007040070015

Aliyev, K., & Gasimov, I. (in press). Fiscal Policy Implementation in Azerbaijan Before-Within-and-After the Oil Boom. *Contemporary Economics, 11*.

Aliyev, K. (2014). *WTO Membership and South Caucasus Economies*. Saarbrucken: LAP Lambert Academic Publishing.

Aliyev, K., & Suleymanov, E. (2015). Macroeconomic Analysis and Graphical Interpretation of Azerbaijan Economy in 1991-2012. *Expert Journal of Economics*, *3*(1), 40–49.

Almaraz, J., Mabood, F., Zhou, X., Gregorich, E., & Smith, D. (2008). Climate Change, Weather Variability and Corn Yield at a Higher Latitude Locale: Southwestern Quebec. *Climatic Change*, *88*(2), 187–197. doi:10.1007/s10584-008-9408-y

Alvarez, J.B., Mackalski, R., Loeb, A., & Mazzanti, L. (2015). Leading Sustainable Change: An Organizational Perspective. *Strategic Management Journal, 34*(13), 1209–1231.

Amadou, S. (2014). *Will There Be an African Economic Community?* Retrieved November 17, 2016, from https://www.brookings.edu/testimonies/hearing-on-will-there-be-an-african-economic-community

Amankwah, K., Klerkx, L., Sakyi Dawson, O., Karbo, N., Oosting, S. J., Leeuwis, C., & Zijpp, A. J. (2014). Institutional Dimensions of Veterinary Services Reforms: Responses to Structural Adjustment in Northern Ghana. *International Journal of Agricultural Sustainability*, *12*(3), 296–315. doi:10.1080/14735903.2014.909635

Ambrose-Oji, B. (2009). Urban Food Systems and African Indigenous Vegetables: Defining the Spaces and Places for African Indigenous Vegetables in Urban and Peri-Urban Agriculture. In C. M. Shackleton, M. W. Pasquini, & A. W. Drescher (Eds.), *African Indigenous Vegetables in Urban Agriculture* (pp. 1–33). London: Earthscan.

Analytical Center at the Government of the Russian Federation. (2016). *The Food Embargo: The Results of 2015*. Moscow: Government of the Russian Federation.

Anderson, D. P., & Hanselka, D. (2009). *Adding Value to Agricultural Products*. Texas AgriLife Extension Service Publication No. E-573. Retrieved February 7, 2017, from https://oaktrust.library.tamu.edu/bitstream/handle/1969.1/86940/pdf_1302.pdf?sequence=1&isAllowed=y

Anderson, J. C., & Gerbing, D. W. (1982). Some Methods for Respecifying Measurement Models to Obtain Unidimensional Construct Measurement. *JMR, Journal of Marketing Research*, *19*(4), 453–460. doi:10.2307/3151719

Anderson, J. E., & Marcouiller, D. S. J. (2005). Autarky and Anarchy: Endogenous Predation and Trade. *International Economic Review*, *46*(1), 189–213. doi:10.1111/j.0020-6598.2005.00315.x

Anderson, K., Jha, S., Nelgen, S., & Strutt, A. (2013). Re-examining Policies for Food Security in Asia. *Food Security*, *5*(2), 195–215. doi:10.1007/s12571-012-0237-5

Anderson, S. A. (1990). Core Indicators of Nutritional State for Difficult-to-Sample Populations. *The Journal of Nutrition*, *120*, 1557–1600. PMID:2243305

Apata, T. G., Samuel, K. D., & Adeola, A. O. (2009). *Analysis of Climate Change Perception and Adaptation among Arable Food Crop Farmers in South Western Nigeria.* Paper presented at the International Association of Agricultural Economists' 2009 Conference, Beijing, China.

Applanaidu, S. D., Bakar, N. A., & Baharudin, A. H. (2014). An Econometric Analysis of Food Security and Related Macroeconomic Variables in Malaysia: A Vector Autoregressive Approach (VAR). *UMK Procedia, 1*, 93–102. doi:10.1016/j.umkpro.2014.07.012

Aras, N., Suleymanov, E., & Mammadov, K. (2016). *Economy of Azerbaijan 25 Years of Independence*. Baku: Sharg-Garb Publishing House.

Arellano, M., & Bond, S. (1991). Some Tests of Specification for Panel Data: Monte Carlo Evidence and an Application to Employment Equations. *The Review of Economic Studies, 58*(2), 277–297. doi:10.2307/2297968

Armisen, R. (1995). Worldwide Use and Importance of Gracilaria. *Journal of Applied Phycology, 7*(3), 231–243. doi:10.1007/BF00003998

Arnell, N. W., Brown, S., Gosling, S. N., Hinkel, J., Huntingford, C., Lloyd-Hughes, B., & Zelazowski, P. et al. (2016). Global-Scale Climate Impact Functions: The Relationship between Climate Forcing and Impact. *Climatic Change, 134*(3), 475–487. doi:10.1007/s10584-013-1034-7

ASEAN Statistics. (2015). *Top Ten ASEAN Trade Partner Countries/Regions*. Retrieved December 13, 2016, from http://www.asean.org/storage/2016/01/statistic/table20_asof121Dec15.pdf

ASEAN-India Relations. (2016). In *ASEAN India Progress & Prosperity*. Retrieved October 17, 2016, from http://www.mea.gov.in/aseanindia/20-years.htm

Asfaw, S., Shiferaw, B., Simtowe, F., & Lipper, L. (2012). Impact of Modern Agricultural Technologies on Smallholder Welfare: Evidence from Tanzania and Ethiopia. *Food Policy, 37*(3), 283–295. doi:10.1016/j.foodpol.2012.02.013

Asian Development Bank Institute. (2016). *Connecting Central Asia with Economic Centers*. Tokyo: Brookings Institution Press.

Asian Development Bank. (2016). *Agriculture and Food Security Issues in Asia and the Pacific*. Retrieved October 6, 2016, from https://www.adb.org/sectors/agriculture/issueshttps://www.adb.org/sectors/agriculture/issues

Audsley, E., Pearn, K. R., Simota, C., Cojocaru, G., Koutsidou, E., Rousevell, M. D. A., & Alexandrov, V. et al. (2006). What Can Scenario Modelling Tell Us about Future European Scale Agricultural Land Use, and What Not? *Environmental Science & Policy, 9*(2), 148–162. doi:10.1016/j.envsci.2005.11.008

Ayinde, O. E., Ajewole, O. O., Ogunlade, I., & Adewumi, M. O. (2010). Empirical Analysis of Agricultural Production and Climate Change: A Case Study of Nigeria. *Journal of Sustainable Development in Africa, 12*(6), 345–253.

Azerbaijan National Fund for Entrepreneurship Support. (2016). *Annual Reports*. Retrieved October 19, 2016, from http://anfes.gov.az/en/show.page.php?guid=d2f7fbbc-269c-11e0-8e86-0022190362dd

Babu, S. C., & Tashmatov, A. (1999). Attaining Food Security in Central Asia – Emerging Issues and Challenges for Policy Research. *Food Policy, 24*(4), 357–362. doi:10.1016/S0306-9192(99)00052-4

Balambal, R. V. (2015). Ethical Values of Food Safety. In S. Hongladarom (Ed.), *Food Security and Food Safety for the XXI Century* (pp. 33–44). Heidelberg, Germany: Springer.

Barenklau, K. E. (2001). *Agricultural Safety*. Boca Raton, FL: Lewis Publishers, CRC Press LLC. doi:10.1201/9781420032772

Barraclough, S., & Utting, P. (1987). *Food Security Trends and Prospects in Latin America*. Notre Dame, IN: The Hellen Kellogg Institute of International Studies.

Bartlett, C., & Ghoshal, S. (1987). Managing Across Borders: New Strategic Requirements. *Sloan Management Review*, *29*, 7–17.

Beloglazov, G. (2007). Food Security of the People's Republic of China and its Russian Vector. *Russia and Pacific RIM*, *3*, 75–83.

Belz, F. M., & Binder, J. K. (2015). Sustainable Entrepreneurship: A Convergent Process Model. *Business Strategy and the Environment*. Retrieved November 26, 2016, from http://onlinelibrary.wiley.com/doi/10.1002/bse.1887/abstract

Bensassi, S., & Martínez-Zarzoso, I. (2012). How Costly is Modern Maritime Piracy for the International Community? *Review of International Economics*, *20*(5), 869–883. doi:10.1111/roie.12000

Berčík, J. (2013). The Modern Forms of Marketing Research in Selected Agri-Food Businesses. In *Proceedings of the International Scientific Conference "Business Management – Practice and Theory in the 21st Century"*. Nitra: Slovak University of Agriculture.

Bhandari, D. P., & Wollen, T. S. (2008). Community-Based Animal Health Care: A Sustainable Solution for Small-Scale Farmers around the World. *Annals of the New York Academy of Sciences*, *1149*(1), 9–11. doi:10.1196/annals.1428.054 PMID:19120163

Biloslavo, F., & Quercia, P. (2013). *Il tesoro dei pirati. Sequestri, riscatti, riciclaggio*. Rome: Ministero della Difesa.

Bioversity International. (n.d.). *Nutritious Underutilized Species*. Retrieved February 15, 2017, from http://www.bioversityinternational.org//Nutritious_underutilized_species_-_Amaranth_1682.pdf

Blázquez-Lidoy, J., Rodríguez, J., & Santiso, J. (2006). *Angel or Devil? China's Trade Impact on Latin American Emerging Markets*. Working paper No. 252. Paris: OECD Development Centre.

Bloch, P. C. (2002). *Agrarian Reform in Uzbekistan and Other Central Asian Countries*. Madison, WI: University of Wisconsin-Madison, Land Tenure Center.

Blomberg, S., & Hess, G. (2006). How Much Does Violence Tax Trade? *The Review of Economics and Statistics*, *88*(4), 599–612. doi:10.1162/rest.88.4.599

Blunden, G. (1991). Agricultural Uses of Seaweeds and Seaweed Extracts. In M. D. Guiry & G. Blunden (Eds.), *Seaweed Resources in Europe: Uses and Potential* (pp. 65–81). Chichester, UK: John Wiley & Sons Ltd.

Bobokulov, I. (2006). Central Asia: Is There an Alternative to Regional Integration? *Central Asian Survey*, *25*(1–2), 75–91. doi:10.1080/02634930600903056

Boehringer, C., & Rutherford, T. (1999). *Decomposing General Equilibrium Effects of Policy Intervention in Multi-Regional Trade Models: Method and Sample Application*. Mannheim: Center for European Economic Research.

Bokusheva, R., Hockmann, H., & Kumbhakar, S. (2012). Dynamics of Productivity and Technical Efficiency in Russian Agriculture. *European Review of Agriculture Economics*, *39*(4), 611–637. doi:10.1093/erae/jbr059

Bondarenko, L. (2011). Employment in Rural Areas and Diversification of Rural Economics. *Economics of Agriculture of Russia*, *1*, 71–76.

Bonnet, P., Lancelot, R., Martinez, D., & Seegers, H. (2011). *When Veterinary Surgeons Start Talking about Food Security!* Retrieved February 21, 2017, from www.cirad.fr/en/news/all-news-items/press-releases/2011/food-security

Boulanger, P., & Messerin, P. (2009). *The 2020 European Agriculture: Long Term Challenges, New Public and Private Policies*. Paper presented at the conference "The 2020 European Agriculture: Long Term Challenges, New Public and Private Policies", Paris, France.

Božić, D., Bogdanov, N., & Ševarlić, M. (2011). *Ekonomika Poljoprivrede*. Belgrade: University of Belgrade.

Braun, J., & Kennedy, E. T. (1994). *Agricultural Commercialization, Economic Development, and Nutrition*. Baltimore, MD: Johns Hopkins University Press.

Breitung, J. (2000). The Local Power of Some Unit Root Tests for Panel Data. In B. H. Baltagi (Ed.), *Nonstationary Panels, Panel Cointegration, and Dynamic Panels* (pp. 161–177). Amsterdam: Elsevier. doi:10.1016/S0731-9053(00)15006-6

Briones, R. M., Chavez, E. C., Durand-Morat, A., & Wailes, E. J. (2012). *Climate Change and Price Volatility: Can We Count on the ASEAN Plus Three Emergency Rice Reserve*. Manila: Asian Development Bank.

Brockett, P. L., Wang, M., & Yang, C. (2005). Weather Derivatives and Weather Risk Management. *Risk Management & Insurance Review*, *8*(1), 127–140. doi:10.1111/j.1540-6296.2005.00052.x

Brookes, G. (2007). *Economic Impact Assessment of the Way in which the EU Novel Foods Regulatory Approval Procedures Affect the EU Food Sector*. Retrieved November 26, 2016, from http://www.pgeconomics.co.uk/archive.php

Brooks, N., Adger, W. N., & Kelly, P. M. (2005). The Determinants of Vulnerability and Adaptive Capacity at the National Level and Implications for Adaptation. *Global Environmental Change*, *15*(2), 151–163. doi:10.1016/j.gloenvcha.2004.12.006

Broom, D. M. (2010). Animal Welfare: An Aspect of Care, Sustainability, and Food Quality Required by the Public. *Journal of Veterinary Medical Education*, *37*(1), 83–88. doi:10.3138/jvme.37.1.83 PMID:20378884

Building Nigeria's Response to Climate Change. (2008). *The Recent Global and Local Action on Climate Change*. Paper presented at the 2008 Annual Workshop of Nigerian Environmental Study Team (NEST), Abuja.

Burlingame, B., Charrondiere, R., & Mouille, B. (2009). Food Composition is Fundamental to the Cross-Cutting Initiative on Biodiversity for Food and Nutrition. *Journal of Food Composition and Analysis*, *22*(5), 361–365. doi:10.1016/j.jfca.2009.05.003

Buschmann, A. H., Correa, J. A., Westermeier, R., del Carmen Hernandez-Gonzalez, M., & Norambuena, R. (2001). Red Algal Farming in Chile: A Review. *Aquaculture (Amsterdam, Netherlands)*, *194*(3), 203–220. doi:10.1016/S0044-8486(00)00518-4

Buschmann, A. H., Troell, M., & Kautsky, N. (2001). Integrated Algal Farming: A Review. *Cahiers de Biologie Marine*, *42*(1–2), 83–90.

Busse, M., & Großmann, H. (2004). Assessing the Impact of ACP/EU Economic Partnership Agreements on West African Countries. Hamburg: Hamburgisches Welt-Wirtschafts-Archiv (HWWA), Hamburg Institute of International Economics.

Caceres, S. B. (2012). The Roles of Veterinarians in Meeting the Challenges of Health and Welfare of Livestock and Global Food Security. *Veterinary Research Forum*, *3*(3), 155–157.

Cannon, A., & Major, R. (2015). *Signature of the Tripartite Free Trade Agreement (TFTA) on 10 June 2015: The First Step Towards a United African Free Trade Bloc?* Retrieved November 17, 2016, from http://hsfnotes.com/publicinternationallaw/2015/07/07/signature-of-the-tripartite-free-trade-agreement-tfta-on-10-june-2015-the-first-step-towards-a-united-african-free-trade-bloc/

Caparros Megido, R., Sablon, L., Geuens, M., Brostaux, Y., Alabi, T., Blecker, C., & Francis, F. et al. (2013). Edible Insects Acceptance by Belgian Consumers: Promising Attitude for Entomophagy Development. *Journal of Sensory Studies*, *29*(1), 14–20. doi:10.1111/joss.12077

Caruso, D., Lusiastuti, A. M., Taukhid, , Slembrouck, J., Komarudin, O., & Legendre, M. (2013). Traditional Pharmacopeia in Small Scale Freshwater Fish Farms in West Java, Indonesia: An Ethnoveterinary Approach. *Aquaculture (Amsterdam, Netherlands)*, *416–417*, 334–345. doi:10.1016/j.aquaculture.2013.09.048

Center for Economic and Social Development. (2015). *Evaluation and Modeling of Comparative Advantages in Agriculture in Azerbaijan*. Baku: CESD.

Central Bank of Nigeria. (2010). *Annual Report and Financial Statement for the Year Ended 31 December 2009*. Abuja: Central Bank of Nigeria.

Central Bank of the Republic of Azerbaijan. (2016). *Statistical Bulletins*. Retrieved October 19, 2016, from http://en.cbar.az/pages/publications-researches

Cernansky, R. (2015). Super Vegetables Long Overlooked in Parts of Africa, Indigenous Greens Are Now Capturing Attention for their Nutritional and Environmental Benefits. *Nature*, *522*, 146–148. doi:10.1038/522146a PMID:26062494

Chagomoka, T., Afari-Sefa, V., & Pitoro, R. (2014). Value Chain Analysis of Traditional Vegetables from Malawi and Mozambique. *International Food and Agribusiness Management Review*, *17*(4), 59–86.

Chen, J., & Fleisher, B. M. (1996). Regional Income Inequality and Economic Growth in China. *Journal of Comparative Economics*, *22*(2), 141–164. doi:10.1006/jcec.1996.0015

Chen, Z., Xiangzheng, D., Jianing, D., & Peiying, H. (2016). Urbanization and Sustainability: Comparison of the Processes in "BIC" Countries. *Sustainability*, *400*(8), 1–18.

Choi, I. (2001). Unit Root Tests for Panel Data. *Journal of International Money and Finance*, *20*(2), 249–272. doi:10.1016/S0261-5606(00)00048-6

Choulamany, X. (n.d.). *Issues and Strategies to Assure Food Security in ASEAN: ASEAN Level Initiative on Food Security* [PowerPoint slides]. Retrieved September 12, 2016, from https://www.oecd.org/tad/events/Issues%20and%20Strategies%20to%20Assure%20Food%20Security%20in%20ASEAN%20-%20%20%20by%20Xaypradeth%20Update.pdf

Christensen, J. H., Hewitson, B., Busuioc, A., Chen, A., Gao, X., Held, I., … Whetton, P. (2007). Regional Climate Projections. In S. Solomon, D. Qin, M. Manning, Z. Chen, M. Marquis, K.B. Averyt, … H.L. Miller (Eds.), Climate Change 2007: The Physical Science Basis (pp. 847–940). Cambridge, UK: Cambridge University Press.

Christiaensen, L., Demery, L., & Kuhl, J. (2011). The (Evolving) Role of Agriculture in Poverty Reduction. *Journal of Development Economics*, *96*(2), 239–254. doi:10.1016/j.jdeveco.2010.10.006

Churchill, G. A. Jr. (1979). A Paradigm for Developing Better Measures of Marketing Constructs. *JMR, Journal of Marketing Research*, *16*(1), 64–73. doi:10.2307/3150876

Clapp, J. (2015). *Food Self-Sufficiency and International Trade: A False Dichotomy?* Rome: Food and Agriculture Organization of the United Nations.

Clay, E. J., Shaw, D. J., & Singer, H. W. (1987). *Poverty, Development, and Food: Essays in Honour of H.W. Singer on his 75th Birthday*. Houndmills, UK: Macmillan. doi:10.1007/978-1-349-09214-7

Colombian Institute of Familiar Wellness. (2009). *Bienestarina: Distribución, cuidado y uso de un recurso sagrado.* Bogotá: ICBF.

Colombian Institute of Familiar Wellness. Instituto Nacional de Salud, Profamilia, Ministerio de Protección Social, & DANE. (2010). Encuesta Nacional de La Situación Nutricional En Colombia 2010 – ENSIN. Bogotá: Ministerio de Protección Social.

Cooperation Union of Vojvodina. (2013). *Pricelist of Services with Mechanisation in 2013.* Novi Sad: Cooperation Union of Vojvodina.

Cotton Australia. (2017). *Sydney, Australia: Properties and Cotton Products.* Retrieved February 7, 2017, from http://cottonaustralia.com.au/cotton-library/fact-sheets/cotton-fact-file-cotton-properties-and-products

Cuevas García, R. (2005). *El diseño de los programas de alimentación escolar y la función de la industria alimentaria.* Retrieved November 26, 2016, from http://www.fao.org/docrep/008/y5906m/Y5906M05.htm

D'Aleo, F., & D'Aleo, V. (2015a). *Preliminary Study in Open Raceway for Production of Spirulina Biomass in Southern Italy. First Experience in Calabria.* Poster presented at the 110° Congresso della Società Botanica Italiana, Pavia.

D'Aleo, F., & D'Aleo, V. (2015b). *Start-up Company for Production of Spirulina Biomass in Southern Italy. First Experience in Calabria.* Poster presented at the 43° Congresso Società Italiana di Microbiologia, Napoli.

D'Aleo, V., & Morabito, G. (2015). Verso il prototipo di start-up mediterranea. Confronto fra modelli di politiche pubbliche, incubatori e acceleratori. *Proceedings of the BIC-Italia Net Congress & EBN AGM.* Milan: BIC; EBN AGM.

Da Silva, R. B. T. R., & Naas, I. D. A. (2012). *A Survey to Investigate Awareness of People about Animal Welfare in Brazil.* Paper presented at the 9th International Livestock Environment Symposium, Valencia.

DAleo, V. (2016). Supply Chain Management: Overview, Competition and Competences, How to Exploit the Hidden Capabilities. *International Journal of Management and Network Economics, 3*(4), 336–346. doi:10.1504/IJMNE.2016.079878

Danes, J. E., & Mann, O. K. (1984). Unidimensional Measurement and Structural Equation Models with Latent Variables. *Journal of Business Research, 12*(3), 337–351. doi:10.1016/0148-2963(84)90016-X

Danilowska, A., Ismayilov, A., & Aliyev, K. (2014). The Comparative Analysis of Agricultural Financial Systems in Poland and Azerbaijan. *Acta Scientiarum Polonorum Oeconomia, 13*(3), 41–53.

Dani, S. (2015). *Food Supply Chain Management and Logistics: From Farm to Fork.* London: Kogan Page Publishers.

Darkow, I. L., Foerster, B., & von der Gracht, H. A. (2015). Sustainability in Food Service Supply Chains: Future Expectations from European Industry Experts Toward the Environmental Perspective. *Supply Chain Management: An International Journal, 20*(2), 163–178. doi:10.1108/SCM-03-2014-0087

Daudu, S., & Ajayi, A. R. (2009). Assessment of the Performance of the National Special Food Security Programme in Benue State, Nigeria. *Journal of Social Sciences, 21*(1), 25–32.

Davies, A. E. (2009). *Food Security Initiatives in Nigeria: Prospects and Challenges.* University of Ilorin.

Davis, D. K. (2008). Brutes, Beasts and Empire: Veterinary Medicine and Environmental Policy in French North Africa and British India. *Journal of Historical Geography, 34*(2), 242–267. doi:10.1016/j.jhg.2007.09.001

Dawes, C. P. (1995). Suspended Cultivation of Gracilaria in the Sea. *Journal of Applied Phycology, 7*(3), 303–313. doi:10.1007/BF00004004

De Jonge, J., Van Trijp, J. C. M., Van der Lans, I. A., Renes, R. J., & Frewer, L. J. (2008a). *A Scale for Consumer Confidence in the Safety of Food.* Discussion Paper no. 44. Wageningen: Mansholt Graduate School of Social Sciences.

De Jonge, J., Frewer, L., Van Trijp, H., Renes, R. J., De Wit, W., & Timmers, J. (2004). Monitoring Consumer Confidence in Food Safety: An Exploratory Study. *British Food Journal, 106*(10/11), 837–849. doi:10.1108/00070700410561423

De Jonge, J., Van Trijp, J. C. M., Van der Lans, I. A., Renes, R. J., & Frewer, L. J. (2008b). How Trust in Institutions and Organizations Builds General Consumer Confidence in the Safety of Food: A Decomposition of Effects. *Appetite, 51*(2), 311–317. doi:10.1016/j.appet.2008.03.008 PMID:18450326

De Sousa, J., Mirza, D., & Verdier, T. (2009). Trade and the Spillovers of Transnational Terrorism. *Swiss Journal of Economics and Statistics, 145*(IV), 453–461.

De Wit, M., & Stankiewicz, J. (2006). Changes in Surface Water Supply across Africa with Predicted Climate Change. *Science, 311*(5769), 1917–1921. doi:10.1126/science.1119929 PMID:16513946

De-Magistris, T., Pascucci, S., & Mitsopoulos, D. (2015). Paying to See a Bug on my Food: How Regulations and Information Can Hamper Radical Innovations in the European Union. *British Food Journal, 117*(6), 1777–1792. doi:10.1108/BFJ-06-2014-0222

Deressa, T. (2007). *Measuring the Economic Impact of Climate Change on Ethiopian Agriculture: Ricardian Approach.* World Bank Policy Research Paper No. 4342. Washington, DC: World Bank.

Deressa, T., Hassan, R., Alemu, T., Yesuf, M., & Ringler, C. (2008). *Analyzing the Determinants of Farmers' Choice of Adaptation Measures and Perceptions of Climate Change in the Nile Basin of Ethiopia.* Discussion Paper No. 00798. Washington, DC: International Food Policy Research Institute (IFPRI).

Deressa, T., & Hassan, T. (2010). Economic Impact of Climate Change on Crop Production in Ethiopia: Evidence from Cross-Section Measures. *Journal of African Economies, 18*(4), 529–554. doi:10.1093/jae/ejp002

Devereux, M. B. (1999). Growth and the Dynamics of Trade Liberalization. *Journal of Economic Dynamics & Control, 23*(5-6), 773–795. doi:10.1016/S0165-1889(98)00043-8

di Commercio d'Italia, C. (2015). *Infocamere.* Retrieved December 12, 2016, from http://www.camcom.gov.it/

Dieye, C. T. (2015). *What Is Africa Worth in the International Trading System?* Retrieved November 18, 2016, from http://www.ictsd.org/bridges-news/bridges-africa/news/what-is-africa-worth-in-the-international-trading-system

Diogo Mateus, S. (2014, July 11). 'Worrying' Rise in Piracy Attacks Around Malacca Strait. *Deutsche Welle.* Retrieved from www.dw.com/en/worrying-rise-in-piracy-attacks-around-malacca-strait/a-17780275

Dirección Nacional de Planeación. (2006). *Conpes 3443. Contratación del operador para la producción y distribución del componente nutricional (Bienestarina) en el Instituto Colombiano de Bienestar Familiar.* Bogotá: ICBF, Dirección Nacional de Planeación.

Dirección Nacional de Planeación. (2007). *Conpes 113. Política Nacional de Seguridad Alimentaria Y Nutricional (PSAN).* Bogotá: Dirección Nacional de Planeación.

Dobbs, R., Ramaswamy, S., Stephenson, E., & Viguerie, P. (2014). *Management Intuition for the Next 50 Years.* Retrieved November 24, 2016, from http://www.mckinsey.com/business-functions/strategy-and-corporate-finance/our-insights/management-intuition-for-the-next-50-years

Đorđević, V., & Nenadić, N. (1980). *Soja.* Belgrade: Nolit.

Dorosh, P. A. (2004). Trade, Food Aid and Food Security: Evolving Rice and Wheat Markets. *Economic and Political Weekly, 36*(39), 4033–4042.

Downing, T. E. (Ed.). (1996). *Climate Change and World Food Security*. Berlin: Springer-Verlag. doi:10.1007/978-3-642-61086-8

Dragović, S. (2008). Effect of Irrigation on Field Crops Yield under the Variable Agro-Climatic Conditions of Serbia. *Agriculture & Forestry*, *54*(1), 25–40.

Dreher, A. (2006). Does Globalization Affect Growth? Evidence from a New Index of Globalization. *Applied Economics*, *38*(10), 1091–1110. doi:10.1080/00036840500392078

Economic and Commercial Counsellor's Office in Nigeria. (2014). *Brief Introduction of Bilateral Relations*. Retrieved November 17, 2016, from http://nigeria2.mofcom.gov.cn/article/bilateralcooperation/inbrief/201407/20140700669579.shtml

Economist Intelligence Unit. (2010). *Country Commerce Nigeria*. New York, NY: Economist Intelligence Unit.

Edoumiekumo, S. G., & Audu, N. P. (2009). The Impact of Agriculture and Agro-based Industries on Economic Development in Nigeria: An Econometric Assessment. *Journal of Research in National Development*, *7*(1), 1–8.

Egunsola, O. (2014). *Should Nigeria Sign the Economic Partnership Agreement?* Retrieved November 18, 2016, from http://nigeriannet.com/should-nigeria-sign-the-economic-partnership-agreement/

Eicher, C., & Witt, L. (1964). *Agriculture in Economic Development*. New York, NY: McGraw Hill.

Ekboir, J. M. (1999). The Role of the Public Sector in the Development and Implementation of Animal Health Policies. *Preventive Veterinary Medicine*, *40*(2), 101–115. doi:10.1016/S0167-5877(99)00015-X PMID:10384947

El Niño in Asia. (2016). In *The United Nations Office for the Coordination of Humanitarian Affairs*. Retrieved September 11, 2016, from http://www.unocha.org/el-nino-asia

El Nuevo Siglo. (2013). *Lupa oficial a la Bienestarina*. Retrieved March 2, 2017, from http://www.elnuevosiglo.com.co/articulos/5-2013-lupa-oficial-bienestarina.html

Emeka, A. (2015). *Experts Debate Proposed EU-ECOWAS Partnership Agreement*. Retrieved from 30 September 2016, from https://www.tralac.org/news/article/7793-experts-debate-proposed-eu-ecowas-partnership-agreement.html

Emerging Market Economy. (n.d.). In *InvestingAnswers*. Retrieved October 2, 2016, from http://www.investinganswers.com/financial-dictionary/world-markets/emerging-market-economy-1518

Encyclopedia Britannica Online. (2016). *Central Asia*. Retrieved February 7, 2017, from https://www.britannica.com/place/Central-Asia

Engelberth, I. (2015). Változó kínai gazdasági és politikai jelenlét Afrikában (The Changing Chinese Economic and Political Presence in Africa). In K. Solt (Ed.), *Alkalmazott tudományok II. Fóruma* (pp. 121–145). Budapest: Budapest Business School.

Enticott, G. (2014). Relational Distance, Neoliberalism and the Regulation of Animal Health. *Geoforum*, *52*, 42–50. doi:10.1016/j.geoforum.2013.12.004

Erokhin, V. (2015). Structural Changes in International Trade in Food: Competitive Growth Models for Economies in Transition. In *Proceedings of the 3rd International Conference "Economic Scientific Research – Theoretical, Empirical and Practical Approaches"*. Bucharest: Academia Romana.

Erokhin, V. (2014). Approaches to Sustainable Rural Development in a Predominantly Non-Rural Region. *Economics of Agriculture*, *2*, 291–306.

Erokhin, V. (2016). Rising Economic Powers among Economies in Transition: Is There Enough Space on the Global Market? In V. Erokhin (Ed.), *Global Perspectives on Trade Integration and Economies in Transition* (pp. 1–25). Hershey, PA: IGI Global. doi:10.4018/978-1-5225-0451-1.ch001

Erokhin, V., & Ivolga, A. (2012). How to Ensure Sustainable Development of Agribusiness in the Conditions of Trade Integration: Russian Approach. *International Journal of Sustainable Economies Management, 2*(1), 12–23. doi:10.4018/ijsem.2012040102

Erokhin, V., Ivolga, A., & Heijman, W. (2014). Trade Liberalization and State Support of Agriculture: Effects for Developing Countries. *Agricultural Economics, 60*(11), 524–537.

Estevadeordal, A., Freund, C., & Ornelas, E. (2008). Does Regionalism Affect Trade Liberalization Toward Nonmembers? *The Quarterly Journal of Economics, 123*(4), 1531–1575. doi:10.1162/qjec.2008.123.4.1531

European Commission. (2007). *Adapting to Climate Change in Europe – Options for EU Action.* Brussels: European Commission.

European Commission. (n.d.). *ACP – The Cotonou Agreement.* Retrieved November 17, 2016, from http://ec.europa.eu/europeaid/regions/african-caribbean-and-pacific-acp-region/cotonou-agreement_en

European Commission. International Monetary Fund, Organization for Economic Co-operation and Development, United Nations, & World Bank. (2009). System of National Accounts 2008. New York, NY: The United Nations.

European Parliament. (1997). *Regulation No 258/97 on Novel Foods and Novel Food Ingredients.* Retrieved December 12, 2016, from http://eur-lex.europa.eu/legal-content/EN/TXT/PDF/?uri=CELEX:31997R0258&from=EN

Eurostat. (2006). *Database.* Retrieved January 5, 2006, from http://epp.eurostat.cec.eu.int/portal/page?_pageid=1090,30070682,1090_33076576&_dad=portal&_schema=PORTAL

Ewing, M., & Msangi, S. (2009). Biofuels Production in Developing Countries: Assessing Tradeoffs in Welfare and Food Security. *Environmental Science & Policy, 12*(4), 520–528. doi:10.1016/j.envsci.2008.10.002

Falke, M. (2002). *Community Interests: An Insolvency Objective in Transition Economies?* F.I.T. Discussion Paper No. 01/02. Frankfurt: Frankfurt Institute for Transformation Studies, European University Viadrina.

FAOSTAT. (2016): *Soybean Production at the World Level.* Retrieved November 21, 2016, from http://faostat.fao.org/site/567/DesktopDefault.aspx?PageID=567#ancor

Feige, E. L. (1994). The Transition to a Market Economy in Russia: Property Rights, Mass Privatization and Stabilization. In G. S. Alexander & G. Skąpska (Eds.), *A Fourth Way?: Privatization, Property, and the Emergence of New Market Economics* (pp. 57–78). London: Routledge.

Filippov, R.V. (2014). International Experience of Subsidizing of Agriculture as Basis of Food Security of Russia. *Naukovedenie, 20*(1).

Food and Agriculture Organization of the United Nations. (1992). *Food, Nutrition, and Agriculture.* Rome: Food and Agriculture Organization of the United Nations.

Food and Agriculture Organization of the United Nations. (1999). *Implications of Economic Policy for Food Security: A Training Manual.* Retrieved February 3, 2017, from http://www. fao.org/docrep/004/x3936e/x3936e03.htm

Food and Agriculture Organization of the United Nations. (2001). *The State of Food Insecurity in the World.* Rome: Food and Agriculture Organization of the United Nations.

Food and Agriculture Organization of the United Nations. (2003). *Trade Reforms and Food Security: Conceptualising the Linkages*. Rome: Food and Agriculture Organization of the United Nations.

Food and Agriculture Organization of the United Nations. (2006). *Food Security*. Policy Brief. Retrieved February 1, 2017, from ftp://ftp.fao.org/es/ESA/policybriefs/pb_02.pdf

Food and Agriculture Organization of the United Nations. (2010). *La Iniciativa América Latina y Caribe sin Hambre*. Rome: Food and Agriculture Organization of the United Nations.

Food and Agriculture Organization of the United Nations. (2011). *Global Food Losses and Food Waste*. Study Conducted for the International Congress SAVE FOOD! Rome: Food and Agriculture Organization of the United Nations.

Food and Agriculture Organization of the United Nations. (2012). Assessment of the Agriculture and Rural Development Sectors in the Eastern Partnership Countries. The Republic of Azerbaijan. Budapest: FAO Regional Office for Europe and Central Asia (REU).

Food and Agriculture Organization of the United Nations. (2014). *Europe and Central Asia Food and Agriculture*. Budapest: Food and Agriculture Organization of the United Nations, Regional Office for Europe and Central Asia.

Food and Agriculture Organization of the United Nations. (2015). *Animal Production and Health*. Retrieved August 8, 2015, from http://www.fao.org/ag/againfo/themes/ru/animal_production.html

Food and Agriculture Organization of the United Nations. (2015). *The State of Food Insecurity in the World 2015. Meeting the 2015 International Hunger Targets: Taking Stock of Uneven Progress*. Rome: Food and Agriculture Organization of the United Nations.

Food and Agriculture Organization of the United Nations. (2016). *FAOSTAT*. Retrieved February 7, 2017, from http://www.faostat.org

Forsyth, T., & Evans, N. (2013). What is Autonomous Adaption? Resource Scarcity and Smallholder Agency in Thailand. *World Development*, *43*, 56–66. doi:10.1016/j.worlddev.2012.11.010

Fortucci, P. (2002). *The Contribution of Cotton to Economy and Food Security in Developing Countries*. Retrieved March 2, 2017, from https://www.icac.org/meetings/cgtn_conf/documents/11_fortucci.pdf

Frewer, L. J., Risvik, E., & Schifferstein, H. (2001). Food, People and Society: A European Perspective of Consumers' Food Choices. Dordrecht: Springer Science+Business Media B.V. doi:10.1007/978-3-662-04601-2

Frewer, L. J., Bergmann, K., Brennan, M., Lion, R., Meertens, R., Rowe, G., & Vereijken, C. et al. (2011). Consumer Response to Novel Agri-Food Technologies: Implications for Predicting Consumer Acceptance of Emerging Food Technologies. *Trends in Food Science & Technology*, *22*(8), 442–456. doi:10.1016/j.tifs.2011.05.005

Frewer, L. J., De Jonge, J., & Van Kleef, E. (2010). Consumer Perceptions of Food Safety. In B. P. Mansourian, A. Wojtczak, & B. McA. Sayers (Eds.), *Medical Sciences* (pp. 244–264). Oxford, UK: EOLSS Publishers.

Friedlander, M., & Levy, I. (1995). Cultivation of Gracilaria in Outdoor Tanks and Ponds. *Journal of Applied Phycology*, *7*(3), 315–324. doi:10.1007/BF00004005

Fuhrer, J., Beniston, M., Fischlin, A., Frei, C., Goyette, S., Jasper, K., & Pfister, C. (2006). Climate Risks and Their Impact on Agriculture and Forests in Switzerland. *Climatic Change*, *79*(1–2), 79–102. doi:10.1007/s10584-006-9106-6

Gami, B., Naik, A., & Patel, B. (2011). Cultivation of Spirulina Species in Different Liquid Media. *Journal of Algal Biomass Utilization*, *2*(3), 15–26.

Garelli, P. S. (2007). *The Competitiveness Roadmap: 2007–2050*. Lausanne: IMD World Competitiveness Center.

Gavin, B., & Sindzingre, A. (2009). EU Trade Relations with Emerging Asia: Identifying the Issues. *Asia Europe Journal*, *7*(1), 9–22. doi:10.1007/s10308-008-0220-0

Geoatlas. (2010). *Map of the Federal Republic of Nigeria.* Retrieved December 12, 2016, from http://www.geoatlas.com/en/maps/countries-4/nigeria-1090

Gerber, P. J., Mottet, A., Opio, C. I., Falcucci, A., & Teillard, F. (2015). Environmental Impacts of Beef Production: Review of Challenges and Perspectives for Durability. *Meat Science*, *109*, 2–12. doi:10.1016/j.meatsci.2015.05.013 PMID:26117397

Ghose, B. (2014). Food Security and Food Self-Sufficiency in China: From Past to 2050. *Food and Energy Security*, *3*(2), 86–95. doi:10.1002/fes3.48

Giannakopoulos, C., Le Sager, P., Bindi, M., Moriondo, M., Kostopoulou, E., & Goodess, C. M. (2009). Climatic Changes and Associated Impacts in the Mediterranean Resulting from a 2C Global Warming. *Global and Planetary Change*, *68*(3), 209–224. doi:10.1016/j.gloplacha.2009.06.001

Gilbert, C. L., & Morgan, C. W. (2010). Food Price Volatility. *Philosophical Transactions of the Royal Society of London. Series B, Biological Sciences*, *365*(1554), 3023–3034. doi:10.1098/rstb.2010.0139 PMID:20713400

Gilder, G. (2012, October 27). Urban Life: Open-Air Computers. *The Economist.* Retrieved March 5, 2017, from http://www.economist.com/news/special-report/21564998-cities-are-turning-vast-data-factories-open-air-computers

Gillson, I., & Fouad, A. (2016). *Trade Policy and Food Security: Improving Access to Food in Developing Countries in the Wake of High World Prices.* Washington, DC: World Bank.

Girard, M. (2015). *Trade and Transition: Russia's Agricultural Pitfalls and Prospects for the 21st Century.* Orono, ME: University of Maine.

Goletti, F., & Samman, E. (2002). Post-Harvest Systems in World Agriculture. In P. Golob, G. Farrell, & J.E. Orchard (Eds.), Crop Post-Harvest: Science and Technology: Vol. 1. Principles and Practice (pp. 1–34). Oxford, UK: Blackwell Science. doi:10.1002/9780470751015.ch1

Gonzalez-Feliu, J., Malhéné, N., Morganti, E., & Morana, J. (2014). The Deployment of City and Area Distribution Centers in France and Italy: Comparison of Six Representative Models. *Supply Chain Forum: An International Journal*, *15*(4), 84–99.

Gonzalez-Feliu, J., & Morana, J. (2014). Assessing Urban Logistics Pooling Sustainability via a Hierarchic Dashboard from a Group Decision Perspective. In C. Macharis, S. Melo, J. Woxenius, & T. Van Lier (Eds.), *Sustainable Logistics* (pp. 113–135). Bingley, UK: Emerald Group Publishing. doi:10.1108/S2044-994120140000006004

Gonzalez-Feliu, J., Taniguchi, E., & d'Arcier, B. F. (2014). Financing Urban Logistics Projects. In J. Gonzalez-Feliu, F. Semet, & J.-L. Frédéric (Eds.), *Sustainable Urban Logistics: Concepts, Methods and Information Systems* (pp. 245–265). Berlin: Springer. doi:10.1007/978-3-642-31788-0_13

Gouel, C., & Jean, S. (2013). Optimal Food Price Stabilization in a Small Open Developing Country. *The World Bank Economic Review*, *29*(1), 72–101. doi:10.1093/wber/lht018

Government of India. (2016). *State of Indian Agriculture 2015-16.* Retrieved October 10, 2016, from http://agricoop.nic.in/imagedefault/state_agri_1516.pdf

Government of Italy. (2012). *Testo del decreto-legge 18 ottobre 2012. Pubblicato nel supplemento ordinario n. 194/L alla 19 ottobre 2012, No. 245. Coordinato con la legge di conversione 17 dicembre 2012, No. 221. Recante: "Ulteriori misure urgenti per la crescita del Paese" (Publication No. 179).* Rome: Gazzetta Ufficiale.

Government of the Republic of Serbia. (2009). *Law on Genetically Modified Organisms (Publication no. 41/09)*. Belgrade: Official Gazette of the Republic of Serbia.

Government of the Republic of Serbia. (2015a). *Regulation on Subsidies Allocation in Agriculture and Rural Development in 2015 (Publication no. 19/15)*. Belgrade: Official Gazette of the Republic of Serbia.

Government of the Republic of Serbia. (2015b). *Regulation on Water Fees for 2015 (Publication no. 15/15)*. Belgrade: Official Gazette of the Republic of Serbia.

Greene, W. H. (2008). *Econometric Analysis*. New York, NY: Prentice Hall.

Grubor, A., & Djokic, N. (2016). Organic Food Consumer Profile in the Republic of Serbia. *British Food Journal*, *118*(1), 164–182. doi:10.1108/BFJ-06-2015-0225

Grubor, A., Djokic, N., Milicevic, N., & Kovac-Znidersic, R. (2014). Acceptance of Food with "Health Claims" in Serbia: Consumers' Profile Research and Recommendations for Promotion. *Custos e Agronegócio Online*, *10*(4), 199–220.

Grunert, K. G. (2005). Food Quality and Safety: Consumer Perception and Demand. *European Review of Agriculture Economics*, *32*(3), 369–391. doi:10.1093/eurrag/jbi011

Guiry, M. D., & Blunden, G. (Eds.). (1991). *Seaweed Resources in Europe: Uses and Potential*. Chichester, UK: John Wiley & Sons Ltd.

Habib, M. A. B., Parvin, M., Huntington, T. C., & Hasan, M. R. (2008). *A Review on Culture, Production and Use of Spirulina as Food for Humans and Feeds for Domestic Animals*. Rome: Food and Agriculutre Organization of the United Nations.

Hadri, K. (2000). Testing for Stationarity in Heterogeneous Panel Data. *The Econometrics Journal*, *3*(2), 148–161. doi:10.1111/1368-423X.00043

Hair, J. H., Tatham, R. L., Anderson, R. E., & Black, W. (1998). *Multivariate Data Analysis*. Upper Saddle River, NJ: Prentice Hall.

Hallström, E., Carlsson-Kanyama, A., & Börjesson, P. (2015). Environmental Impact of Dietary Change: A Systematic Review. *Journal of Cleaner Production*, *91*, 1–11. doi:10.1016/j.jclepro.2014.12.008

Hartzenberg, T. (2011). *Regional Integration in Africa*. Retrieved January 26, 2017, from https://www.wto.org/english/res_e/reser_e/ersd201114_e.pdf

Headey, D., & Fan, S. (2008). Anatomy of a Crisis: The Causes and Consequences of Surging Food Prices. *Agricultural Economics*, *39*, 375–391. doi:10.1111/j.1574-0862.2008.00345.x

Heilig, G. K. (2003). *Sustainable Regional and Rural Development in China: Where Do We Stand?* Laxenburg: IIASA.

Henneberry, S. R., & Diaz, C. C. (2014). Global Food Security: Emerging Economies and Diverging Food Markets. *Journal of Food Distribution Research*, *45*(1), 29–34.

Hepburn, P., Howlett, J., Boeing, H., Cockburn, A., Constable, A., Davi, A., & Wal, J. M. et al. (2008). The Application of Post-Market Monitoring to Novel Foods. *Food and Chemical Toxicology*, *46*(1), 9–33. doi:10.1016/j.fct.2007.09.008 PMID:17950974

Heribert, D. (2004). The Stability of International Financial Markets: A Global Public Good? In S. Schirm (Ed.), *New Rules for Global Markets: Public and Private Governance in the World Economy* (pp. 23–43). London: Palgrave Macmillan.

Hermann, M. (2009). The Impact of the European Novel Food Regulation on Trade and Food Innovation Based on Traditional Plant Foods from Developing Countries. *Food Policy*, *34*(6), 499–507. doi:10.1016/j.foodpol.2009.08.005

Hijioka, Y., Lin, E., Pereira, J. J., Corlett, R. T., Cui, X., Insarov, G. E., … Surjan, A. (2014). Asia. In V.R. Barros, C.B. Field, D.J. Dokken, M.D. Mastrandrea, K.J. Mach, T.E. Bilir, … L.L. White (Eds.), Climate Change 2014: Impacts, Adaptation, and Vulnerability. Part B: Regional Aspects. Contribution of Working Group II to the Fifth Assessment Report of the Intergovernmental Panel on Climate Change (pp. 1327–1370). Cambridge, UK: Cambridge University Press.

Hofstede, G. (n.d.). *Geert Hofstede.* Retrieved November 24, 2016, from https://geert-hofstede.com/

Holguin-Veras, J., Jaller, M., Van Wassenhove, L. N., Pérez, N., & Wachtendorf, T. (2012). On the Unique Features of Post-Disaster Humanitarian Logistics. *Journal of Operations Management*, *30*(7), 494–506. doi:10.1016/j.jom.2012.08.003

Holguin-Veras, J., Wachtendorf, T., Jaller, M., & Jefferson, T. (2013). Logistics and the Management of Critical Supplies Following Catastrophes. In R. Bissell (Ed.), *Preparedness and Response for Catastrophic Disasters* (pp. 131–150). Boca Raton, FL: CRC Press.

Holt-Giménez, E. (2008). *The World Food Crisis: What Is behind It and What We Can Do?* Oakland, CA: Institute for Food and Development Policy.

Hooper, D., Coughlan, J., & Mullen, M. (2008). Structural Equation Modelling: Guidelines for Determining Model Fit. *Electronic Journal of Business Research Methods*, *6*(1), 53–60.

Hopewell, K. (2014). The Transformation of State-Business Relations in an Emerging Economy. The Case of Brazilian Agribusiness. *Critical Perspectives on International Business*, *10*(4), 291–309. doi:10.1108/cpoib-03-2014-0019

Hsieh, Y. P., Leong, F. M., & Rudloe, J. (2001). Jellyfish as Food. *Hydrobiologia*, *451*(1–3), 11–17. doi:10.1023/A:1011875720415

Hu, L., & Bentler, P.M. (1999). Cutoff Criteria for Fit Indexes in Covariance Structure Analysis: Conventional Criteria versus New Alternatives. *Structural Equation Modelling: A Multidisciplinary Journal*, *6*(1), 1–55.

Huh, H., & Park, C. (2013). *Examining the Determinants of Food Prices in Developing Asia*. Manila: Asian Development Bank.

Hurlin, C., & Dumitrescu, E. (2012). *Testing for Granger Non-causality in Heterogeneous Panels*. Retrieved November 17, 2016, from https://halshs.archives-ouvertes.fr/halshs-00224434/document

Hurlin, C., & Venet, B. (2001). *Granger Causality Tests in Panel Data Models with Fixed Coefficients.* Paris: EURIsCO.

Huseyn, R. (2014). An Analysis of the Agricultural Subsidy Policy in Azerbaijan. *Azerbaijan Journal of Economics and Social Studies*, *1*(1), 6–18.

Ibadoglu, G. (2014). *Azerbaijan's Economic Model and Its Development since Independence.* Retrieved November 16, 2016, from http://azerireport.com/index.php?option=com_content&task=view&id=2981&Itemid=55

Ihle, R., von Cramon-Taubadel, S., & Zorya, S. (2009). Markov-switching Estimation of Spatial Maize Price Transmission Processes between Tanzania and Kenya. *American Journal of Agricultural Economics*, *91*(5), 1432–1439. doi:10.1111/j.1467-8276.2009.01360.x

Ijirshar, V. U. (2015). The Empirical Analysis of Agricultural Exports and Economic Growth in Nigeria. *Journal of Development and Agricultural Economics*, *7*(3), 113–122. doi:10.5897/JDAE2014.0615

Ikoku, E. U. (1980). *Self-Reliance: African Survival.* Enugu: Fourth Dimension Publishers.

Ilyinykh, P. A. (2013). Economy under Conditions of Veterinary Service Optimization in Sverdlovsk Region. *Agricultural Bulletin of the Urals, 115*(9), 91–94.

Im, K. S., Pesaran, M. H., & Shin, Y. (2003). Testing for Unit Roots in Heterogenous Panels. *Journal of Econometrics, 115*(1), 53–74. doi:10.1016/S0304-4076(03)00092-7

Ingram, M., Nabhan, G., & Buchmann, S. L. (1996). Our Forgotten Pollinators: Protecting the Birds and Bees. *Global Pesticide Campaigner, 6*(4), 1–8.

Institute of Agricultural Economics. (2015). *Results of In-depth Interview with Members of Selected Agricultural Holdings. Internal documentation.* Belgrade: Institute of Agricultural Economics.

Intergovernmental Panel on Climate Change. (2007). *Climate Change: Impacts, Adaptation and Vulnerability. Contribution of Working Group II to the Fourth Assessment Report of the Intergovernmental Panel on Climate Change.* Cambridge, UK: Cambridge University Press.

Intergovernmental Panel on Climate Change. (2012). *Managing the Risks of Extreme Events and Disasters to Advance Climate Change Adaptation. Special Report of the Intergovernmental Panel on Climate Change.* Cambridge, UK: Cambridge University Press.

International Chamber of Commerce. (2015). *Piracy and Armed Robbery against Ships. Report for the Period of 1 January – 30 June 2015.* Retrieved February 11, 2017, from www.hellenicshippingnews.com/wp-content/uploads/2015/07/2015-Q2-IMB-Piracy-Report-ABRIDGED.pdf

International Cocoa Organization. (2010). *SPS Capacity Building in Africa to Mitigate the Harmful Effect of Pesticides Residues in Cocoa and to Maintain Market Access.* Retrieved November 18, 2016, from http://www.icco.org/projects/10-projects/193-sps-capacity-building-in-africa-to-mitigate-the-harmful-effects-of-pesticides%20residues-in-cocoa-and-to-maintain-market-access.html

International Trade Center. (2016). *Trade Map.* Retrieved February 7, 2017, from http://www.trademap.org

International, C. A. B. (2010). *Climate Change and Crop Production.* Oxfordshire, UK: CAB International.

Investopedia. (2016). *Value Added Definition.* Retrieved February 6, 2017, from http://www.investopedia.com/terms/v/valueadded.asp

Ipema, A. H., Bleumer, E. J. B., & Lokhorst, C. (2011). *Precision Livestock Farming Claims for Sustainable Dairy Production.* Paper presented at the 5th European Conference on Precision Livestock Farming, Prague.

Irungu, C., Mburu, J., Maundu, P., Grum, M., & Hoeschle-Zeledon, I. (2011). The Effect of Market Development On-farm Conservation of Diversity of African Leafy Vegetables around Nairobi. *International Journal of Humanities and Social Science, 8*(1), 198–207.

Ismayilov, A., & Aliyev, K. (2013). Innovation Based Economic Development and the Importance of Government Promotion. *Journal of Qafqaz University – Economics and Administration, 1*(2), 231–241.

Ismayilov, A., & Aliyev, K. (2016). Government Support to the Agriculture for Promotion and Application of Innovations in Azerbaijan: Broker-State Framework. In *Proceedings of the Scientific Conference "Expanding Innovation Activities in Azerbaijan's Non-oil Sector"*. Baku: Azerbaijan Technical University.

Iversen, E. S., & Hale, K. K. (1992). Species for Human Food. In E. S. Iversen & K. K. Hale (Eds.), *Aquaculture Sourcebook* (pp. 35–204). New York, NY: Springer US. doi:10.1007/978-1-4684-1428-8_3

Ivolga, A., & Erokhin, V. (2011). Rural Development and Agrarian Reform: Russian Experience during the Transition Period. In *Proceedings of the International Conference "Rural Development Policies from the EU Enlargement Perspective"*. Ecka: Institute of Agricultural Economics.

Ivolga, A. (2014). Overview of Contemporary Issues of Sustainable Rural Development in Russia in Terms of Existing Differences between Regions. *Economics of Agriculture*, *2*, 331–345.

Jaffee, S., & Gordon, P. (1993). *Exporting High-Value Food Commodities: Success Stories from Developing Countries.* Washington, DC: World Bank Publications.

Jančić, M. (2015). *Uticaj klimatskih promena na biljnu proizvodnju* (Doctoral dissertation). University of Novi Sad, Novi Sad.

Jha, S., Kubo, K., & Ramaswami, B. (2013). *International Trade and Risk Sharing in the Global Rice Market: The Impact of Foreign and Domestic Supply Shocks.* Manila: Asian Development Bank.

Jianhua, Z. (2011). China's Success in Increasing Per Capita Food Production. *Journal of Experimental Botany*, *32*, 1–5.

Jongema, Y. (2015). *List of Edible Insects of the World.* Retrieved November 26, 2016, from http://www.wur.nl/en/Expertise-Services/Chair-groups/Plant-Sciences/Laboratory-of-Entomology/Edible-insects/Worldwide-species-list.htm

Josling, T., Anderson, K., Schmitz, A., & Tangerman, S. (2010). Understanding International Trade in Agricultural Products: One Hundred Years of Contributions by Agricultural Economists. *American Journal of Agricultural Economics*, *92*(2), 424–446. doi:10.1093/ajae/aaq011

Juma, C., & Mangeni, F. (2015). *The Benefits of Africa's New Free Trade Area.* Retrieved November 18, 2016, from http://belfercenter.ksg.harvard.edu/publication/25444/benefits_of_africas_new_free_trade_area.html

Kaiser, H. F. (1960). The Application of Electronic Computers to Factors Analysis. *Educational and Psychological Measurement*, *20*(1), 141–151. doi:10.1177/001316446002000116

Kaiser, H. F. (1970). A Second Generation Little Jiffy. *Psychometrika*, *35*(4), 401–415. doi:10.1007/BF02291817

Kaiser, H. F. (1974). An Index of Factorial Simplicity. *Psychometrika*, *39*(1), 31–36. doi:10.1007/BF02291575

Kalugina, Z. (2014). Agricultural Policy in Russia: Global Challenges and the Viability of Rural Communities. *International Journal of Sociology of Agriculture and Food*, *1*(21), 115–131.

Kane, S., Reilly, J., & Tobey, J. (1992). An Empirical Study of the Economic Effects of Climate Change on World Agriculture. *Climatic Change*, *21*(1), 17–35. doi:10.1007/BF00143251

Kaplan, R. S., & Norton, D. P. (2007). Using the Balanced Scorecard as a Strategic Management System. *Harvard Business Review*, *85*(7–8), 150–161.

Kaplinsky, R. (2001). Globalization and Economic Insecurity. Institute of Development Studies Bulletin (Sussex), 32(2), 13–24.

Kapraun, D. F. (1999). Red Algal Polysaccharide Industry: Economics and Research Status at the Turn of the Century. In *Proceedings of the Sixteenth International Seaweed Symposium*. Cebu City: Springer-Science+Business Media, B.V. doi:10.1007/978-94-011-4449-0_2

Karimov, I. A. (1993). *Building the Future: Uzbekistan-Its Own Model for Transition to a Market Economy.* Tashkent: Uzbekistan Publishers.

Karimov, R. (2015). Development of Non-Oil Sector in Azerbaijan: Tendencies and Opportunities. *Journal of Business & Economic Policy*, *2*(2), 39–52.

Katayama, N., Ishikawa, Y., Takaoki, M., Yamashita, M., Nakayama, S., Kiguchi, K., & Force, S. A. T. et al. (2008). Entomophagy: A Key to Space Agriculture. *Advances in Space Research*, *41*(5), 701–705. doi:10.1016/j.asr.2007.01.027

Kawashima, S. (1993). Cultivation of the Brown Alga, Laminaria "Kombu". In M. Ohno & A. T. Critchley (Eds.), *Seaweed Cultivation and Marine Ranching* (pp. 25–40). Yokosuka: Japan International Cooperation Agency.

Kearney, A. T. (n.d.). *Global Business Policy Council*. Retrieved November 23, 2016, from https://www.atkearney.com/gbpc

Kellar, J. A. (2012). Animal Health Surveillance: Navigation amidst the Flotsam of Human Frailty and Fiscal Inertia. *Preventive Veterinary Medicine*, *105*(3), 169–175. doi:10.1016/j.prevetmed.2011.12.009 PMID:22310236

Khalilov, H. (2015). Sustainable Agricultural Production in Azerbaijan: Current Situation and Main Tasks. *Azerbaijan Journal of Economics and Social Studies*, *3*, 6–23.

Khalilov, H., Shalbuzov, N., & Huseyn, R. (2015). *Country Report: Azerbaijan*. Baku: Research Institute of Agricultural Economics.

Khokhlov, A. V. (2014). *Reference Materials on Geography of World's Economy*. Retrieved May 30, 2015, from http://vlant-consult.ru/projects/materials/

Khoury, C. K., Bjorkmanc, A. D., Dempewolfd, H., Ramirez-Villegasa, J., Guarinof, L., Jarvisa, A., & Struik, P. C. et al. (2014). Increasing Homogeneity in Global Food Supplies and the Implications for Food Security. *Sustainability Science Agricultural Sciences*, *111*(11), 4001–4006. PMID:24591623

Kim, W. C., & Mauborgne, R. (2015). *Blue Ocean Strategy, Expanded Edition: How to Create Uncontested Market Space and Make the Competition Irrelevant*. Brighton, MA: Harvard Business Review Press.

Kitamura, M., & Omori, M. (2010). Synopsis of Edible Jellyfishes Collected from Southeast Asia, with Notes on Jellyfish Fisheries. *Plankton and Benthos Research*, *5*(3), 106–118. doi:10.3800/pbr.5.106

Kleinová, K., & Ürgeová, J. (2011). Country Image and Branding of Slovakia. *Polityki europejskie, 54*(5), 50–58.

Klein, R. J., Midgley, G. F., Preston, B. L., Alam, M., Berkhout, F. G. H., Dow, K., & Shaw, M. R. (2014). Adaptation Opportunities, Constraints, and Limits. In C. B. Field et al. (Eds.), *Climate Change 2014: Impacts, Adaptation, and Vulnerability, Part A: Global and Sectoral Aspects. Contribution of Working group II to V Assessment Report of IPCC* (pp. 899–943). Cambridge, UK: Cambridge University Press.

Kneller, R., Morgan, C. W., & Kanchanahatakij, S. (2008). Trade Liberalisation and Economic Growth. *World Economy*, *31*(6), 701–719. doi:10.1111/j.1467-9701.2008.01101.x

Kolodko, G. W. (1998). *Equity Issues in Policymaking in Transition Economies*. Washington, DC: International Monetary Fund.

Kovačević, V. (2014). *Razvoj robnih berzi u funkciji upravljanja rizikom poslovanja poljoprivrednih preduzeća u Srbiji* (Doctoral dissertation). University of Belgrade, Belgrade, Serbia.

Kovács, G., & Spens, K. M. (2007). Humanitarian Logistics in Disaster Relief Operations. *International Journal of Physical Distribution & Logistics Management*, *37*(2), 99–114. doi:10.1108/09600030710734820

Kovács, G., & Spens, K. M. (2009). Identifying Challenges in Humanitarian Logistics. *International Journal of Physical Distribution & Logistics Management*, *39*(6), 506–528. doi:10.1108/09600030910985848

KPMG. (2011). *Confronting Complexity*. Amsterdam: KPMG International Cooperative.

Krivonos, E. (2015). *Russian Restrictions on Agri-food Imports: Effects on Domestic and World Markets*. Paper presented at the International Conference of Agricultural Economists, Milan.

Kurukulasuriya, P., & Mendelsohn, R. (2006). *A Ricardian Analysis of the Impact of Climate Change on African Cropland*. CEEPA Discussion Paper No. 8. Pretoria: Centre for Environmental Economics and Policy in Africa, University of Pretoria.

Lalić, B., Stričević, R., Janković, D., Jaćimović, G., & Firanj, A. (2015). *Prilagođavanje na nove klimatske uslove: Zagrevanje useva*. Report presented at Dialogue on Climate Change in Serbia: Towards the Conference in Paris, Belgrade. Retrieved March 28, 2016, from http://www.klimatskepromene.rs/uploads/useruploads/Documents/Zagrevanje-useva_B.Lalic.pdf

Lambert, D. M. (2001). The Supply Chain Management and Logistics Controversy. In A. M. Brewer, K. M. Button, & D. A. Hensher (Eds.), *Handbook of Logistics and Supply Chain Management* (pp. 99–126). Oxford, UK: Pergamon Press.

Lambert, D. M., Cooper, M. C., & Pagh, J. D. (1998). Supply Chain Management: Implementation Issues and Research Opportunities. *The International Journal of Logistics Management*, *9*(2), 1–20. doi:10.1108/09574099810805807

Laudicina, P. (2012). *Beating the Global Odds: Successful Decision-making in a Confused and Troubled World*. Hoboken, NJ: John Wiley & Sons, Inc.

Lavy, V. (1992). Alleviating Transitory Food Crisis in Sub-Saharan Africa: International Altruism and Trade. *The World Bank Economic Review*, *6*(1), 125–138. doi:10.1093/wber/6.1.125

Lawrence, G., & McMichael, P. (2012). The Question of Food Security. *International Journal of Sociology of Agriculture and Food*, *2*(19), 135–142.

Le Pere, G. (2005). Emerging Markets – Emerging Powers: Changing Parameters for Global Economic Governance. *IPG*, *2*, 36–51.

Lee, H., & Park, C. (2013). *International Transmission of Food Prices and Volatilities: A Panel Analysis*. Manila: Asian Development Bank.

Lee, R. (2007). *Food Security and Food Sovereignty*. Newcastle upon Tyne, UK: Centre for Rural Economy, University of Newcastle upon Tyne.

Leonard, D. K. (1987). The Supply of Veterinary Services: Kenyan Lessons. *Agricultural Administration and Extension*, *26*(4), 219–236. doi:10.1016/0269-7475(87)90061-4

Lerman, Z. (2001). Agriculture in Transition Economies: From Common Heritage to Divergence. *Agricultural Economics*, *26*(2), 96–114. doi:10.1111/j.1574-0862.2001.tb00057.x

Levin, A., Lin, C. F., & Chu, C.-S. J. (2002). Unit Root Tests in Panel Data: Asymptotic and Finite Sample Properties. *Journal of Applied Econometrics*, *108*(1), 1–22. doi:10.1016/S0304-4076(01)00098-7

Liefert, W. (2004). *Food Security in Russia: Economic Growth and Rising Incomes are Reducing Insecurity*. Washington, DC: Economic Research Service, USDA.

Liefert, W., & Swinnen, J. (2002). *Changes in Agricultural Markets in Transition Economies*. Washington, DC: Economic Research Service, USDA.

Lin, J. H., Kaphle, K., Wu, L. S., Yang, N. Y. J., Lu, G., Yu, C., & Rogers, P. A. M. et al. (2003). Sustainable Veterinary Medicine for the New Era. *OIE Revue Scientifique et Technique*, *22*(3), 949–964. doi:10.20506/rst.22.3.1451 PMID:15005552

Lioubimtseva, E., & Henerbry, G. (2009). Climate and Environmental Change in Arid Central Asia: Impacts, Vulnerability, and Adoptions. *Journal of Arid Environments*, *73*(11), 963–977. doi:10.1016/j.jaridenv.2009.04.022

Lipton, M., & Heald, C. (1984). *African Food Strategies and the EEC's Role: An Interim Review* (1st ed.). Brussels: Centre for European Policy Studies.

Liss, C. (2007). *The Roots of Piracy in Southeast Asia*. Retrieved February 11, 2017, from http://nautilus.org/apsnet/the-roots-of-piracy-in-southeast-asia/

Liu, H., Guenther, F., & Laixiang, S. (2004). Study on the Impacts of Climate Change on Chinas Agriculture. *Climatic Change*, *65*(1/2), 125–148. doi:10.1023/B:CLIM.0000037490.17099.97

Liu, J., & Diamond, J. (2005). Chinas Environment in a Globalizing World. *Nature*, *435*(7046), 1179–1186. doi:10.1038/4351179a PMID:15988514

Liu, M., Heijman, W., Zhu, X., Dries, L., & Huang, J. (2016). Individual and Social Optima of Rural Land Allocation by Stakeholders: A Case Study on Eco-Fragile Areas of Northern China. *The Rangeland Journal*, *38*, 57–72.

Li, W. J., Ali, S. H., & Zhang, Q. (2007). Property Rights and Grassland Degradation: A Study of the Xilingol Pasture, Inner Mongolia, China. *Journal of Environmental Management*, *85*(2), 461–470. doi:10.1016/j.jenvman.2006.10.010 PMID:17129663

Lobell, D. B., & Burke, M. B. (2008). Why are Agricultural Impacts of Climate Change so Uncertain? The Importance of Temperature Relative to Precipitation. *Environmental Research Letters*, *3*(3), 034007. doi:10.1088/1748-9326/3/3/034007

Lobell, D., & Field, C. (2007). Global Scale Climate–Crop Yield Relationships and the Impacts of Recent Warming. *Environmental Research Letters*, *2*(1), 1–7. doi:10.1088/1748-9326/2/1/014002

Long, H., Liu, Y., Li, X., & Chen, Y. (2010). Building New Countryside in China: A Geographical Perspective. *Land Use Policy*, *27*(2), 457–470. doi:10.1016/j.landusepol.2009.06.006

Luan, Y., Cui, X., & Ferrat, M. (2013). Historical Trends of Food Self-Sufficiency in Africa. *Food Security*, *3*(5), 393–405. doi:10.1007/s12571-013-0260-1

Luckmann, J., Ihle, R., Kleinwechter, U., & Grethe, H. (2014). World Market Integration of Vietnamese Rice Markets during the 2008 Food Price Crisis. *Food Security. The Science. Sociology and Economics of Food Production and Access to Food*, *6*(6), 1–17.

Lüning, K., Yarish, C., & Kirkman, H. (1990). *Seaweeds: Their Environment, Biogeography, and Ecophysiology*. Hoboken, NJ: John Wiley & Sons.

Lunze, K., Yurasova, E., Idrisov, B., Gnatienko, N., & Migliorini, L. (2015). Food Security and Nutrition in the Russian Federation – a Health Policy Analysis. *Global Health Action*, *8*(1), 10. doi:10.3402/gha.v8.27537

Ma, L. (2011). Sustainable Development of Rural Household Energy in Northern China. *Journal of Sustainable Development*, *5*(4), 115–124.

Maldonado, M., & Moya, S. (2013). The Challenge of Implementing Reverse Logistics in Social Improvement: The Possibility of Expanding Sovereignty Food in Developing Communities. In J. Cheung & H. Song (Eds.), *Logistics: Perspectives, Approaches and Challenges* (pp. 1–34). New York, NY: Nova Science Publishers.

Malikov, N., Qineti, A., & Pulatov, A. (2016). *Agriculture and Economic Development in Uzbekistan.* Paper presented at the International Scientific Days, Nitra.

Malozemov, S. I. (2014). Experience of State Support of Agriculture in Foreign Countries. *Science and Modernity*, *32*(1), 136–141.

Marcotty, T., Thys, E., Conrad, P., Godfroid, J., Craig, P., Zinsstag, J., & Boelaert, M. et al. (2013). Intersectoral Collaboration between the Medical and Veterinary Professions in Low-Resource Societies: The Role of Research and Training Institutions. *Comparative Immunology, Microbiology and Infectious Diseases*, *36*(3), 233–239. doi:10.1016/j.cimid.2012.10.009 PMID:23260374

Markovic, I., & Markovic, M. (2014). Agricultural Protectionism of the European Union in the Conditions of International Trade Liberalization. *Economics of Agriculture*, *61*(2), 423–440.

Marsden, T., & Smith, E. (2005). Ecological Entrepreneurship: Sustainable Development in Local Communities through Quality Food Production and Local Branding. *Geoforum*, *36*(4), 440–451. doi:10.1016/j.geoforum.2004.07.008

Martínez-Zarzoso, I., & Bensassi, S. (2013). The Price of Modern Maritime Piracy. *Defence and Peace Economics*, *24*(5), 397–418. doi:10.1080/10242694.2012.723156

Masudul Haq, M., Kaitibie, S., Huda, S., Vock, N., Coughlan, K., Abdella, K., … Moustafa, A. (2014). *Using a Gravity Approach to Explain Food Imports for Improved Food Security in Qatar.* Paper presented at Qatar Foundation Annual Research Conference, Doha.

Matsunaga, M. (2010). How to Factor-Analyse your Data Right: Do's Dont's, and How-To's. *International Journal of Psychological Research*, *3*(1), 97–110.

Maundu, P., Achigan-Dako, E., & Morimoto, Y. (2009). Biodiversity of African Vegetables. In C. M. Shackleton, M. W. Pasquini, & A. W. Drescher (Eds.), *African Indigenous Vegetables in Urban Agriculture* (pp. 65–104). London: Earthscan.

Maxwell, S. (1990). Food Security in Developing Countries: Issues and Options for the 1990s. *IDS Bulletin*, *21*(3), 2–13. doi:10.1111/j.1759-5436.1990.mp21003002.x

McCarthy, J., Canziani, O. F., Leary, N. A., Dokken, D. J., & White, C. (Eds.). (2001). *Climate Change 2001: Impacts, Adaptation, and Vulnerability. Contribution of Working Group II to the Third Assessment Report of the Intergovernmental Panel on Climate Change*. Cambridge, UK: Cambridge University Press.

McLean, M., Watson, H. K., & Muswema, A. (2007). Veterinary Waste Disposal: Practice and Policy in Durban, South Africa (2001–2003). *Waste Management (New York, N.Y.)*, *27*(7), 902–911. doi:10.1016/j.wasman.2006.05.004 PMID:16876398

Mefor, L. (2007). *Senator Ken Nnamani Portrait and the New Nigerian Senate*. Abuja: Psykom Publishing Ltd.

Mendelsohn, R., & Dinar, A. (Eds.). (2009). *Climate Change and Agriculture – An Economic Analysis of Global Impacts, Adaptation and Distributional Effect, New Horizons in Environmental Economics*. Cheltenham, UK: Edward Elgar.

Mendelsohn, R., Nordhaus, W. D., & Shaw, D. (1994). The Impact of Global Warming on Agriculture: A Ricardian Analysis. *The American Economic Review*, *84*(4), 753–771.

Mendelsohn, R., Nordhaus, W. D., & Shaw, D. (1996). Climate Impacts on Aggregate Farm Values: Accounting for Adaptation. *Agricultural and Forest Meteorology*, *80*(1), 55–67. doi:10.1016/0168-1923(95)02316-X

Merzlov, A., Ovchintseva, L., & Popova, O. (2012). *Regional Experience of Elaboration of Programs of Sustainable Rural Development*. Moscow: Rosinformagrotekh.

Meskhia, I. E. (2016). Food Security Problems in Post Soviet Georgia. *Annals of Agrarian Science*, *2*(14), 46–51. doi:10.1016/j.aasci.2016.03.004

Miladinović, J., Hrustić, M., & Vidić, M. (2011). *Soybean*. Novi Sad: Institute of Field and Vegetable Crops.

Ministerio de Desarrollo e Inclusión Social. (2013). *Programa Vaso de Leche*. Lima: Gobierno de la República del Perú.

Ministry of Agriculture and Environmental Protection of the Republic of Serbia. (2016). *Balance of Soybean and Basic Soybean Products Production/Consumption at the Level of the Republic of Serbia*. Retrieved November 21, 2016, from www.mpzzs.gov.rs/dokumenti/

Ministry of Agriculture of the Republic of Azerbaijan. (2016). *Tax Incentives*. Retrieved October 19, 2016, from http://www.agro.gov.az/vergi-gztlri

Mirzabaev, A., Goedecke, J., Dubovyk, O., Djanibekov, U., Le, Q. B., & Aw-Hassan, A. (2016). Economics of Land Degradation in Central Asia. In E. Nkonya, A. Mirzabaev, & J. von Braun (Eds.), *Economics of Land Degradation and Improvement – A Global Assessment for Sustainable Development* (pp. 261–290). Cham: Springer International Publishing. doi:10.1007/978-3-319-19168-3_10

Mirzabaev, A., & Tsegai, D. (2012). *Effects of Weather Shocks on Agricultural Commodity Prices in Central Asia. ZEF Discussion Papers on Development Policy No. 171*. Bonn: Center for Development Research.

Mohamed Arshad, F., & Hameed, A.A.A. (2010). Global Food Prices: Implications for Food Security in Malaysia. *Journal of the Consumer Research and Resource Centre*, 21–38.

Mohring, S. A. I., Strzysch, I., Fernandes, M. R., Kiffmeyer, T. K., Tuerk, J., & Hamscher, G. (2009). Degradation and Elimination of Various Sulfonamides during Anaerobic Fermentation: A Promising Step on the Way to Sustainable Pharmacy? *Environmental Science & Technology*, *43*(7), 2569–2574. doi:10.1021/es802042d PMID:19452918

Morana, J., & Gonzalez-Feliu, J. (2015). A Sustainable Urban Logistics Dashboard from the Perspective of a Group of Operational Managers. *Management Research Review*, *38*(10), 1068–1085. doi:10.1108/MRR-11-2014-0260

Morgan Stanley Capital International. (n.d.). *Market Classification*. Retrieved September 6, 2016, from https://www.msci.com/market-classification

Morgounov, A. (1998). *Spring Wheat in Kazakstan: Current Status and Future Directions*. CIMMYT.

Morikawa, T. (1984). Jellyfish. *FAO Infofish Marketing Digest*, *1*, 37–39.

Morozova, E., Glushakova, O., & Fadeikina, N. (2016). Food Consumption as an Indicator of the Quality of Life of the Population in Regions. *Foods and Raw Materials*, *4*(1), 171–180. doi:10.21179/2308-4057-2016-1-171-180

Mpala, C., Dlamini, M., & Sibanda, P. (2013). The Accessibility, Utilisation and Role of Indigenous Traditional Vegetables in Household Food Security in Rural Hwange District. *International Open & Distance Learning Journal International Research Conference Special Edition*, *3*(1), 40–48.

Mumbi, K., Karanja, N., Njenga, M., Kamore, M., Achieng, C., & Ngeli, P. (2006). *Investigative Market Research: Viable Market Opportunities and Threats for Urban and Peri-Urban Farmers, Farm Concern International*. Nairobi: Urban Harvest and International Potato Centre.

Nastić, L., Jeločnik, M., & Subić, J. (2014). Analysis of Calla Lily and Cucumber Production in Greenhouse. *Ekonomika*, *60*(4), 209–217.

Natale, F., Borrello, A., & Molotova, A. (2015). Analysis of the Determinants of International Seafood Trade Using a Gravity Model. *Marine Policy*, *60*, 98–106. doi:10.1016/j.marpol.2015.05.016

Nath, R., Luan, Y., Yang, W., Yang, C., Chen, W., Li, Q., & Cui, X. (2015). Changes in Arable Land Demand for Food in India and China: A Potential Threat to Food Security. *Sustainability*, *7*(5), 5371–5397. doi:10.3390/su7055371

National Bureau of Statistics of the People's Republic of China. (n.d.). *Database*. Retrieved December 3, 2016, from http://www.stats.gov.cn/

National Cotton Council of America. (2016). *Rankings*. Retrieved February 7, 2017, from http://www.cotton.org/econ/cropinfo/cropdata/rankings.cfm

National Drought Mitigation Centre. (2016). *Types of Drought Impacts*. Lincoln, NE: National Drought Mitigation Centre, University of Nebraska.

National Planning Commission. (2009). *Nigeria Vision 20:2020*. Abuja: Government of the Federal Republic of Nigeria.

National Planning Commission. (2016). *The Transformation Agenda – Summary of Federal Government's Key. Priority Policies, Programmes and Projects 2011–2015*. Abuja: Government of the Federal Republic of Nigeria.

National Population Commission of Nigeria. (2006). *Population and Housing Census of the Federal Republic of Nigeria*. Retrieved November 7, 2016, from http://www.population.gov.ng/images/Priority%20Tables%20Volume%20I-update.pdf

Nazarbayev, N. A. (1997). *Kazakhstan-2030*. Almaty: Bilim.

Nechaev, V. (2008). *Conceptual Foundations of Ensurance of Food Security in Russia*. Moscow: All-Russian Research Institute of Agricultural Economics.

Nekesa, P., & Meso, B. (1997). Traditional African Vegetables in Kenya: Production, Marketing and Utilization. In *Proceedings of the IPGRI International Workshop on Genetic Resources of Traditional Vegetables in Africa: Conservation and Use*. Rome: International Plant Genetic Resources Institute.

Nelson, G. C., Rosegrant, M. W., Koo, J., Robertson, R., Sulser, T., Zhu, T., … Lee, D. (2009). Climate Change: Impact on Agriculture and Costs of Adaptation. Washington, DC: International Food Policy Research Institute.

Nelson, G. C., Van der Mensbrugghe, D., Ahammad, H., Blanc, E., Calvin, K., Hasegawa, T., & Willenbockel, D. et al. (2014). Agriculture and Climate Change in Global Scenarios: Why dont the Models Agree. *Agricultural Economics*, *45*(1), 85–101. doi:10.1111/agec.12091

Neori, A., Chopin, T., Troell, M., Buschmann, A. H., Kraemer, G. P., Halling, C., & Yarish, C. et al. (2004). Integrated Aquaculture: Rationale, Evolution, and State of the Art Emphasizing Seaweed Biofiltration in Modern Mariculture. *Aquaculture (Amsterdam, Netherlands)*, *231*(1), 361–391. doi:10.1016/j.aquaculture.2003.11.015

Neszmélyi, G. (2014). The Issues of Sustainability and Self-sufficiency in the Nigerian Food Economy. In G. Neszmélyi (Ed.), *Socio-economic and Regional Processes in the Developing Countries* (pp. 98–116). Gödöllő: Szent István University Publishing.

Nicita, A. (2015). *Key Statistics and Trends in International Trade 2014*. Geneva: *United Nations Conference on Trade and Development*.

Nielsen, L. (2011). *Classifications of Countries Based on Their Level of Development: How It Is Done and How It Could Be Done*. IMF Working Paper, WP/11/31. Washington, DC: International Monetary Fund.

Nigerian Export Promotion Council. (2011). *Expanding Nigeria's Exports of Sesame Seeds and Sheanut/Butter through Improved SPS Capacity Building for Private and Public Sector*. Retrieved November 18, 2016, from http://www.aflasafe.com/c/document_library/get_file?uuid=9adeb84f-67c2-4de8-9082-110a27b94489&groupId=524500

Nikitin, I. N., Akmullin, A. I., & Trofimova, E. N. (2012). Pricing of Paid Veterinary Services in Volgograd Region. *Scientific Notes of Kazan State Academy of Veterinary Medicine*, *211*, 423–429.

Nowak, V., Persijn, D., Rittenschober, D., & Charrondiere, U. R. (2014). Review of Food Composition Data for Edible Insects. *Food Chemistry*, *193*, 36–46. PMID:26433285

Nunnally, J. C. (1978). *Psychometric Theory*. New York, NY: McGraw-Hill.

Ochieng, J., Afari-Sefa, V., Karanja, D., Kessy, R., Rajendran, S., & Samali, S. (2016). How Promoting Consumption of Traditional African Vegetables Affects Household Nutrition Security in Tanzania. *Renewable Agriculture and Food Systems*, 1–11. doi:10.1017/S1742170516000508

Ohmae, K. (1989). The Global Logic of Strategic Alliances. *Harvard Business Review*, (March-April), 143–154.

Ohno, T. (1988). *Toyota Production System: Beyond Large-Scale Production*. Boca Raton, FL: CRC Press.

Ojiewo, C., Tenkouano, A., Oluoch, M., & Yang, R. (2010). The Role of AVRDC – the World Vegetable Centre in Vegetable Value Chains. *African Journal of Horticultural Science*, *3*, 1–23.

Olajide, O. T., Akinlabi, B. H., & Tijani, A. A. (2012). Agriculture Resource and Economic Growth in Nigeria. *European Scientific Journal*, *22*(8), 103–115.

Oliveira, E. C., Alveal, K., & Anderson, R. J. (2000). Mariculture of the Agar-Producing Gracilarioid Red Algae. *Reviews in Fisheries Science*, *8*(4), 345–377. doi:10.1080/10408340308951116

Olson, D. R. (2003). *Towards Food Sovereignty: Constructing an Alternative to the World Trade Organization's Agreement on Agriculture*. Minneapolis, MN: Institute for Agriculture and Trade Policy.

Oluwasanmi, H. A. (1966). *Agriculture and Nigeria's Economic Development*. Ibadan: Oxford University Press.

Orbell, J. M., & Dawes, R. M. (1993). Social Welfare, Cooperators Advantage, and the Option of not Playing the Game. *American Sociological Review*, *58*(6), 787–800. doi:10.2307/2095951

Organization for Economic Cooperation and Development. (2015). *OECD-FAO Agricultural Outlook 2015*. Paris: OECD Publishing.

Organization for Economic Cooperation and Development. (2016). *National Accounts*. Retrieved November 24, 2016, from http://www.oecd.org/std/na/

Orindi, V., Ochieng, A., Otiende, B., Bhadwal, S., Anantram, K., Nair, S., … Kelkar, U. (2006). Mapping Climate Vulnerability and Poverty in Africa. In P.K. Thornton, P.G. Jones, T. Owiyo, R.L. Kruska, M. Herrero, P. Kristjanson, … A. Omolo (Eds.), Report to the Department for International Development. Nairobi: ILRI.

Orumgbe, R. (n.d.). *The Effect of World Trade Organisation on Nigeria Maritime Industry*. Retrieved November 16, 2016, from http://www.academia.edu/11962449/THE_EFFECT_OF_WORLD_TRADE_ORGANISATION_ON_NIGERIA_MARITIME_INDUSTRY

Osipova, N. I. (2008). Sanitary and Bacteriological Examination of Organic Wastes of Livestock Enterprises. *Veterinary Medicine*, *1*, 4–5.

Otsuka, K., Larson, D. F., & Hazell, P. B. R. (2013). An Overview. In K. Otsuka & D. F. Larson (Eds.), *An African Green Revolution. Finding Ways to Boost Productivity of Small Farms* (pp. 1–14). Amsterdam: Springer. doi:10.1007/978-94-007-5760-8_1

Overview. (n.d.). In *Association of Southeast Asian Nations*. Retrieved November 2, 2016, from http://asean.org/asean/about-asean/overview/

Palacios, L., Morana, J., Gonzalez-Feliu, J., & Devia, J. (2016). Traceability for a More Sustainable Supply Chain Management: The Case of Bienestarina in Colombia. In *Proceedings of the Third International Conference on Green Supply Chain Management*. London: Loughborough University.

Pannhausen, C. (2006). Economic Partnership Agreements and Food Security: What Is at Stake for West Africa? Bonn: German Development Institute (DIE).

Papić Brankov, T., & Lovre, K. (2011). Investiranje istraživanja i razvoja u oblasti poljoprivredne biotehnologije. *Ekonomika*, *57*(1), 30–36.

Papić Brankov, T., & Milovanović, M. (2015). Measuring Food Security in the Republic of Serbia. *Economics of Agriculture*, *62*(3), 801–812.

Pappaioanou, M. (2004). Veterinary Medicine Protecting and Promoting the Publics Health and Well-Being. *Preventive Veterinary Medicine*, *62*(3), 153–163. doi:10.1016/j.prevetmed.2003.11.001 PMID:15041202

Parry, M., Arnell, N., Berry, P., Dodman, D., Fankhauser, S., & Hope, C. … Wheeler, T. (2009). Assessing the Costs of Adaptation to Climate Change – A Review of UNFCCC and other Recent Studies. London: International Institute for Environment and Development and Grantham Institute for Climate Change, Imperial College London.

Parry, M. L., Rosenzweig, C., Iglesias, A., Fischer, G., & Livermore, M. (1999). Climate Change and World Food Security: A New Assessment. *Global Environmental Change: Human and Policy Dimensions*, *9*(supplement 1), 51–67. doi:10.1016/S0959-3780(99)00018-7

Parry, M., Rosenzweig, C., Iglesias, A., Livermore, M., & Fisher, G. (2004). Effects of Climate Change on Global Food Production under SRES Emissions and Socio-Economic Scenarios. *Global Environmental Change*, *14*(1), 53–67. doi:10.1016/j.gloenvcha.2003.10.008

Pearce, D., Cline, W., Achanta, A., Fankhauser, S., Pachauri, R., Tol, R., & Vellinga, P. (1996). The Social Costs of Climate Change: Greenhouse Damage and Benefits of Control. In J. Bruce, H. Lee, & E. Haites (Eds.), *Climate Change 1995: Economic and Social Dimensions of Climate Change* (pp. 179–224). Cambridge, UK: Cambridge University Press.

Peñaloza, C., Palacios, L., & Gonzalez-Feliu, J. (2016). Characterization of the Collection and Distribution Processes of Bienestarina in Bogotá. In *Proceedings of the International Workshop on Franchising & Distribution Networks in Emerging Countries*. Saint-Etienne: Jean Monet University.

Perkins, W. A., & Stembridge, J. H. (1966). *Nigeria – A Descriptive Geography* (3rd ed.). Ibadan: Oxford University Press.

Pesaran, M. H., Shin, Y., & Smith, R. J. (1999). *Bounds Testing Approaches to the Analysis of Long-run Relationships*. Cambridge Working Papers in Economics, 9907. Cambridgeshire, UK: University of Cambridge.

Petrick, M. (2015). *The Prospects for Russian Food Self-Sufficiency*. Paper presented at the International Conference of Agricultural Economists, Milan, Italy.

Petrick, M., Oshakbaev, D., & Wandel, J. (2014). *Kazakhstan's Wheat, Beef and Dairy Sectors: An Assessment of Their Development Constraints and Recent Policy Responses*. Halle: Leibniz Institute of Agricultural Development in Central and Eastern Europe.

Petrikov, A. V. (2012). It Is Necessary to Increase Adaptation of Russian Agrarian Sector to WTO Conditions. *Economy of Agricultural and Processing Enterprises*, *6*, 6–8.

Petrikov, A. V. (2016). Major Directions of Implementation of Modern Agrifood and Rural Policy. *International Agricultural Journal, 1*, 3–9.

Pomfret, R. (2001). *The Economics of Regional Trading Arrangements*. Oxford, UK: Oxford University Press. doi:10.1093/0199248877.001.0001

Pomfret, R. (2010). Trade and Transport in Central Asia. *Global Journal of Emerging Market Economies, 2*(3), 237–256. doi:10.1177/097491011000200302

Popkova, E. G., Dubova, Y. I., Yakovleva, E. A., Azarova, N. A., & Titova, E. V. (2014). Role of Ecological Marketing in Formation and Development of Ecological Cluster. *Asian Social Science, 23*(10), 1–8.

Power, E. M. (2005). Determinants of Healthy Eating Among Low-Income Canadians. *Canadian Journal of Public Health, 96*, 537–542. PMID:16042163

Presidencia de la República. (2013). *Bienestarina Más' llegará gratuitamente este año a 6,5 millones de beneficiarios, especialmente niños y niñas de la primera infancia.* Retrieved November 23, 2016, from http://wsp.presidencia.gov.co/Prensa/2013/Febrero/Paginas/20130228_07.aspx

President of Azerbaijan. (2015). *Decree on Declaring 2015 the Year of Agriculture in the Republic of Azerbaijan*. Baku: Administration of the President of the Republic of Azerbaijan.

President of Azerbaijan. (2016a). *Decree on Additional Actions on Promotion of Non-oil Products Export*. Baku: Administration of the President of the Republic of Azerbaijan.

President of Azerbaijan. (2016b). *Decree on Establishment of the Open Joint Stock Company on Food Products Supply and Procurement*. Baku: Administration of the President of the Republic of Azerbaijan.

President of Azerbaijan. (2016c). *Law of the Republic of Azerbaijan on Agricultural Cooperatives*. Baku: Administration of the President of the Republic of Azerbaijan.

Rakhimov, M. (2010). Internal and External Dynamics of Regional Cooperation in Central Asia. *Journal of Eurasian Studies, 1*(2), 95–101. doi:10.1016/j.euras.2010.04.002

Ramonet, I. (2008). *Africa Says No – And Means It.* Retrieved November 18, 2016, from https://mondediplo.com/2008/01/01africa

Ramos-Elorduy, J., Moreno, J. M. P., & Camacho, V. H. M. (2009). Edible Aquatic Coleoptera of the World with an Emphasis on Mexico. *Journal of Ethnobiology and Ethnomedicine, 5*(11). Retrieved from https://ethnobiomed.biomedcentral.com/articles/10.1186/1746-4269-5-11 PMID:19379486

Reidsma, P., Wolf, J., Kanellopoulos, A., Schaap, B. F., Mandryk, M., Verhagen, J., & Van Ittersum, M. K. (2015). Climate Change Impact and Adaptation Research Requires Integrated Assessment and Farming Systems Analysis: A Case Study in the Netherlands. *Environmental Research Letters, 10*(4). doi:10.1088/1748-9326/10/4/045004

Reilly, J. (1999). Climate Change – Can Agriculture Adapt? Choices. In C. Rosenzweig & A. Iglesias (Eds.), *Implications of Climate Change for International Agriculture: Crop Modeling Study* (pp. 4–8). Washington, DC: US EPA Office of Policy, Planning and Evaluation, Climate Change Division, Adaptation Branch.

Republican Hydrometeorological Service of Serbia. (2006). *Agrometeorological Conditions in Production Cycle 2005/2006 at the Territory of the Republic of Serbia*. Belgrade: Republican Hydrometeorological Service of Serbia.

Republican Hydrometeorological Service of Serbia. (2008a). *Agrometeorological Conditions in Production Cycle 2006/2007 at the Territory of the Republic of Serbia*. Belgrade: Republican Hydrometeorological Service of Serbia.

Republican Hydrometeorological Service of Serbia. (2008b). *Agrometeorological Conditions in Production Cycle 2007/2008 at the Territory of the Republic of Serbia*. Belgrade: Republican Hydrometeorological Service of Serbia.

Republican Hydrometeorological Service of Serbia. (2009). *Agrometeorological Conditions in Production Cycle 2008/2009 at the Territory of the Republic of Serbia*. Belgrade: Republican Hydrometeorological Service of Serbia.

Republican Hydrometeorological Service of Serbia. (2010). *Agrometeorological Conditions in Production Cycle 2009/2010 at the Territory of the Republic of Serbia*. Belgrade: Republican Hydrometeorological Service of Serbia.

Republican Hydrometeorological Service of Serbia. (2011). *Agrometeorological Conditions in Production Cycle 2010/2011 at the Territory of the Republic of Serbia*. Belgrade: Republican Hydrometeorological Service of Serbia.

Republican Hydrometeorological Service of Serbia. (2012). *Agrometeorological Conditions in Production Cycle 2011/2012 at the Territory of the Republic of Serbia*. Belgrade: Republican Hydrometeorological Service of Serbia.

Republican Hydrometeorological Service of Serbia. (2013). *Agrometeorological Conditions in Production Cycle 2012/2013 at the Territory of the Republic of Serbia*. Belgrade: Republican Hydrometeorological Service of Serbia.

Republican Hydrometeorological Service of Serbia. (2014). *Agrometeorological Conditions in Production Cycle 2013/2014 at the Territory of the Republic of Serbia*. Belgrade: Republican Hydrometeorological Service of Serbia.

Republican Hydrometeorological Service of Serbia. (2015). *Agrometeorological Conditions in Production Cycle 2014/2015 at the Territory of the Republic of Serbia*. Belgrade: Republican Hydrometeorological Service of Serbia.

Reutlinger, S. (1986). *Poverty and Hunger: Issues and Options for Food Security in Developing Countries. A World Bank Policy Study*. Washington, DC: World Bank.

Reutlinger, S., & Knapp, K. (1980). *Food Security in Food Deficit Countries* (1st ed.). Washington, DC: World Bank.

Ritson, C. (1980). *Self-sufficiency and Food Security*. Reading, UK: Centre for Agricultural Strategy, University of Reading.

Ritson, C., & Mai, L. W. (1998). The Economics of Food Safety. *Nutrition & Food Science*, *98*(5), 253–259. doi:10.1108/00346659810224163

Roljević, S., Subić, J., & Jeločnik, M. (2009). Importance of Soybean Production as Safe Food in Conditions of Economic Crisis. In *Proceedings of Competitivitatea economiei agrolimentare si rurale in conditile crizei mondiale*. Bucharest: Academiei de Studii Economice.

Rosenzweig, C., Parry, M.L., Fischer, G., & Frohberg, K. (1993). *Climate Change and World Food Supply*. Oxford, UK: Environmental Change Unit, University of Oxford.

Rosenzweig, C., & Parry, M. L. (1994). Potential Impacts of Climate Change on World Food Supply. *Nature*, *13*(6459), 133–138. doi:10.1038/367133a0

Rovný, P. (2016). The Analysis of Farm Population with Respect to Young Farmers in the European Union. *Procedia: Social and Behavioral Sciences*, *220*, 391–398. doi:10.1016/j.sbspro.2016.05.513

Rozo, C. (2000). Complementary Foods in Colombia. *Food and Nutrition Bulletin*, *21*(1), 55–61. doi:10.1177/156482650002100109

Rubyogo, J. C., Murithii, P. M., Agumbah, G. J. O., & Obhai, G. (2005). Sustainability of a Privatized Community-Based Animal Health Worker System in Mwingi District, Kenya. *Tropical Animal Health and Production*, *37*(4), 253–266. doi:10.1007/s11250-005-4164-8 PMID:15934633

Rumpold, B. A., & Schlüter, O. K. (2013). Potential and Challenges of Insects as an Innovative Source for Food and Feed Production. *Innovative Food Science & Emerging Technologies*, *17*, 1–11. doi:10.1016/j.ifset.2012.11.005

Rutabanzibwa, A. P. (2011). *Veterinary Legal Reform in Tanzania.* Retrieved February 12, 2017, from http://sites.tufts.edu/capeipst/files/2011/03/Rutabanzibwa-Mombasa.pdf

Ryder, I. (2005). Issues and Patterns in Global Branding. In C. Distler & B. Nivolett (Eds.), *Securing the Business Benefits of Globalization: A European Perspective* (pp. 205–226). Blue Bell, PA: Unisys.

Sally, R. (2001). *International Trade and Emerging Markets since the Crisis.* Johannesburg: South African Institute of International Affairs.

Salvatici, L. (2013). *The Gravity Model in International Trade.* Washington, DC: African Growth and Development Policy Modelling Consortium.

Samaratunga, P. A. (2011). Multiple Facets of Food (In) Security in Sri Lanka: An Input to Food Policy. In S. Mittal & D. Sethi (Eds.), *Policy Options to Achieve Food Security in South Asia* (pp. 51–76). New Delhi: Foundation Books.

Sargeant, J. M. (2008). The Influence of Veterinary Epidemiology on Public Health: Past, Present and Future. *Preventive Veterinary Medicine*, *86*(3–4), 250–259. doi:10.1016/j.prevetmed.2008.02.011 PMID:18375003

Sarker, R., & Jayasinghe, S. (2007). Regional Trade Agreements and Trade in Agri-food Products: Evidence for the European Union from Gravity Modelling Using Disaggregated Data. *Agricultural Economics*, *37*(1), 93–104. doi:10.1111/j.1574-0862.2007.00227.x

Sasson, A. (2012). Food Security for Africa: An Urgent Global Challenge. *Agriculture & Food Security*, *2*(1), 1–16.

Schippers, R. R. (2000). *African Indigenous Vegetables. An Overview of the Cultivated Species.* Chatham: Natural Resources Institute/ACP-EU Technical Centre of Agricultural and Rural Cooperation Book.

Schlenker, W., & Roberts, M. J. (2009). Nonlinear Temperature Effects Indicate Severe Damages to us Crop Yields under Climate Change. *Proceedings of the National Academy of Sciences of the United States of America*, *106*(37), 15594–15598. doi:10.1073/pnas.0906865106 PMID:19717432

Schmitz, A., Moss, C., Schmitz, T., Furtan, W., & Schmitz, H. (2010). *Agricultural Policy, Agribusiness, and Rent-Seeking Behaviour.* Toronto: University of Toronto Press.

Schwab, K. (2015). *World Economic Forum's Global Competitiveness Report, 2014-2015.* Retrieved February 7, 2017, from http://www3.weforum.org/docs/WEF_GlobalCompetitivenessReport_2014-15.pdf

Secretariat of the Convention on Biological Diversity. (2005). *The Impact of Trade Liberalization on Agricultural Biological Diversity, Domestic Support Measures and their Effects on Agricultural Biological Diversity.* Montreal: SCBD.

Sedik, D., & Lerman, Z. (in press). *Food Security – Prospects of the Farming Sector and Rural Development across the Eurasian Plain: The Case of Wheat Production Potential in the Belarus, Uzbek and Turkmen Republic.* Brussels: Joint Research Centre.

Sen, A. (1983). *Poverty and Famines: An Essay on Entitlement and Deprivation.* Oxford, UK: Oxford University Press. doi:10.1093/0198284632.001.0001

Shackleton, C. M., Pasquini, M. W., & Drescher, A. W. (Eds.). (2009). *African Indigenous Vegetables in Urban Agriculture.* London: Earthscan.

Shagaida, N., & Uzun, V. (2015). *Food Security in Russia: Monitoring, Tendencies, and Threats.* Moscow: Russian Academy of National Economy and Public Administration under the President of the Russian Federation.

Shah, A. (2010). *Deregulation or Protectionism?* Retrieved June 21, 2016, from http://www.globalissues.org/article/43/deregulation-or-protectionism

Sharma, R. (2014). The Ever-Emerging Markets. Why Economic Forecasts Fail. *Foreign Affairs*. Retrieved November 23, 2016, from https://www.foreignaffairs.com/articles/2013-12-06/ever-emerging-markets

Shelton, A. M., Zhao, J. Z., & Roush, R. T. (2002). Economic, Ecological, Food Safety, and Social Consequences of the Deployment of Bt Transgenic Plants. *Annual Review of Entomology*, *47*(1), 845–881. doi:10.1146/annurev.ento.47.091201.145309 PMID:11729093

Shengming, H. (2000). Present Situation, Policies and Outlook for China's Food Industry Development. In *The Agro-food Processing Sector in China: Developments and Policy Challenges* (pp. 75–82). Paris: OECD Publishing.

Sherman, R. A., & Pechter, E. A. (1988). Maggot Therapy: A Review of the Therapeutic Applications of Fly Larvae in Human Medicine, Especially for Treating Osteomyelitis. *Medical and Veterinary Entomology*, *2*(3), 225–230. doi:10.1111/j.1365-2915.1988.tb00188.x PMID:2980178

Shtaltovna, A. (2016). Knowledge Gaps and Rural Development in Tajikistan: Agricultural Advisory Services as a Panacea? *Journal of Agricultural Education and Extension*, *22*(1), 25–41. doi:10.1080/1389224X.2014.997257

Siamwalla, A., & Valdes, A. (1980). Food Insecurity in Developing Countries. *Food Policy*, *5*(4), 258–272. doi:10.1016/0306-9192(80)90055-X

Simon, J. Y. (2008). Detoxification Mechanisms in Insects. In J.L. Capinera (Ed.), Encyclopedia of Entomology (pp. 1187–1201). Dordrecht: Springer Science+Business Media B.V.

Smit, B., & Skinner, M. W. (2002). Adaptation Options in Agriculture to Climate Change: A Typology. *Mitigation and Adaptation Strategies for Global Change*, *7*(1), 85–114. doi:10.1023/A:1015862228270

Smith, J. W. (2005). *Economic Democracy: The Political Struggle for the 21st Century*. Sun City, AZ: Institute for Economic Democracy.

Snowden, A. L. (2010). Post-Harvest Diseases and Disorders of Fruits and Vegetables: Vol. 1. *General Introduction and Fruits*. Boca Raton, FL: CRC Press.

Socha, T., Zahaf, M., Chambers, L., Abraham, R., & Fiddler, T. (2012). Food Security in a Northern First Nations Community: An Exploratory Study on Food Availability and Accessibility. *Journal of Aboriginal Health*, *2*(8), 5–14.

Song, B., Marchant, M. A., & Xu, S. (2006). Competitive Analysis of Chinese Soybean Import Suppliers: U.S., Brazil, and Argentina. In *Proceedings of the American Agricultural Economics Association Annual Meetings*. Long Beach, CA: American Agricultural Economics Association.

Sonnino, R., Faus, A. M., & Maggio, A. (2014). Sustainable Food Security: An Emerging Research and Policy Agenda. *International Journal of Sociology of Agriculture and Food*, *1*(21), 173–188.

Soyak, A., & Nesirova, Z. (2003). The Consequences of IMF's Policies in the Globalization Process: The Case of Azerbaijan. In *Proceedings of the Caucasus and Central Asia in the Globalization Process International Conference*. Baku: Qafqaz University.

Spiertz, H. (2012). Avenues to Meet Food Security. The Role of Agronomy on Solving Complexity in Food Production and Resource Use. *European Journal of Agronomy*, *43*, 1–8. doi:10.1016/j.eja.2012.04.004

Spoor, M., & Robbins, M. J. (Eds.). (2012). *Agriculture, Food Security, and Inclusive Growth*. The Hague: Institute of Social Sciences.

Stamenković, I., Bošnjak, D., & Stamenković, A. (2007). Geographic Characteristics of Soya Production in Agricultural Areas of Vojvodina Region. *Geographical Journal – Geographica Pannonica, 10*, 78–84.

Starck, S. (2012). *The Theoretical Foundation of Gravity Modeling: What are the Developments that have Brought Gravity Modeling into Mainstream Economics?* (Unpublished master dissertation). Copenhagen Business School, Copenhagen, Denmark.

State Council of the Russian Federation. (2014). *Report on Sustainable Development of Rural Territories of the Russian Federation.* Moscow: Council under the President of the Russian Federation on Implementation of Priority National Projects and Demographic Policy.

State Statistical Committee of the Republic of Azerbaijan. (2016). *Statistical Database*. Retrieved October 19, 2016, from http://www.stat.gov.az/menu/13/indexen.php

Statistical Office of the Republic of Serbia. (2008). *Statistical Yearbook of the Republic of Serbia – 2008*. Belgrade: Statistical Office of the Republic of Serbia.

Statistical Office of the Republic of Serbia. (2011). *Statistical Yearbook of the Republic of Serbia – 2011*. Belgrade: Statistical Office of the Republic of Serbia.

Statistical Office of the Republic of Serbia. (2014). *Statistical Yearbook of the Republic of Serbia – 2014*. Belgrade: Statistical Office of the Republic of Serbia.

Statistical Office of the Republic of Serbia. (2015). *Statistical Yearbook of the Republic of Serbia – 2015*. Belgrade: Statistical Office of the Republic of Serbia.

Statistical Office of the Republic of Serbia. (2016a). *Basic Data of Soybean Production in 2015*. Retrieved November 21, 2016, from http://webrzs.stat.gov.rs/WebSite/public/ReportView.aspx

Statistical Office of the Republic of Serbia. (2016b). Monthly Statistical Bulletin, vol. LXV, no. 12/2015. Belgrade: Statistical Office of the Republic of Serbia.

Steele, J. H. (2008). Veterinary Public Health: Past Success, New Opportunities. *Preventive Veterinary Medicine, 86*(3–4), 224–243. doi:10.1016/j.prevetmed.2008.02.014 PMID:18417229

Steenkamp, J. E. M., & van Trijp, H. C. M. (1991). The Use of LISREL in Validating Marketing Constructs. *International Journal of Research in Marketing, 8*(4), 283–299. doi:10.1016/0167-8116(91)90027-5

Stern, N. (2007). *The Economics of Climate Change: The Stern Review*. Cambridge, UK: Cambridge University Press. doi:10.1017/CBO9780511817434

Stige, L. C., Stave, J., Chan, K., Ciannelli, L., Pettorelli, N., Glantz, M., & Stenseth, N. C. et al. (2006). The Effect of Climate Variation on Agro-pastoral Production in Africa. *Proceedings of the National Academy of Sciences of the United States of America, 103*(9), 3049–3053. doi:10.1073/pnas.0600057103 PMID:16492727

Stiglitz, J. (2003). *The Roaring Nineties: A New History of the World's Most Prosperous Decade*. New York, NY: W.W. Norton & Co.

Stiglitz, J., & Greenwald, B. (2006). Helping Infant Economies Grow: Foundations of Trade Policies for Developing Countries. *The American Economic Review, 96*(2), 141–146. doi:10.1257/000282806777212206

Stock, J. R. (1990). Logistics Thought and Practice: A Perspective. *International Journal of Physical Distribution & Logistics Management, 20*(1), 3–6. doi:10.1108/09600039010140845

Subasinghe, S. (1992). Jellyfish Processing. *INFOFISH International, 4*, 63–65.

Subić, J., Jeločnik, M., & Zubović, J. (2015). Primena navodnjavanja kao agrotehničke mere – analiza marže pokrića u proizvodnji kukuruza. *Ecologica, 78*(22), 245–251.

Sulakshin, S. S., Bardasaryan, V. E., Vilisov, M. V., Zachesova, Y. A., Pak, N. K., Seredkina, O. A., & Chirva, A. N. (2007). *State Competition Policy and Competition Stimulation in the Russian Federation.* Retrieved February 12, 2017, from http://rucont.ru/file.ashx?guid=af71340c-4bee-41a2-8723-7514a217fb50

Supply Chain Council. (2008). *Supply-Chain Operations Reference-Model. Overview of SCOR version, 5(0).* Chicago, IL: Supply Chain Council.

Swiss Federal Institute of Technology Zurich. (2016). *KOF Index of Globalization.* Retrieved November 24, 2016, from http://globalization.kof.ethz.ch/

Tabor, S., Ginting, E., & Aji, P. (2015). *Preparing for El Niño: Policy Options.* Manila: Asian Development Bank.

Taranov, P. (2015). The Russian Economy in the Context of Trends in International Trade and Economic Policy. *Economics and Entrepreneurship, 10*(63), 94–98.

Taranov, P., & Panasyuk, A. (2010). Food Dependence of Russia in the Light of Joining the World Trade Organization (WTO). *Bulletin of Orenburg State Agrarian University, 26*, 106–109.

Taranov, P., & Panasyuk, A. (2012). Competitiveness of Agri-food System of Russia in the Light of Eurasian Economic Integration. *National Interests: Priorities and Security, 40*(181), 56–60.

Tarrósy, I. (2009). Politikai elméletek és módszertanok a mai afrikai problémák megértéséhez. In S. Csizmadia & I. Tarrósy (Eds.), Afrika Ma – Tradíció, átalakulás, fejlődés (pp. 13–17). Pecs: IDResearch Kft. & Publikon Kiadó.

Tarrósy, I. (2014). Nigéria választ – Újrázik-e Goodluck Jonathan? (Nigeria will Elect – Will Goodluck Jonathan Come Again). *Afrika Tanulmányok, 4*, 5–12.

Taylor, I. (2014). *Africa Rising? BRICS – Diversifying Dependency.* Oxford, UK: James Currey.

Termini, M. J. (1996). *The New Manufacturing Engineer: Coming of Age in an Agile Environment.* Dearborn, MI: Society of Manufacturing Engineers.

Territorial Authority of the Federal State Statistics Service of the Russian Federation in Stavropol Region. (2014). *Livestock and Poultry Population – Livestock Production.* Stavropol: Territorial Authority of the Federal State Statistics Service of the Russian Federation in Stavropol Region.

The Economist Intelligence Unit. (2016). *Global Food Security Index 2016. An Annual Measure of the State of Global Food Security.* London: The Economist Intelligence Unit Limited.

The Economist. (2016). *The Global Food Security Index.* Retrieved September 10, 2016, from http://foodsecurityindex.eiu.com/

The Global Economy. (2016). *Countries.* Retrieved February 7, 2017, from http://www.theglobaleconomy.com

The Korean Post. (2013). *Uzbekistan Uses Uzbek Model of Transition to Socially Oriented Market Economy.* Retrieved February 7, 2017, from http://www.koreapost.com/news/articleView.html?idxno=377

The Observatory of Economic Complexity. (n.d.). *Nigeria.* Retrieved October 11, 2016, from http://atlas.media.mit.edu/en/profile/country/nga/

The Swiss Re Group. (2013). *New Swiss Re Sigma Study Puts the Spotlight on the Role that Insurance Can Play in Improving Food Security for over 850 Million People Globally.* Retrieved February 1, 2017, from http://www.swissre.com/media/news_releases/nr_20130116_improving_food_security.html

Thow, A. M., & Hawkes, C. (2009). The Implications of Trade Liberalization for Diet and Health: A Case Study from Central America. *Globalization and Health, 28*(1), 5. doi:10.1186/1744-8603-5-5 PMID:19638196

Tinbergen, J. (1962). *Shaping the World Economy: Suggestions for an International Economic Policy* (1st ed.). New York, NY: Twentieth Century Fund.

Tol, R. S. J. (2002). Estimates of the Damage Costs of Climate Change. Part 1: Benchmark Estimates. *Environmental and Resource Economics, 21*(1), 47–73. doi:10.1023/A:1014500930521

Torero, M., & von Braun, J. (2010). *Alternative Mechanisms to Reduce Food Price Volatility and Price Spikes.* Paper presented at the International Food and Agricultural Trade Policy Council, Barcelona, Spain.

Torzillo, G., Carlozzi, P., Pushparaj, B., Montaini, E., & Materassi, R. (1993). A Two-Plane Tubular Photobioreactor for Outdoor Culture of Spirulina. *Biotechnology and Bioengineering, 42*(7), 891–898. doi:10.1002/bit.260420714 PMID:18613137

Trading Economics. (2016). Statistics Database [Data file]. Retrieved from http://ru.tradingeconomics.com/countries

Traverse, M., & Aceto, H. (2015). Environmental Cleaning and Disinfection. *The Veterinary Clinics of North America. Small Animal Practice, 45*(2), 299–330. doi:10.1016/j.cvsm.2014.11.011 PMID:25555560

Tregubov, V. I., Efimov, Y. G., Kononov, A. N., Zaerko, V. I., Svetlakova, E. V., & Ozheredova, N. A. (2012). *Procedural and Institutional Issues of Veterinary Service.* Retrieved February 12, 2017, from http://www.science-education.ru/102#61

Trevisi, E., Zecconi, A., Cogrossi, S., Razzuoli, E., Grossi, P., & Amadori, M. (2014). Strategies for Reduced Antibiotic Usage in Dairy Cattle Farms. *Research in Veterinary Science, 96*(2), 229–233. doi:10.1016/j.rvsc.2014.01.001 PMID:24508188

Trukhachev, V., Ivolga, A., & Lescheva, M. (2015). Enhancement of Land Tenure Relations as a Factor of Sustainable Agricultural Development: Case of Stavropol Krai, Russia. *Sustainability, 7*(1), 164–179. doi:10.3390/su7010164

Turaeva, R., & Hornidge, A. K. (2013). From Knowledge Ecology to Innovation Systems: Agricultural Innovations and Their Diffusion in Uzbekistan. *Innovation (Abingdon), 15*(2), 183–193.

Turkson, P. K., Slenning, B. D., & Brownie, C. F. (1999). Perceptions of Veterinarians Regarding Privatization of Veterinary Services Delivery in Ghana and Jamaica. *Preventive Veterinary Medicine, 40*(3–4), 221–232. doi:10.1016/S0167-5877(99)00023-9 PMID:10423776

United Nations Comtrade. (2006). *UN Comtrade Database.* Retrieved December 6, 2016, from https://comtrade.un.org/

United Nations Conference on Trade and Development. (2002). *The Least Developed Countries Report.* Geneva: United Nations Conference on Trade and Development.

United Nations Conference on Trade and Development. (2013). *Nigeria Bilateral Investment Treaties (BITs).* Retrieved November 17, 2016, from http://investmentpolicyhub.unctad.org/IIA/CountryBits/153#iiaInnerMenu

United Nations Conference on Trade and Development. (2016). Statistics Database [Data file]. Retrieved from http://unctad.org/en/Pages/Statistics.aspx

United Nations Department of Economic and Social Affairs. (2015). *World Population Prospects: The 2015 Revision.* New York, NY: The United Nations.

United Nations Development Programme. (2016). *Support to the Implementation of the 2030 Agenda for Sustainable Development*. Retrieved July 27, 2016, from http://www.undp.org/content/undp/en/home/librarypage/sustainable-development-goals/undp-support-to-the-implementation-of-the-2030-agenda/

United Nations Environment Programme. (2006). *Challenges to International Waters – Regional Assessments in a Global Perspective*. Nairobi: UNEP.

United Nations Framework Convention on Climate Change. (2007). *Report on the Workshop on Climate Related Risks and Extreme Events*. Note by the Secretariat. FCCC/SBSTA/2007/7. Bonn: UNFCCC.

United Nations Industrial Development Organization. (2015). *Industrial Development Report 2016: The Role of Technology and Innovation in Inclusive and Sustainable Industrial Development*. Vienna: United Nations Industrial Development Organization.

United Nations Office for Disaster Risk Reduction, & World Bank. (2009). *Central Asia and Caucasus Disaster Risk Management Initiative: Risk Assessment for Central Asia and Caucasus*. Geneva: ISDR.

US Department of State. (2016). *U.S. Relations with Nigeria*. Retrieved November 18, 2016, from http://www.state.gov/r/pa/ei/bgn/2836.htm

Ushachev, I. (2013). *Food Security of Russia in View of the Global Partnership*. Moscow: Russian Academy of Agricultural Sciences, All-Russian Research Institute of Agricultural Economics.

Uye, S. I. (2008). Blooms of the Giant Jellyfish Nemopilema Nomurai: A Threat to the Fisheries Sustainability of the East Asian Marginal Seas. *Plankton and Benthos Research*, *3*(Supplement), 125–131. doi:10.3800/pbr.3.125

Valdes, A. (1981). *Food Security for Developing Countries*. Boulder, CO: Westview Press.

Van Dijk, M. (2011). *African Regional Integration: Implications for Food Security*. Retrieved October 18, 2016, from http://edepot.wur.nl/168845

Van Huis, A. (2003). Medical and Stimulating Properties Ascribed to Arthropods and Their Products in Sub-Saharan Africa. In E. Motte-Florac & J. M. C. Thomas (Eds.), *Insects in Oral Literature and Traditions* (pp. 367–382). Paris: Peeters.

Van Huis, A., van Itterbeeck, J., Klunder, H., Mertens, E., Halloran, A., Muir, G., & Vantomme, P. (2013). *Edible Insects: Future Prospects for Food and Feed Security*. Rome: Food and Agriculture Organization of the United Nations.

Van Rijswijk, W., & Frewer, L. J. (2008). Consumer Perceptions of Food Quality and Safety and Their Relation to Traceability. *British Food Journal*, *110*(10), 1034–1046. doi:10.1108/00070700810906642

Vantomme, P. (2015). Way forward to Bring Insects in the Human Food Chain. *Journal of Insects as Food and Feed*, *1*(2), 121–129. doi:10.3920/JIFF2014.0014

Vasiljević, Z., & Subić, J. (2010). Importance of the Costs Calculation at the Family Farms in Serbia. In D. Tomic & M. Sevarlic (Eds.), *Agriculture in Late Transition – Experience of Serbia* (pp. 123–138). Belgrade: Serbian Association of Agricultural Economists.

Vassileva, B., & Nikolov, M. (2016). Market Entry Strategies to Emerging Markets: A Conceptual Model of Turnkey Project Development. *Serbian Journal of Management*, *11*(2), 291–310. doi:10.5937/sjm11-10177

Velarde, A., Fàbrega, E., Blanco-Penedo, I., & Dalmau, A. (2015). Animal Welfare towards Sustainability in Pork Meat Production. *Meat Science*, *109*, 13–17. doi:10.1016/j.meatsci.2015.05.010 PMID:26013042

Veninga, W., & Ihle, R. (in press). Import Vulnerability in the Middle East: Effects of the Arab Spring on Egyptian Wheat Trade. *Food Security: The Science, Sociology and Economics of Food Production and Access to Food.*

Verbeke, W., Frewer, L. J., Scholderer, J., & De Brabander, H. F. (2007). Why Consumers Behave as They Do with Respect to Food Safety and Risk Information. *Analytica Chimica Acta, 586*(1/2), 2–7. doi:10.1016/j.aca.2006.07.065 PMID:17386689

Visser, O., Mamonova, N., Spoor, M., & Nikulin, A. (2015). Quiet Food Sovereignty as Food Sovereignty without a Movement? Insights from Post-socialist Russia. *Globalizations, 12*(4), 1–16. doi:10.1080/14747731.2015.1005968

Von Braun, J., & Tadesse, G. (2012). *Global Food Price Volatility and Spikes: An Overview of Costs, Causes, and Solutions.* ZEF Discussion Papers, 161.

Von Braun, J., Ahmed, A., Asenso-Okyere, K., Fan, S., Gulati, A., Hoddinot, J., & Von Grebmer, K. (2008). *High Food Prices: The What, Who, and How of Proposed Policy Actions.* Washington, DC: International Food Policy Research Institute.

Von Braun, J., Serova, E., Seeth, H., & Melyukhina, O. (1996). *Russia's Food Economy in Transition: What Do Reforms Mean for the Long-term Outlook?* Washington, DC: International Food Policy Research Institute.

Walker, M. (2007). Globalization 3.0. Prospects for a New World Order. *The Wilson Quarterly*, (Autumn): 16–24.

Wang, J. (2016). Analysis and Comparison of the Factors Influencing Worldwide Four Kinds of Vegetable Oil Trade: Based on Gravity Model. *Modern Economy, 7*(2), 173–182. doi:10.4236/me.2016.72019

Warr, P. (2011). *Food Security vs. Food Self-Sufficiency: The Indonesian Case.* Canberra: The Australian National University.

Wegren, S. K. (2013). *Return to Putin's Russia. Past Imperfect, Future Uncertain.* Plymouth: Rowman & Littlefield Publishers.

Weinberger, K., & Msuya, J. (2004). *Indigenous Vegetables in Tanzania- Significance and Prospects.* Shanhua: AVRDC – The World Vegetable Center.

Weinberger, K., & Pichop, G. N. (2009). Marketing of African Indigenous Vegetables along Urban and Peri-Urban Supply Chains in Sub-Saharan Africa. In C. M. Shackleton, M. W. Pasquini, & A. W. Drescher (Eds.), *African Indigenous Vegetables in Urban Agriculture* (pp. 225–244). London: Earthscan.

Welton, G., Astaryan, A., & Jijeleva, D. (2013). *Comparative Analysis of Agriculture in the South Caucasus.* Tbilisi: United Nations Development Programme.

Wikipedia. (2013). *Climate Change and Agriculture.* Retrieved November 6, 2016, from http://en.wikipedia.org/wiki/Climate_change_and_agriculture

Wittman, H., Desmarais, A., & Wiebe, N. (2010). *Food Sovereignty: Reconnecting Food, Nature and Community.* Halifax, Canada: Fernwood Publishing and Food First Books.

Wolfe, D. W., Schwartz, M. D., Lakso, A. N., Otsuki, Y., Pool, R. M., & Shaulis, N. J. (2005). Climate Change and Shifts in Spring Phenology of Three Horticultural Woody Perennials in Northeastern USA. *International Journal of Biometeorology, 49*(5), 303–309. doi:10.1007/s00484-004-0248-9 PMID:15592880

Wolf, J. (2000). Modelling Climate Change Impacts at the Site-Scale on Soybean. In T. E. Downing, P. A. Harrison, R. E. Butterfield, & K. G. Lonsdale (Eds.), *Climate Change, Climatic Variability and Agriculture in Europe* (pp. 103–116). Oxfordshire, UK: Environmental Change Unit, Oxford University.

Wollina, U., Karte, K., Herold, C., & Looks, A. (2000). Biosurgery in Wound Healing – The Renaissance of Maggot Therapy. *Journal of the European Academy of Dermatology and Venereology*, *14*(4), 285–289. doi:10.1046/j.1468-3083.2000.00105.x PMID:11204517

Woolf, S. S., & Jones, E. I. (1969). *Agrarian Change and Economic Development: The Historical Problem*. London: Methuen.

World Bank. (2000). *Economic Review, 11*. Washington, DC: World Bank.

World Bank. (2009). *Adapting to Climate Change in Europe and Central Asia*. Washington, DC: World Bank.

World Bank. (2016). *World Bank Open Data*. Retrieved November 24, 2016, from http://data.worldbank.org/

World Bank. (2016). *World Development Indicators*. Retrieved February 7, 2017, from http://data.worldbank.org

World Development Report. (2010). *Development and Climate Change*. Washington, DC: The World Bank. Retrieved November 7, 2016, from https://openknowledge.worldbank.org/bitstream/handle/10986/4387/9780821379875_fm.pdf?sequence=104&isAllowed=y

World Factbook. (2016). *Central Intelligence Agency*. Retrieved February 8, 2017, from https://www.cia.gov/library/publications/the-world-factbook

World Food Programme. (2009). *Emergency Food Security Assessments (EFSAs). Technical Guidance Sheet No. 5*. Retrieved September 11, 2016, from http://documents.wfp.org/stellent/groups/public/documents/manual_guide_proced/wfp197295.pdf?_ga=1.244370689.508075766.1479356271

World Health Organization. (2014). *World Malaria Report 2013*. Geneva: World Health Organization.

World Trade Organization. (2007). *Council for Trade in Services – Special Session – Nigeria – Initial Offer – Document TN/S/O/NGA*. Retrieved November 18, 2016, from https://www.wto.org/english/docs_e/docs_e.htm

World Trade Organization. (2010). *Committee on Trade and Development – Sixteenth Session on Aid for Trade – Nigeria's Trade and Productive Capacity Reform – Document WT/COMTD/AFT/W/23*. Retrieved November 18, 2016, from https://www.wto.org/english/docs_e/docs_e.htm

World Trade Organization. (2011). *Trade Policy Review Body – Trade Policy Review – Report by the Secretariat – Nigeria – Document for Nigeria WT/TPR/S/247/Rev.1*. Retrieved November 18, 2016, from https://www.wto.org/english/docs_e/docs_e.htm

World Trade Organization. (2014a). *International Trade Statistics 2014*. Retrieved January 10, 2017, from https://www.wto.org/english/res_e/statis_e/its2014_e/its2014_e.pdf

World Trade Organization. (2014b). *Trade and Development: Recent Trends and the Role of the WTO*. Retrieved December 15, 2016, from https://www.wto.org/english/res_e/booksp_e/world_trade_report14_e.pdf

World Vegetable Center. (2006). *Empowering Small Scale and Women Farmers through Sustainable Production, Seed Supply and Marketing of African Indigenous Vegetables in Eastern Africa*. Bangkok: AVRDC – The World Vegetable Center and Family Concern.

Yamanaka, R., & Akiyama, K. (1993). Cultivation and Utilization of Undaria Pinnatifida (Wakame) as Food. *Journal of Applied Phycology*, *5*(2), 249–253. doi:10.1007/BF00004026

Yang, R., & Keding, G. B. (2009). Nutritional Contributions of Important African Indigenous Vegetables. In C. M. Shackleton, M. W. Pasquini, & A. W. Drescher (Eds.), *African Indigenous Vegetables in Urban Agriculture* (pp. 105–143). London: Earthscan.

Yesuf, M., Di Falco, S., Deressa, T., Ringer, C., & Kohlin, G. (2008). *The Impact of Climate Change and Adaptation on Food Production in Low-Income Countries: Evidence from Nile Basin, Ethiopia*. International Food Policy Research Institute (IFPRI) Discussion Paper, 00828.

Yeung, R. M. W., & Morris, J. (2001). Food Safety Risk – Consumer Perception and Purchase Behavior. *British Food Journal, 103*(3), 170–186. doi:10.1108/00070700110386728

Yong-Woo, L. (1999). *Silk Reeling and Testing Manual*. Rome: Food and Agriculture Organization of the United Nations.

Yu, W., Elleby, C., & Zobbe, H. (2015). Food Security Policies in India and China: Implications for National and Global Food Security. *Food Security, 7*(2), 405–414. doi:10.1007/s12571-015-0432-2

Yu, Y., Feng, K., Hubacek, K., & Sun, L. (2016). Global Implications of Chinas Future Food Consumption. *Journal of Industrial Ecology, 20*(3), 593–602. doi:10.1111/jiec.12392

Zhou, Z. (2010). Achieving Food Security in China: Past Three Decades and Beyond. *China Agricultural Economic Review, 2*(3), 251–275. doi:10.1108/17561371011078417

Zhou, Z., Liu, H., Cao, L., Tian, W., & Wang, J. (2014). *Food Consumption in China: The Revolution Continues*. Cheltenham, UK: Edward Elgar Publishing. doi:10.4337/9781782549208

Zhou, Z., Wu, W., Chen, Q., & Chen, F. (2007). Study on Sustainable Development of Rural Household Energy in Northern China. *Renewable & Sustainable Energy Reviews, 12*(8), 2237–2239.

Zhuplei, I. (2015). Sustainable Development of Agriculture: Regional and Applied Aspect. *Almanac of Modern Science and Education, 99*(9), 64–67.

Zhu, Y. (2016). International Trade and Food Security: Conceptual Discussion, WTO and the Case of China. *China Agricultural Economic Review, 8*(3), 399–411. doi:10.1108/CAER-09-2015-0127

Zinsstag, J., Schelling, E., Waltner Toews, D., & Tanner, M. (2011). From One Medicine to One Health and Systemic Approaches to Health and Well-Being. *Preventive Veterinary Medicine, 101*(3–4), 148–156. doi:10.1016/j.prevetmed.2010.07.003 PMID:20832879

Zsarnóczai, J. S., Lakner, Z., & Mughram, Y. A. (2011). Hungarian Participation in Modernisation of the Third World. In I. Szűcs & J. S. Zsarnóczai (Eds.), *Economics of Sustainable Agriculture* (pp. 87–108). Gödöllő: Szent István University Publishing.

Zubovic, J., & Domazet, I. (2009). Organic Food Market Development: The Case of Serbia. In J.-P. Guichard (Ed.), *Challenges of Economic Sciences in the 21st Century* (pp. 179–189). Belgrade: Institute of Economic Sciences.

About the Contributors

Vasilii Erokhin is an Associate Professor, School of Economics and Management, Harbin Engineering University, China. He holds a Ph.D. degree in Economics and Management (2006) and a Specialist degree in International Economics (2003) from Stavropol State Agrarian University, Russia. Spheres of scientific interests include international trade, globalization, trade integration and liberalization, sustainable development, food security issues with a focus on emerging markets, developing countries, and economies in transition. Vasilii Erokhin is an author of over 150 scientific works in international economics, international trade, sustainable rural development, and regionalization, including 6 books, 10 monographs, over 40 papers in peer-reviewed journals and proceedings (the USA, the Czech Republic, Slovakia, Hungary, Romania, Serbia, Ukraine, Russia). Vasilii Erokhin is an editor of the IGI Global book Global Perspectives on Trade Integration and Economies in Transition (2016). He is also an author of the Series "International Trade from a Russian Perspective" (2014), published online by HSTalks, London. Vasilii Erokhin is a Guest Editor at MDPI Economies, special issue "Macroeconomics and Urban Labour Markets Paradigms"; an Academic Editor at SCIENCEDOMAIN Asian Journal of Economics, Business and Accounting; an Editorial Board Member at SPG Journal of World Economic Research; an Editorial Board Member at IKP Journal of Economics and Trade; a member of the Editorial Review Board at IGI International Journal of Sustainable Economies Management (IJSEM); a member of the Editorial Review Board at IGI International Journal of Social Ecology and Sustainable Development (IJSESD); an Editorial Board Member at SPRINGER Journal of the Knowledge Economy; an Editorial Board Member at SPRINGER Journal of Innovation and Entrepreneurship: A Systems View across Time and Space; an Editorial Board Member at SPRINGER Journal of Technology, Innovation and Education; a member of the International Editorial Review Board at Economics of Agriculture (Ekonomika poljoprivrede); a member of the Editorial Advisory Board at IGI book "Green Economic Structures in Modern Business and Society"; a member of the Editorial Advisory Board at IGI book "Research on Food Science, Production and Engineering in Green Economies"; a member of the Review Board at Applied Research Conference, Koforidua Polytechnic, Ghana. He is a holder of the honorary awards from the Ministry of Agriculture of the Russian Federation (2010), Ministry of Education and Science of the Russian Federation (2012). Vasilii Erokhin is a winner of the 3rd All-Russian competition of business ideas and projects "Creation of Future Russia", project "Efficiency Increasing of the Regional Entrepreneurship in the Conditions of the World Trade Organization Accession" and a scholarship holder of the Vladimir Potanin's Charity Fund among young teachers (2009). Since 2016, Vasilii Erokhin is also a researcher at the Polar Development and Northeast Asian Economic Research Center (PAERC), China and a Specially Invited Expert at the Association of International Economics and Trade of Heilongjiang Province, China.

* * *

Khatai Aliyev is a Ph.D. candidate in Econometrics, Institute of Control Systems, Azerbaijan National Academy of Sciences, Azerbaijan. Since February 2014, he is employed at the Institute of Social Sciences and Humanities within Qafqaz University. Khatai Aliyev is currently employed at the newly established Baku Engineering University, Azerbaijan. He obtained his Bachelor degree in Economics from Qafqaz University, Azerbaijan in 2010 and Master degree in Economics from Eastern Mediterranean University, North Cyprus in 2014. Khatai Aliyev is an author of one book and one book chapter, author or co-author of 16 journal articles, and 8 conference publications. His research focuses on fiscal policy effectiveness in resource-rich economies, particularly Azerbaijan.

Goodness C. Aye is a Senior Lecturer and Researcher, University of Agriculture Makurdi, Nigeria. She obtained her Ph.D. in Agricultural Economics from the University of Pretoria, South Africa. Her M.Sc and Bachelor's degree in the same field of study were respectively obtained from the University of Nigeria Nsukka and the University of Agriculture Makurdi, Nigeria. She has over 50 publications in the form of books, peer reviewed local and international journals, and conference proceedings covering micro and macroeconomic policy analysis, agricultural development and policy analysis, production economics, environmental economics, energy economics, and poverty studies. She is an editor and has reviewed articles for several high ranking journals.

Anzhelika Baicherova is an Assistant Professor, Department of Entrepreneurship and International Economics, Stavropol State Agrarian University, Russia. Research areas include issues of agricultural economics, an efficiency of veterinary service, food security, and import substitution.

Natalia Bannikova is a Professor, Doctor in Economics, Head of the Department of Entrepreneurship and International Economics, Stavropol State Agrarian University, Russia. Research areas include issues of agricultural economics, food markets, and agribusiness management.

Roberta Bonanno obtained Master degree in Biology from the University of Messina, Italy. NHS in Microbiology and Virology from the University of Catania. Skilled in bacteriology and virology. Nutritionist and expert of algae and microbic applications in foods. Member of the "Società Italiana di Microbiologia".

Francesco D'Aleo obtained Master degree in Biology from the University of Messina, Italy. NHS in Microbiology and Virology from the University of Catania. Skilled in bacteriology and forensic microbiology. Practiced to microalgae and their applications in foods and biotechnology. Member of the American Society of Microbiology. Manager of young microbiologists for "Società Italiana di Microbiologia". President of the "Associazione Microbiologi Italiani".

Vittorio D'Aleo is a Ph.D. Student in Economics and Management at the University of Messina, Italy. After graduation in Political Science, he obtained an MBA degree from Luiss Business School, Rome, Italy. During his MBA studies, he successfully participated in the International Activities Program in Japan, where he deepened his studies in business planning and Asian public policy at the Nagoya University of Commerce and Business.

Davide Di Fatta is a Ph.D. student in Economics, University of Messina, Italy and teaching assistant in marketing, University of Palermo, Italy. Junior member referent for the Business System Laboratory and associate editor for many international journals such as IJMABS and IJEMR.

Nenad Djokic, Ph.D., teaching assistant, Faculty of Economics in Subotica (University of Novi Sad, Republic of Serbia), teaching Marketing Principles and Global Marketing. In 2015, he defended his Ph.D. thesis "Profile of Organic and Functional Food Consumers in Chosen Emerging Economies". He is an author and co-author of over 30 papers published in Serbian journals and abroad (the UK, the Czech Republic, Lithuania, and Brazil), or presented at scientific conferences in Serbia and abroad (Spain, the Czech Republic, Bulgaria, and Bosnia and Herzegovina). He is also a member of the Serbian Marketing Association.

Ivana Domazet is a Senior Research Associate, Institute of Economic Sciences, Republic of Serbia and Associate Professor, Faculty of Banking, Insurance, and Finance, Union University, Republic of Serbia. She teaches courses on Marketing Management, Market Research and Competitiveness Enhancement (postgraduate studies). Her scientific interest refers to improving competitiveness, marketing research and strategy, CRM and strategic management. She published numerous papers related to foregoing topics. She is a President of Scientific Council at the Institute of Economic Sciences, Vice-president of Board member of the Institute of Economics Sciences, Board member of the Institute of Social Science, and a member of the Serbian Scientific Association of Economists and Serbian Marketing Association.

Praxedis Dube is a Seed Technologist with M.Sc. in Plant Sciences specializing in Greenhouse Horticulture from Wageningen University, The Netherlands (2010) and B.Sc. (Hons.) in Agriculture specializing in Horticulture from Women's University in Africa, Zimbabwe (2006). She possesses vast expertise in seed potato production and certification, seed crop production and certification for field crops, agricultural crops management, seed and agricultural policy having attended various courses seed production and certification at Danish Plant Directorate, Copenhagen, Denmark, Wageningen International, Wageningen, The Netherlands, Corpoica, Bogota, Colombia and DLF-Trifolium, Odense, Denmark. Praxedis is currently studying her Ph.D. with Wageningen University in the Agricultural Economics and Rural Policy Chair Group since April 2014, working on the Supply Chain of Traditional Leafy Vegetables in Zimbabwe; a seed quality and socio-economic analysis. She is interested in the seed and vegetables supply markets for economic development and change values challenging the use of Traditional Leafy Vegetables for food and nutritional security and poverty reduction in Africa.

Jana Galova is an Assistant Professor and Project Manager, Faculty of Economics and Management, Slovak University of Agriculture in Nitra, Slovakia. She has a doctoral degree in the field of Economics and Management of Agriculture and Food Processing and a Master degree in Agrarian Trade and Marketing. Her doctoral studies were devoted to international marketing strategies for emerging markets, where she applied knowledge obtained during research and study stays in Ukraine, Austria, Hungary, Czech Republic, Poland, and Crete. She has published several scientific papers in journals and chapters in international scientific monographs and textbooks released in Slovakia and abroad. She also has several years of experience from working on various projects, such as Erasmus+ (KA2 Capacity Building in Higher Education, KA2 Strategic Partnerships), TEMPUS, CEEPUS, Erasmus Mundus, International Visegrad Fund, Leonardo da Vinci, and other domestic and international projects.

Ilkin Gasimov is a Senior researcher, Institute for Scientific Researches on Economic Reforms, Azerbaijan. He graduated from the Faculty of Journalism, Baku Slavian University, Azerbaijan (Bachelor degree) in 2009 and from the Faculty of Business and Economics, Eastern Mediterranean University, North Cyprus (Master) in 2014.

Ishita Ghosh is an Assistant Professor at the Symbiosis School of Economics, Symbiosis International University. With fifteen years of combined work experience in academia and industry, she has a Masters in Economics – specialized in Trade. Ishita has worked on urban policy and planning for Maharashtra (India) with special emphasis on Energy, Economic Policy, and Governance. Her Ph.D. focuses on trade in services and the latter day new-new trade theories. Currently, Ishita teaches Research Methodology, International Trade, Macroeconomics, Economic Journalism, and Econometrics". Her research interests include trade in services, applications of the gravity model, aid for trade, trade and gender, etc.

Ishita Ghoshal obtained her B.Sc. in Economics (Hons) with Mathematics and Statistics from Lady Brabourne College, University of Calcutta and M.A in Economics from Gokhale Institute of Politics and Economics, Pune. She is currently pursuing Ph.D. from Gokhale Institute of Politics and Economics. Her areas of interest are macroeconomics, applied econometrics, and international trade. She is an Assistant Professor with Symbiosis School of Economics and teaches Macro-Economics and Econometrics to M.Sc. and B.Sc. Economics students. She has published in reputed journals like Journal of Quantitative Economics, Global Economy Journal, and Procedia - Economics and Finance.

Jesus Gonzalez-Feliu is an Associate Professor of Industrial Engineering at Mines Saint-Etienne, France, assigned to Institut Henri Fayol. He holds a Master degree in Civil and Urban Engineering from INSA Lyon, France obtained in 2003, a Master degree in Sustainable Transport and Mobility from Politecnico di Torino and Universitá di Torino, Italy obtained in 2004, and a Ph.D. degree in Computer and System Sciences, option operations research at Politecnico di Torino, Italy obtained in 2008. His main research interests are urban logistics (demand estimation, demand-supply interactions, freight and land use relations, and evaluation and assessment), collaborative freight transport systems, food logistics, and non-emergency humanitarian and social welfare logistics, among others. He is an author of more than 40 papers in international peer-review journals, more than 20 book chapters, and more than 50 communications on national and international conferences. He is a co-editor of two books on urban logistics and commercial transport.

Ruth F. Haruna obtained her Bachelor's degree in Agricultural Economics from the University of Agriculture Makurdi, Nigeria. She is currently serving the one year compulsory National Youth Service Corps. She has an interest in climate issues and their impacts on the agricultural sector as well as policies that can help mitigate those impacts.

Wim Heijman is a Professor of Regional Economics, Wageningen University, the Netherlands. Wim Heijman studied Economics at Tilburg University and Human Geography at the University of Utrecht in the Netherlands. He received his Ph.D. degree from Wageningen University.

Elena Horska is a Professor, Dean of the Faculty of Economics and Management, Slovak University of Agriculture in Nitra, Slovakia, and a Professor of Marketing and Management at the Department of

Marketing and Trade. She has been Vice-dean for International Relations at the Faculty of Economics and Management between 2003-2010 and Vice-rector for International Relations of the Slovak University of Agriculture in Nitra between 2010-2014. In 2011, she was awarded the Professor Honoris Causa title by the Delhi School of Professional Studies and Research, New Delhi, India. Her main research areas include internationalization within and beyond the EU borders, identifying opportunities for doing business in neighboring (Visegrad), European, less developed and emerging markets (focus on agribusiness and agri-food market), food marketing, consumer studies, neuromarketing, and visual merchandising. She has initiated three double degree programs and coordinated many national and international research and educational projects. She has more than 200 publications, including national and international student books and monographs published in Slovakia and abroad, as well as conference papers presented at international conferences.

Rico Ihle is an Assistant Professor, Agricultural Economics and Rural Policy Group, Wageningen University, the Netherlands.

Anna Ivolga is an Associate Professor at the Department of Tourism and Service, Stavropol State Agrarian University, Russia. She published more than 90 papers in the sphere of business analysis, sustainable development, rural development, management of land relations and effectiveness of business in agriculture. She participated in the project Tempus RUDECO "Vocational Training in Rural Development and Ecology" as a co-coordinator and in the research project "Sustainability in Agrisector of Visegrad countries and cooperating regions" as a researcher in the sphere of international trade in agricultural products and influences of market integration and globalization.

Marko Jeločnik is a Ph.D. candidate in Agro-economics, Faculty of Agriculture, University of Novi Sad, Republic of Serbia. He is currently working as a full-time employee in a rank of a research assistant at the Institute of Agricultural Economics (IAE) Belgrade within the sector for scientific-research work. He holds a B.Sc. degree in agro-economics (2006) from the Faculty of Agriculture, University of Belgrade, Serbia as well as M.Sc. degree in the same field (2009), earned at the Faculty of Agriculture, University of Novi Sad, Serbia. Marko Jeločnik is a member of several professional associations: Balkan Environment Association (BENA), European Association of Agrarian Economists (EAAE), Balkan Scientific Association of Agrarian Economists (BSAAE), and Serbian Association of Agricultural Economists (SAAE). He is a member of the international editorial review board of the International Journal of Sustainable Economies Management (published by IGI Global, the USA). He is mainly oriented to agro-economy and economy, with a special focus on the economic effectiveness of investments, rural development, and strategic planning of local communities. He is an author/co-author of over 75 bibliographic units and a member of the research teams at many national and international projects.

Shikha Jha is a Principal Economist at the Asian Development Bank (ADB). She received her Ph.D. from the Indian Statistical Institute, Delhi Centre for her thesis on pricing of food grains and edible oils in India and completed post-doctoral research on food prices and commodity tax reforms at the London School of Economics. She has published extensively on the topic of food security and has been a consultant on food and agricultural policy issues for organizations such as the World Bank, USDA, UNDP and the International Food Policy Research Institute. As a Global Associate of the International Service for National Agricultural Research (CGIAR group) she was part of the expert group designing its 5-year strategy. Shikha also served on many Government of India expert committees including the Tenth-Plan Working Groups of the Planning Commission on 'Agricultural Marketing' and 'Public Distribution System and Food Security'. Following the food price crisis in 2008, she led ADB's corporate flagship publications 'Food Prices and Inflation in Developing Asia' and 'Global Food Price Inflation and Developing Asia'. Her work has informed ADB's policy dialogue and development assistance for food security in the region.

József Káposzta is an Associate Professor, Szent István University, Hungary. He received Ph.D. degree almost 20 years ago. Since then, he has had experience in teaching at bachelor, master and Ph.D. levels in both Hungarian and English languages, as well as had participated in national and international research projects at the Szent István University, Hungary. He has been the Dean of the Faculty of Economics and Social Sciences since 2013. He has over 320 scientific publications, including 55 journal articles and nearly 50 books/book chapters. He is a member of editorial boards for several Hungarian and foreign journals and has been a member of the organization committee of numerous national and international scientific conferences. He has been the supervisor of hundreds of students, including more than 20 Ph.D. candidates. He has been the supervisor of 10 Ph.D. successfully defended dissertations. His major research topic is regional economics. He is a member of the Hungarian Academy of Science.

Oleksandr Labenko is an Associate Professor, Faculty of Economics, National University of Life and Environmental Sciences of Ukraine, Ukraine. Skills and expertise: ecological economics, environmental issues, politics, and analysis, and environmental management.

Marina Lescheva is a Professor, Doctor in Economics, Head of the Department of Economic Analysis and Audit, Stavropol State Agrarian University, Russia. Specialization areas: agrarian economy, state support of agricultural production, sustainable development of agriculture, integration processes in the agricultural and industrial complex.

Giacomo Morabito is a Ph.D. student in Economics at the University of Messina, Italy. He graduated with a Master's degree in International Relations and European Studies. He is also a Researcher on Macroeconomics Desk at Wikistrat. His main research topics are economics, politics, and security, with a focus on maritime piracy.

Roberto Musotto is a Ph.D. student in Economics, University of Messina, Italy. Corporate and Civil Suit Lawyer in Palermo.

Henrietta Nagy is a Habilitated Associate Professor, Szent István University, Hungary. She received Ph.D. degree ten years ago. Since then, she has had experience in teaching at bachelor, master, and Ph.D. levels, as well as in the research projects at the Szent István University. In addition to academic activities, she has been responsible for the development of international relations at the Faculty of Economics and Social Sciences. She has more than 120 scientific publications, including 29 journal articles. She was the editor or author for over 10 books and wrote 20 book chapters. She has been the supervisor for more than 150 students and the organizer of over 25 national and international scientific conferences. Her major research fields are regional/cohesion policy, local economic development, and enterprise development. She is the member of the Hungarian Academy of Science.

György Iván Neszmély is an Associate Professor, Budapest Business School, Hungary. He received a Ph.D. degree in economic sciences in 1997, and then he worked as the Head of Department in the Research and Information Institute of Agricultural Economics (AKII). He worked as a diplomat between 1998 and 2012, served in several countries in Asia and Africa altogether around a decade, while as a part-time lecturer he had continuously been involved in various research and academic activities of Szent István University and Budapest Business School. He has over ten years of teaching experience at bachelor and master levels. He has been an honorary professor at Szent István University for over 15 years and since 2013, he has been the full-time Associate Professor at Budapest Business School. He has around 90 scientific publications, including 29 journal articles. He is an author and an editor of three books and wrote 11 book chapters. Many of his articles and papers were published in English, and some of them in the United States, Canada, and the Republic of Korea. His major research field is the examination of economic, political, and social factors in the development of Asian and African countries. He earned the post-doctoral title "Dr. habil" in regional sciences in 2016. He is a member of the Public Body of the Hungarian Academy of Science and several other scientific and civilian organizations, like the Hungarian Economic Association, the Hungarian Foreign Policy Association, and the Hungarian-Korean Association.

Omokheka Gregory Obozuwa is a Master student in Rural Development and Agribusiness, Faculty of Economics and Social Sciences, Szent István University, Hungary. Stipendium Hungaricum scholar.

Justus Ochieng is an Agricultural Economist, World Vegetable Center, Eastern and Southern Africa office in Arusha, Tanzania. He has experience in economic analysis of smallholder agricultural and livestock systems in sub-Saharan Africa, with a focus on production efficiency, technology adoption, market development, climate change and household welfare impacts. Before joining Worldveg, he worked as a Research Fellow for Tegemeo Institute of Agricultural Policy and Development, Nairobi, Kenya and the International Institute of Tropical Agriculture (IITA) in the Consortium for Improving Agricultural-based Livelihoods in Central Africa (CIALCA), which covered Burundi, Rwanda, and Democratic Republic of Congo (DRC).

Carlos Osorio-Ramirez obtained Master degree in Industrial Engineering and research associate at grupo SEPRO, National University of Colombia. He is currently in charge of the project "Agro-industrial Technological Corridor of Bogotá Region" and finishing his Ph.D. thesis on coordination strategies to improve humanitarian logistics operations. His main research subjects are humanitarian logistics, distribution systems, agro-industrial systems, and urban logistics.

Laura Palacios-Arguello is a Ph.D. student at Institut Fayol, Department of Environment and Organization Engineering, Mines Saint-Etienne, France. She holds Bachelor and Master degrees in Industrial Engineering at National University of Colombia and a Master in Management of Information Systems (Supply Chain Management) at the University of Grenoble, France. Her main research interests are food supply chain management, urban logistics systems, demand modeling, and information systems.

Khabibullo Pirmatov is a Ph.D. student, Faculty of Economics and Management, Slovak University of Agriculture in Nitra, Slovakia. He obtained his MSc degree from the Peoples' Friendship University of Russia with magna cum laude in 2013. He also holds an MSc degree in International Management.

His research interests are international tourism, management, global marketing with a focus on emerging markets, international economics, and agricultural economics. Currently, he is a grantee of TIMUR Erasmus Mundus project. At the Faculty of Economics and Management, Slovak University of Agriculture in Nitra he conducts his doctorate research on Innovations and Value Added Processes at Food Market System in Uzbekistan.

Andrew Schmitz is the Ben Hill Griffin, Jr., Eminent Scholar and Professor of Food and Resource Economics, University of Florida; Research Professor, University of California, Berkeley; and Adjunct Professor, University of Saskatchewan. He received awards for both his Master's thesis (University of Saskatchewan) and his Ph.D. dissertation (University of Wisconsin-Madison). He is a Fellow of the American Agricultural Economics Association. His publications have won six major research awards and three research of enduring quality awards from the American Agricultural Economics Association, and two major awards from the Canadian Agricultural Economics Association. He was awarded the Who's Who in America in 2004, the Southern Agricultural Economics Association Lifetime Achievement Award in 2003, and the Earned Doctor of Letters degree from the University of Saskatchewan in 1999. He has been a consultant to hundreds of private and public organizations. Some of his recent books include Government Policy and Farmland Markets (Iowa State Press); The Welfare Economics of Public Policy (Elgar Publishing); Agricultural Policy, Agribusiness, and Rent-Seeking Behavior (University of Toronto Press); Agricultural Policy, Agribusiness, and Rent-Seeking Behavior, Second Edition (University of Toronto Press); Sugar and Related Sweetener Markets (CABI Publishing), and The Economics of Alternative Energy Sources and Globalization (Bentham Publishing).

Carlos Alberto Talamantes is a student in Industrial Engineering at the Autonomous University of Ciudad Juarez, Mexico. His participation in this work took place during his internship at the National University of Colombia, grupo SEPRO.

Gao Tianming is an Associate Professor, School of Economics and Management, Harbin Engineering University, China. He obtained his Bachelor degree in Industrial External Trade from Harbin Engineering University, Master degree in Management from the Russian Presidential Academy of National Economy and Public Administration under the President of the Russian Federation, and Ph.D. degree in International Economics from Moscow State Institute of International Relations under the Ministry of Foreign Affairs of the Russian Federation. He has over 15 years of professional experience as a general representative of Chinese transnational corporation in Russia and the CIS. He is a Director of the Polar Development and Northeast Asian Economic Research Center (PAERC), Deputy Head of the Heilongjiang International Economic and Trade Association, leading consultant of governmental bodies and commercial organizations in the sphere of economic collaboration between China, Russia, and South Korea.

Vladimir Trukhachev is a Rector, Stavropol State Agrarian University, Russia. Doctor of Science in Economics, Doctor of Science in Agriculture, Professor, member of the Russian Academy of Sciences, Meritorious worker of science of Russia, Honorable worker of agricultural complex of Russia, Honorable worker of science and technics of Russia, Honorable worker of higher professional education of Russia, Hero of labor of Stavropol region, Honorable citizen of Stavropol Krai, Honorable worker of agricultural complex of Stavropol Krai. Awards: Order for Merit to the Fatherland, Order of Friendship, Diploma of the President of the Russian Federation "For success in labor, long work, and public activities". Member of the Academy of Agrarian Education, International Academy of Agrarian Education, International Academy of Authors of Scientific Discoveries, International Academy of Sciences in Higher Education, International Academy of Security Issues and Law Enforcement. The author of over 870 research, academic, and methodical works, including over 660 papers in journals and conference proceedings and 41 monographs. Co-author of a scientific discovery on the link between the live weight of endothermic organisms and optimal air temperature. Sphere of research interests: post-crisis development of the agricultural sector of Russia and its regions; biological and physiological specifics of fine-wool sheep and pigs in relation to their feed rations.

Bistra Vassileva is an Associate Professor of Marketing, University of Economics–Varna, Bulgaria. She graduated as Master of Science of Commodities followed by a specialization in marketing and management and Ph.D. study in International Marketing. Since 1992, she is lecturing and consulting in the fields of Marketing Research, International Marketing, Marketing Communications, TQM, Marketing Management. Dr. Vassileva was a visiting professor in Portugal, France, Germany, Spain, the UK and a guest lecturer in Belgium. Dr. Bistra Vassileva implemented more than 12 international and national projects of different donors. As a Director of the Centre for Innovation and Development at the University of Economics-Varna, she is responsible for the organization of research and ICT projects with the different scope. Her scientific interests are focused on applied mathematics (nonlinear dynamics and network theory) in the field of marketing as well as on marketing analytics and digital marketing. Dr. Vassileva is a member of CIM, ESOMAR, and EMAC and a national expert on Reference Budgets for Bulgaria.

Walter Vesperi is a Ph.D. student in Economics, University of Messina, Italy. He is an assistant teaching in human resource management and organization behavior. Walter Vesperi is a reviewer for the international scientific journal. His research focuses on the management of entrepreneurship, spinoffs, and knowledge transfer. He is a member of ASSIOA and AIDP.

Jovan Zubović is a Director of the Institute of Economic Sciences in Belgrade, Republic of Serbia. He holds a Ph.D. degree (2009) with a doctoral thesis "Investments in Human Resources as a Factor for Attracting Foreign Direct Investments Focusing on Serbian Financial Industry". Jovan Zubović is a member of the European Association of Labour Economists (EALE) as of 2011 and the European Association of Agricultural Economists (EAAE) as of 2009. He actively participates in editing of scientific journals: since 2013, an editor of the Journal Industrija; since 2013, a member of the editorial board of International Journal of Young Researchers in Business Economics published by Mind Reader Publications, New Delhi, India; since 2012, a member of the editorial board of Journal of Sustainable Economies Management published by IGI Global, the USA; in the period of 2009-2012, a member of the editorial board of the journal Economic Analysis; in the period of 2006-2008, an editor-in-chief of the journal Economic Analysis; in the period of 1996-2006, a business editor of the journal Marketing.

Index

M

N

O

P

R

S

T

U

V

W

Purchase Print, E-Book, or Print + E-Book

IGI Global books can now be purchased from three unique pricing formats: Print Only, E-Book Only, or Print + E-Book. Shipping fees apply.

www.igi-global.com

Recommended Reference Books

ISBN: 978-1-5225-0717-8
© 2017; 296 pp.
List Price: $190

ISBN: 978-1-5225-0320-0
© 2016; 502 pp.
List Price: $205

ISBN: 978-1-4666-8358-7
© 2015; 2,307 pp.
List Price: $2,150

ISBN: 978-1-5225-1674-3
© 2017; 1,536 pp.
List Price: $2,200

ISBN: 978-1-5225-0053-7
© 2016; 618 pp.
List Price: $300

ISBN: 978-1-5225-0731-4
© 2017; 609 pp.
List Price: $300

Looking for free content, product updates, news, and special offers?

Join IGI Global's mailing list today and start enjoying exclusive perks sent only to IGI Global members. Add your name to the list at **www.igi-global.com/newsletters.**

Publishing Information Science and Technology Research Since 1988

IGI Global
DISSEMINATOR OF KNOWLEDGE

www.igi-global.com Sign up at www.igi-global.com/newsletters facebook.com/igiglobal twitter.com/igiglobal

www.ingramcontent.com/pod-product-compliance
Lightning Source LLC
LaVergne TN
LVHW060300011026
849161LV00011B/2741